Fitting the Task to the Human

A Textbook of Occupational Ergonomics

BY

K. H. E. KROEMER and E. GRANDJEAN

Fifth Edition

TAYLOR & FRANCIS
ALERE FLAMMAM
· Founded 1798 ·

UK   Taylor & Francis Ltd., 11 New Fetter Lane, London, EC4P 4EE
USA   Taylor & Francis Inc., 325 Chestnut Street, 8th Floor, Philadelphia, PA 19106

Reprinted 1999, 2000

Taylor & Francis is an imprint of the Taylor & Francis Group

British Library Cataloguing in Publication Data

A catalogue record for this book is available from the British Library

ISBN 0-7484-0664-6 (cased)
ISBN 0-7484-0665-4 (paperback)

Library of Congress Cataloging Publication Data are available

Cover design by Jim Wilkie
Typeset in Times 10/12pt by Keyset Composition, Colchester
Printed in Great Britain by T.J. International Ltd., Padstow, Cornwall

Fitting the Task to the Human

Contents

About the authors   ix

1 Muscular work   **1**

    1.1   Physiological principles   1
    1.2   Static muscular efforts   7

2 Nervous control of movements   **17**

    2.1   Physiological principles   17
    2.2   Reflexes and skills   20

3 Improving work efficiency   **25**

    3.1   Optimal use of muscle strength   25
    3.2   Practical guidelines for work layout   28

4 Body size   **33**

    4.1   Variations in body dimensions   33
    4.2   National and international data   47
    4.3   Hand size   48
    4.4   Angles of rotation in joints   49

5 The design of workstations   **53**

    5.1   Working heights   53

5.2   Neck and head postures   63
5.3   Room to grasp and move things   66
5.4   Sitting at work   69
5.5   The design of computer workstations   83
5.6   The design of keyboards   95

6 Heavy work   101

6.1   Physiological principles   101
6.2   Energy consumption at work   103
6.3   Upper limits of heavy work   105
6.4   Energy efficiency of heavy work   105
6.5   Heart rate as a measure of workload   115
6.6   Combined effects of work and heat   120
6.7   Case histories involving heavy work   121

7 Handling loads   129

7.1   Back troubles   129
7.2   Intervertebral disc pressure   131
7.3   Biomechanical models of the lower back   134
7.4   Intra-abdominal pressure   136
7.5   Subjective judgements   137
7.6   Recommendations   137

8 Skilled work   147

8.1   Acquiring skill   147
8.2   Maximal control of skilled movements   149
8.3   Design of tools and equipment   152

9 Human-machine systems   157

9.1   Introduction   157
9.2   Displays   158
9.3   Controls   160
9.4   Relationship between controls and displays   172

10 Mental activity   177

10.1   Elements of 'brain work'   177
10.2   Uptake of information   178
10.3   Memory   180
10.4   Sustained alertness (vigilance)   183

11 Fatigue   191

11.1   Muscular fatigue   191
11.2   General fatigue   194
11.3   Fatigue in industrial practice   201
11.4   Measuring fatigue   203

12 Occupational stress   211

12.1   What is stress?   211
12.2   The measurement of stress   214
12.3   Stress among VDT operators   215

13 Boredom   219

13.1   Causes   219
13.2   The physiology of boredom   221
13.3   Field studies and laboratory experiments   225

14 Job design to avoid monotonous tasks   231

14.1   The fragmented work organisation   231
14.2   Principles of job design   233

15 Working hours and eating habits   241

15.1   Daily and weekly working time   241
15.2   Rest pauses   246
15.3   Nutrition at work   250

16 Night work and shift work   259

16.1   Day- and night-time sleep   259
16.2   Night work and health   264
16.3   The organisation of shift work   269
16.4   Recommendations   271

17 Vision   275

17.1   The visual system   275
17.2   Accommodation   278
17.3   The aperture of the pupil   281
17.4   Adaptation of the retina   282
17.5   Eye movements   284

17.6   Visual capacities   285
17.7   Physiology of reading   288
17.8   Visual strain   291

18 Ergonomic principles of lighting   295

18.1   Light measurement and light sources   295
18.2   Physiological requirements of artificial lighting   299
18.3   Appropriate arrangement of lights   306
18.4   Lighting for fine work   306
18.5   Lighting in computerised offices   310

19 Noise and vibration   319

19.1   Perception of sound   319
19.2   Noise   325
19.3   Damage to hearing through noise   329
19.4   Physiological and psychological effects of noise   332
19.5   Protection against noise   340
19.6   Vibrations   346

20 Indoor climate   355

20.1   Thermoregulation in humans   355
20.2   Comfort   359
20.3   Dryness of the air   364
20.4   Field studies on indoor climate   365
20.5   Recommendations for comfort indoors   368
20.6   Heat in industry   369
20.7   Air pollution and ventilation   377

21 Daylight, colours, and music for a pleasant work environment   383

21.1   Daylight   383
21.2   Colour in the workplace   385
21.3   Music and work   390

References   393
Index   411

About the authors

Professor Etienne Grandjean was one of the leading figures in ergonomics in Europe for over 30 years. Born in 1914 in Bern, Switzerland, he obtained his MD in 1939 and became Director of the Department of Hygiene and Ergonomics at the ETH, the Swiss Federal Institute of Technology, Zürich in 1950, where he remained until his retirement in 1983. His main research interests were the sitting posture, fatigue and working conditions in industries, and for the last decade VDT workstations.

Professor Grandjean was granted several international rewards and received Honorary Doctorates from the Universities of Surrey, Stuttgart and Geneva. He was one of the founders of the International Ergonomics Association and its General Secretary from 1961 to 1970. He published some 300 scientific papers and edited the first edition of *Fitting the Task to the Man* in 1963, which has since been translated into 10 different languages. He also wrote two other books in English – *Ergonomics of the Home* (Taylor & Francis, 1973) and *Ergonomics of Computerized Offices* (Taylor & Francis, 1987).

Karl H. E. Kroemer was born in 1933 near Berlin, Germany. He obtained Dipl.-Ing. and Dr.-Ing. degrees in 1959 and 1965, respectively, from the Technical University Hannover, Germany. Starting in 1960, he worked as a research engineer at the Max-Planck Institute for Work Physiology in Dortmund and from 1966 to 1973 he worked at the USAF Aerospace Laboratories in Dayton, Ohio. Then, after three years as Director of the Ergonomics Division of the Federal Institute of Occupational Safety and Accident Research in Dortmund, Germany, he was appointed Professor of Ergonomics at Wayne State University in Detroit, Michigan, in 1976. Since 1981 he has been Professor of Industrial and Systems Engineering at Virginia Tech where he directs the Industrial Ergonomics Laboratory. In his student years he almost moved to Zürich to work under Etienne Grandjean.

Although he did not move, they stayed in private and professional contact, often meeting at scientific conferences.

On 11 November 1991, Etienne Grandjean died. His death was a great loss to the ergonomic community. He was a pioneering researcher, a person who had great rapport with industry, an excellent teacher and an author who could write easily and simply about topics of great complexity. The first four editions of this book attest to this ability.

When talking about the fifth edition, Richard Steele, Publisher at Taylor & Francis, and Dr Kroemer decided to maintain Professor Grandjean's approach and style while editing and updating as needed. This fifth revised edition, *Fitting the Task to the Human*, is their homage to Etienne Grandjean.

1 May 1997
ISE Dept., Virginia Tech
Blacksburg, VA 24061-0118
USA
email: KARLK@VT.EDU

CHAPTER ONE

Muscular work

1.1 Physiological principles

Structure of muscle

The human body is able to move because it has a widely distributed system of muscles, which together make up approximately 40 per cent of the total body weight. Each muscle consists of a large number of muscle fibres, which can be between 5 mm and 140 mm long, according to the size of the muscle. The diameter of a muscle fibre is about 0.1 mm. A muscle contains between 100 000 and 1 million such fibres. The fibres of long muscles are sometimes bound together in bundles. At each end of the muscle the sinews are combined into a tough, nearly non-elastic tendon, which ends firmly attached to the bony skeleton.

Muscular contraction

The most important characteristic of a muscle is its ability to shorten to about half its normal resting length, a phenomenon we call muscular contraction. The work done by a muscle in such a complete contraction increases with its length: for this reason, we often try to pre-stretch a muscle before we contract it, such as bringing the arm back before we throw something.

Each muscle fibre contains proteins, including actin and myosin which have special importance for muscular contraction: they are the filaments which can slide over each other. During the process of contraction the actin filaments curl around and slide along the stationary myosin rods, as illustrated in Figure 1.1, thus shortening the muscle.

Muscle strength

Each muscle fibre contracts with a certain force and the strength of the whole muscle is the sum of these muscle fibre forces. The maximal strength of a human muscle lies *between 0.3 and 0.4 N*

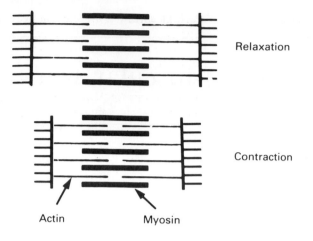

Relaxation

Contraction

Actin   Myosin

Figure 1.1   Model of muscular contraction. *The actin fibres slide between the myosin fibres and the two ends of the muscle section are drawn closer together.*

per mm² of the cross-section: thus a muscle of 100 mm² cross-section can develop a force of 30 to 40 N. Hence a person's inherent muscular strength depends, in the first instance, on the cross-section of his or her muscles. Given equal training, men and women can become equally strong per cross-section but women have, as a group, narrower muscles; they exert, on average, about two-thirds the force of men – but there are some strong women and some weak men. A muscle produces its greatest active strength at the beginning of its contraction, when it is still near its relaxed length. As the muscle shortens, its ability to produce force declines. Many of the recommendations in Chapter 3 for improving working efficiency are based on this relationship.

Regulation of muscular effort

The number of actively contracting muscle fibres determines how strength is developed during the period of contraction. As we shall see later, a muscle fibre is made to contract by incoming nervous impulses; hence the amount of muscle strength produced is determined by the number of nervous impulses, that is by the number of motor nerve cells in the brain that have been excited. The speed of a muscular contraction depends upon how quickly force is developed during a given interval of time, so the rapidity of a movement is governed by the number of actively contracting muscle fibres.

When muscular contraction is slow, or kept constant for a long time (static muscular effort), muscle fibres are brought into active contraction in succession, and they may alternate. This allows individual fibres some resting periods, which permits some 'recuperation from fatigue' to take place.

Sources of energy

During contraction, mechanical energy is developed at the expense of the reserves of chemical energy in the muscle. Muscular work involves the transformation of chemical into mechanical energy. The energy released by chemical reaction acts on the protein molecules of the actin and myosin filaments, causing them to change position and so bring about contraction. The immediate sources of energy for contraction are energy-rich phosphate compounds which change from a high-energy to a low-energy state in the course of chemical reactions. The source of energy most widely used by living organisms is *adenosine triphosphate* (ATP), which releases considerable amounts of energy when it is broken down into adenosine diphosphate. Moreover, ATP is present not only in muscles but in nearly every kind of tissue, where it acts as a reservoir of readily available energy. Another source of chemical energy in muscle fibres is phosphocreatine (phosphagen) which releases an equally significant amount of energy when it is broken down into phosphoric acid and creatine.

These low-energy phosphate compounds are continuously converted back to these high-energy state in the muscles so that the reserves of energy available to muscle remain undiminished. However, this conversion requires energy gained from digested foodstuffs.

The roles of glucose, fat and protein

This regeneration of high-energy phosphate compounds itself consumes energy, which is obtained from glucose and components of fat and protein. Glucose, the most important of the sugars circulating in the blood, is the main energy supply in intensive physical work. It is immediately available and easily converted. For maintained physical work the components of fat (fatty acids) and protein (amino acids) are the dominant energy supplies. *These nutritive substances, glucose, fat and protein, are, therefore, the indirect energy sources for the continuous replenishment of energy reserves in the form of ATP or other energy-rich phosphate compounds.* The glucose passes out of the bloodstream into the cells, where it is converted by various stages into *pyruvic (pyro-racemic) acid*. Further breakdown can take two directions, depending on whether oxygen is available (aerobic glycolysis) or the oxygen supply is deficient (anaerobic glycolysis).

The role of the oxygen

If oxygen is present, then the pyruvic acid is further broken down by oxidation (i.e. under continuous oxygen consumption), the end-products being water and carbon dioxide. This releases enough energy to reconstitute a large amount of ATP.

If oxygen is lacking, then the normal breakdown of pyruvic acid cannot take place. Instead, it is converted into *lactic acid*, a form of metabolic waste product which plays a vital part in symptoms

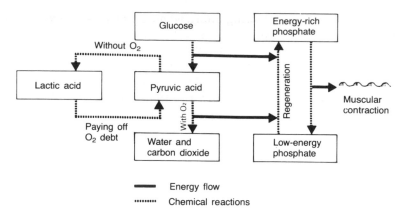

Figure 1.2   Diagram of the metabolic processes which take place during muscular work.

of muscle fatigue and 'muscular hangover'. This process releases a lesser amount of energy for the reconstitution of energy-rich phosphate compounds but allows a higher muscular performance under low-oxygen conditions, at least for a short time.

Oxygen debt

After heavy muscular effort a person is said to be 'out of breath'. This means making up for a shortage of oxygen by breathing more heavily; *paying off the oxygen debt.* This oxygen debt arises from previous consumption of energy; the extra oxygen is needed to convert lactic acid back into pyruvic acid and to reconstitute energy-rich phosphate compounds. After that, energy can again be obtained by oxidative breakdown of pyruvic acid. Figure 1.2 shows a simple diagram of the energy supply of a muscle.

Protein and fat

As already mentioned, fats and proteins are also involved in these metabolic changes. When breakdown of these substances has reached a certain stage, a *common metabolic pool* comes into being. The remaining fragments of fatty acids (from the break-down of fats) and amino acids (from the breakdown of proteins) undergo the same final breakdown as the pyruvic acid releasing energy and end up as water and carbon dioxide. This last stage, therefore, makes its own contribution of energy to the muscular effort.

The blood supply

The substances that are so important for energy production – glucose and oxygen – are only stored in small amounts in the muscles themselves. Both must therefore be transported to the muscles by the blood. For this reason, the blood supply often is the limiting factor in the efficiency of the muscular machinery.

During effort, a muscle increases its need for energy and thus for blood severalfold, and to supply this the most important adaptations of the blood system are more active pumping by the heart, raising of the blood pressure and enlargement of the blood vessels that lead to the muscles.

The following increases in circulation can be expected:

Muscle at rest:	4 ml/min/100 g of muscle
Moderate work:	80 ml/min/100 g muscle
Heavy work:	150 ml/min/100 g muscle
After a restriction of blood circulation:	50 to 100 ml/min/100 g muscle.

Heat production

According to the first law of thermodynamics, a muscle must be supplied with at least the same amount of energy that it uses. In practice, the incoming energy is transformed into (a) work performed, (b) heat and (c) energy-rich chemical compounds. As will be explained, only a small percentage of the incoming energy is converted into performed work.

Energy stored in the form of phosphate compounds is the smallest component; in contrast, generation of heat is by far the biggest: in terms of energy use for muscular work, the human is a very inefficient converter. If heat generation within the muscles is measured with delicate instruments, the following constituent parts are recognisable:

1 *Resting heat production* amounts to 1.3 kJ/min for a man weighing 70 kg. This serves to maintain the molecular structure and the electrical potentials in the muscle fibres.

2 *Initial heat* is considerably greater than the resting heat. This is the heat produced during the whole course of contraction of the muscle and is proportional to the work done.

3 *Recovery heat* is produced after the contraction is finished and over a longer period, up to 30 min. This is obviously the output from the oxidative processes of the recovery phase and is of the same order of magnitude as the initial heat.

Associated electrical phenomena

It has been known for a long time that muscular contraction is accompanied by electrical phenomena, which are similar to the processes of transmission of impulses along a nerve. In recent decades these electrical processes have been studied in much detail, using a delicate electrophysiological method called electromyography, discussed in more detail in Chapter 2.

In simple terms we can make the following points:

1 The resting muscle fibre exhibits an electrical potential – the so-called *resting membrane potential* – of about 90 mV. The

interior of the fibre is negatively charged in relation to the exterior.

2 The beginning of contraction is associated with a collapse of the resting potential and an overriding positive charge in the interior. This reversal of potential is called the *action potential*, so-called because it arises 'during the action' of the nerve. The action potential in the muscle lasts about 2–4 ms and spreads along the muscle fibre at a speed of about 5 m/s.

3 The action potential involves depolarisation and repolarisation of the membrane of the muscle fibre. During this period the muscle fibre is no longer capable of being excited; this period is called the absolute refractory period and lasts 1–3 ms. By analogy with the processes that go on in nerves, in the muscle fibre de- and repolarisation are also manifestations of reciprocal streams of potassium and sodium ions through the membrane of the muscle fibre.

**Electromyo-
graphy**

The electrical activity of a muscle can be recorded with the help of electrodes and amplifiers in an *electromyogram*, usually abbreviated as EMG. To record and interpret an EMG is nearly as complicated as doing an electrocardiogram, ECD (spelled EKD in German), which records the electrical signals of the heart muscle.

The electric current can be picked up via electrodes either inserted into the muscle or attached to the surface of the skin directly over the muscle to be studied. To insert needle or wire electrodes into muscles and thereby monitor individual muscle fibres is more exact but, obviously, also more intrusive and therefore not done so often.

An electromyogram registering the output of surface electrodes records the total electrical activity of the entire muscle. For this purpose two electrodes, each about 100 mm^2 in area, are applied a few centimetres apart. The output of the electrodes is usually amplified and then squared, averaged and the square root of the result is calculated (Basmajian and DeLuca, 1985; Soderberg, 1992). The use of computers has facilitated this procedure very much over manual techniques used before. However, EMGs are strictly valid only for one particular set of electrodes in one particular experiment and are often limited to maintained isometric muscle contractions. (Such static efforts are discussed below.)

Nevertheless, electromyography has shown that electrical activity increases with the level of muscular force developed. Changes in EMG signals, especially in frequency, can indicate muscle fatigue; changes in intensity are indicative of differing tensions (forces) in the muscle. Thus, although not easy to use, electromyography is especially suited for investigating the involve-

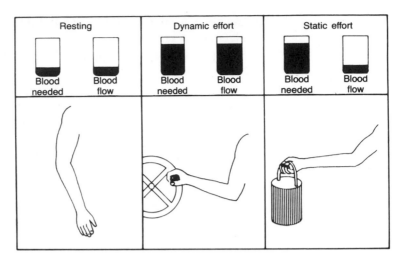

Figure 1.3   Diagram of dynamic and static muscular effort.

ment of muscles, and of their individual contributions, in efforts expended to maintain body postures.

1.2   Static muscular efforts

There are two kinds of muscular effort: *dynamic (motion) effort and static (posture) effort.*

Dynamic and static forms of work

Figure 1.3 illustrates these two kinds of muscular activity. The dynamic example is cranking a wheel and the static example is supporting a weight at arm's length.

The two forms of muscular effort can be described as follows:

1 Dynamic effort is characterised by an alternation of contraction and extension, tension and relaxation; muscle length changes, often rhythmically.

2 Static effort, in contrast, is characterised by a prolonged state of contraction of the muscles, which usually maintains a postural stance.

In a dynamic situation the muscular effort can be expressed as the product of the force developed and the shortening of the muscle (work = force × distance; here: weight × height it is raised). During static effort the muscle does not change its length but remains in a state of heightened tension, with force exerted over the duration of the effort. (Since muscle length does not change, this effort is called 'isometric' in physiology.) During such static effort no useful work is externally visible nor can this be defined by a formula such as force × distance. It rather resembles an electromagnet, which has a steady consumption of energy while it is supporting a given weight, but does not appear to be doing useful work.

Blood supply

There are certain basic differences between static and dynamic muscular efforts.

During a strong static effort the blood vessels are compressed by the internal pressure of the muscle tissue so that blood no longer flows through the muscle. During dynamic effort, on the other hand, as when walking, the muscle acts as a pump in the blood system: compression squeezes blood out of the muscle and the subsequent relaxation releases a fresh flow of blood into it. By this means the blood supply can increase; in fact, the muscle may receive up to 20 times more blood than when it is resting. A muscle performing dynamic work is therefore flushed with blood and retains the energy-rich sugar and oxygen contained in it, while at the same time waste products are removed.

In contrast, a muscle that is performing heavy static effort is receiving no fresh blood and no sugar or oxygen and must depend upon its own reserves. Moreover – and this is a serious disadvantage – waste products are not being removed. Quite the reverse: these waste products are accumulating and produce the acute pain of muscular fatigue. Figure 1.4 shows how the two kinds of muscular effort affect the blood supply to the working muscle.

For this reason we cannot continue a static muscular effort for very long; the pain will compel us to relax. On the other hand, a dynamic effort can be carried on for a very long time without fatigue, provided that we choose a suitable rhythm for it. There is one muscle that is able to work dynamically throughout our lives without interruption and without tiring: the muscle of the heart.

Examples of static effort

Our bodies must often perform static effort during everyday life. Thus, when we keep standing, a whole series of muscle groups in the legs, hips, back and neck are tensed. It is thanks to these static efforts that we can hold our bodies in any desired attitude. If, however, we keep standing, our strained muscles start hurting. When we sit down, the static effort in the legs is relieved and the total muscular strain of the body is reduced. When we lie down, nearly all static muscular effort is avoided; that is why a recumbent posture is the most restful. *There is no sharp line between dynamic and static effort. Often one particular task is partly static and partly dynamic.* Since static effort is much more arduous than dynamic, the static component of mixed effort assumes the greater importance.

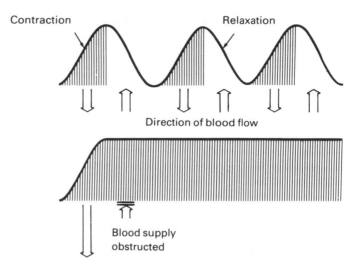

Figure 1.4 Flow of blood through muscles during dynamic and static effort. *The curves show the variation of muscular tension (internal pressure). (Upper) dynamic effort operates like a pumping action, which ensures a flow of blood through the muscle. (Lower) static effort obstructs the flow of blood.*

In general terms, static effort can be said to be considerable under the following circumstances:

1 If a high level of effort is maintained for 10 s or more.

2 If moderate effort persists for 1 min or more.

3 If slight effort (about one-third of maximum force) lasts for 5 min or more.

There is a static component in nearly every form of factory work or any other occupation. The following are some common examples:

1 Jobs which involve bending the back forward or sideways.

2 Holding things in the hands.

3 Manipulations which require the arms to be held stretched out or raised above shoulder height.

4 Putting the weight on one leg while the other works a pedal.

5 Standing in one place for long periods.

6 Pushing and pulling heavy objects.

7 Tilting the head strongly forwards or backwards.

8 Raising the shoulders for long periods.

Constrained postures are certainly the most frequent form of static muscular work. The main cause of constrained postures is carrying the trunk, head or limbs in unnatural positions. Figure 1.5 shows examples of static loads.

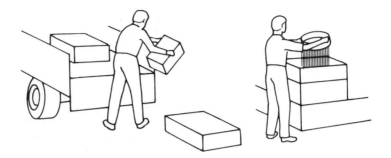

Figure 1.5 Examples of static muscular effort. *(Left) loading parcels. (Right) sieving sand into a mould in a foundry. In both examples there are high static loads on the muscles of the back, shoulders and arms.*

Effects of static work

During static effort the flow of blood is constricted in proportion to the force exerted. If the effort is 60 per cent of the maximum, the flow is almost completely interrupted, but a certain amount of circulation of blood is possible during lesser efforts because the tension in the muscles is less. When the effort is less than 15–20 per cent of the maximum, blood flow should be normal.

Obviously, therefore, the onset of muscular fatigue from static effort will be more rapid the greater the force exerted, i.e. the greater the muscular tension. This can be expressed in terms of the relation between the maximal duration of a muscular contraction and the force expended. This was systematically studied by Moltech (1963), Rohmert (1960) and Monod (1967). Figure 1.6 shows the results obtained by Monod for four muscles. It appears from their work that a static effort which requires 50 per cent of maximum force can last no more than 1 min whereas, if the force expended is less than 20 per cent of maximum, the muscular contraction can continue for some time. Field studies as well as general experience show that a static force of 15–20 per cent of the maximum force will induce painful fatigue if such loads have to be kept up for very long periods of time (van Wely, 1970; Nemecek and Grandjean, 1975). Many experts are therefore of the opinion that work can be maintained for several hours per day without symptoms of fatigue if the force exerted does not exceed about 10 per cent of the maximum force of the muscle involved.

Static muscular effort is strenuous

Under roughly similar conditions a static muscular effort, compared with dynamic work, leads to:

A higher energy consumption.

Raised heart rate.

Longer rest periods needed.

This is easy to understand if we bear in mind that on the one hand the metabolism of sugar with an inadequate supply of oxygen

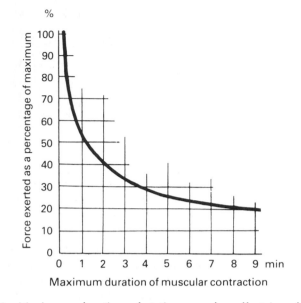

%

Figure 1.6 Maximum duration of static muscular effort in relation to the force exerted. *After Monod* (1967).

releases less energy for regeneration of energy-rich phosphates and on the other hand produces a large amount of lactic acid, which interferes with muscular effort. Oxygen deficiency, which is unavoidable during static muscular effort, inevitably lowers the effective working level of the muscle.

Figure 1.7 shows a good example of this, repeating the research results of Malhotra and Sengupta (1965). These authors showed that schoolchildren who carried their satchel in one hand needed more than twice as much energy as when they carried the satchel on their back. This increased energy consumption must be attributed to the high static loads on the arms, shoulders and trunk. Figure 1.8 shows another example from Hettinger (1970) involving potato planting. In one case the basket of potatoes was carried in one hand; in the other the basket hung from a harness in front of the body. Carrying the basket by hand raised the heart rate by 45 pulses per minute, compared with only 31 pulses per minute when the basket was carried in a harness. Carrying the basket required a muscular effort in the left arm amounting to 38 per cent of its maximum. Hettinger drew the conclusion that the increased strain on the heart was entirely caused by the heavy static effort needed to carry the basket in one hand.

Combination of dynamic and static efforts

In many cases no sharp distinction between dynamic and static effort can be made. A particular task can be partly static and partly dynamic. Keyboard operating is an example of a combination of both types of muscular work: muscles in the back, shoulders and

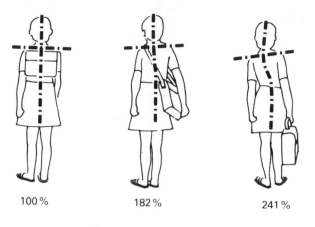

100%   182%   241%

Oxygen consumption

Figure 1.7 Effect of static effort on energy consumption (measured by oxygen consumption) for three ways of carrying a school satchel. *After Malhotra and Sengupta* (1965).

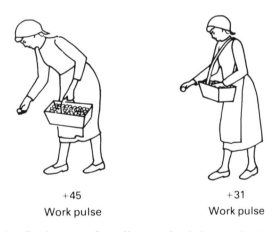

+45   +31
Work pulse   Work pulse

Figure 1.8. Static muscular effort in the left arm during potato planting. *The use of a sling avoids static effort in the left arm. Over a period of 30 min the heart rate rose by up to 45 pulses/min (left) or 31 pulses/min (right). After Hettinger* (1970).

arms do mainly static work to hold the hands in position over the keyboard while the hands' digits perform mainly dynamic work when operating the keys. The static component of the combined effort assumes great importance in terms of postural fatigue while the muscles and tendons moving the fingers and thumbs may experience repetitive motion strain. There is a static component in almost every form of physical work.

Localised fatigue and musculoskeletal troubles

As already explained, even moderate static work might produce troublesome localised fatigue in the muscles involved, which can build up to intolerable pain. If excessive efforts, whether static or dynamic, are repeated over a long period, first light and then more intense aches and pains will appear and may involve not only the muscles but also the joints, tendons and other tissues. Thus long-lasting and often repeated efforts can lead to damage of joints, ligaments and tendons. These impairments are usually summarised under the term 'musculoskeletal disorders'.

Field studies as well as general experience have shown that maintained and repetitive loads are associated with a higher risk of:

arthritis of the joints due to mechanical stress;
inflammation of the tendons and/or tendon sheaths (tendinitis or tenosynovitis);
inflammation of the attachment-points of tendons;
symptoms of arthrosis (chronic degeneration of the joints);
painful muscle spasms; and
intervertebral disc troubles.

Persistent musculoskeletal troubles

These symptoms of overstress can be divided into two groups: *reversible* and *persistent musculoskeletal troubles*.

The *reversible symptoms* are short-lived. The pains are mostly localised to the muscles and tendons and disappear as soon as the load is relieved. *These troubles are the pains of weariness.*

Persistent troubles are also localised to strained muscles and tendons, but they affect the joints and adjacent tissues as well. The pains do not disappear when the work stops but continue. *These persistent pains are attributable to inflammatory and degenerative processes in the overloaded tissues.* Elderly employees are more prone to such persistent troubles. If musculoskeletal troubles persist over years, they may get worse and lead to chronic inflammations especially of tendons and tendon sheaths, even to deformations of joints. The health problems that may be expected to follow from certain forms of static load are set out in Table 1.1.

Examples of morbid symptoms

During a training course lasting 12 weeks, Tichauer (1976) compared the effects of using a wire-cutter that was shaped to the hand with those of a normal pair of pliers. The normal pliers called for an unnatural grip, with the hand rotated bent sideways at the wrist, and hence a pronounced static muscular strain. During those 12 weeks, 25 of the 40 workers who used the unnatural grip showed overstress symptoms, in the form of inflammation of the tendon attachments or of the tendon sheaths. In contrast, there were only four cases of inflamed tendon sheaths among an equal

Table 1.1 Static load and bodily pains

Work posture	Possible consequences affecting
Standing in one place	Feet and legs; possibly varicose veins
Sitting erect without back support	Extensor muscles of the back
Seat too high	Knee; calf of leg; foot
Seat too low	Shoulders and neck
Trunk curved forward when sitting or standing	Lumbar region; deterioration of intervertebral discs
Arm outstretched, sideways, forwards or upwards	Shoulders and upper arm; possibly periarthritis of shoulders
Head excessively inclined backwards or forwards	Neck; deterioration of intervertebral discs
Unnatural grasp of hand grip or tools	Forearm; possibly inflammation of tendons

number of workers who used the curved wire-cutters which permitted a natural grip. The two forms of pliers are shown in Figure 1.9.

Figure 1.10 shows two postures at a machine which could lead to a danger of aches and other unhealthy symptoms.

Standing in one place

Standing in one place involves static effort through prolonged immobility of the joints of the feet, knees and hips. The force involved is not great and falls below the critical level of 15 per cent of the maximum. All the same, standing in one place for a long time is wearisome and painful. This is not entirely the effect of static muscular effort; pain is also partly caused by increased hydrostatic pressure of the blood in the veins of the legs and general restriction of lymphal circulation in the lower extremities. In practice the hydrostatic pressure in the veins, when standing motionless, is increased as follows:

1 At the level of the feet by 80 mm Hg.

2 At the level of the thighs by 40 mm Hg.

During walking the muscles of the legs act as a pump, which compensates for the hydrostatic pressure of the veins by actively propelling fluid back towards the heart.

Thus, prolonged standing in one place causes not only fatigue of the muscles which are under static load but also discomfort which is attributable to insufficient return flow of the venous blood.

This unhealthy state of the blood circulation is the cause of cumulative ailments of the lower extremities in occupations which

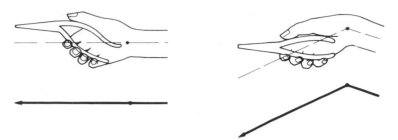

Figure 1.9 (Left) the curved pliers are shaped to fit the hand, which remains in line with the forearm. (Right) pliers of the traditional shape require sustained static effort, with the wrist bent; the hand is no longer in line with the forearm. *Modified from Tichauer* (1976).

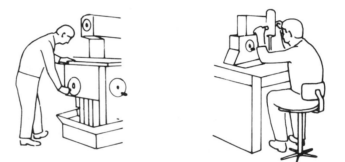

Figure 1.10 Unnatural postures at work which involve static loads can lead to physical disabilities. *(Left) risk of back troubles. (Right) risk of aches in shoulders and upper arms.*

call for prolonged standing without movement. Such occupations involve increased liability to:

1 Dilation of the veins of the legs (varicose veins).

2 Swelling of the tissues in the calves and feet (oedema).

3 Ulceration of the oedematous skin.

Cumulative trauma disorders

The musculoskeletal system may be overloaded by a succession of small traumas (conveniently called microtraumas) which, taken one by one, do not injure but, by their cumulative effects, can lead to overexertions. They have long been reported to occur in jobs with repetitive work elements, for example in milking cows, hand-wringing of washed cloths, clerical handwriting, hammering nails, operating telegraphs and playing musical instruments – the composer Robert Schumann reportedly lost the use of his right hand to 'repetitive strain injury' related to his piano playing. With stereotyped jobs in production and assembly in modern industry, the same movements are repeated over and over: for example, when meat cutting or at check-out counters in stores, while operating keyboards in typing and today in computer word processing, or when playing tennis or golf. The occurrence of musculoskeletal disorders due to accumulated microtrauma effects has become a widespread problem in manual jobs. Human engineering actions to avoid such overuse disorders have been discussed by Grandjean (1987), Putz-Anderson (1988), Kroemer et al. (1989, 1994, 1997) and Kuorinka and Forcier (1995). These interventions include the reduction of frequency of manual activities and of their energy content (muscle force) as well as the incorporation of proper body motions and postures.

Summary

Muscles are the engines that propel our bodies. They convert chemical energy extracted from food and drink into mechanically useful force and travel of body limbs. For their work they depend on the circulatory, respiratory and metabolic subsystems of the body. Muscles can perform well-organised dynamic work easily but they fatigue quickly in static efforts. Therefore avoiding static efforts, including standing or sitting still over long periods of time, is an important human engineering task.

CHAPTER TWO

Nervous control of movements

2.1 Physiological principles

Structure of the nervous system

The central nervous system consists of the brain and the spinal cord. The peripheral nerves either run outwards from the spinal cord to end in the muscles (motor nerves) or inwards from the skin, muscles or eyes and ears to the spinal cord or brain (sensory nerves). The sensory and motor nerves, together with their associated tracts and centres in the spinal cord and brain, comprise the somatic nervous system, which links the organism with the outside world through perception, awareness and reaction.

Complementary to this is the visceral or autonomic nervous system, which controls the activities of all the internal organs: blood circulation, breathing organs, digestive organs, glands and so on. The visceral nervous system therefore governs those internal mechanisms that are essential to the life of the body.

The complete nervous system is made up of millions of nerve cells, or neurones, each of which has basically a cell body and a comparatively long nerve fibre. The cell body is a few thousandths of a millimetre thick whereas the nerve fibre can be more than a metre long. Figure 2.1 shows a diagrammatic picture of a neurone.

The function of nerves

The nervous system is essentially a control system, which regulates external and internal activities as well as monitoring a variety of sensations. The working of a neurone depends upon its being sensitive to stimuli and being able to transmit a stimulus along the length of the nerve fibre. When a nerve cell is stimulated, the resulting impulse travels along the nerve fibre to the operative organ, which may be a muscle fibre among others.

Nervous impulses are of an electromechanical nature. Nerves are not just 'telephone wires' transmitting impulses passively. A

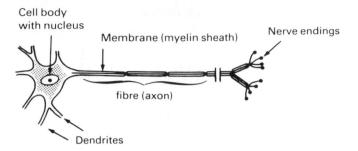

Figure 2.1   Diagram of a motor neurone, consisting of cell-body, dendrites, a nerve fibre (axon) and nerve endings.

nervous impulse is an active process, self-generating and energy-consuming, like a spark running the length of a fuse. Unlike a fuse, however, a nerve fibre is not dead once it has been used but is regenerated in a fraction of a second, becoming receptive once again after the so-called refractory period. The nerve fibre cannot transmit a continuous 'direct current' but only single impulses, with brief interruptions between them.

The speed of transmission varies in different types of nerve: motor fibres transmit at 70–120 m/s; other fibres in the region of 12–70 m/s.

The nature of nervous impulses

What are these nervous impulses? Like muscle fibres, nerve fibres have a resting membrane potential. *In the resting phase the membrane of the neurone is polarised; positive charges predominate on the outer surface, while negative charges predominate internally. Depolarisation of the membrane produces the nervous impulse.* The resting potential, −70 mV, collapses completely and depolarisation continues until a reverse peak of +35 mV is reached. Then comes a repolarisation with a quick return to the resting potential of −70 mV. Figure 2.2 shows the trace of an action potential as a nervous impulse passes along a nerve fibre. As mentioned in Chapter 1, *this electrical oscillation between de- and repolarisation of the membrane is called the action potential.*

Causes of the action potential

The action potential is therefore the electrical manifestation of a wave of de- and repolarisations, which travels along the nerve fibre at a speed of between 12 and 120 m/s. Depolarisation, the breakdown of the resting potential, is brought about by a sudden change in the permeability of the membrane, allowing a stream of positively charged sodium ions to penetrate the interior of the fibre. Almost simultaneously, positively charged potassium ions move from inside to outside the membrane but these are fewer

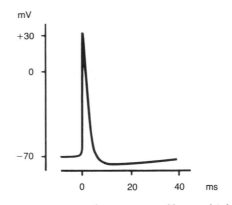

Figure 2.2 The action potential in a nerve fibre, which indicates the passage of a nervous impulse.

than the sodium ions moving inwards so the net result is a sharp increase in the total positive charge in the interior. *It is this displacement of electrical charges which produces depolarisation and the associated action potential.*

In the subsequent repolarisation, these charged ions move in the reverse directions: sodium outwards, potassium inwards. We are back at the starting-point, with the membrane potential restored. This mechanism is called the *sodium–potassium pump* for short. It is not confined to nerve fibres but occurs in nearly all living cells. The energy for this pumping mechanism is derived from adenosine triphosphate (ATP). The sodium–potassium pump is an essential requirement before a nerve can react to stimuli and transmit impulses.

The nervous system requires a supply of energy, principally to maintain the membrane potential, and derives this energy from ATP. The metabolic energy need of a nerve approximately doubles when it is active; a small increase compared with that of skeletal muscle, which increases its metabolism about one hundredfold when it is working.

Innervation of muscles

Every muscle is connected to the brain, the overriding control centre, by nerves of two kinds: efferent or motor nerves, and afferent or sensory nerves.

Motor nerves carry impulses, movement orders, from the brain to the skeletal muscles where they bring about a contraction and collectively control muscular activity. Within a muscle the nerve divides into its constituent fibres, each nerve fibre serving to innervate several muscle fibres. Each single motor neuron, together with the muscle fibres that it innervates, constitutes a *motor unit*. In muscles which carry out fine and precise movements, as in skilled work, there are only 3–6 muscle fibres per motor unit whereas muscles which do heavy work may have 100 or more muscle fibres innervated by one neuron.

The bundle of motor nerve fibres shown in Figure 2.1 ends in so-called *motor end plates* where the cell membrane of the nerve fibre is thickened. This is where the motor impulse leaps across from the nerve fibre to the muscle fibre and where the action potential finally provokes muscular contraction.

Sensory nerves conduct impulses from the muscles into the central nervous system, either to the spinal cord or to the brain. *Sensory impulses are bearers of 'signals'*, which will be utilised in the central nervous system in part to direct muscular work and in part to store as information.

Receptor organs of a special kind are the *muscle spindles* which are parallel to the muscle fibres and end in the tendons. The muscle spindles are sensitive to stretching of the muscle and send signals about this to the spinal cord (proprioceptor system).

Further organs of sensory response are the *organs of Golgi* (Golgi–Mazzoni corpuscles), which are composed of a network of nodular nerve endings and are embedded in the tendons. These organs also transmit sensory impulses to the nerve cord whenever the tendon is under tension.

Within the spinal cord, the sensory impulses pass by means of an intermediate neuron to the motor nerve so that new impulses flow back to the muscles. This system of an afferent sensory nerve and an efferent motor nerve running back to the same muscle is called a *reflex arc*. Reflexes keep muscle tension and muscle length continually adjusted to each other; *the muscle spindle and the Organ of Golgi are the 'detectors' in this regulatory system.*

Other sensory nerves conduct impulses from the muscles across a first intermediate neuron in the spinal cord and over a second in the medulla of the brain (brain stem) up to the cerebral cortex *where the incoming pulses are finally experienced as a sensation* This is how pains are felt which arise in the muscles. The innervation of the skeletal muscles with sensory and motor nerve fibres is sketched in Figure 2.3.

2.2 Reflexes and skills

One special way in which movement and activity are controlled is by means of reflexes. Because these are not consciously directed they are *'automatic' in a physiological sense*. A reflex consists of three parts: (a) an impulse which travels along a sensory nerve, carrying the information to the spinal cord or to the brain; (b) intermediate neurons, which pass the impulse across to a motor nerve; and (c) a final impulse along the motor nerve which activates the appropriate muscle. An example is reflex blinking of the eyelids. When anything unexpected moves close to either eye, both eyes close automatically. The unexpected movement in the visual field provides the initial stimulus; the sensory impulse travels to a particular centre in the brain, which itself acts as

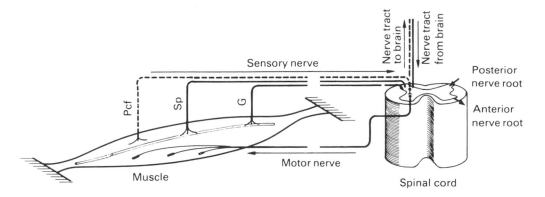

Figure 2.3 Innervation of a muscle. *The sensory nerve comprises pain-conducting fibres (Pcf) as well as fibres from the muscle spindle (Sp) and from the Golgi receptors of the tendon (G). The fibres of the motor nerve terminate in the motor end plates in the membranes of the muscle fibres.*

intermediate neuron and passes on the message to a motor nerve; the motor nerve in turn operates the muscles of the eyelid. Reflex blinking is thus an automatic protective mechanism, safeguarding the eyes against damage. The body makes use of thousands of such reflexes, not only for protection but as part of normal control functions.

Reflexes also play an essential part in muscular activity. One example has been described above in connection with skeletal muscle and the reflex arc involved is shown diagrammatically in Figure 2.3. Another example of a more complicated and important reflex is the antagonistic control of a muscular movement. When the lower arm is bent the bending muscles are caused to contract by stimulating their motor nerves and if this is to proceed smoothly the opposing muscles behind the arm must be simultaneously relaxed by exactly the right amount. This co-contraction is a reflex automatism by which movement is carried out in a well-controlled manner.

Skilled work

To give an idea of the complexity of nervous control, the most important stages in a piece of skilled work are shown diagrammatically in Figure 2.4. During a simple grasping operation such as that illustrated, the first step is to make use of visual information to direct the movements of the arms, hands and fingers towards the object to be grasped. For this purpose nervous impulses travel along the optic nerve from the retina of the eye to the brain and are integrated there into a hand–finger–object picture. These impulses are then passed over to other centres, in the medulla and cerebellum, which control muscular activity. On the basis of the visual signals it has received, the brain thus decides what the next

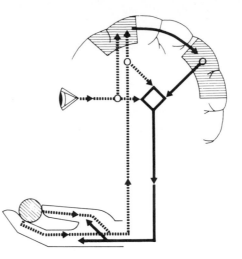

Figure 2.4   Diagram illustrating nervous controls in skilled work.

move will be. When the object has been grasped, pressure-sensitive nerves in the skin send new signals to the brain and the operator can adjust finger pressure accordingly.

Conditioned reflexes

Becoming skilled is largely a matter of developing new reflexes which go on without conscious control; they are called conditioned reflexes. Figure 2.4 indicates new reflex pathways as arrows passing directly from the synapses (intermediate neurones) of the sensory pathways into the muscular control centres†, where *combinations of movements are retained as a sort of template*. In other words, whenever a sequence of movements is practised for a long time, the *complete movement pattern becomes 'engraved' in the brain*. Coordination and delicate adjustment of individual muscular movements are achieved when a continuous stream of sensory information reaches these motor coordination centres. Skill reaches a maximum when learning has eliminated conscious control and movements have become automatic. The task of the conscious mind is to concentrate all nervous activity upon the job at hand and to give overriding 'orders' to the control centres.

An example: writing

The whole process of automatism can be illustrated by the example of learning to write. First the child learns to control the movement of hand and fingers so delicately that the correct shapes appear on the paper. Then he or she begins to make the necessary movements at will. After a very long period of practice the sequence of movements necessary to write each letter has become an 'engram', a pattern in the motor control centres of the brain. The writing process itself becomes more and more automatic, and eventually the conscious mind concerns itself only with finding the right words and making them into sentences.

Summary

The central nervous system in the brain controls our activities. It receives information about events outside and inside the body through sensors which send signals via the sensory pathways (afferent part of the peripheral nervous system) to the command centre. Here, decisions are made and motor control signals are transmitted via the efferent pathways to muscles. Automation of such actions by repetition into reflexes is an important part of acquiring skills.

CHAPTER THREE

Improving work efficiency

3.1 Optimal use of muscle strength

Strength, length and leverage

When any form of body activity calls for a considerable expenditure of effort the necessary movements must be organised in such a way that the muscles are developing as much power as possible with the least effort feasible. In this way the muscles will be at their most efficient and skilful.

Where the work involves holding something statically, a posture must be taken up which will allow as many strong muscles as possible to contribute. This is the quickest way to bring the loading of each muscle down below 15 per cent of its maximum (see Chapter 1).

Since a muscle is most powerful at the beginning of its contraction it is a good idea, in principle, to start from a posture in which the muscle is fully extended. There are so many exceptions to this general rule, however, that it has more theoretical than practical value. One must also take into account the leverage effect of the bones and, if several muscles join forces, their total effect. In the last case the force exerted is usually at its greatest when as many muscles as possible contract simultaneously. The maximal force a muscle, or group of muscles, is capable of depends upon:

age
gender
constitution
state of training
momentary motivation

Age and gender

Figure 3.1 sets out the effects of age and gender on muscle strength, according to data obtained by Hettinger (1960). The peak of muscle strength for both men and women is reached

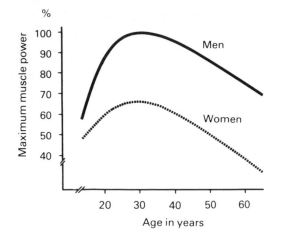

Figure 3.1 Muscle strength in relation to age and sex. *After Hettinger* (1960).

Table 3.1 Maximal muscular force for men and women. According to Hettinger (1960). *SD* = standard deviation of the individual values

Function	Max. force Men (N)	*SD*	Max. force Women (N)	*SD*
Hand clasp	460	120	280	70
Kicking (with knee bent at 90°)	400	60	320	50
Stretching the back	1100	160	740	160

between the ages of 25 and 35 years. Most older workers aged between 50 and 60 can produce only about 75–85 per cent as much muscular strength.

As has already been pointed out in Chapter 1 it can be assumed that on average, women will be only about two-thirds as powerful as men.

Hettinger studied the maximum force developed by both men and women when using three groups of muscles which he considered have special significance in the assessment of human muscular power. His results are summarised in Table 3.1.

Maximum power when sitting at work

The following rules, deduced from studies made by Caldwell (1959), apply to test subjects who are sitting with their backs against a back rest:

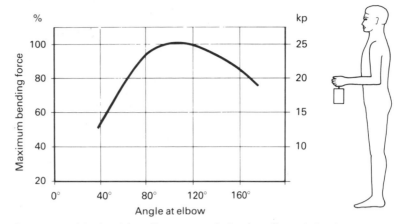

Figure 3.2 Maximal bending strength in the elbow joint in relation to the angle at the elbow. *After Clarke* et al (1950) *and Wakim* et al (1950).

1 The hand is significantly more powerful when it turns inwards (pronation of the forearm, 180 N) than when it turns outwards (supination, 110 N). Note that these force values are given only to indicate orders of magnitude. Different persons have different strengths.

2 The rotation force is greatest if the hand is grasping 30 cm in front of the body.

3 The hand is significantly more powerful when it is pulling downwards (370 N) than when pulling upwards (160 N).

4 The hand is more forceful when pushing (600 N) than when pulling (360 N).

5 Pushing strength is greatest when the hand is grasping 50 cm in front of the body.

6 Pulling strength is greatest at a grasping distance of 70 cm.

Maximal force for bending elbows

The maximum force exerted by the muscles which bend the elbow (biceps muscles) seems to be particularly dependent upon leverage in the arm. The results of studies by Clarke et al. (1950) and Wakim et al. (1950) and their co-workers, as set out in Figure 3.2, show that *the greatest bending moment is achieved at angles between 90 and 120°.*

Maximal force for standing work

The maximal pulling and pushing forces of the hand during standing work have been thoroughly studied by Rohmert (1966) and by Rohmert and Jenik (1972). Selections of their results for men are set out in Figure 3.3.

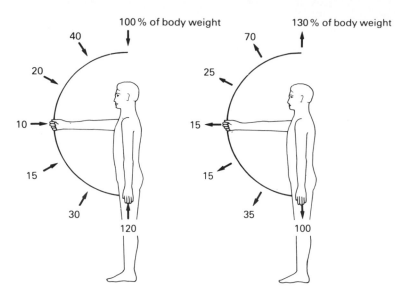

Figure 3.3   Maximal force of pulling (left) and pushing (right) for a man. *Feet 300 mm apart. The values at various angles are given as a percentage of the body weight. Simplified, after Rohmert (1966).*

The following conclusions may be drawn from Rohmert's (1966) studies:

1 At most positions of the arms, while standing up, the pushing force is greater than the pulling force.
2 Both pulling and pushing forces are greatest in the vertical plane and lowest in the horizontal plane.
3 Pulling and pushing forces are of the same order of magnitude whether the arms are held out sideways or forwards (in the sagittal plane).
4 Pushing force in the horizontal plane may amount to:
160–170 N for men
80–90 N for women.

3.2 Practical guidelines for work layout

Most important principles

As has already been emphasised, static loads on the muscles lead to painful fatigue; they are wasteful and exhausting. For this reason, a major objective in the design and layout of jobs, workplaces, machines, instruments and tools should be to *minimise or abolish altogether the need to grasp and hold things.*

Unavoidable static effort should be reduced to not more than 15 per cent of the maximum and to 10 per cent for long-lasting loads.

Figure 3.4 Strongly bent posture while cleaning a casting in a foundry. *If the casting was placed vertically it could be cleaned with substantially less static load on the muscles of the back.*

According to van Wely (1970), dynamic effort of a repetitive nature should not exceed 30 per cent of the maximum, although it may rise to 50 per cent as long as the effort is not prolonged for more than 5 min.

Seven guidelines

1 *Avoid any kind of bent or unnatural posture* (Figure 3.4). Bending the trunk or the head sideways is more harmful than bending forwards.

2 *Avoid keeping an arm outstretched either forwards or sideways.* Such postures not only lead to rapid fatigue but also markedly reduce the precision and general level of skill of operations using the hands and arms (Figure 3.5).

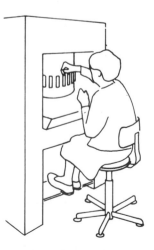

Figure 3.5 Fitting components onto a machine with the arm continuously outstretched. *The static loads on the right arm and on the shoulder muscles are tiring and reduce skill. The machine should be redesigned so that the operator can work with the elbow lowered and bent at right angles.*

Figure 3.6 The work level should be at such a height that the body takes up a natural posture, slightly inclined forwards, with the eyes at the best viewing distance from the work. *This workbench is a model of its kind, with the elbows supported in a natural position without static effort.*

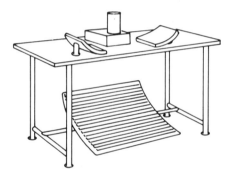

Figure 3.7 The forearms and elbows can be supported by adjustable, padded rests. *Restful support is also given to the legs by a footrest that is adjustable for height and which allows all normal leg movements to be made. For women the bars of the footrest should be closer together, to prevent shoe heels becoming trapped.*

3 *Work sitting down as much as possible.* Workplaces where the operator can choose to either stand or sit are recommended.

4 *Arm movements should be either in opposition to each other or otherwise symmetrical.* Moving one arm by itself sets up static loads on the trunk muscles. Furthermore, symmetrical movements of the arms facilitate nervous control of the operation.

5 *The working area should be located so that it is at the best distance from the eyes of the operator.* See Figure 3.6 and Chapter 5 for more information.

6 *Hand grips, operating levers, tools and materials should be arranged in such a way that the most frequent movements are carried out with the elbows bent and near to the body.* The best position for both strength and skill in the hands is to have them 25–50 cm from the eyes, with the elbows lowered and bent at right angles.

7 *Manipulation can be made easier by using supports under the elbows, forearms or hands.* These supports should be padded with felt or some other soft, warm material and should be adjustable to suit people of different sizes (Figure 3.7).

Summary

People have differing muscle strengths, depending on individual training, age, gender, health and other factors. However, every human body follows the same biomechanical layout. This allows us to derive principles of work and workplace design which enable the exertion of muscular strength with most efficiency and least effort.

CHAPTER FOUR

Body size

Size to fit

Since natural postures – attitudes of the trunk, arms and legs which do not involve static effort – and natural movements are a necessary part of efficient work, it is essential that *the workplace should be suited to the body size and mobility of the operator*.

4.1 Variations in body dimensions

The enormous variations in body size among individuals pose a great challenge for the designer of equipment and workstations. It is not acceptable, as a rule, to design a workplace to suit the phantom of the 'average person'. Often it is necessary to take into account the tallest persons (e.g. to determine the necessary leg room under a table) or the shortest persons (e.g. to make sure they can reach high enough). If doorways' heights were fixed to suit people of average height, many persons passing through them would have bloodied foreheads as they would strike their heads on the lintels.

Design range and percentiles

It is usually not possible to design workplaces to suit the very biggest or the very smallest workers so we must be content with meeting the requirements of the majority. A selection is therefore made, usually so that the extreme body sizes are disregarded. If we decide to design for the 'central 90 per cent' of a group, we fit the persons bigger than the smallest 5 per cent (in the body dimension considered) and smaller than the biggest 5 per cent; in other words, we exclude the smallest 5 per cent and the largest 5 per cent. *The design cut-off points must be carefully selected depending upon the design purpose.*

A particular percentage point on the distribution is called a *percentile (p)*. Thus, in the present example only the percentiles

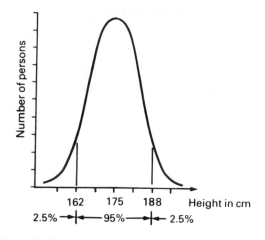

Figure 4.1 Typical distribution of anthropometric data: body heights of male Americans.

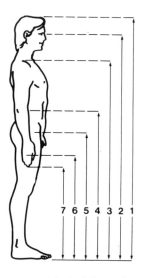

Figure 4.2 Static anthropometric heights of standing persons. *From Pheasant* (1986, 1996).

lying between 5 per cent and 95 per cent (including therefore 90 per cent of the sample) are being considered.

The most striking differences in body size are related to ethnic diversity, gender and age. As a rule, females are smaller than males except in hip dimensions. With increasing age, many adults become shorter, but heavier.

Unless designed specifically to fit individuals, regular workplaces should allow for the bodily dimensions of all users,

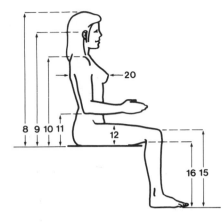

Figure 4.3 Static anthropometric heights of sitting persons. *From Pheasant* (1986, 1996).

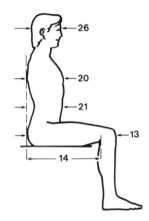

Figure 4.4 Static anthropometric depths of sitting persons. *From Pheasant* (1986, 1996).

female or male, between about 20 and 65 years. This often requires adjustment capabilities, with their lower and upper limits being selected to fit given percentiles, such as the 5th and 95th.

Figure 4.1 shows a typical distribution of anthropometric measurements. In statistical terms it is called a Gauss or 'normal' distribution that can be described by the mean (average) and the standard deviation.

Figures 4.2 through 4.7 illustrate the tabulated dimensions. Their numbering corresponds to that used in Tables 4.1 to 4.6.

If we know the mean (m) and the standard deviation (SD) of

Table 4.1 Anthropometric (measured) data in mm of US adults aged 19–60 years according to Gordon et al. (1989). *The reference numbers of the dimensions are shown in Figures 4.2 to 4.7*

Dimension	Men				Women			
	5th percentile	50th percentile	95th percentile	SD	5th percentile	50th percentile	95th percentile	SD
1 Stature	1647	1755	1867	67	1528	1628	1737	64
2 Eye height	1528	1633	1743	66	1415	1515	1621	63
3 Shoulder height (acromion)	1342	1442	1546	62	1241	1332	1432	58
4 Elbow height	995	1072	1153	48	926	997	1074	45
5 Hip height (trochanter)	853	927	1009	48	789	860	938	45
6 Knuckle height	na	na	na	na	na	na	na	na
7 Fingertip height	591	653	716	40	531	610	670	36
8 Sitting height	855	914	972	36	795	851	910	35
9 Sitting eye height	735	792	848	34	685	738	794	33
10 Sitting shoulder height (acromion)	549	598	646	30	509	555	604	29
11 Sitting elbow height	184	232	274	27	176	221	264	27
12 Thigh height (thickness)	149	168	190	13	140	158	180	12
13 Buttock–knee length	569	615	667	30	542	588	640	30
14 Buttock–popliteal length	458	500	546	27	440	481	528	27
15 Knee height	514	558	606	28	474	514	560	26

16	Popliteal height	395	433	476	25	351	389	429	24
17	Shoulder breadth (bideltoid)	450	491	535	26	397	431	472	23
18	Shoulder breadth (biacromial)	367	397	426	18	333	363	391	17
19	Hip breadth (sitting)	329	365	412	25	343	383	432	27
20	Chest (bust) depth	210	242	280	22	209	237	279	21
21	Abdominal depth (sitting)	199	236	291	28	185	219	271	26
22	Shoulder–elbow length	340	369	399	18	308	335	365	17
23	Elbow–fingertip length	448	483	524	23	406	442	483	23
24	Vertical upper limb length	729	788	856	39	662	723	788	38
25	Vertical shoulder–grip length	612	665	722	33	557	609	664	33
26	Head length	185	197	209	7	176	187	198	6
27	Head breadth	143	152	161	5	137	144	153	5
28	Hand length	179	193	211	10	165	180	197	10
29	Hand breadth	84	90	98	4	73	79	86	4
30	Foot length	249	269	292	13	224	244	265	12
31	Foot breadth	92	101	110	5	82	90	98	5
32	Span	1693	1821	1960	82	1542	1670	1809	81
33	Elbow span	na	na	na	na	na	na	na	na
34	Vertical grip reach (standing)	1958	2106	2260	92	1808	1945	2094	87
35	Vertical grip reach (sitting)	1221	1309	1401	55	1127	1213	1296	51
36	Forward grip reach	693	750	813	37	632	685	744	34
	Weight (kg), estimated by Kroemer	58	78	99	13	39	62	85	14

na = not available

Table 4.2 Anthropometric (estimated) data in mm of British adults aged 19–35 years according to Pheasant (1986, 1996). *The reference numbers of the dimensions are shown in Figures 4.2 to 4.7*

Dimension	Men				Women			
	5th percentile	50th percentile	95th percentile	SD	5th percentile	50th percentile	95th percentile	SD
1 Stature	1625	1740	1855	70	1505	1610	1710	62
2 Eye height	1515	1630	1745	69	1405	1505	1610	61
3 Shoulder height (acromion)	1315	1425	1535	66	1215	1310	1405	58
4 Elbow height	1005	1090	1180	52	930	1005	1085	46
5 Hip height (trochanter)	840	920	1000	50	740	810	885	43
6 Knuckle height	690	755	825	41	660	720	780	36
7 Fingertip height	590	655	720	38	560	625	685	38
8 Sitting height	850	910	965	36	795	850	910	35
9 Sitting eye height	735	790	845	35	685	740	795	33
10 Sitting shoulder height (acromion)	540	595	645	32	505	555	610	31
11 Sitting elbow height	195	245	295	31	185	235	280	29
12 Thigh height (thickness)	135	160	185	15	125	155	180	17
13 Buttock–knee length	540	595	645	31	520	570	620	30
14 Buttock–popliteal length	440	495	550	32	435	480	530	30
15 Knee height	490	545	595	32	455	500	540	27

16	Popliteal height	395	440	490	29	355	400	445	27
17	Shoulder breadth (bideltoid)	420	465	510	28	355	395	435	24
18	Shoulder breadth (biacromial)	365	400	430	20	325	355	385	18
19	Hip breadth (sitting)	310	360	405	29	310	370	435	38
20	Chest (bust) depth	215	250	285	22	210	250	295	27
21	Abdominal depth (sitting)	220	270	325	32	205	255	305	30
22	Shoulder–elbow length	330	365	395	20	300	330	360	17
23	Elbow–fingertip length	440	475	510	21	400	430	460	19
24	Horizontal upper limb length	720	780	840	36	655	705	760	32
25	Horizontal shoulder–grip length	610	665	715	32	555	600	650	29
26	Head length	180	195	205	8	165	180	190	7
27	Head breadth	145	155	165	6	135	145	150	6
28	Hand length	175	190	205	10	160	175	190	9
29	Hand breadth	80	85	95	5	70	75	85	4
30	Foot length	240	265	285	14	215	235	255	12
31	Foot breadth	85	95	110	6	80	90	100	6
32	Span	1655	1790	1925	83	1490	1605	1725	71
33	Elbow span	865	945	1020	47	780	850	920	43
34	Vertical grip reach (standing)	1925	2060	2190	80	1790	1905	2020	71
35	Vertical grip reach (sitting)	1145	1245	1340	60	1060	1150	1235	53
36	Forward grip reach	720	780	835	34	650	705	755	31
	Weight (kg)	55	75	94	12	44	63	81	11

Table 4.3 Anthropometric (estimated) data in mm of West German adults according to Pheasant (1986). *The reference numbers of the dimensions are shown in Figures 4.2 to 4.7*

Dimension	Men				Women			
	5th percentile	50th percentile	95th percentile	SD	5th percentile	50th percentile	95th percentile	SD
1 Stature	1645	1745	1845	62	1520	1635	1750	69
2 Eye height	na	na	na	na	na	na	na	na
3 Shoulder height	1370	1465	1560	58	1240	1320	1400	50
4 Elbow height	1020	1095	1170	46	925	1000	1075	46
5 Hip height	840	910	980	44	760	840	920	48
6 Knuckle height	na	na	na	na	na	na	na	na
7 Fingertip height	na	na	na	na	na	na	na	na
8 Sitting height	865	920	975	32	800	865	930	39
9 Sitting eye height	750	800	850	31	680	740	800	37
10 Sitting shoulder height	na	na	na	na	na	na	na	na
11 Sitting elbow height	195	235	275	25	165	205	245	23
12 Thigh height (thickness)	135	150	265	70	125	155	185	19
13 Buttock–knee length	560	600	640	25	525	580	635	33
14 Buttock–popliteal length	na	na	na	na	na	na	na	na
15 Knee height	500	545	590	28	455	505	555	30

16	Popliteal height	415	455	495	25	355	395	435	23
17	Shoulder breadth (bideltoid)	425	465	505	23	355	400	445	27
18	Shoulder breadth (biacromial)	na	na	na	na	na	na	na	na
19	Hip breadth (sitting)	na	na	na	na	na	na	na	na
20	Chest (bust) depth	na	na	na	na	na	na	na	na
21	Abdominal depth (sitting)	na	na	na	na	na	na	na	na
22	Shoulder–elbow length	na	na	na	na	na	na	na	na
23	Elbow–fingertip length	na	na	na	na	na	na	na	na
24	Horizontal upper limb length	735	785	835	31	660	720	780	36
25	Shoulder–grip length	na	na	na	na	na	na	na	na
26	Head length	na	na	na	na	na	na	na	na
27	Head breadth	na	na	na	na	na	na	na	na
28	Hand length	na	na	na	na	na	na	na	na
29	Hand breadth	na	na	na	na	na	na	na	na
30	Foot length	na	na	na	na	na	na	na	na
31	Foot breadth	na	na	na	na	na	na	na	na
32	Span	na	na	na	na	na	na	na	na
33	Elbow span	na	na	na	na	na	na	na	na
34	Vertical grip reach (standing)	na	na	na	na	na	na	na	na
35	Vertical grip reach (sitting)	na	na	na	na	na	na	na	na
36	Forward grip reach	na	na	na	na	na	na	na	na
	Weight (kg)	na	na	na	na	na	na	na	na

na = not available

Table 4.4 Anthropometric (measured) data in mm of East German adults aged 18–59 years according to Fluegel (1986). *The reference numbers of the dimensions are shown in Figures 4.2 to 4.7*

Dimension	Men				Women			
	5th percentile	50th percentile	95th percentile	SD	5th percentile	50th percentile	95th percentile	SD
1 Stature	1607	1715	1825	66	1514	1606	1707	59
2 Eye height	1498	1600	1705	64	1415	1501	1597	57
3 Shoulder height (acromion)	1320	1412	1512	60	1232	1319	1403	53
4 Elbow height	na	na	na	na	na	na	na	na
5 Hip height	na	na	na	na	na	na	na	na
6 Knuckle height	682	748	819	42	643	702	764	37
7 Fingertip height	588	651	717	39	557	615	672	35
8 Sitting height	846	904	958	34	804	855	905	31
9 Sitting eye height	719	775	831	34	684	733	782	30
10 Sitting shoulder height	552	602	650	31	517	562	609	29
11 Sitting elbow height	198	243	293	29	190	234	282	28
12 Thigh height (thickness)	126	151	176	15	125	146	175	15
13 Buttock–knee length	560	603	648	27	541	584	630	27
14 Buttock–popliteal length	444	485	527	25	437	479	521	26
15 Knee height	490	530	575	27	458	496	538	24

16	Popliteal height	410	452	496	26	380	415	455	23
17	Shoulder breadth (bideltoid)	432	472	510	24	393	436	481	27
18	Shoulder breadth (biacromial)	365	399	430	20	336	365	393	17
19	Hip breadth (sitting)	334	368	406	22	346	400	460	35
20	Chest (bust) depth	na	na	na	na	na	na	na	na
21	Abdominal depth (sitting)	na	na	na	na	na	na	na	na
22	Shoulder–elbow length	432	464	500	20	394	425	556	19
23	Elbow–fingertip length	704	762	820	35	650	702	758	33
24	Vertical upper limb length	na	na	na	na	na	na	na	na
25	Shoulder–grip length	179	190	201	7	170	181	191	6
26	Head length	148	158	168	6	141	151	160	6
27	Head breadth	174	189	205	9	161	174	189	9
28	Hand length	81	88	96	5	71	78	85	4
29	Hand breadth	243	264	285	13	222	241	260	12
30	Foot length	91	102	113	6	83	93	104	6
31	Foot breadth	1640	1761	1885	75	1503	1614	1735	70
32	Span	833	895	911	39	757	816	881	38
33	Elbow span	1975	2121	2267	89	1843	1973	2103	79
34	Vertical grip reach (standing)	na	na	na	na	na	na	na	na
35	Vertical grip reach (sitting)	704	763	824	37	650	706	767	35
36	Forward grip reach	na	na	na	na	na	na	na	na
	Weight (kg)								

na = not available

Table 4.5 Anthropometric data in mm of Japanese adults according to Pheasant (1986). *The reference numbers of the dimensions are shown in Figures 4.2 to 4.7*

Dimension	Men				Women			
	5th percentile	50th percentile	95th percentile	SD	5th percentile	50th percentile	95th percentile	SD
1 Stature	1560	1655	1750	58	1450	1530	1610	48
2 Eye height	na	na	na	na	na	na	na	na
3 Shoulder height	1250	1340	1430	54	1075	1145	1215	44
4 Elbow height	965	1035	1105	43	895	955	1015	36
5 Hip height (trochanter)	765	830	895	41	700	755	810	33
6 Knuckle height	na	na	na	na	na	na	na	na
7 Fingertip height	na	na	na	na	na	na	na	na
8 Sitting height	850	900	950	31	800	845	890	28
9 Sitting eye height	735	785	835	31	690	735	780	28
10 Sitting shoulder height	na	na	na	na	na	na	na	na
11 Sitting elbow height	220	260	300	23	215	250	285	20
12 Thigh height (thickness)	110	135	160	14	105	130	155	14
13 Buttock–knee length	500	550	600	29	485	530	575	26
14 Buttock–popliteal length	na	na	na	na	na	na	na	na
15 Knee height	450	490	530	23	420	450	480	18

16 Popliteal height	360	400	440	24	325	360	395	21
17 Shoulder breadth (bideltoid)	405	440	475	22	365	395	425	18
18 Shoulder breadth (biacromial)	na	na	na	na	na	na	na	na
19 Hip breadth	na	na	na	na	na	na	na	na
20 Chest (bust) depth	na	na	na	na	na	na	na	na
21 Abdominal depth	na	na	na	na	na	na	na	na
22 Shoulder–elbow length	na	na	na	na	na	na	na	na
23 Elbow–fingertip length	na	na	na	na	na	na	na	na
24 Horizontal upper limb length	665	715	765	29	605	645	685	25
25 Shoulder–grip length	na	na	na	na	na	na	na	na
26 Head length	na	na	na	na	na	na	na	na
27 Head breadth	na	na	na	na	na	na	na	na
28 Hand length	na	na	na	na	na	na	na	na
29 Hand breadth	na	na	na	na	na	na	na	na
30 Foot length	na	na	na	na	na	na	na	na
31 Foot breadth	na	na	na	na	na	na	na	na
32 Span	na	na	na	na	na	na	na	na
33 Elbow span	na	na	na	na	na	na	na	na
34 Vertical grip reach (standing)	na	na	na	na	na	na	na	na
35 Vertical grip reach (sitting)	na	na	na	na	na	na	na	na
36 Forward grip reach	na	na	na	na	na	na	na	na
Weight (kg)	na	na	na	na	na	na	na	na

na = not available

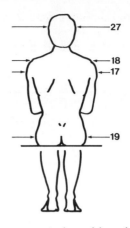

Figure 4.5 Static anthropometric breadths of sitting persons. *From Pheasant* (1986, 1996).

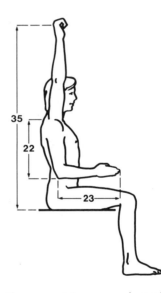

Figure 4.6 Static anthropometric arm and reach dimensions of sitting persons. *From Pheasant* (1986, 1996).

any group measurement then we can calculate percentiles. For example:

$$2.5p = m - 1.96SD \quad 97.5p = m + 1.96SD$$
$$5p = m - 1.65SD \quad 95p = m + 1.65SD$$
$$10p = m - 1.28SD \quad 90p = m + 1.28SD$$
$$16.5p = m - 1.00SD \quad 83.5p = m + 1.65SD$$
$$50p = m +/- 0 \; SD$$

Table 4.6 Average anthropometric data in mm estimated for 20 regions of the earth. *Adapted from Juergens et al.* (1990).

	1. Stature see Fig. 4.2		8. Sitting height see Fig. 4.3		15. Knee height, sitting see Fig. 4.3	
	Females	Males	Females	Males	Females	Males
North America	1650	1790	880	930	500	550
Latin America						
Indian population	1480	1620	800	850	445	495
European and Negroid population	1620	1750	860	930	480	540
Europe						
North	1690	1810	900	950	500	550
Central	1660	1770	880	940	500	550
East	1630	1750	870	910	510	550
Southeast	1620	1730	860	900	460	535
France	1630	1770	860	930	490	540
Iberia	1600	1710	850	890	480	520
Africa						
North	1610	1690	840	870	500	535
West	1530	1670	790	820	480	530
Southeast	1570	1680	820	860	495	540
Near East	1610	1710	850	890	490	520
India						
North	1540	1670	820	870	490	530
South	1500	1620	800	820	470	510
Asia						
North	1590	1690	850	900	475	515
Southeast	1530	1630	800	840	460	495
South China	1520	1660	790	840	460	505
Japan	1590	1720	860	920	395	515
Australia						
European extraction	1670	1770	880	930	525	570

Statistical procedures

Anthropometric data can yield a wealth of information to the skilled statistician, but even a 'layperson' such as an engineer or designer can calculate percentiles, ranges, of homogeneous or composite population groups using the advice given in books written for ergonomists and human factors engineers, such as those by Kroemer *et al.* (1994, 1997), Pheasant (1986, 1996) and Roebuck (1995).

4.2 National and international data

Tables 4.1 to 4.7 show body measurements of several national populations. Unfortunately, rather few large-scale anthropometric

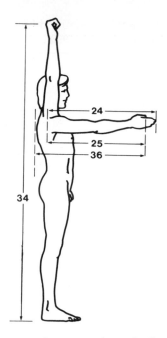

Figure 4.7 Static anthropometric arm and reach dimensions of standing persons. *From Pheasant* (1986, 1996).

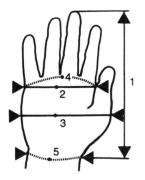

Figure 4.8 Static anthropometric hand dimensions listed in Table 4.7: length (1), breadth across knuckles (2), maximal breadth (3), circumferences at knuckles (4) and at wrist (5).

surveys have been conducted, and often these were done on military samples, not on civilians. Therefore, data must often be interpreted and estimated from surveys on soldiers. Kroemer *et al.* (1994), Pheasant (1986, 1996) and Roebuck (1995) discuss in detail the use of available anthropometric information. The data in Tables 4.1 through 4.6 are taken from these authors' publications, where all the references to original sources or measurements can be looked up.

Table 4.7 Hand and wrist sizes (in mm)

		Males		Females	
		Mean	*SD*	Mean	*SD*
1 Length	US soldiers	194	10	181	10
	US Vietnamese	177	12	165	9
	Japanese	–	–	–	–
	Chinese, Hong Kong	–	–	–	–
	British	189	10	174	9
	Germans	189	9	175	9
2 Breadth at knuckles	US soldiers	90	4	80	4
	US Vietnamese	79	7	71	4
	Japanese	–	–	90	5
	Chinese, Hong Kong	–	–	92	5
	British	87	5	76	4
	Germans	88	5	78	4
3 Maximal breadth	US soldiers	–	–	–	–
	US Vietnamese	100	6	87	6
	Japanese	–	–	–	–
	Chinese, Hong Kong	–	–	–	–
	British	105	5	92	5
	Germans	107	6	94	6
4 Circumference at knuckles	US soldiers	214	10	186	9
	US Vietnamese	–	–	–	–
	Japanese	–	–	–	–
	Chinese, Hong Kong	–	–	–	–
	British	–	–	–	–
	Germans	–	–	–	–
5 Wrist circumference	US soldiers	174	8	151	7
	US Vietnamese	163	15	137	18
	Japanese	–	–	–	–
	Chinese, Hong Kong	–	–	–	–
	British	–	–	–	–
	Germans	–	–	–	–

4.3 Hand size

Hand size is particularly important when designing controls. Figure 4.8 and Table 4.7 show relevant information reported by Courtney (1984), Fluegel et al. (1986), Garrett and Kennedy (1971), Greiner (1991), Imrhan et al. (1993) and Pheasant (1996).

4.4 Angles of rotation in joints

The operating range of a limb is a product of its length and the angle of rotation about its proximal joint. Figure 4.9 and Table

Table 4.8   Mobility of wrist and forearm (in degrees). (*From Kroemer et al., 1997*).

Direction (see Figure 4.9)	Males			Females		
	5th percentile	50th percentile	95th percentile	5th percentile	50th percentile	95th percentile
Wrist flexion F	51	68	85	54	72	90
Wrist extension E	47	62	76	57	72	88
Wrist radial deviation (adduction) R	14	22	30	17	27	37
Wrist ulnar deviation (abduction) U	22	31	40	19	28	37
Forearm supination S	86	108	135	87	109	130
Forearm pronation P	43	65	87	63	81	99

Table 4.9   Mobility of ankle (in degrees). (*From Kroemer et al., 1997*).

Direction (see Figure 4.10)	Males			Females		
	5th percentile	50th percentile	95th percentile	5th percentile	50th percentile	95th percentile
Ankle flexion F	18	29	34	13	23	33
Ankle extension E	21	36	52	31	41	52

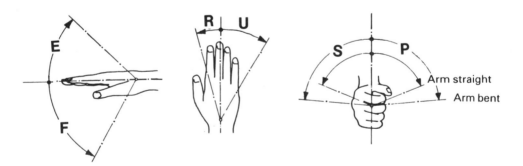

Figure 4.9   Movements of the hand and wrist listed in Table 4.8.

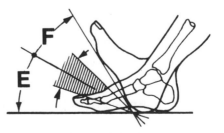

Figure 4.10   Movements of the ankle listed in Table 4.9.

4.8 show some angular motion ranges of the hand and forearm. Figure 4.10 indicates motion ranges of the foot, listed in Table 4.9. However, the actual mobility may be strongly influenced by a person's individual flexibility and by the strength or precision of motion to be exerted. The 'most convenient' range is usually in the middle of the operating range.

Summary

To fit equipment to persons of various body dimensions requires (a) data and (b) proper procedures. Data on many populations are available; for other groups they may be estimated or should be measured according to standard procedures. The design procedures involve the selection of percentile values to be fitted.

CHAPTER FIVE

The design of workstations

**Recommendations
are compromises**

The ergonomic recommendations for the dimensions of workstations are to some extent based on anthropometric data but behavioural patterns of people and specific requirements of the work itself must also be considered. Thus, the recommended dimensions given in textbooks or in various standard works are compromise solutions which may be quite arbitrary. Another critical remark is necessary. Most standard specifications for ergonomic workstations were worked out by committees in which many interested goups were represented: manufacturers, industry associations, unions, employers and ergonomists. The resulting recommendations seem reasonable and suitable in most cases but they are seldom ideal in the eyes of the human factors specialist under practical conditions. It is therefore not surprising that field studies or practical experience do not always confirm recommended standard dimensions.

5.1 Working heights

**Working heights
when standing**

Working height is of critical importance in the design of workplaces. If work is raised too high the shoulders must frequently be lifted up to compensate, which may lead to discomfort, even painful cramp in the neck and shoulders. If the working height is too low the back must be excessively bowed, which often causes backache. Hence, the work surface must be of such a height that it suits the stature of the operator, whether standing or sitting at the work.

The most favourable working height for handwork while standing is 50–100 mm below elbow level. The average elbow height (distance from floor to underside of elbow when it is bent at right angles with the upper arm vertical) is almost 1070 mm for

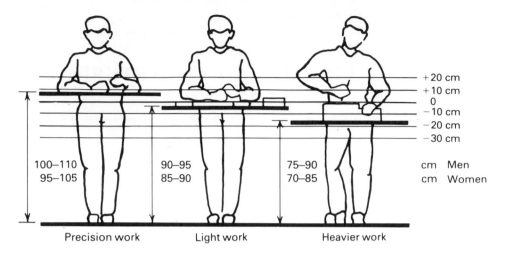

Figure 5.1 Recommended heights of benches for standing work. *The reference line (±0) is the height of the elbows above the floor, which averages 1050 mm for men and 980 mm for women.*

men and 1000 mm for women both in Europe and North America.

Hence, we may conclude that, on average, working heights of 970–1020 mm will be convenient for 'western' men and 900–950 mm for women when standing. Of course, for populations with distinctly different body dimensions or working habits, other working heights may be desirable. In addition to these anthropometric considerations, we must allow for the nature of the work:

1 For delicate work (e.g. drawing) it is desirable to support the elbow to help reduce static loads in the muscles of the back. A good working height is about 50–100 mm *above* elbow height.

2 During manual work an operator often needs space for tools, materials and containers of various kinds, and a suitable height for these is 100–150 mm *below* elbow height.

3 During standing work, if it involves much effort and makes use of the weight of the upper part of the body (e.g. woodworking or heavy assembly work), the working surface needs to be lower; 150–400 mm below elbow height is adequate.

Recommended working heights when standing are set out in Figure 5.1.

The dimensions recommended in Figure 5.1 are merely general guidelines, since they are based on average body 'western' measurements and make no allowances for individual variation. The table heights quoted are too high for short people, who will need to use a platform, footrests or some similar support. On the

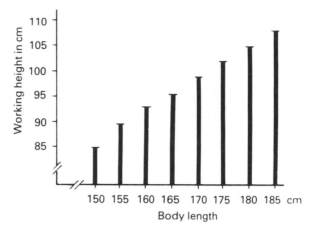

Figure 5.2 Working heights for light work while standing, in relation to body length.

other hand, tall people will have to bend over the work table, which can be a cause of back strain.

Fully adjustable work tables when standing

Ergonomically speaking, therefore, it is often desirable to be able to adjust the working height to suit the individual. Instead of improvisations such as foot supports or lengthening the legs of the work table, a fully adjustable bench is recommended. Figure 5.2 shows the recommended working heights for light standing work, in relation to the stature of the operators.

If one is unable to provide fully adjustable benches, or if the operating level at a machine cannot be varied, then in principle *working heights should be set to suit the tallest operators*: smaller people can be accommodated by giving them platforms to stand on.

Workplaces for sedentary work

A classic study carried out in 1951 is often cited when working heights are discussed. Ellis (1951) was able to confirm an old empirical rule: *the maximal speed of operation for manual jobs carried out in front of the body is achieved by keeping the elbows down at the sides and the arms bent at a right angle*. This elbow height is a generally accepted basis for the assessment of working heights for sedentary activities.

Since the work may also require fine or precise manipulation, working heights must also allow for good visual distance and angle, as shown in Figure 3.6. In such cases the work level must be raised until the operator can see clearly while keeping his or her back in a natural posture. The opposite, that is lowering of the working surface, is necessary when handwork calls for great force or much freedom of movement. However, the need for a low table conflicts with the necessity for enough knee room under the table and this can be limiting factor. If we extract the measurement 'knee height' (which is the distance between ground and the upper surface of knee) for large individuals (95th percentile) from Table 4.1 and if we add 50 mm to allow for heels and minimal clearance to facilitate movement, we arrive (after rounding to convenient numbers) at the following recommendations for *free knee room*

> *men (600 + 50 mm) = 650 mm*
> *women (560 + 50 mm) = 610 mm*

If we allow 40 mm for the thickness of the tabletop, then *the lowest table level* is:

> *men 690 mm*
> *women 650 mm*

This lowest working surface is recommended for heavy assembly work, for working over any kind of container and for preparatory work in the kitchen. It should also be noted here that the maximum distance from seating surface to the underside of the table should be 190 mm (95 per cent of the thickness of the thighs, from Table 4.1).

How office employees sit

A survey carried out by Grandjean and Burandt (1962) on 261 men and 117 women engaged in traditional office work revealed interesting links between desk height and musculoskeletal troubles. The work-sampling analysis gave particulars about the different sitting postures shown in Figure 5.3.

An upright trunk posture was observed for only about 50 per cent of the time, the trunk leaning against the backrest about 40 per cent of the time, although most of the chairs were provided with rather poor back supports.

Figure 5.4 presents the results of the survey on musculoskeletal complaints.

Links between desk height, sitting behaviour and pains

The principal anthropometric data and desk heights were assessed and compared with the reports on musculoskeletal troubles. From a large number of results and calculated correlations, the following conclusions emerged:

1 Twenty-four per cent reported pains in neck and shoulders which most of the subjects, especially the typists, blamed on a desktop that was too high.

2 Twenty-nine per cent reported pains in the knees and feet, most of them short people who had to sit on the front edge of their chairs, probably because they had no footrests.

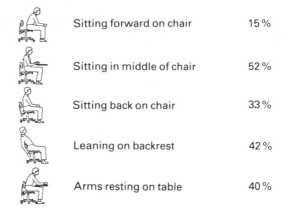

	Sitting forward on chair	15%
	Sitting in middle of chair	52%
	Sitting back on chair	33%
	Leaning on backrest	42%
	Arms resting on table	40%

Figure 5.3 Sitting postures of 378 office employees, as shown by a multi-moment observation technique. *There were 4920 observations. The percentages quoted indicate how much of the working period was spent in that posture. The two lower observations were seen simultaneously with the three upper postures; that is why the sum of all five characteristics exceeds 100 per cent. After Grandjean and Burandt* (1962).

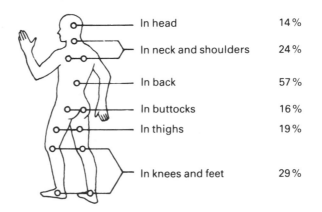

	In head	14%
	In neck and shoulders	24%
	In back	57%
	In buttocks	16%
	In thighs	19%
	In knees and feet	29%

Figure 5.4 Incidence of bodily aches among 246 employees engaged in traditional sedentary office jobs. *Multiple answers were possible. After Grandjean and Burandt* (1962).

3 *A desktop 740–780 mm high gave employees the most scope for adaptation to suit themselves,* provided that a fully adjustable seat and footrests were available.

4 *Regardless of their stature, the great majority of workers preferred the seat to be 270–300 mm below the desktop. This seems to permit a natural position of the trunk, obviously a point of major importance to these employees.*

5 The incidence of backache (57 per cent) and the frequent use of the backrest (42 per cent of the time) indicate the need to

relax the back muscles periodically and may be quoted as evidence of the importance of a well-constructed backrest.

Electromyo-graphy of shoulder muscles

Several authors have recorded the electrical activity of the shoulder muscles while subjects worked at different desk heights. It should be remembered here that the electrical activity of a muscle is an indicator of the contraction of muscle fibres (to be exact, of motor units) which can be used, after careful calibration, to estimate the exerted muscular force, as described by Basmajian and de Luca (1985) and Soderberg (1992). The procedure is called electromyography. As far back as 1951, Lundervold investigated the electrical activity of shoulder and arm muscles of subjects operating typewriters at high and lower levels. (At that time electrical impulses were obtained with ink-writer records.) Lundervold concluded that the least action potentials (electrical impulses) were recorded when the person undergoing the experiment was sitting in a relaxed and well-balanced state of muscular equilibrium or was using a backrest. Intensive typewriting was associated with raised shoulders and a strongly increased electrical activity of the trapezius and deltoid muscles. (The trapezius muscle lifts the shoulders, the deltoid the upper arms.) More recently, Hagberg (1982) made a quantitative analysis of the electromyograms of shoulder and arm muscles when typing at different heights. These results are shown in Figure 5.5.

Keeping the shoulders lifted is strenuous static work

A working level that is too high can be compensated for either by lifting the shoulders (mostly through contraction of the trapezius muscle) or by lifting the upper arm (with the deltoid muscle). When the maintained contraction force of the lifting muscle is substantive, such as reaching 20 per cent of the maximal force, this would suffice to eventually generate great pains.

Most important: seat-to-desk distance

A slight forward stoop, with the arms on the desk, is only minimally tiring when reading or writing, but in order to relax the back *the distance from seat surface to desktop must be between 270 and 300 mm* for most 'westerners'. As mentioned above, an employee sitting at an office desk first looks for a comfortable and relaxed trunk position, and often accepts a seat height that is bad for the legs or buttocks rather than sacrifice a comfortable trunk posture. This is described in books edited by Grandjean (1973, 1987) and recently by Lueder and Noro (1995).

The height of tables which are not adjustable is primarily based on average body measurements and makes no allowance for individual variation. Therefore, all table heights recommended on the basis of 'average' body dimensions are too high for short people, who will need some kind of footrest. Yet tall people will

A

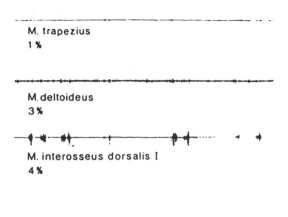

M. trapezius
1 %

M. deltoideus
3 %

M. interosseus dorsalis I
4 %

B

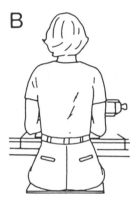

M. trapezius
20 %

M. deltoideus
3 %

M. interosseus dorsalis I
6 %

C

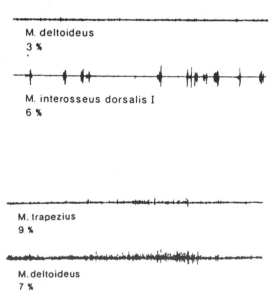

M. trapezius
9 %

M. deltoideus
7 %

M. interosseus dorsalis I
4 %

Figure 5.5 Electromyographic recording of shoulder muscle activity. *The figures in brackets refer to percentage of maximum voluntary contraction position. A = optimal height of the typewriter (i.e. home row at elbow height); B = too high, resulting in elevation of shoulders by the trapezius muscle; C = too high, compensated by a sideward elevation of the upper arms by the deltoid muscle. According to Hagberg (1982).*

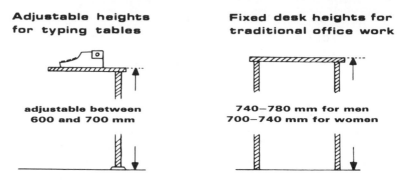

Figure 5.6 Recommended desk-top heights for traditional office jobs. *Left: range of adjustability for typing-desks. Right: desktop heights for reading and writing without typewriter.*

have to bend their necks over the work table which may cause musculoskeletal troubles in the neck and back. Table heights for traditional office work (with the exception of typing) thus follow the rule that it is more practical to choose a height to suit the tall rather than the short person; the latter can always be given a footrest to raise the seat to a suitable level. On the other hand, a tall person given a table that is too low can do nothing about it except fix the seat so low that it might be uncomfortable for their legs.

Conclusion for an office desk without keyboards

As illustrated in Figure 5.6, *office desks to be used without a typewriter or other keyboard equipment should have a height of 740–780 mm for men and 700–740 mm for women (of European extraction) assuming that the chairs are fully adjustable and footrests are available.* These figures are slightly higher than most standard specifications recommending desktop heights between 720 and 750 mm, which are certainly not ideal for tall male employees.

Vertical and horizontal leg room

It is important that office desks allow plenty of space for leg movement and it is an advantage if the legs can be crossed without difficulty. For this reason there should be no drawers above the knees and no thick edge to the desktop. *The tabletop should not be thicker than 30 mm and the space for legs and feet under the table should be at least 680 mm wide and 690 mm high. (As discussed above and below, these recommendations apply to most Europeans and North Americans. For populations of different body sizes, or working habits, these recommendations probably are not appropriate.)*

Many sitting persons, especially those who lean back, periodically like to stretch their legs under the table or desk. It is therefore necessary to leave enough space in the horizontal plane as well. *At knee level the distance from table front edge to back wall should not be less than 600 mm and at the level of the feet be at least 800 mm.* These recommendations are also valid for workplaces with typewriters or other keyboard equipment.

Typing tables shall be height-adjustable

The classical guideline, mentioned above, assumes a straight upright trunk position with the elbows at the sides and bent at right angles, and has for a long time been the basis for the design of typing tables. Since the height of the keyboard defines the working level, the middle row (or so-called home row) should be at about elbow level. However, the need for such a low table often conflicts with the necessity for enough knee space under the table. This can be a limiting factor and *calls for height-adjustable typing tables.* Indeed, *today most experts recommend the height of such tables to be adjustable between 600 and 700 mm* (see Figure 5.6).

Recommendations for fixed typing tables are problematic

Until recently many 'western' experts have recommended a fixed top height of 650 mm. This recommendation is based on two assumptions which are questionable for the following reasons:

1 The assumed straight upright trunk posture is normally not adopted by keyboarders if their work lasts for hours. They often lean back or forward in order to relax the back muscles or to get a suitable viewing distance.

2 Old mechanical typewriters required considerable key displacements and stroke forces of more than 5 N. Modern electrical typewriters and electronic computer keyboards need much less key travel and lower stroke forces of 0.4 to 1.0 N. Their keys can be operated with little muscular effort from the upper arms and shoulders and the operator may assume any posture that he or she likes. Such posture can include standing, at least for a while, or various ways of sitting, and the keyboard may be placed on a traditional table or on one's lap.

Concluding, one can recommend the heights reported in Table 5.1 as general rules but many people would be served best if they could choose their working heights according to their own body sizes, posture preferences and working habits; adjustability allows many choices.

Alternate sitting and standing

A workplace which allows the operator to sit or stand, as he or she wishes, is highly recommended from a physiological and orthopaedic point of view. There is less to hold up when one is sitting down than when standing; on the other hand, sitting is the

Table 5.1 Desk and table heights (in mm) for seated work

Kind of work	Men	Women
Precision work with close visual range	900–1100	800–1000
Reading and writing	740–780	700–740
Manual work requiring strength or space for containers	680	650
Adjustable range for typing tables	600–700	600–700

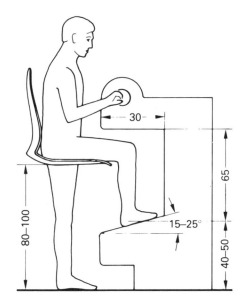

Figure 5.7 Alternately sitting and standing at work (Dimensions are in cm).

cause of many aches and pains which can be relieved by standing up and moving about. Standing and sitting impose stresses upon different muscles so that each changeover relaxes some muscles and stresses others. Furthermore, we have good grounds for believing that each change from standing to sitting (and vice versa) is accompanied by variations in the supply of nutrients to the intervertebral discs so that the change is beneficial in this respect too.

Figure 5.7 shows a machine workstation which permits alternate sitting and standing, and the relevant measurements are as follows:

Horizontal knee room	300×650 mm
Height of working area above seat	300–600 mm
Height of working area above floor	1000–1200 mm
Range of adjustment of seat	800–1000 mm

Inclined desktop and trunk posture

The question of work surfaces for special purposes arises in relation to school desks and tables where much reading and writing are carried on. Eastman and Kamon (1976) made a contribution to this problem. Six male subjects carried out writing and reading exercises for 2.5 h each at a flat table and at tables which were tilted to 12 and 24°, respectively. Inclinations of the body (measured as the angle between the horizontal plane and a straight line from the twelfth thoracic vertebra to the eyes) were as follows:

Horizontal table   35–45°
Table sloped 12°   37–48°
Table sloped 24°   40–50°

So, tilting the tabletop leads to a more erect posture, as well as to fewer pulses on the electromyogram. Subjectively, there was an impression of fewer aches and pains.

Bendix and Hagberg (1984) examined the effects of inclined desktops on 10 reading subjects. With increasing desk inclination the cervical as well as the lumbar spine were extended and the head and trunk assumed a more upright posture. The electromyography of the trapezius muscle (shoulder-lifting muscle) showed no change in muscle strain. A rating of acceptability for both reading and writing with either desk inclination favoured a steep slope of the desk for reading but a flatter slope for writing. The authors conclude that reading material should be placed on a sloping desk and paper for writing on a horizontal tabletop. A separate sloping desk ('lectern') placed on a level table may be preferred to inclining the whole tabletop since a slope of more than 10° usually causes paper and pencils to slide off.

As a general rule it seems that a sloped work table is an improvement over a flat one, posturally as well as visually. These advantages have certain drawbacks, however, notably that it is difficult to keep things put on an inclined surface from rolling or sliding off.

5.2 Neck and head postures

The posture of the neck and head is not easy to assess since seven joints determine the mobility of this part of the body. In fact, it is possible to combine an erect or even backward flexed (lordotic) neck with a downward bent head, or a forward flexed neck with an upwards directed head.

Looking at a subject from the (right) side, some authors define the neck/head posture by measuring an angle between a line along the neck related either to a horizontal (or vertical) or to a line along the trunk. Most authors consider such a neck/head angle of 15° to be acceptable. Chaffin (1973) determined the neck/head angle related to the horizontal and observed on five subjects that

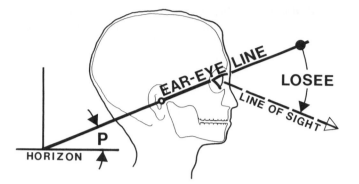

Figure 5.8 The EE line is easy to determine using the earhole and the junction of the eyelids. The EE line describes the posture of the head and serves as reference for the angle of the line of sight.

the average time to reach a marked muscle fatigue was shorter with increasing neck/head angles. The author concludes that localised muscle fatigue in the neck area can be a preliminary sign of other more serious and chronic musculoskeletal disorders and that the inclination angle of the head should not exceed 30° for any prolonged period of time. Let us note here that neck pains can vanish overnight when documents are presented on raised reading stands instead of laying them flat on the desktop.

Another way of assessing neck/head posture is by measuring the angle between a line running from the seventh cervical vertebra to the earhole and a vertical line. However, this is difficult to do – just as it is difficult to locate the so-called 'Frankfurt Line' that connects the earhole and the lower rim of the orbit, the opening in the skull in which our eyeballs are located. For very practical reasons, it is much easier to establish the 'Ear–Eye Line' (EE) which runs from the earhole to the meeting point of the eyelids (Figure 5.8).

The EE Line can be used for two purposes: to describe the tilt posture of the head and as the reference for the angle of the 'line of sight'. If the EE Line is tilted approximately 15° above the horizon so that the eyes are higher than the earholes, the head is held 'erect' or 'upright'.

Line of sight

The line of sight connects the pupil with the visual target. Assuming an erect head, as just defined, the preferred direction of sight is approximately 'straight ahead' for distant targets but more and more declined the closer the object that the eye must focus upon. At 'reading distance' (about 400 to 700 mm from the pupil), the preferred declination below the EE line is about 45°. However, there are large differences in the preferred angles from

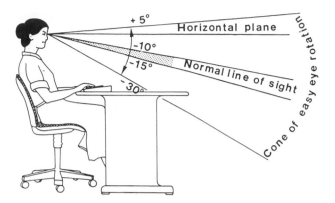

Figure 5.9   The cone of easy eye rotation.

person to person. Eye movement within about 15° above and below the average line-of-sight angle is still comfortable. That means that regular viewing tasks should be within a 30° cone around the principal line of sight as shown in Figure 5.9.

This arrangement accomodates the resting condition of the eye and avoids 'visual fatigue'. *For this reason one should be able to position a display or any other close visual target so that the viewing angle is about 45° below the EE line.*

Differing opinions

The human engineering literature contains recommendations for the angle of the line of sight to be 'normally' at 0 to 15° below the horizontal but this applies only if the head is held erect and when distant objects must be viewed. Lehmann and Stier (1961) reported that seated subjects preferred to view objects placed on a table at an average viewing angle of 38° below the horizontal. Approximately half of this angle was attributed to bending the head downward and the other half to rotation of the eyeballs inside the skull.

Preferred line of sight related to head position

In a more recent study, Hill and Kroemer (1986) investigated the preferred line of sight by measuring its angle to a plane affixed to the head so that it moved with the tilting of the head. Thirty-two subjects were tested in several experiments, with a total duration of about 2 h. The seat was provided with a high backrest and could be inclined backwards from 90° to 105 and 130°. The subject's head was placed against the headrest and was not flexed. The results showed an overall mean angle of 45° below the EE Line (34° below the Frankfurt line) with a large range of between 3° above and 82° below EE; the standard deviation was about 12°.

Relating the line of sight to the horizon (and not to the EE or Frankfurt planes) yielded the following results:

The line-of-sight angle was
29° below horizontal with a backrest angle of 90° (upright posture),
19° below horizontal with a backrest angle of 105° (leaning back),
8° above horizontal with a backrest angle of 130° (steeply reclining)

These results indicate that one cannot recommend any angle of the line of sight against the horizontal without considering the body position, especially the head posture. Using the EE line as reference allows us to define both the head tilt angle and the angle of the line of sight.

Conclusions for posture of neck and head

The present state of knowledge suggests that the head and neck should not be bent forward by more than 30° (meaning that the EE line should not be more than 15° below the horizon) when the trunk is erect – otherwise fatigue and troubles are likely to occur. For persons keeping their trunks and heads erect, the average preferred line of sight lies horizontal if the visual target is far away, but distinctly declined below the horizon if the object of focus is close. This also applies to the preferred viewing angles of VDU operators watching their computer screens.

5.3 Room to grasp and move things

Grasp and reach envelopes

An understanding of how much room the hands and arms need to take hold of things and to move them about is an important factor in the planning of controls, tools, accessories of various kinds and workplaces on which to put these down, such as in assembly or inspection. Reaching too far to pick things up leads to excessive movements of the trunk, making the operation itself less accurate and more energy consuming, and increasing the risks of pains in the back and shoulders.

The grasp or reach envelope is determined by the sweep radius of the arms, with the hands in a grasping or reaching attitude. Decisive factors are the location of the person's shoulder joint and the distance from this joint to the hand. This is a case in which we need to consider persons with short arms. Fifth percentile measurements, taken on Americans, are as follows:

Shoulder height when standing
Small men 1342 mm
Small women 1241 mm

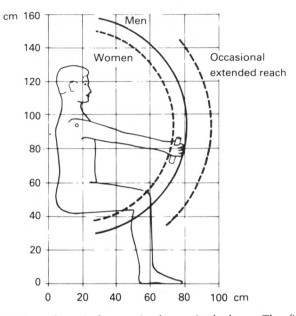

Figure 5.10 Arc of vertical grasp in the sagittal plane. *The figures include the 5th percentile and so apply to small men and women. The grasp can be extended occasionally by reaching out another 150 mm or so by stretching the feet, legs and shoulders.*

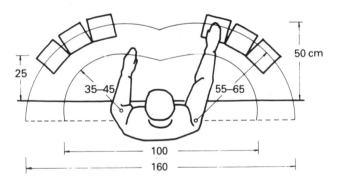

Figure 5.11 Horizontal arc of grasp and working area at tabletop height. *The grasping distance takes account of the distance from shoulder to hand; the working distance only elbow to hand. The values include the 5th percentile and so apply to men and women of less than average size.*

Shoulder height above seat
Small men 549 mm
Small women 509 mm

Arm length (thumbtip)
Small men 739 mm
Small women 677 mm

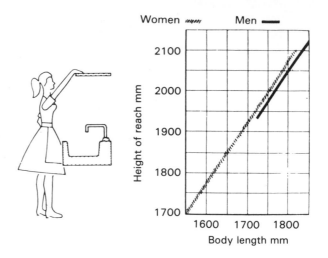

Figure 5.12 Height to which a free-standing person can reach and place a hand flat on a shelf. *Drawn from data given by Thiberg (1965–70)*.

Table 5.2 Maximal height of reach for American men and women (standing)

	Percentile	To fingertip (mm) 1	Grasping height (mm) 2
Men			
Tall	95	2393	2260
Short	5	2073	1958
Women			
Tall	95	2217	2094
Short	5	1914	1808

1 Measurement 83, 2 Measurement D42 in Gordon et al (1989)

With these data, the work envelope of the hands is within an arc that has a radius of about 730 mm for men and 670 mm for women about their shoulders. Of course, 'convenient' reaches are not as far out but close to the body. For each arm the working space becomes a nearly semi-circular shell. The shells overlap in front of the body. Figure 5.10 shows this in the sagittal plane through the shoulder Figure 5.11 depicts a top view.

An occasional stretch to reach beyond this range is permissible since the momentary effect on trunk and shoulders is transient – in fact, it might be desirable to change the body posture.

Table 5.2 summarises maximal vertical reach and grasp heights for American men and women, standing.

It is often necessary to be able to look to the back of a shelf

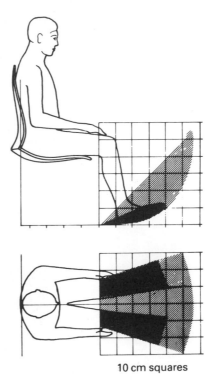

10 cm squares

Figure 5.13 Range of operation of the feet. *The optimal range for delicately controlled pedal work, where only slight force is needed, is double-shaded. After Kroemer (1971).*

in order to see the object to be grasped: a bottle, a box or some other object. Under these conditions the highest shelf should not be located higher than about 1500–1600 mm for men and 1400–1500 mm for women as shown in Figure 5.12. At these heights shelves may be up to 600 mm deep.

Horizontal grasp at tabletop level

Figure 5.11 illustrates grasping and working space over a tabletop. All materials, tools, controls and containers should be deployed within this space but remember that an occasional stretch up to distances of 700–800 mm is not harmful.

Operating space for the legs

Room to operate the feet is essential for all forms of pedal control. This is illustrated in Figure 5.13. The placement and design of pedals are discussed in Chapter 9.

5.4 Sitting at work

Historical background

Some people have no need for seats because they are used to crouching, kneeling or squatting. So where did seats come from? The anthropological answer is that they originated as status

Figure 5.14 The seat of the mayor of Berne (Switzerland) created by the cabinet maker M. Funk in 1735. *This beautiful chair was certainly, above all, a status symbol for the distinguished mayor who was allowed to use it. (Historisches Museum Berne, Switzerland.)*

symbols: only the chief had the right to be raised up. Hence, the gradual development of ceremonial stools which indicated status by their size and decoration. The highest point of this development in cultural history is the splendour of a throne. Figure 5.14 shows such an example.

This status function persists to the present day. Anyone who doubts this should look inside a factory making office furniture and notice that there is a type of chair associated with salary levels. Not long ago there was, for example, a wooden chair for typists, a thinly upholstered one for the senior clerk, a thickly upholstered chair for the office manager and a swivel armchair upholstered in leather for the director.

At the end of the nineteenth century the idea gradually emerged that well-being and efficiency are improved and fatigue reduced if people can sit, and sit well, while doing their work. The reason is physiological. As long as a person is standing an outlay of static muscular effort to keep the joints of the feet, knees and hips in fixed positions is required; this muscular effort ceases when a person sits down.

This realisation led to a greater application of orthopaedic and ergonomic ideas to the design of seats for work. This development gained in importance as more and more people sat down at their work until today about three-quarters of all operatives in industrial countries have sedentary jobs.

Pros and cons

The advantages of sedentary work are:

1 Taking the weight off the legs.

2 Stability of upper body posture.

3 Reduced energy consumption.

4 Fewer demands on the circulatory system.

These advantages must be set against certain drawbacks. Prolonged sitting leads to a slackening of the abdominal muscles ('sedentary tummy') and to curvature of the spine, which in turn is bad for the organs of digestion and breathing.

Main problem is the back

The most severe problem involves the spine and the muscles of the back, which in many sitting positions are not merely not relaxed but positively stressed in various ways.

About 80 per cent of adults have backache at least once in their lives and the commonest cause of this is disc trouble.

Intervertebral discs

The intervertebral disc is a sort of cushion which separates two vertebrae. Collectively, the discs give flexibility to the spine. A disc consists internally of a viscous fluid enclosed in a tough, fibrous ring which encircles the disc. A schematic representation of a disc between two vertebrae and its location close to the spinal cord and its nerve roots is shown in Figure 5.15.

Disc injuries

For various reasons, often related to wear and tear with age and use, intervertebral discs may degenerate and lose their strength. They can become flattened and in advanced cases may even be so deformed that the fibrous ring is damaged. The degenerative processes impair the mechanics of the vertebral column and allow

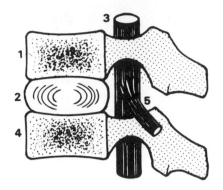

Figure 5.15 Diagram of a section of the spine. *The disc (2) lies between two vertebrae, (1) and (4); behind, the spinal cord (3) and a nervous tract (5). The disc is like a cushion which gives flexibility to the spine.*

tissues and nerves to be strained and pinched, leading to various back troubles, most commonly lumbago (painful muscle cramps) and sciatic troubles, and even in severe instances to leg paralysis.

Unnatural postures, heavy lifting and bad seating can speed up the deterioration of the discs, resulting in all the ailments mentioned above. For this reason, during the middle of the twentieth century many orthopaedists concerned themselves with the medical aspects of the sitting posture, for example Akerblom (1948), Andersson and Ortengren (1974a), Keegan (1953), Krämer (1973), Nachemson (1974), Schoberth (1962) and Yamaguchi *et al.* (1972).

Orthopaedic research

A very important contribution was made by the Swedish orthopaedists Nachemson (1974), Nachemson and Elfström (1970), and Andersson and Ortengren (1974b) who employed a sophisticated method to measure the pressure inside a disc during a variety of standing and sitting postures. They emphasised that increased disc pressure means that the discs are being overloaded and will wear out more quickly. Therefore, disc pressure is a criterion for evaluating the risk of disc injuries and backaches.

Disc pressure for four postures

The effects of four different postures on nine healthy subjects are shown in Figure 5.16.

The results disclose *that the disc pressure can be greater when sitting (without use of a backrest) than when standing.* The explanation lies in the mechanism of the pelvis and sacrum during the transition from standing to sitting:

The upper edge of the pelvis is rotated backwards
The sacrum turns upright
The vertebral column changes from a lordosis to either a straight or a kyphotic shape.

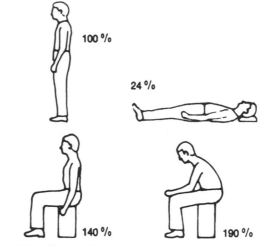

Figure 5.16 The effect of four postures on the intervertebral disc pressure between the 3rd and 4th lumbar vertebrae. *The pressure measured when standing is taken as 100 per cent. According to Nachemson and Elfström* (1970).

The spine when standing and sitting

It should be recalled here that lordosis means that the spine is curved forward, as it normally is in the lumbar region when standing erect. Kyphosis describes backward curving, which is normal in the thoracic region when standing upright.

These effects of the standing and sitting posture are illustrated in Figure 5.17.

Erect or relaxed sitting posture

Many orthopaedists still recommend an upright sitting posture as this holds the spine in a shape of an elongated 'S' with a lordosis of the lumbar spine. They believe disc pressure is lower in such a posture than when the body is curved forwards with kyphoses in the lumbar and thoracic sections. One of the recent advocates for an upright trunk at working desks is Mandal who in 1984 recommended higher seats and higher sloping desks which automatically lead to a more upright posture with reduced forward bending of the back.

In the same line are the *Balans* seats from Norway, which induce a half-sitting/half-kneeling posture. The seat surface is tilted forward strongly and a support for the knees prevents a forward sliding of the buttocks. The result is an opening of the hip angle (between legs and trunk) and a pronounced lumbar lordosis with a straight upright trunk posture. Krueger (1984) tested four models and found that the load on knees and lower legs is too high and sitting becomes painful after a while. (Some subjects even refused to sit longer than 2 h.) With desks of 720–780 mm in height the effect of a lumbar lordosis drops since the subjects cannot avoid

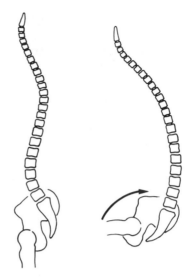

Figure 5.17   Rotation of the pelvis when changing from a standing to a relaxed sitting posture. *(Left) Standing upright. (Right) Sitting down. Sitting down involves a backward rotation of the pelvis (indicated by the arrow), bringing the sacrum to an upright position and turning the lumbar lordosis into a kyphosis.*

bending the trunk forward. Drury and Francher (1985) tested a similar forward-tilting chair. It elicited mixed responses, with complaints of leg discomfort from VDT users. Overall, the chair was no better than conventional ones and could be worse than well-designed office seats. Looking at these forward-tilting chairs, the question arises whether they would not be of greater use in a physiotherapeutical exercise than in an office.

The orthopaedic advice of an upright trunk posture conflicts with the fact that a slightly forward or reclined sitting posture relieves strain on the back muscles and makes sitting more comfortable. This was established, in part, through electromyographic studies made in the 1950s by Lundervold (1951, 1958). Some of his results are shown in Figure 5.18.

Slightly forward-bending the trunk holds upper body weight in balance

A relaxed posture, with a slightly forward bent trunk, holds the weight of the body in balance. This is the posture that many people adopt when they make notes or read in a sitting position because it is relaxing and exerts a minimum of strain on the muscles of the back. The visual distance for good readability might in some cases be another reason for a slightly forward-bent trunk. Thus there may be a 'conflict of interests' between the demands of the muscles and those of the intervertebral discs. While the discs prefer an erect posture, the muscles prefer a slight forward

Figure 5.18 Electrical activity in the back muscles when sitting upright and in a relaxed posture, slightly bent forward. *Sitting upright involves a considerable electrical activity, revealing the static effort imposed upon the muscles of the back. According to Lundervold* (1951).

bending. Of course, leaning against a well-designed backrest also relieves the spine and connective tissues (especially the muscles) of the back, as will be discussed below.

'Feeding' the intervertebral discs

Here we refer again to the interesting work of Krämer (1973) who made a study of the nutritional needs of intervertebral discs. The interior of a disc has no blood supply and must be fed by diffusion through the fibrous outer ring. Krämer produced evidence that pressure on the disc creates a diffusion gradient from the interior to the exterior so that tissue fluid leaks out. When the pressure is taken off, this gradient is reversed and tissue fluid diffuses back, taking nutrients with it. It seems from this that, to keep the discs well nourished and in good condition, they need to be subjected to frequent changes of pressure, as a kind of pumping mechanism.

From a medical point of view, therefore, *an occasional change of posture from bent to erect, and vice versa, must be beneficial.*

Increased seat angle reduces disc load

The above-mentioned orthopaedists Andersson and Ortengren (1974a,b) studied the effects of seat angle and different postures at desks on disc pressure. The electrical activity of back muscles was recorded to measure the static load. The effects of different postures are described in Figure 5.19.

The results show that both leaning back and bending forward with supported upper limbs (writing posture) are favourable conditions for reduced disc pressure. The effects of the seat angle are presented in Figure 5.20.

The results are clear: *by increasing the seat angle both disc pressure and muscle strain are reduced.*

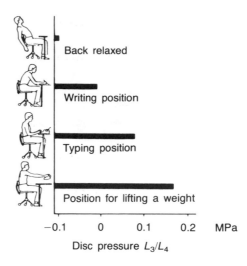

Figure 5.19 Effects of various sitting postures on disc pressure. L_3 and L_4 = third and fourth lumbar vertebrae. Zero (0) on the pressure scale is a relative reference value for a seat angle of 90°. Absolute values at the reference level zero were about 0.5 MPa. According to Andersson and Ortengren (1974a,b).

A proper lumbar pad relieves disc strain

Another study by Andersson and Ortengren (1974b) showed that a proper lumbar support also resulted in a decrease in disc pressure. These results are shown in Figure 5.21.

Further studies on the adjustment of back support of office chairs at different lumbar levels showed that providing back support at the level of the fourth and fifth lumbar vertebrae slightly decreased pressure compared with placing suport at the first and second lumbar vertebrae. The use of armrests always resulted in a decrease in disc pressure. This was less pronounced, however, when the backrest–seat angle was large. A comparison between these findings shows that *the disc load of a person leaning back at an angle between 110 and 120° and supplied with a 50 mm lumbar pad is even lower than that of a standing posture with the advocated lordosis of the lumbar region.*

Conclusion drawn from orthopaedic research

All these studies lead to an important conclusion: *resting the back against an inclined backrest transfers a significant portion of the weight of the upper part of the body to the backrest and reduces the strain on discs and muscles. In view of the design of chairs it is deduced that optimal conditions concerning disc pressure and muscular activity are obtained when the backrest has an inclination of 110 or 120° against horizontal (that is 20 or 30° behind vertical) and a lumbar pad protruding up to 50 mm.*

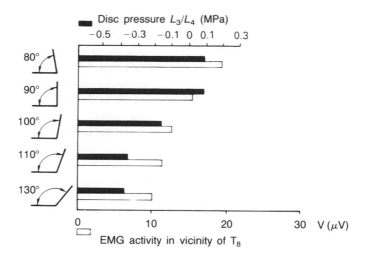

Figure 5.20 Effect of the seat angle (i.e. between seat and backrest) on the disc pressure and electrical activity in the back muscles recorded at the level of the eighth thoracic vertebra (T_8). *For more details see Figure 5.19. According to Andersson and Ortengren (1974b).*

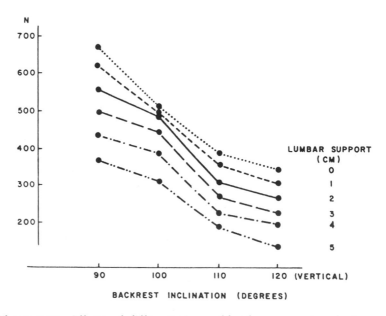

Figure 5.21 Effects of different sizes of lumbar support and of increasing seat angles on disc pressure. *The size of the lumbar support is defined as the distance between the front of the lumbar pad and the plane of the backrest. The backrest inclination is defined as the angle between the seat and the backrest. From Andersson and Ortengren (1974b).*

The cervical spine

Another part of the spine, the cervical spine of the neck consisting of the upmost seven vertebrae, is as important as the lumbar region. Like the lumbar spine, this is a very mobile segment, also showing a lordosis when 'upright'. The cervical spine is delicate and prone to degenerative processes and arthrosis. A great number of adults have neck troubles due to injuries of the cervical vertebrae and discs, generally referred to as cervical syndrome. The most common symptoms of cervical syndrome are painful cramps in the shoulder muscles, pains and reduced mobility in the cervical spine and sometimes pain radiating into the arms, ailments which are also called cervicobrachial syndrome. In Japan these cervicobrachial disorders were considered an occupational disease since they often occurred among key-punchers, assembly-plant workers, typists, cash-register operators and telephone operators (Maeda *et al.*, 1982). Subsequently, several authors have discovered physical troubles among VDT operators which fit the description of cervicobrachial disorders (Laubli *et al.*, 1986; Nishiyama *et al.*, 1984). It is often observed that discomfort in the neck increases with the degree of forward bending of the head.

These findings indicate that it is desirable to lean the neck and head against a tall backrest. For this, the section of the backrest in the shoulder region should be slightly concave; above, convex to accommodate the kyphotic curve of the cervical column. The top of the backrest should be bent backward a bit to provide a 'cushion' for the head. Even if one cannot use the upper part of the backrest all the time, it is there to provide support for relaxing the neck and upper trunk during a break. Thus, the function and shape of a tall backrest of a 'work chair' are much the same as for an 'easy chair', described below.

Ergonomic research

A highly adjustable sitting apparatus and a variety of moulded seat shells of different profiles were tested by Grandjean and co-workers (1967, 1973) on a large number of people, including a group of 68 who complained of back ailments. They were asked to give their subjective impressions of the different seat profiles and their effects upon various parts of the body. The profiles of a multi-purpose seat and of an easy chair, which the test subjects thought produced the fewest aches and pains, are shown in Figure 5.22.

The best chair for relaxing

The result for the easy chair agrees almost exactly with what orthopaedists say. *A seat profile which produces a low pressure in the intervertebral discs and requires very little static muscular effort is also the one that causes the fewest aches and pains. When more discomfort is experienced, it is evidently associated with stresses falling upon the discs and fatigue symptoms in the muscles.*

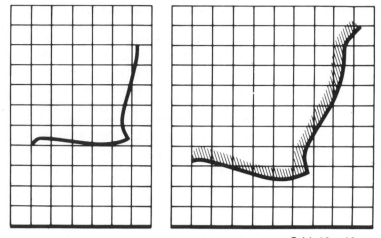

Grid: 10 × 10 cm

Figure 5.22   Seat profiles of a multi-purpose chair (left) and an easy chair (right) both of which caused a minimum of subjective complaints. *Grid 100 × 100 mm. After Grandjean et al.* (1967, 1973).

This was also pointed out by Rosemeyer (1971), who noted that opening the angle between the seat and backrest to 110° resulted in less electrical activity in the involved muscles and in greater comfort.

Taking these research results as a whole, the following recommendations for an easy chair have both orthopaedic/medical and ergonomic backing:

1 *The seat pan should be tilted backward* so that the buttocks will not slide forward. A back tilt of up to 24° below horizontal is recommended.

2 *The backrest should be tilted at the following angles*:
    105 to 110° to the seat
    20 to 30° behind vertical

The backrest should be provided with a lumbar pad, as Akerblom (1948) said when he pioneered the study of seating. The apex of this pad should meet the spine between the third and fifth lumbar vertebrae. This means that its vertical height above the back of the seat should be 100–180 mm. The pad helps reduce kyphosis of the lumbar region and holds the spine in as natural a position as possible. The preferred shape of the multi-purpose seat shown in Figure 5.22 is characterised by a slightly moulded seat surface in order to prevent the buttocks from sliding forward and by a backrest with a lumbar pad.

The work seat   *As far as work seats are concerned, the research performed over the years in Professor Grandjean's laboratory indicates that a high*

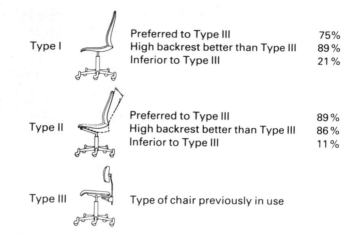

Type I — Preferred to Type III — 75%
High backrest better than Type III — 89%
Inferior to Type III — 21%

Type II — Preferred to Type III — 89%
High backrest better than Type III — 86%
Inferior to Type III — 11%

Type III — Type of chair previously in use

Figure 5.23 Comparative assessment of three experimental chairs which were used by 66 office workers over a period of two weeks. *Type I = fixed moulded chair with high backrest. Type II = moulded chair which tilts 2° forward and 14° backward, freely movable with high backrest. Type III = standard office chair with adjustable backrest. After Hünting and Grandjean (1976).*

backrest formed to follow the contour of the human back is good both medically and ergonomically. Such a profile gives support to the lumbar region when the occupant is leaning forward (in a working attitude) yet relaxes the back muscles thoroughly when leaning backward because it then holds the spine in a natural position.

Office chairs

Hünting and Grandjean (1976) studied office chairs with backrests under practical working conditions. They recorded sitting habits and reports of physical discomfort in different parts of the body. A tiltable chair and a similar model with a fixed seat were compared with a traditional type of chair fitted with an adjustable but low backrest. The subjects carried out their normal work while using each of the three chairs for two weeks. The most interesting results related to the subjects' preferences, as shown in Figure 5.23.

The survey indicated quite clearly that the office workers favoured the two types of chairs with high backrests. This confirms the view expressed earlier that a high backrest is preferable for office work as most employees often want to lean back. It is obvious that a high backrest will be more effective in supporting the weight of the trunk than a chair with a small backrest.

It follows that *the office seat should offer the possibility of leaning back – all the time or only occasionally – by providing a high backrest.*

Figure 5.24 An office chair must be conceived for a forward- as well as a backward-leaning sitting posture. *The lumbar spine must get proper support from the backrest in both sitting postures.*

It is of interest to note that the subjects in the experiments could not arrest the tiltable chair at the desired inclination; thus it did not provide stable enough support for the whole body. This was criticised by many subjects and leads to the conclusion that tiltable chairs or chairs with adjustable backrest inclinations should be fitted with a mechanism that can be fixed to the desired degree of inclination.

General experience as well as a number of studies have yielded the following 'golden rules' for office chairs:

1 *Office chairs must be adapted to both the traditional office job and the modern equipment of information technology, especially to jobs at computer workstations.*

2 *Office chairs must be conceived for forward and reclined sitting postures* (Figure 5.24).

3 *The backrest should have an adjustable inclination. It should be possible to lock the backrest at any desired angle.*

4 *A backrest height of at least about 500 mm vertically above the seat surface is a necessity.*

5 *The backrest must have a well-formed lumbar pad, which should offer good support to the lumbar spine between the third vertebra and the sacrum, for example at a height of 100–200 mm above the lowest point of the seat surface. These recommendations are illustrated in Figure 5.25.*

Tilting chair with high backrest

Backrest:

  height (above seat) 50 cm

  lumbar pad

  slightly concave at thorax level

  adjustable inclination (104–120°)
  with locking device

(do not forget a footrest)

Figure 5.25   A Grandjean-designed office chair.

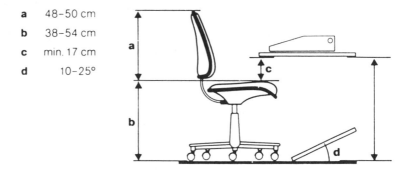

a	48–50 cm
b	38–54 cm
c	min. 17 cm
d	10–25°

Figure 5.26   Recommended dimensions for the design of the seat and table.

6 *The seat surface should measure 400–450 mm across and 380–420 mm from back to front. A slight cavity in the seat pan will prevent the buttocks from sliding forward. A light padding, about 20 mm thick, covered with non-slip, permeable material is a great aid to comfort.*

7 *Footrests are important for people with short lower legs to avoid sitting with hanging feet.*

8 *An office chair must fulfil all the requirements of a modern seat: adjustable height (380–540 mm), swivel, rounded front edge of the seat surface, castors at a five-arm base and user-friendly controls. The most important dimensions of a seat and desk are shown in Figure 5.26. A chair in which the backrest angle changes automatically with the seat pan inclination is shown in Figure 5.27.*

The main objection to a good adjustable office chair is its cost, of course. However, one should bear in mind that the life span

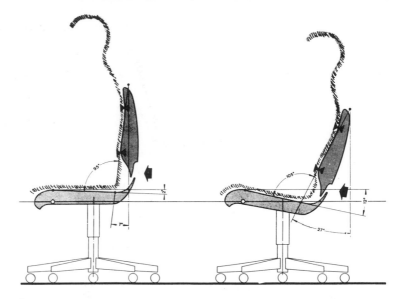

Figure 5.27 A recent office chair design in which the backrest reclines with increasing seat pan declination. *This mechanism allows the back to get adequate support at the correct level for any backrest declination, as indicated by the small arrows. The large arrows show the way the backrest descends with increasing declination.*

of a well-constructed chair is about 10 years or about 2000 working days. The price of a good chair, which reduces physical discomfort and promotes well-being, is for a few coins per day, certainly a good investment.

5.5 The design of computer workstations

The metamorphosis of offices

Computers have invaded all types of offices. They often enter a world where machines have not been used before. The result is a considerable change in offices and their working conditions. To call the present change a metamorphosis, similar to that of caterpillars and butterflies, is hardly an exaggeration.

At the traditional office desk an employee performs a great variety of physical and mental activities and has a large space for various body postures and movements: he or she might look for documents, take notes, file correspondence, use the telephone, read a text, exchange information with colleagues or type for a while. He or she will leave the desk many times during the course of the working day. A desk which is a bit low or high, an unfavourable chair, deficient lighting conditions or other ergonomic shortcomings are not likely to cause annoyance or physical discomfort because the great variety of activities prevents adverse effects of long-lasting invariable physical or mental loads.

The situation is, however, entirely different for an operator working with a computer for hours without interruption or perhaps for a whole day. *Such a VDT operator is tied to the machine system.* His or her movements are restricted: attention is concentrated on the screen and hands are linked to the keyboard. VDT operators are more vulnerable to ergonomic shortcomings: they are more susceptible to the effects of constrained postures, repetitive activities, poor photometric display characteristics and inadequate lighting conditions. This is the reason why the computerised office calls for ergonomics; consequently the VDT workstation has become the launch vehicle for ergonomics in the office world.

Reports on discomfort

As long as engineers and other highly motivated experts operated VDTs, hardly anyone complained about negative effects. However, the situation changed drastically with the expansion of computers into workplaces where traditional working methods had formerly prevailed. Complaints from VDT operators about visual strain and physical discomfort in the back, the neck/shoulder area and in the forearm/wrist/hand area became more and more frequent. This has provoked different reactions: some believe that the complaints are highly exaggerated and mainly a pretext for social and monetary claims, while others consider the complaints to be symptoms of a health hazard requiring immediate measures to protect operators from injuries to their health. *Ergonomics as a science stands between these opposing beliefs; its duty is to analyse the situation objectively and to deduce guidelines for the appropriate design of computer workstations.*

Controversial field studies

Many field studies have been carried out into complaints about physical discomfort. These studies have used self-rating questionnaires and included different types of VDT jobs as well as control groups (Grandjean, 1987). In most cases the non-VDT groups differ from the VDT groups not only in the use of VDTs but also in many other respects.

The problem of control groups is indeed very intricate. The introduction of VDTs is normally accompanied by changes in task design, speed of work and especially by major differences in performance and productivity. It is therefore not surprising that these field studies disclosed controversial results. If the control groups were engaged in traditional office work with low productivity and a great variety of activities they were much less affected by musculoskeletal discomfort than the VDT groups. If, on the

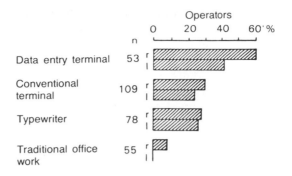

Figure 5.28 Palpation findings in the shoulders of four groups of office workers. *Painful pressure points are at the tendons, joints and muscles: r = right; l = left; n = number of examined operators. Differences between groups were significant at P < 0.01; Kruskal–Wallis test. According to Läubli et al.* (1981).

contrary, the control groups did strenuous work, like full-time typing, the complaints were as frequent as among VDT groups. An example of this phenomenon is shown in Figure 5.28.

We shall discuss here only a few studies related to the proper design of VDT workstations.

Medical findings

In this context an interesting contribution was made by a research group around Läubli (1981) and Läubli *et al.* (1986) who applied investigation methods which are used in rheumatology and include the assessment of joint and back mobility, painful pressure points at tendons or other characteristic locations and painful reactions to muscle palpation. They noticed a high correlation between the medical findings of two examining doctors and between the medical findings and self-rated physical discomfort. The study included 295 subjects engaged in two VDT jobs and in two non-VDT activities. Some of the results are presented in Table 5.3 and in Figure 5.28.

Medical findings indicating musculoskeletal troubles in muscles, tendons and joints were frequent in the groups using data-entry terminals and among full-time typists, whereas the groups performing traditional office work consisting of many different activities and movements showed the least pain. The palpation findings in the shoulders disclose a similar distribution of symptoms. Both the complaints and the medical findings must be taken seriously, especially as 13 to 27 per cent of the examined employees had consulted a doctor because of pains.

Table 5.3 Incidence of medical findings in the neck–shoulder–arm area of office employees

	Data-entry tasks (*n* = 53) (%)	Conversational tasks (*n* = 109) (%)	Full-time typists (*n* = 78) (%)	Traditional office jobs (*n* = 54) (%)
Tendomyotic pressure pains in shoulders and neck	38	28	35	11
Painfully limited head movability	30	26	37	10
Pains during isometric contractions of forearm	32	15	23	6

Note: n = number of subjects.

Physical discomfort related to workstation design

In this field study on VDT operators several significant relationships were discovered between the design of workstations or postures on the one hand and the incidence of complaints or medical findings on the other (Läubli *et al.*, 1986). These results can be summarised as follows.

Physical discomfort and/or the number of medical findings in the neck/shoulder/arm/hand area are likely to increase when:

The keyboard level is too low or too high.
Forearms and wrists cannot rest on an adequate support.
The key tops are too high above the support surface (table).
Operators work with a marked bend (in terms of flexion/extension and/or in lateral deviation) of the wrist.
Operators have a marked head inclination.
Operators adopt a slanting position of the thighs under the table due to insufficient space for the legs. This is illustrated in Figure 5.29.

The frequent complaints about physical discomfort among VDT operators induced office furniture manufacturers to place adjustable VDT workstations on the market. At the same time several experiments with adjustable workstations were carried out both under laboratory conditions and during practical office work. Since the latter gave important indications for the ergonomic design of VDT workstations they will be described below.

Preferred settings of VDT workstations in offices

A study on postures and preferred settings of adjustable VDT workstations during subjects' usual working activities was carried out by Grandjean *et al.* (1983). The experiments were conducted on 68 operators (48 females and 20 males, average age 28 years old) in four companies: 45 subjects had a conversational job in

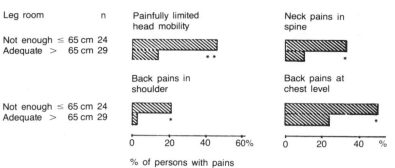

Figure 5.29 Vertical leg room and physical discomfort in 53 VDT operators at conversational terminals. *Leg room = distance from the floor to the lower edge of the desk. n = number of operators.* * *P<0.5, ** P<0.01, Mann–Whitney U test. According to Läubli and Grandjean (1984).*

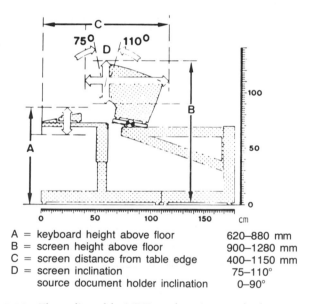

A = keyboard height above floor   620–880 mm
B = screen height above floor   900–1280 mm
C = screen distance from table edge   400–1150 mm
D = screen inclination   75–110°
     source document holder inclination   0–90°

Figure 5.30 The adjustable VDT workstation, with the ranges of adjustability, used in a field study during subjects' usual working activities.

an airline company, 17 subjects had primarily data-entry activities in two banks and six subjects were engaged in word-processing operations. Each subject used the adjustable workstation shown in Figure 5.30 for one week.

The key tops were as high as 80 mm above support level. A chair was provided with a high backrest and an adjustable inclination. For the first two days a forearm/wrist support was used; on the following two days the subjects operated the keyboard without support and on the last day they were given the

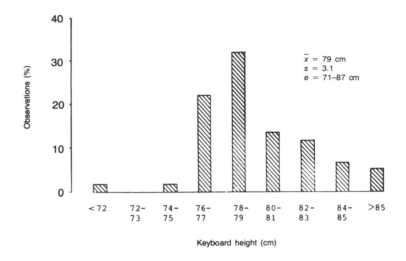

Figure 5.31 Preferred keyboard heights of 59 VDT operators (236 observations) while performing their usual daily jobs. *Keyboard height = home row above floor; x̄ = mean value; s = standard deviation; e = range.*

option of whether to use it or not. Document holders were provided as an optional device for each subject. The preferred settings and postures were assessed and determined every day.

The analysis of the results of preferred settings disclosed no noticeable differences throughout the five days. In other words, the mean values remained practically the same for the whole week, independent of the use of wrist support. Thus the data obtained during the week could be put together for evaluation.

The frequency distribution of all preferred keyboard heights is reported in Figure 5.31.

Range of adjustability for keyboard desk

The 95 per cent confidence interval lies between 730 and 850 mm. A desk level between 630 and 790 mm suits a keyboard height of 80 mm, a level between 680 and 840 mm would be adequate for a keyboard height of 30 mm. Assuming the 95 per cent confidence interval, *the range for the adjustability of desk levels lies between 650 and 820 mm.* This seems to be a reasonable recommendation for workstation manufacturers.

The results obtained in this field study reveal slightly higher keyboard levels than those obtained in comparable laboratory studies. It is assumed that in short-term experiments subjects are less relaxed, sit more upright and try to keep the elbows low and the forearms in a horizontal position, thus giving preference to a slightly lower keyboard height. All the results of preferred settings are assembled in Table 5.4.

Table 5.4 Preferred VDT workstation settings and eye levels during habitual working activities

Adjustable dimensions	n_1	n_2	Mean	Range
Seat height (mm)	58	232	480	430–570
Keyboard height above floor (mm)	59	236	790	710–870
Screen height above floor (mm)	59	236	1030	920–1160
Visual down angle, eye to screen centre (°)	56	224	−9°	+2°– −2°
Visual distance, eye to screen centre (mm)	59	236	760	610–930
Screen upward inclination (°)	59	236	94°	88°–103°
Eye level above floor (mm)	65	65	1150	1070–1270

Note: n_1 = number of subjects; n_2 = number of observations. Visual declination angle and screen inclination are measured to a horizontal plane.

Preferred settings of the screen

The preferred screen heights and screen inclinations are in some cases influenced by the operators' attempts to reduce reflections. In fact, many operators reported less annoyance by reflections if they could adjust the screen.

The capital letters on the screen were 3.4 mm high corresponding to a comfortable visual distance of 680 mm. At the adjustable VDT workstation the operators tended to choose greater viewing distances; 75 per cent of them had visual distances between 710 and 930 mm. No explanation can be found for this preference.

Correlations with anthropometric data

The calculation of Pearson correlation coefficients between anthropometric data and preferred settings revealed only poor relationships: between eye level and screen height $r = 0.25$ ($P = 0.03$) and between stature and keyboard height $r = 0.13$ (not significant). Some of the laboratory studies revealed similar results of poor or no relationships. It can be concluded, therefore, that in this study *the preferred settings of VDT workstations were hardly influenced by anthropometric factors*; individual habits had probably the strongest influence.

Preferred postures

The most striking result of this field study concerned the postures associated with the preferred settings. The operators moved very seldom and did not noticeably change the main postural elements which are obviously determined by the position of keyboard and screen. Figure 5.32 shows the distribution of observed trunk postures expressed as angles of a line 'shoulder articulation to trochanter' related to a horizontal plane.

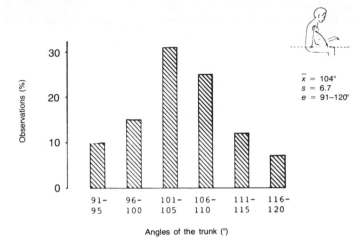

Figure 5.32 Trunk postures of 59 VDT operators (236 observations) while performing their usual jobs. *Trunk posture is assessed as the angle of a line between the hip and shoulder joints to the horizontal. $\bar{x}$ = mean value; s = standard deviation; e = range.*

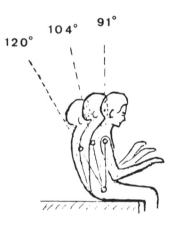

Figure 5.33 Mean and range of observed trunk postures of 59 operators. *Trunk posture is assessed as the angle of a line between the hip and shoulder joints to the horizontal.*

Most operators lean back

The trunk inclinations approximated a normal distribution. The majority of subjects preferred trunk inclinations between 100 and 110°. Only 10 per cent demonstrated an upright trunk posture. Figure 5.33 illustrates the mean and the range of observed trunk postures.

It is obvious that the majority of operators leaned back. This was the basis for all other adopted postural elements: the upper arms were kept higher and the elbow angles slightly opened. The mean values for preferred trunk–arm positions are shown in Figure 5.34.

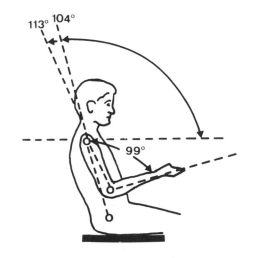

Figure 5.34 The 'average posture' of VDT operators at workstations with preferred settings.

Table 5.5 Means ($\bar{x}$), standard deviations (*SD*) and ranges of postural measurements obtained from VDT operators during their daily work at workstations with preferred settings. 59 operators, 236 observations

Postural element	$\bar{x}$	SD	Range
Trunk inclination (°)	104	6.7	91–120
Head inclination[a] (°)	51	6.1	34–65
Upper arm flexion[b] (°)	113	10.4	91–140
Upper arm abduction[c] (°)	22	7.7	11–44
Elbow angle (°)	99	12.3	75–125
Lateral abduction of hands (°)	9	5.5	0–20
Acromion–home row distance (mm)	510	50	420–620

Notes:
[a]Angle C7–earhole–vertical.
[b]See Figure 5.32.
[c]Abduction = lateral raising of the upper arm.

It must be pointed out here that about 80 per cent of the subjects did rest their forearms or wrists if a proper support was available. If no special support was provided, about 50 per cent of the subjects rested forearms and wrists on the desk surface in front of the 80-mm-high keyboard. The results of all the measured postural elements, expressed as mean values and ranges, are reported in Table 5.5.

The observed postures are not due to the experimental

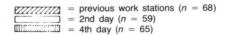

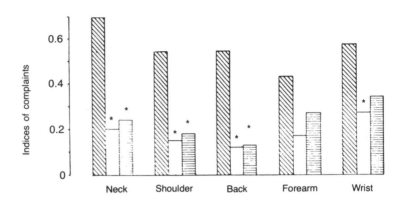

Figure 5.35 Mean indices of complaints at the original workstation and the redesigned workstation with preferred settings. *0 = relaxed; 1 = tense; 3 = impaired.* *P ≤ 0.05.

workstation, for the measurements carried out at the previous workstations had already revealed nearly the same trunk and arm inclinations.

The study of Grandjean et al. (1983) confirms a general impression one gets when observing the sitting posture of many VDT operators in offices: most of them lean back and often stretch out the legs. They seem to put up with having to bend the head forward and lift their arms. In fact, *many VDT operators in offices disclose postures very similar to those of car drivers*. This is understandable: who would like to adopt an upright trunk posture when driving a car for hours?

Preferred settings and physical discomfort

The VDT operators completed a questionnaire relating to feelings of relaxation and physical discomfort, once at the previous workstation and twice at the adjustable workstation with preferred settings. An index was calculated from the answers 'relaxed', 'tense' and 'impaired' for each of the involved parts of the body (neck, shoulders, back, forearms and wrists). In Figure 5.35 the mean indices of complaints from the previous workstation are compared with those reported on the second and on the fourth day. An index of less than 0.5 means that the majority of the subjects rated their postures as relaxed; an index of more than 0.5 means that many subjects indicated that their muscles were tense or even that they experienced impairments.

Discomfort is reduced with preferred settings

From Figure 5.35 it is obvious that the indices were distinctly higher at the previous workstation than with the preferred settings. A chi-squared analysis showed significant differences between the previous and adjustable workstation for the neck, shoulders and back. It must be pointed out that at the previous workstation subjects sat on traditional office chairs with relatively small backrests. At the adjustable workstations, however, they were provided with particularly suitable office chairs, featuring high backrests with adjustable inclinations, which allowed the whole back to relax. It is therefore reasonable to assume that the decrease in physical discomfort reported at the adjustable workstation was due to both the preferred settings and the proper chairs.

These results were confirmed by Shute and Starr (1984). In a first field study telephone operators were provided with an adjustable VDT table and in a second study with an additional conveniently adjustable chair. Subjects used the adjustable workstation for several weeks while doing their normal work. The previous VDT table had a fixed height of 686 mm and the screen was 400 mm above the table. The previous chair was difficult to adjust and had an unsuitable backrest. The main difference to the new advanced chair was its easy adjustability. The results revealed a reduction of discomfort when either conventional component was replaced by an advanced component. However, the reduction of physical discomfort was far greater when the advanced table was used together with the advanced chair. The authors concluded that the benefit of an advanced table can only be fully realised if it is used in combination with an advanced chair.

Preferred settings at CAD workstations

Van der Heiden and Krueger (1984) examined the use and acceptance of an adjustable workstation for computer-aided design (CAD) operations. Height and inclination of the work surface as well as height, inclination, rotation and distance of the monitor could be adjusted with the aid of motorised devices. To study the use of the adjustment fixtures a continuous registration of settings was carried out during one week. The majority of operators had more than six weeks' experience in using the adjustable workstation. In the test week eight women and three men were studied during their normal CAD work, consisting of mechanical design. A total of 67 CAD work sessions were registered. Questionnaire answers and preferred settings were obtained from 11 female and 4 male operators. Of a total of 166 registered adjustments, 142 (= 86 per cent) were made at the beginning of a work session and 24 (= 14 per cent) were readjustments. Short operators used the adjustment device more frequently than tall people. Furthermore, operators who had not received specific instructions adjusted less frequently than others

Table 5.6 Preferred settings of 15 operators at an adjustable CAD workstation

	$\bar{x}$	e
Seat height (mm)	540	500–570
Work surface height (mm)	730	700–800
Monitor centre above floor (mm)	1130	1070–1150
Monitor visual distance (mm)	700	590–780
Work surface tilt(°)	8.6	2–13
Monitor tilt[a] (°)	−7.7	−15–+1

Notes:
$\bar{x}$ = mean values; e = range
[a]Negative tilt = a forward monitor inclination (top of the screen towards the operator).

who had been given such instructions. The preferred settings are presented in Table 5.6.

The mean seat height of 540 mm is quite unusual. Another striking result is the forward tilting of the monitor with a preferred mean angle of −8°. The operators claimed that with this setting reflections resulting from windows behind them could be avoided. For that reason most operators preferred a relatively high monitor setting. All other preferred dimensions are similar to those of the VDT operators shown in Table 5.5.

Wishful thinking of standards versus operators' instinctive behaviour

Let us come back to the backward declination of the trunk observed at VDT workstations. This leaning back does not correspond at all with the often recommended 'upright' postures – even the 1988 American ANSI Standard 100 assumes such erect operators. Figure 5.36 illustrates the great gap between 'wishful thinking' (recommendations) and actual postures. An important question suggests itself here. Is the upright posture healthy and therefore recommendable or is the relaxed position with the reclining trunk to be preferred? As already mentioned, by increasing the backrest declination from 90 to 120° a significant decrease in discal load and muscle strain is achieved. Orthopaedic studies by Andersson and Ortengren (1974) suggest *that resting the back on a sloping backrest transfers a good portion of the trunk weight to the backrest and reduces strain on discs and muscles as compared with sitting straight and upright. It is therefore concluded that VDT operators instinctively do the right thing when they prefer a reclined sitting posture and ignore the recommended upright trunk position.* Of course, there is nothing wrong with an occasional erect posture just to vary the body position. *Variety and motion are the keys to 'healthy sitting'.*

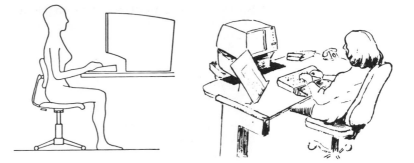

Wishful thinking   Preferred body posture

Figure 5.36 Recommended and actual postures at office VDT workstations. *(Left) The upright trunk posture with elbows down and forearms almost horizontal, postulated in many brochures and standards. (Right) The actual posture most commonly observed at VDT workstations resembles the posture of a car driver.*

Guidelines for the design of computer workstations

From the studies mentioned above, which have been largely reinforced by more recent research, such as compiled by Lueder and Noro (1995), the following guidelines for the design of a VDT workstation can be proposed for users in Europe or North America:

1 *The furniture should, in principle, be conceived to be as flexible as possible. A proper computer workstation should be adjustable in the following dimensions:*

Keyboard height (floor to home row)	*700–850 mm*
Screen centre above floor	*800–1100 mm*
Screen inclination to horizontal	*about 105°*
Screen distance to table edge	*500–750 mm*

2 *A VDT workstation without adjustable keyboard height and adjustable height and distance of the screen is not suitable for a continuous job at a VDT.*

3 *The controls for adjusting the dimensions should be easy to handle, particularly at workstations with several users.*

4 *At knee level the distance between the front table edge and the back wall should not be less than 600 mm and at least 800 mm at the level of the feet.*

5.6 The design of keyboards

Parallel rows require unnatural hand positions

A commercially successful machine for typing letters was invented around 1870. It was a mechanical device with at first two, then four, parallel rows of keys. To operate these keys rapidly, the typist had to hold the hands parallel to the rows. This required

Figure 5.37 Position of wrists and hands operating a traditional keyboard. *The parallel position of the rows requires an inward rotation (ptonation) of the forearms and wrists and a sideways bend (ulnar deviation) of the hands.*

an unnatural position of the wrists and hands, characterised by an inward rotation of the forearms and wrists and a lateral (ulnar) adduction of the hands. These constrained postures often caused physical discomfort and in some cases even inflammation of tendons or tendon sheaths in the hands, wrists or forearms. Figure 5.37 illustrates such constrained postures of the wrists and hands at a keyboard.

Nearly a century later, the mechanical typewriter was replaced by an electrified machine. The mechanical resistance of keys was much reduced and the operation of the keyboard was made easier because the hands no longer had to provide all the energy for generating imprints but the unnatural position of wrists and hands remained.

Keyboards at VDTs

Even at today's computer word-processing workstations the keying activities and the hand/arm postures are similar to the traditional operation of typewriters. There are some differences, however. The number of keys has increased from about 60 to 100 or more on most keyboards with the addition of specially arranged numerical and function keys. While much less energy is needed per keystroke from the operator, the keying speed and the number of people using keyboards have become much larger. For these and other reasons, more and more keyboarders have complained about fatigue, pains and physical ailments in shoulders, arms and hands. Tendonitis, tenosynovitis and carpal tunnel syndrome, disorders from which typists had suffered early in the twentieth century (as reported in the medical and physiological literature), became 'epidemics' among computer keyboard users in the 1980s.

This induced some designers to develop keyboards which allow operators to maintain a 'straight wrist'. For this reason many ergonomists today recommend *a keyboard with a home row not higher than 30 mm above the support table. The operator should be able to place the keyboard freely on an adjustable support table as preferred and needed.*

The next step is an ergonomic design of the keyboard in order to avoid constrained and unnatural hand postures by reducing or eliminating the inward twist (pronation) and lateral bend (ulnar deviation). This is largely a matter of arranging keys and sets of keys so that they comply better with the human body. Such key arrangements should also help to avoid bending the wrist up or down (flexion or extension). A soft wrist rest in front of the keyboard can support the forearm and hand, at least during breaks in keyboarding, and help to keep the wrist straight.

Studies on split keyboards

Studies along this line were carried out in 1926 by Klockenberg and 40 years later by Kroemer (1964, 1965a, 1972) who proposed splitting the keyboard into two sections, one for the left and one for the right hand, and arranging the sections in such a way that the hands could be kept in a more natural position. Kroemer's halved keyboards had an opening angle ('slant' viewed from above) of 30° and could be declined laterally ('tilt-down' viewed from the front) between 0 (horizontal) and 90° (vertical). Experiments disclosed that the experimental keyboard with its halves tilted downward 30 to 45° generated less fatigue and pain in arms and hands than the traditional horizontal arrangement inherited from the typewriter. The test subjects performed about the same number of correct keystrokes on either keyboard but made fewer errors on the Klockenberg–Kroemer design.

EMG of forearms and shoulders

Following the example of Lundervold in the 1950s, Zipp *et al.* in 1983 studied the electrical activity of various muscles in the shoulder/arm area in relation to arm/hand postures according to the characteristics of keyboard operations. With increasing lateral adduction of the hands from a neutral position, an increase in the electrical activity of the muscles involved was recorded. Use of a split keyboard, as proposed by Kroemer (1964, 1965a,b), was associated with decreased electrical activities in the arm/shoulder area, even with lateral keyboard tilt-downs of only 10–30°. The same result was observed when the slant angle between the two keyboard halves was opened. The authors concluded that the static muscle load in the arm/shoulder area is significantly reduced with such a keyboard design.

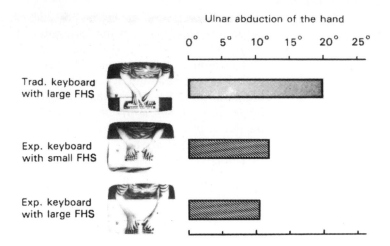

Figure 5.38 Mean angles of sideways twisting (ulnar deviation) of the right hand with three types of keyboards. *(Top) Traditional typewriter with large forearm wrist support (200 mm). (Middle) Split keyboard with a slant angle of 25°, a lateral down tilt of 10° and a small forearm wrist support of 100 mm. (Bottom) The same split keyboard but with a large forearm wrist support of 200 mm. FHS: forearm-hand support.*

Experiments with split keyboards

Following this line of research, Grandjean *et al.* (1981), Hünting *et al.* (1982) and Nakaseko *et al.* (1985) developed an adjustable model of a split keyboard and studied the preferred settings of opening (slant) and lateral tilt-down angles, and distances between the split keyboards with 51 subjects. Typing with the split keyboard with preferred settings decreased the lateral adduction of the hands, as shown in Figure 5.38, reduced discomfort and increased feelings of being relaxed in the neck/shoulder/arm/hand areas.

Effects of a large forearm/wrist support

The use of a large forearm/wrist support was associated with a reclining sitting posture of the subjects and an increased pressure load of the forearm/wrist on the support, reaching mean values of nearly 40 N. Such weight transfers onto the support will strongly decrease the load on the intervertebral discs. Of 51 subjects, 40 preferred the split keyboard with the following characteristics:

Slant angle between the two half-keyboards 25°
Distance between the two half-keyboards
(measured as distance between the keys
G and H)   95 mm
Lateral tilt angle of both half-keyboards   10°
A hand-configured design of the keys

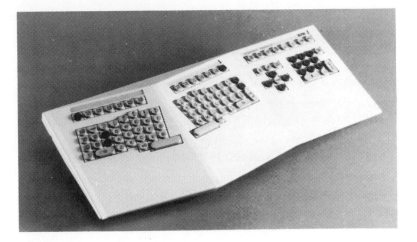

Figure 5.39 A keyboard designed in accordance with ergonomic principles. *The two keyboard halves show an opening (slant) angle of 25° in order to avoid a sideways twisting (ulnar deviation) of the hands. They tilt sideways down at 10° below horizontal to lessen the inward rotation (pronation) of the forearms and wrists. According to Nakaseko et al. (1985).*

A prototype of such a keyboard is shown in Figure 5.39. A commercial model was presented at the Ergodesign 84 conference by Buesen (1984). Since then, a large number of 'ergonomic' keyboards have come onto the market, incorporating an ingenious variety of keys, key sets, angular adjustments and operating principles. While previous design principles seemed to have followed mostly the productivity principle ('more data input per time period is better'), advanced ergonomic design goals keep the human operator in focus and try to alleviate the biomechanical workload by deviating from the old typewriter keyboard. However, while alternative keyboards should bring about some operator relief, Kroemer (1995) suggested that technical alternatives to keyboards such as voice inputs to the computer together with organisational changes ('do we need all this keyboarding?') may open new avenues for computer ergonomics.

Summary

Workstation design should facilitate movement of the body instead of promoting maintained static postures. However, excessive motions, such as in keyboarding or using hand tools, must be avoided as well.

CHAPTER SIX

Heavy work

6.1 Physiological principles

Heavy work is any activity that calls for great physical exertion and is characterised by a high energy consumption and severe stresses on the heart and lungs. Energy consumption and cardiac effort set limits to the performance of heavy work and these two functions are often used to assess the severity of a physical task.

Mechanisation has reduced the demands for strength and energy of the operator; nevertheless, in many industries there are still jobs which rank as heavy work and which not infrequently lead to overstrain. Heavy work is common in mining, building, agriculture, forestry and transport including, for example, baggage handling by airline personnel. It is a major problem for ergonomics in developing countries.

Metabolism

A fundamental biological process is to take in nutrients in the form of food and drink and to convert their chemical energy into mechanical energy and heat. Food is progressively broken down in the intestines until its constituents can pass through the gut wall and be absorbed into the blood. Most of the nutrients then pass to the liver, which controls their storage as glucose and glycogen, and as fat, the body's final energy reserve. When needed, first glucose and then glycogen again pass into the bloodstream as readily usable compounds, mainly sugars. Only a small proportion of the food is used for building up body tissues or reaches the adipose tissues as fat.

Blood carries nutrients to all the cells of the body where they are broken down further by precisely controlled processes to yield needed energy, ending up as the by-products water, carbon dioxide and heat. The collective term for these processes of converting chemical energy is *metabolism*, which can be compared

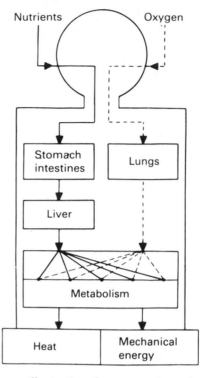

Figure 6.1   Diagram illustrating the conversion of nutrients into heat and mechanical energy in the human body.

to a slow self-regulated combustion. The comparison is quite apt since metabolism, like combustion, needs a supply of oxygen, which it obtains via the lungs and the bloodstream. These metabolic processes liberate heat as well as mechanical energy, depending on what muscular activity is going on. Figure 6.1 shows these processes diagrammatically.

Energy consumption

Energy consumption is measured in kilojoules (kJ). It can be assessed indirectly by recording the uptake of oxygen that is necessary for the oxidation of nutrients. When 1 litre of oxygen is consumed in the human body there is, on average, a turnover of 20 kJ (5 kcal) of energy. Fortunately, this single conversion factor applies whether carbohydrates, fats, proteins or alcohol provide the original chemical energy. If one wants to know which is actually combusted, one must also measure the volume of carbon dioxide in the exhaled air. The ratio between the volumes of CO_2 and O_2 is called the respiratory quotient (RQ). The RQ is as follows:

1.00 for carbohydrate
about 0.8 for protein
about 0.7 for fat and alcohol

Basal metabolism Measurements show that a resting person has a steady consumption of energy, depending on size, weight and gender. When the person is lying down, completely at rest, with the stomach empty, this quantity is known as the *basal metabolism*. For a man weighing 70 kg it amounts to about 7000 kJ per 24 h and for a woman weighing 60 kg, it is about 5900 kJ. Under the conditions of basal metabolism nearly all the chemical energy from nutrients is converted into heat.

6.2 Energy consumption at work

As soon as physical work is performed, energy consumption rises sharply. The greater the demands made on the muscles by one's activity, the more energy is consumed.

Work joules The increased consumption associated with a particular activity is expressed in *work joules* and is obtained by measuring the energy consumption while working and subtracting from this the resting consumption or the basal metabolism.

Work joules indicate the level of bodily stress and, especially in heavy work, they can be used to assess the level of effort, to determine necessary rest periods and to compare the energetic efficiency of different tools and different ways of arranging the work. In this context it should be clearly understood that energy consumption measures only the level of physical effort; it tells us nothing about mental stress, the demands the work may make in alertness, concentration or skill, nor about any special physical problems such as excessive heat or static loads from awkward postures.

Hence, energy consumption should be used as a measure of comparison only for strenuous physical effort and never for studying mental activities or skilled work.

Leisure joules Off-work activities also consume energy, which we may call *leisure joules*. An average consumption would be 2400 kJ daily for a man and 2000–2200 kJ for a woman.

Thus the total energy expenditure is made up as follows:

1 Basal metabolism
2 Work joules
3 Leisure joules

Energy expenditure at work During and just after World War II a number of physiologists systematically studied the energy expenditure in a variety of occupations. At that time work joules were often used as the basis

Table 6.1   Energy demands of some occupational activities

Type of work	Example of occupation	Energy demand in kJ day	
		Men	Women
Light work, sitting	Book-keeper	9 600	8 400
Heavy manual work	Tractor driver	12 500	9 800
Moderate bodily work	Butcher	15 000	12 000
Heavily bodily work	Shunter	16 500	13 500
Extreme bodily effort	Coal miner Lumberjack }	19 000	–

Note: The figures of kJ/day are approximate annual means of daily energy expenditure.
Source: After Lehmann (1962).

Table 6.2   Energy expenditure in work joules during various forms of physical activities in the mid-1900s

Activity	Conditions of work	Energy expenditure in kJ/min
Walking	Level, smooth surface, 4 km/h	8.8
Walking with load	30 kg load, 4 km/h	22.3
Climbing stairs	30° gradient, 17.2 m/min	57.5
Cycling	Speed 16 km/h	22.0
Sawing wood	60 double strokes/min	38.0
Household work	Cleaning, ironing, washing floors	8–20

Note: The figures of kJ/min refer to the net working time.
Source: After Lehmann (1962).

for assessing occupational loads and the severity of tasks. Today many jobs are different, these procedures are no longer in vogue and new methods are used to assess workloads. Nevertheless, some of these results are assembled in Tables 6.1 and 6.2.

Working posture might have a significant influence on the energy expenditure, as the example in Figure 6.2 shows.

Energy consumption and health

Most workers in industrial countries sit down at their work. If we add to this the time they sit while they travel to and from work, and in front of the television screen in the evening, it is obvious that *today's human is clearly on the way to becoming a sedentary animal.* Such a sedentary life leaves many organs of the body under-used. Often more chemical energy is taken into the body

Sitting
3–5%

Standing
8–10%

Stooping
50–60%

Kneeling
30–40%

Figure 6.2 Percentage increase in energy consumption for different bodily postures. *100 per cent = energy consumption lying down. The relative increase, as a percentage of this, is the same for men and women.*

than is consumed, leading to overweight, with increased risk of cardiac and circulatory disease, as well as metabolic troubles. Research has shown that a healthy occupation should involve a daily energy consumption of 12 000–15 000 kJ for a man and 10 000–12 000 kJ for a woman. This category includes, for example, being a postman, mechanic, a shoemaker or having one of many jobs in building or agriculture. People who have sedentary jobs can make up some of the deficiency during their leisure time: some examples are given in Figure 6.3.

6.3 Upper limits of heavy work

Work physiologists today consider an energy consumption of 20 000 kJ per working day (averaged over a year) to be a reasonable maximum for heavy work in Europe and North America.

This corresponds to an average of 10 500 occupational kJ per working day and if these are spread over an 8-h work period they amount to 1300 kJ/h.

Seasonal workers may exceed these values for a few weeks, or even a few months, provided slack periods intervene.

This is true in principle for many heavy workers in forestry, haulage and so on who can reach levels of 22 000–30 000 kJ for a few days without ill effects. This limit of 20 000 kJ per working day is therefore a yearly average which applies only to a heavy worker in a good state of health. Clearly there are individual variations both above and below this mean value, depending particularly upon such factors as constitution, level of training, age and gender.

6.4 Energy efficiency of heavy work

Heavy work is becoming increasingly rare in industrial countries but it is still widespread in developing countries. A major target

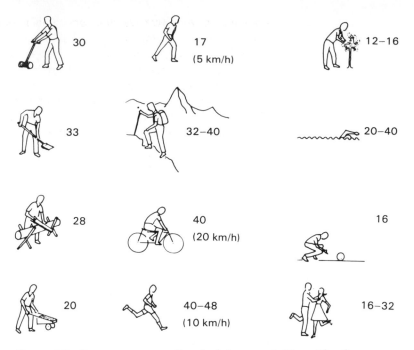

Figure 6.3 Energy consumption in leisure activities. *The figures are the average overall consumptions in kJ/min for men. Slightly lower figures apply to women, about 10–20 per cent less. Drawn from figures given by Durnin and Passmore* (1967).

of ergonomics in developing countries is therefore the achievement of a high level of efficiency in heavy work.

Efficiency

In terms of energy, a person doing physical work can be compared to a combustion engine. The engine converts chemical energy of gasoline or oil into mechanical performance but with certain losses. Similarly, in the human body, a large part of the energy it receives is 'wasted' by being converted into heat and only a small proportion becomes useful mechanical energy. In both engine and human the term *efficiency* denotes the ratio between the externally measurable useful effort and the energy consumption that was necessary to produce it.

Under most favourable conditions, human physical effort can be 30 per cent efficient, turning 30 per cent of the energy consumed into mechanical work and the remaining 70 per cent into heat. Heat is not the only form in which energy is wasted. Unproductive static or dynamic effort is equally important. Hence, the highest level of efficiency is possible only by converting as much as possible of the mechanical effort into a useful form, with little or none of it being dissipated in holding or supporting things. *The greater the proportion of mechanical energy to go into static effort, the lower the efficiency*. This is particularly true when work is carried out with a bent back.

In every kind of heavy work it is important to aim at maximal physiological efficiency, not only to use energy economically but to minimise stresses upon the operator. For this reason work physiologists have made a great many attempts to measure the physiological efficiency of various kinds of working methods and the use of different tools and other equipment. Their results make it possible to formulate guidelines for the layout of work and the design of equipment, which are of particular importance where strenuous work is concerned.

Some examples are given in Table 6.3.

Shovelling

Shovelling is a common form of manual work, which has been thoroughly studied at the Max-Planck-Institut für Arbeitsphysiologie in Dortmund, Germany (Lehmann, 1962) and, more recently, at Pennsylvania State University in the USA by Freivalds (1987). Both studies demonstrated that certain tool forms are better suitable than others to perform maximal work with minimal effort.

The highest level of efficiency was attained when *a load of 8–10 kg was shovelled 12–15 times per minute*. In addition to the

Table 6.3 Maximum efficiency in various physical tasks

Activity	Per cent efficiency
Shovelling in stooped posture	3
Screw driving	5
Shovelling in normal posture	6
Lifting weights	9
Turning a handwheel	13
Using a heavy hammer	15
Carrying a load on the back on the level, returning without load	17
Carrying a load on the back up an incline, returning without load	20
Going up and down ladders, with and without load	19
Turning a handle or crank	21
Going up and down stairs, without load	23
Pulling a cart	24
Cycling	25
Pushing a cart	27
Walking on the level, without load	27
Walking uphill on a 5° slope, without load	30

Efficiency (in per cent) = useful work $\times 100$ divided by energy consumption.

load, one must take into account the weight of the shovel itself so a big shovel should be used when the material is light and a small one for heavy materials. For fine-grained materials the shovel should be slightly hollow, spoon-shaped with a pointed tip for penetration. For coarse materials, the cutting edge should be straight and the blade flat, with a rim round the back and sides. For stiff materials such as clay, the cutting edge can be either straight or pointed but the blade should be flat. The handles of spades and shovels should be 600–650 mm long.

Sawing

In a study by Grandjean et al. (1952) the physiological efficiency of various types of timber saws was investigated. Oxygen consumption was measured before and during the use of five different types of saw and the number of work joules calculated. This energy consumption, related to cutting slices or discs 1 m^2 in area, is shown in Figure 6.4.

The results show that the best saws were the two ripsaws with a broad blade and either two or four cutting teeth in each group, and these saws were preferred. A more extensive study showed that the best results were obtained from a sawing rate of 42 double strokes per minute and a vertical force of about 100 N.

Hoeing

Figure 6.5 shows the results of loosening the soil of a vegetable plot using hoes of two different types. In soft soil a swivel hoe is much more efficient than the ordinary chopping hoe but if the soil is hard and dry the two types are almost equally good.

Walking

A pleasant and not too strenuous walking pace is 75–110 steps per minute, with a length of pace between 0.5 and 0.75 m, but this is not the most efficient pace when the work performed is compared with the energy consumed. Figure 6.6 shows the efficiency of walking, expressed in terms of the energy consumed per kg/m of work performed.

From studies by Hettinger and Müller (1953) it appears that the *most efficient walking speed is 4–5 km/h, reduced to 3–4 km/h in heavy shoes.*

Carrying loads

Work physiologists have given special attention to all kinds of jobs which involve carrying heavy loads since these are rightly considered to be among the most strenuous forms of work. (Perhaps even more taxing is 'lifting', discussed in Chapter 7.) While Lehmann (1962) found 50 to 60 kg to be the most efficient load to carry, smaller ones are more convenient and are 'safer' for the body (as discussed below) but require more journeys and carrying one's own body weight to and fro increases the total consumption

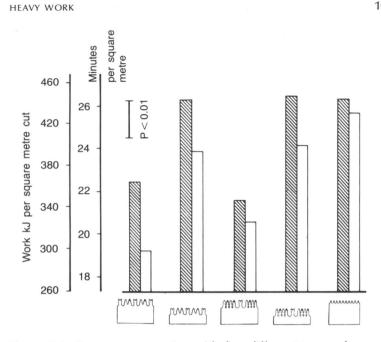

Figure 6.4 Energy consumption with five different types of saws. *The shaded columns represent the average energy consumption per square metre of cut surface; the white columns show the average time needed to saw 1 m². The vertical marker (P < 0.01) indicates a difference in energy consumption that is statistically significant. After Grandjean et al.* (1952).

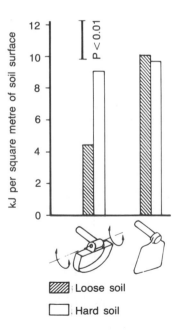

Figure 6.5 Energy consumption while using a swivel hoe (left) and a traditional hoe (right) in loose and hard soil. *After Egli et al.* (1943).

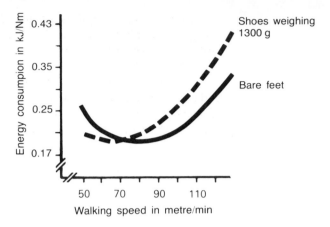

Figure 6.6 Best walking speed. *Efficiency expressed as the energy consumption in kJ per unit of walking effort (in Nm). Curves shown for bare feet (solid line) and for shoes weighing 1300 g (broken line). The heavier the shoes, the lower the best speed.*

of energy. Of course, the carried weight might be broken into smaller portions which can be attached to the back, the shoulders, the chest or even be put on the head. Carrying a medium load (of about 30 kg) distributed on the back and chest is the least energy consuming, as Table 6.4 shows.

In most modern industrial jobs that involve carrying the work conditions are rather well controlled. In such circumstances, the data published by Snook and Ciriello (1991) provide good guidelines about the loads that experienced American workers are willing to carry (Table 6.5).

Nature of the ground surface

The nature of the ground surface has a great effect on energy consumption. Walking on smooth and solid ground is least demanding. The effort increases on a dirt road, gets heavier through light brush, on hard-packed snow, through heavy brush or in a swampy bog. Compared with walking on black top, walking in loose sand takes twice as much energy, in soft snow 20 cm deep it takes three times as much and in 35 cm, four times as much. (Data collected by Kroemer *et al.*, 1994.) Use of a wheelbarrow or other kind of cart for carrying loads can be helpful. The following are the distances that can be covered with a load of 1 tonne on a four-wheeled hand-truck and using up 1050 work kJ:

On a light railway track	850 m
On a good road	700 m
On a bad road	400 m
On a dirt road	150 m

Pushing a cart requires about 15 per cent less effort than dragging it behind. Snook and Ciriello (1991) published extensive tables of the forces that experienced material handlers were willing to push or pull.

The handles of the cart should be close to 1 m above the ground and about 40 mm thick. A two-wheeled cart should have its centre of mass as low and as close to the axle as possible for better balance and to reduce the amount of weight to be supported.

Slopes

When a task involves climbing, a slope of 10° gives the best efficiency. The following heights can be reached for an outlay of 42 work kJ:

On a ladder at 90°	11.5 m
On a ladder at 70°	14.4 m
On a staircase at 30°	13.2 m
On a path at 25°	13.1 m
On a path at 10°	15.5 m

Climbing ladders is most efficient when the ladder is at an angle of 70° and the rungs are 260 mm apart but if heavy loads are being carried, 170 mm between rungs is better. In some cases, alternating rungs may be advantageous, as Jorna *et al*. demonstrated in 1989.

Staircases

Climbing stairs is one of the best forms of exercise in everyday life. *From the preventive medicine standpoint, sedentary people are strongly advised to take every opportunity to climb the stairs as a 'gymnastic exercise'.* At the same time it is sensible to have the stairs designed in such a way that they can be climbed as efficiently as possible. This is particularly important when the stairs are in frequent use or used by infirm or old people.

Lehmann (1962) found that least energy was consumed when climbing stairs with a slope of 25–30° and he recommended the following empirical norms:

Tread height (riser)	*170 mm*
Tread depth	*290 mm*

Stairs of these dimensions are not only the most efficient but also seem to cause the fewest accidents. This recommendation can be expressed as a formula:

2h + d = 630 mm

where h = height of riser and d = depth of tread, both in mm. Figure 6.7 gives optimum dimensions for the design of flights of stairs.

Table 6.4 Techniques of carrying loads totalling about 30 kg in various ways affects energy consumption, fatigue, pressure on the body and stability while walking. *Adapted with permission from Kroemer et al. (1994) who compiled this information from many sources*

	Estimated energy expenditure on a straight flat path (kJ/min)	Estimated muscular fatigue	Local pressure and ischemia	Stability of the carrying person
In one hand	Very high	Very high	Very high	Very poor
In both hands	Very high (30; equal weights)	High	High	Poor
Clasped between arms and trunk	(Not measured)	(Unknown)	(Unknown)	(Unknown)
On head	Fairly low (22; supported with one hand)	(High if stabilized by hand)	(Unknown)	Very poor
On neck	Medium (23; therpa-type strap around forehead)	(Unknown)	(Unknown)	Poor
On one shoulder	(Not measured)	High	Very high	Very poor
Across both shoulders	High (26; yoke held with one hand)	(Unknown)	High	Poor
On back	Medium (22; backpack) High (25; bag held in place with hands)	Low	(Unknown)	Poor
On chest	(Not measured)	Low	(Unknown)	Poor
On chest and back	20, lowest	Lowest	(Unknown)	Good
At waist, on buttocks	(Not measured)	(Unknown)	(Unknown)	Very good
On hip	(Not measured)	Low	(Unknown)	Very good
On legs	(Not measured)	High	(Unknown)	Good
On foot	Highest	Highest	(Unknown)	Poor

Special aspects

Load easily manipulated and released Load easily manipulated and released Comprise between hand and trunk use	Suitable for quick pick-up and release; for short-term carriage even of heavy loads.
May free hand(s); strongly limits body mobility; determines posture; pad is needed May free hand(s); affects posture	If accustomed to this technique, suitable for heavy and bulky loads.
May free hands; strongly affect posture	Suitable for short-term transport of heavy and bulky loads.
May free hand(s); affects posture	Suitable for bulky and heavy loads; pads and means of attachment must be carefully designed
Usually frees hands; forces forward trunk bend; skin cooling problem	Suitable for large loads and long-time carriage. Packaging must be done carefully, attachment means should not generate areas of high pressure on body.
Frees hands, easy hand access; reduces trunk mobility; skin cooling problem	Very advantageous for small loads that must be accessible
Frees hands; may reduce trunk mobility; skin cooling problem	Very advantageous for loads that can be divided/distributed; suitable for long-time carriage
Frees hands; may reduce trunk mobility	Around waist for smaller items, distributed in pockets or by special attachments; superior surface of buttocks often used to partially support backpacks.
Frees hands; may affect mobility	Often used to prop large loads temporarily.
Easily reached with hands; may affect walking	Requires pockets in garments and/or special attachments.
Usually not useful	

Table 6.5 Maximally acceptable weights (kg) of carry (North American workers)

	(a)	(b)	One 2.1 m carry every							One 4.3 m carry every							One 8.5 m carry every						
			6	12	1	2	5	30	8	10	16	1	2	5	30	8	18	24	1	2	5	30	8
			s		min				h	s		min				h	s		min				h
Males	111	90	10	14	17	17	19	21	25	9	11	15	15	17	19	22	10	11	13	13	15	17	20
		75	14	19	23	23	26	29	34	13	16	21	21	23	26	30	13	15	18	18	20	23	27
		50	19	25	30	30	33	38	44	17	20	27	27	30	34	39	17	19	23	24	26	29	35
		25	23	30	37	37	41	46	54	20	25	33	33	37	41	48	21	24	29	29	32	36	43
		10	27	35	43	43	48	54	63	24	29	38	39	43	48	57	24	28	34	34	38	42	50
Males	79	90	13	17	21	21	23	26	31	11	14	18	19	21	23	27	13	15	17	18	20	22	26
		75	18	23	28	29	32	36	42	16	19	25	25	28	32	37	17	20	24	24	27	30	35
		50	23	30	37	37	41	46	54	20	25	32	33	36	41	48	22	26	31	31	35	39	46
		25	28	37	45	46	51	57	67	25	30	40	40	45	50	59	27	32	38	38	42	48	56
		10	33	43	53	53	59	66	78	29	35	47	47	52	59	69	32	38	44	45	50	56	65
Females	105	90	11	12	13	13	13	13	18	9	10	13	13	13	13	18	10	11	12	12	12	12	16
		75	13	14	15	15	16	16	21	11	12	15	15	16	16	21	12	13	14	14	14	14	19
		50	15	16	18	18	18	18	25	12	13	18	18	18	18	24	14	15	16	16	16	16	22
		25	17	18	20	20	21	21	28	14	15	20	20	21	21	28	15	17	18	18	19	19	25
		10	19	20	22	22	23	23	31	16	17	22	22	23	23	31	17	19	20	20	21	21	28
Females	72	90	13	14	16	16	16	16	22	10	11	14	14	14	14	20	12	12	14	14	14	14	19
		75	15	17	18	18	19	19	25	11	13	16	16	17	17	23	14	15	16	16	17	17	23
		50	17	19	21	21	22	22	29	13	15	19	19	20	20	26	16	17	19	19	20	20	26
		25	20	22	24	24	25	25	33	15	17	22	22	22	22	30	18	19	21	21	22	22	30
		10	22	24	27	27	28	28	37	17	19	24	24	25	25	33	20	21	24	24	25	25	33

(a) Vertical distance from floor to hands (cm).
(b) Acceptable to 90, 75, 50, 25, 10 per cent of industrial workers.
Source: Adapted from Snook and Ciriello (1991).

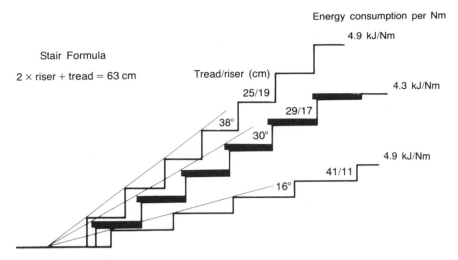

Figure 6.7 Physiological recommendations for the slope and dimensions of stairs.

6.5 Heart rate as a measure of workload

Energy consumption and heart rate

Into the mid-1900s energy consumption was usually the means by which the severity of physical stress was estimated but it is evident that energy consumption alone may not be a sufficient measure. The degree of physical stress depends not only on the number of kJ consumed but also on the number of muscles involved and on the extent to which they are under static load. A given level of energy consumption is much more strenuous if it is achieved by using only a few muscles than if many others are employed. Similarly, the same energy consumption by static muscular effort is distinctly more tiring than if it is applied to dynamic work.

A further argument against the sole use of energy consumption as a measure of workload is that it may not respond to certain conditions at work, such as the heat that often prevails. This may be a trivial part of the energy consumption yet it may cause a sharp rise in the heart rate.

The various ways in which a rise in heart rate is related to workload are shown diagrammatically in Figure 6.8. This figure shows that work at a given energy consumption can make different demands on the heart according to the circumstances.

Summarising, it can be said *the rise in heart rate with increasing workload is the steeper*:

(a) the higher the ambient temperature

(b) the greater the proportion of static to dynamic effort

(c) the smaller the number of muscles involved.

For these reasons, heart rate has been used increasingly in

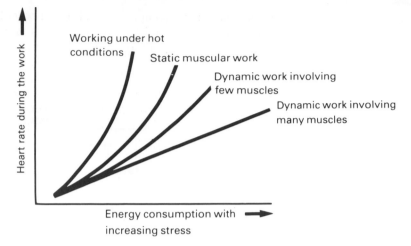

Figure 6.8   Increase in heart rate associated with various types of stress.

recent years as an index of workload. Before going into details about heart rate and pulse we must first consider the relationships between blood circulation and respiration.

Blood and respiration

Physical work demands adjustments and adaptations which affect nearly all the organs, tissues and fluids of the body. The most important adjustments are as follows:

1 Deeper and more rapid breathing.
2 Increased heart rate, accompanied by an initial rise in cardiac output per heart beat.
3 Vasomotor adaptations, with dilatation of the blood vessels in the organs involved (muscles and heart), while other blood vessels are constricted. This diverts blood from the organs not immediately concerned into those which need more oxygen and nutrients.
4 Rise in blood pressure, increasing the pressure gradients from the main arteries into the dilated vessels of the working organs and so speeding up the flow of blood.
5 Increased supply of sugars (glucose and glycogen) released into the blood by the liver.
6 Rise in body temperature and increased metabolism. The increase in temperature speeds up the chemical reactions of metabolism and ensures that more chemical energy is converted into mechanical energy (for this reason athletes 'warm up' before a contest).

As work continues, secondary metabolic effects arise, particularly in the chemical composition of the body fluids. There is an accumulation of metabolic waste products, notably lactic acid, and the kidneys have more waste products to excrete. Muscular activity generates additional heat in the body and, to restore the balance, more heat must be dissipated through the skin by sweating. Some of the heat is also transported in the bloodstream to the lungs where it is released into the air to be exhaled, together with some of the water liberated in the metabolic process. Obviously, the circulatory system is intensively involved in all of this: in the transport of nutrients and oxygen to the working muscles, and in the transport of heat, water and carbon dioxide to the lungs and of heat and water to the skin.

Within certain limits, the changes – ventilation of the lungs, sweating, body temperature and especially heart rate – show a linear relationship with energy consumption, or the work performed. Since these changes can be measured while a person is at work, they can be used to assess the physical effort involved. Table 6.6 shows reactions measured at various workloads.

Measuring the heart rate (pulse)

Measuring the heart rate ('taking the pulse') is one the most useful ways of assessing the workload because it can be done so easily.

One can easily feel the pulses (the blood pressure in waves) at the radial artery in the wrist or at the carotid artery at the neck but to do this while a person is at work is an intrusion that may disturb his or her work, thereby producing a false result. Instruments have been devised which give a continuous record of heart rate during work. One kind uses the changes in tissue volume with each pulse; the most modern method uses the electrical current associated with the actions of the heart muscle (as can be seen in the electrocardiogram) and the heart rate is counted by the number of R peaks (the strongest action potential) per unit of time, usually per minute.

Heart rate during physical activity

As already mentioned, and within certain limits, the heart rate increases linearly with the work performed, provided this is dynamic not static and is performed with a steady rhythm.

When the work is comparatively light, the pulse rate increases quickly to a level appropriate to the effort, then remains constant for the duration of the work. When work ceases, the heart returns to its normal beat after a few minutes.

With more strenuous work, however, the heart rate goes on increasing until either the work is interrupted or the operator is forced to stop from exhaustion. Figure 6.9 shows diagrammatically the behaviour of the pulse during certain work studies.

Table 6.6   Metabolism, respiration, temperature and heart rate as indications of workload

Assessment of workload	Oxygen consumption (litre/min)	Lung ventilation (litre/min)	Rectal temperature (°C)	Heart rate (pulses/min)
Very low (resting)	0.25–0.3	6–7	37.5	60–70
Low	0.5–1.0	11–20	37.5	75–100
Moderate	1.0–1.5	20–31	37.5–38.0	100–125
High	1.5–2.0	31–43	38.0–38.5	125–150
Very high	2.0–2.5	43–56	38.5–39.0	150–175
Extremely high (e.g. sport)	2.4–4.0	60–100	over 39	over 175

Source: Christensen (1964).

Scales of heart rate

Müller (1961) proposed the following definitions:

Resting pulse. Average heart rate before the work begins.
Working pulse. Average heart rate during the work.
Work pulse. Difference between the resting and working pulses.
Total recovery pulse (recovery cost). Sum of heart beats from the cessation of work until the pulse returns to its resting level.
Total work pulse (cardiac cost). Sum of heart beats from the start of the work until resting level is restored.

Müller believed that the total recovery pulse is a way of measuring fatigue and recovery. Since 'fatigue' can be a subjective term it would be appropriate to regard heart rate, and particularly total recovery pulse, as a measure of the physical workload of an individual.

Acceptable limits

Karrasch and Müller (1951) made use of their studies to define as an acceptable upper limit of workload that below which the working pulse does not continue to rise indefinitely and, when the work is stopped, returns to the resting level within about 15 min. This limit seems to ensure that energy is being used up at the same rate as it is being replaced, that is to maintain a 'steady state' of the body. *The maximal output under these conditions is the limit of continuous performance throughout an 8-h working day.*

The limit of continuous performance for men is reached when the average working pulse is 30 beats/min above the resting pulse; both of these are measured in the same posture (for example, both standing up), so that the static loads are the same. Rohmert and Hettinger (1965) made a systematic study of limits of workload, during which the heart rate remained steady, using a bicycle

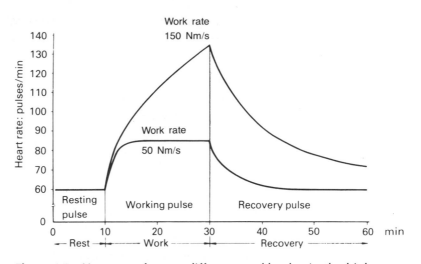

Figure 6.9 Heart rate for two different workloads. *At the higher level of stress the heart rate increases as long as the stress lasts, whereas at the lower rate it levels off at a 'plateau' or 'steady state'.*

ergometer and a hand-crank for eight hours at a time. They came to the conclusion that this limit was still valid up to a work pulse of 40/min, provided that *the resting pulse was measured when the operator was lying down.* The authors show that for dynamic work involving a moderate number of muscles 4 work kJ/min are equivalent to 10 work pulses.

Several studies undertaken in factories have shown that it is easier to measure the resting pulse when the subject is sitting rather than lying down so we suggest that, *for men,* we should start from *a resting pulse taken when seated* and fix *35 work pulses as the limit for continuous performance.* There are no corresponding studies of *women* but on physiological grounds it seems reasonable to postulate *30 work pulses as the limit for continuous performance,* again taking the resting pulse in a seated position.

Recovery pulse

In the USA, Brouha (1967) made detailed studies of heart rate as an index of workload and the most important of his findings include:

1 *A workload of 3500 Nm/min produced work pulses of 50 for women and 40 for men.*

2 The total recovery pulse (recovery cost) was as good as, or even better than, the total work pulse (cardiac cost) as a measure of workload.

3 The following procedure seems to be suitable for recording the recovery process. After work has stopped, take the pulse at the wrist during the following 30-s intervals:

> from 30 s to 1 min;
> from 1½ min to 2 min;
> from 2½ min to 3 min.

Take the average of these three readings as the heart rate during the recovery phase as well as an indication of the preceding workload.

4 The following criterion was recommended for determination of acceptable limits of workload: *the first reading should not exceed 110 pulses/min with a fall of at least 10 pulses between the first and third readings*. Given this condition the workload could be sustained during an 8-h working day.

6.6 Combined effects of work and heat

As we have already noted, heart rate can be used as a measure of heat load as well as of workload. This is understandable since the heart is merely a 'pump' which supplies the muscles with blood and suffuses the skin and lungs with blood when it is necessary to eliminate excess heat. So when work is being performed under hot conditions, the heart and the circulatory system that it powers have two functions:

(1) Transporting energy to the muscles.

(2) Transporting heat energy from the interior of the body to the skin.

This double burden on heart and circulatory system is a common occurrence in industry, forestry and agriculture. When heavy work has to be carried out in ambient temperatures of 25°C, the elimination of excess heat throws an additional load upon the heart. An example of such a double loading is drop forging. Hünting and his colleagues (1974) were able to show that holding the forging tongs during the operation caused considerable loads on the back and arms, both static and dynamic. Working conditions were certainly strenuous and deliberate pauses of up to 60 per cent of the working time were noted. At the same time the operator was subjected to intensive radiation of heat. Figure 6.10 gives a simple sketch of the drop forge and results from measurements of the workload on a 47-year-old operator are shown in Figure 6.11.

The mean work pulse of 41 beats/min during forging rates as a high level. Figure 6.11 also indicates the limit of continuous performance set by Müller (1961), which is valid in this case because the resting pulse of the operator was measured when he was standing up.

The net working time was related to the number of forgings/min made and to the ambient temperature. The heart rate followed in step with these two quantities.

Static work and heart rate

Chapter 1 gave an example of 'potato planting' where the 'static effort' produced a work pulse of 40 beats/min, which fell to 31 beats/min when the need to support the potato basket was eliminated (Figure 1.8), although the energy consumption remained the same. This shows that *static effort can cause a rise in heart rate, even though there is no increase seen in the total consumption of energy. This rise must be interpreted as an increased physical stress*. This phenomenon is also shown in Figure 6.9.

In the laboratory, the effects of static muscular effort have been studied mainly by using loads which had to be either supported or dragged along. Figure 6.12 shows the results of experiments by Lind and McNicol (1968). It was found that in spite of considerable static loads the heart rate rose only a little above 100 beats/min and that the initial resting pulse was quickly restored afterwards.

6.7 Case histories involving heavy work

Heavy work in the iron and steel industry

Hettinger (1970) and Scholz (1963) carried out intensive studies of workloads in the German iron and steel industry. In Figure 6.13 the maximum heart rate of 380 workers, measured over periods of 2 and 4 min, is displayed as a frequency distribution curve. It is evident from this that the most frequent peak value lies in the range 130–140 beats/min (mean 132.6) and that extreme increases of up to 180 beats/min occur.

Hettinger (1970) compared the energy expenditure (work kJ) with the work pulses of 552 workers in the German iron and steel industry. In proportion to their respective maximal limits, values for work kJ were generally lower than the values for work pulses, a discrepancy which was attributed to the effects of static effort and heat. In about one-third of the workplaces work pulses were above the limit of 40 beats/min.

Heavy work in agriculture

Despite mechanisation heavy work still exists in agriculture and studies by Brundke (1973) are mentioned in this connection.

A fruit grower, a farmer and a dairyman were studied, and their pulses taken throughout a work week. Resting pulse was measured when they were sleeping at night. Figure 6.14 shows the results of recording their work pulses.

Figure 6.10 Drop forging. *The operator manipulates a piece of glowing metal weighing 18.5 kg with tongs and is exposed to a strong radiant heat.*

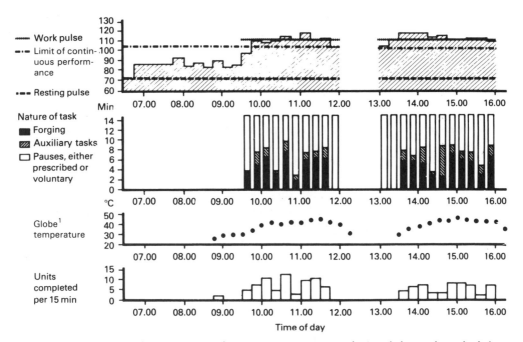

Figure 6.11 Pulse measurements, analysis of the task and globe temperatures of an operator of a drop forge. *Top heart rate (averaged over 15 min). Average work pulse during drop forging is 41/min. Middle: distribution in time of the various phases of the job. Pure drop forging occupies 28 per cent of the total time. Globe temperatures for each phase are given in °C. Bottom: number of completed units per 15 min period. After Hünting et al. (1974).*[1] *Globe temperature measures the Mean Radiant Temperature – see Chapter 20.*

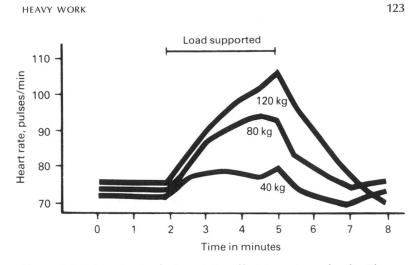

Figure 6.12 Heart rate during static effort (carrying a load). *The heavier the load, the greater the total of both work pulses and recovery pulses. After Lind and McNicol* (1968).

It is clear from Figure 6.14 that the critical limit of 40 work pulses/min was exceeded on several occasions, especially by the dairyman and the farmer at harvest time. These studies confirm that *in spite of mechanisation, heavy work is still part of the routine for the agricultural worker.*

Heavy work by women in the textile industry

There are many jobs in industry that involve heavy work without appearing to do so at first glance. One such example may be taken from the textile industry, where women have the job of examining spools of artificial fibre and packing them into boxes (Nemecek and Grandjean, 1975). The three key movements in this task are illustrated in Figure 6.15.

Reach to the left varied between 500 and 900 mm while that to the right was about 900 mm. The movement across from left to right required an effort that was partly static (supporting a weight of 3 kg) and partly dynamic, and each turn was accompanied by slight bending and twisting of the trunk. The factory introduced a '*new method*' of knotting the ends of the fibres but this did not produce the improved performance that was expected. Therefore, the question arose of whether the operators might be physically overstressed.

Therefore, the researchers selected five workers and took their pulse, counted how many spools they handled in 15 min and investigated the extent of bodily aches and subjective impressions of fatigue.

To assess the bodily aches and the fatigue an open-ended questionnaire was applied, both before and after the task.

The resulting pulse measurements are shown in Figure 6.16. Assuming that the desirable upper limit for continuous effort for

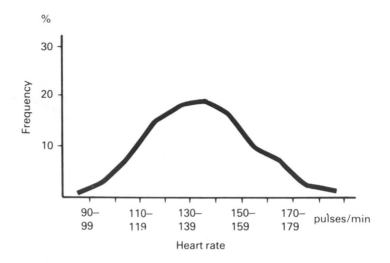

Figure 6.13 Frequency distribution of maximal heart rates. *Measured on 380 factory workers during the course of a working day in the German iron and steel industry during the years 1961–1969. After Hettinger (1970).*

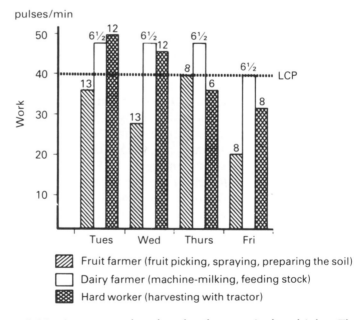

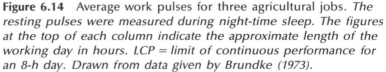

Figure 6.14 Average work pulses for three agricultural jobs. *The resting pulses were measured during night-time sleep. The figures at the top of each column indicate the approximate length of the working day in hours. LCP = limit of continuous performance for an 8-h day. Drawn from data given by Brundke (1973).*

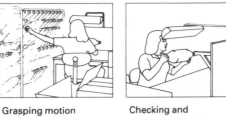

Grasping motion   Checking and   Packing to the right 2 s
to the left 1–2 s   tying ends 10–12 s

Figure 6.15 Inspection of spools for artificial fibre. *The three drawings show the three operations involved. Weight of a spool = 3 kg. Performance: 730–960 × 3 kg = 2200–2900 kg per arm per shift. After Nemecek and Grandjean* (1975).

women is 30 work pulses/min, we can see that this has been exceeded in six out of 10 cases since twice the number of work pulses of more than 40/min were recorded. *The cause for the higher effort is in the increased demand for muscular work of the arms.* According to Rohmert (1960), on average a woman is able to hold 6 ± 1.5 kg with an outstretched arm so that to lift a spool weighing 3 kg she is exerting about half her maximal strength. This is too high even for dynamic effort and obviously so for static effort, which should not exceed 15–20 per cent of maximal force capability. Moreover, we have not allowed for the weight of the outstretched arm itself, which must be supported as well. So these studies have shown that even *apparently light forms of work can involve operators in too much muscular effort.*

The assessment of bodily aches and fatigue disclosed that pains in the right arm, back and neck greatly increased during the working shifts. There is no doubt that these complaints are related to the postures that the workers had to adopt.

Finally, Table 6.7 summarises the effects of the *new method*, with its demand for increased performance, on the output, the work pulse and the level of aches and fatigue. The 'self-assessment' scheme made use of an average of the daily variations in replies to the questions involving fatigue and aches in arms, hands, back and neck.

Although these results were not statistically significant, probably because of the small number of observations, the trend is obvious. The workload was already excessive under the old procedure and the new method – by reducing the inspection time and involving more arm work to left and right – increased the workload still further, which tended to raise the work pulse and apparently also caused more fatigue and aches and pains in the workers.

Conclusions   From the evidence of these results, the work could be lightened in the following ways:

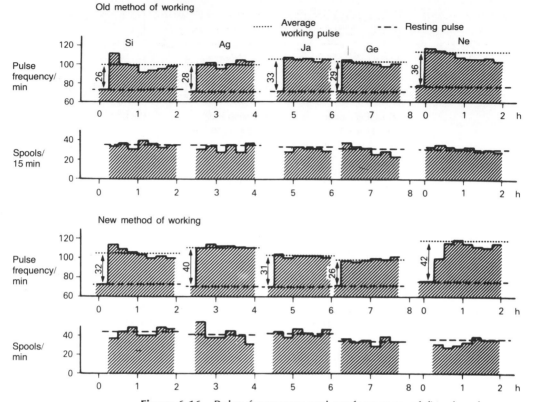

Figure 6.16 Pulse frequency and performance of five female workers checking spools (bobbins) in a textile factory. *Si, Ag, Ja, Ge, Ne = abbreviation of the worker's name. Broken line = resting pulse while sitting. Dotted line = average working pulse. Vertical arrows = work pulse averaged over 2 h. After Nemecek and Grandjean (1975).*

1 Shortening the distance to be reached at each side.

2 Lowering the working level.

3 Introducing mechanical aids to reduce the load on the hands, e.g. a swivelling support for the spool.

4 Reorganising the work, with a rotation between different operations.

Summary

The human is not an 'energy-efficient animal' as measured by joules, heart rate or fatigue but the ergonomist can employ many techniques to make heavy physical work.

Table 6.7 Effect of the increased performance of the new method on output, work pulse, fatigue and aches and pains in five female textile workers

Worker[a]	No. of spools handled per minute		Work pulses		Increases in daily aches and fatigue	
	Old M	New M	Old M	New M	Old M	New M
Ne	730	959	36	42	1	18
Ag	720	912	28	41	22	27
Ge	732	840	29	26	7	19
Ja	732	960	33	31	3	31
Si	852	960	26	32	26	30

Note: M = method. [a]Abbreviation of worker's name.
Source: After Nemecek and Grandjean (1975).

CHAPTER SEVEN

Handling loads

7.1 Back troubles

Handling loads (lifting, lowering, pushing, pulling, carrying, holding and dragging) often involves a good deal of static and dynamic effort, enough to be classified as heavy work.

Loss of work through back troubles

The main problem with these forms of work, however, is not the heavy loading of muscles but more *the wear and tear on the back, especially on the lumbar intervertebral discs*, with the increased risk of injury. That is why the handling of loads deserves a chapter to itself.

Back troubles can be painful and reduce one's mobility and vitality. They often lead to long absences from work and in modern times are among the main causes of early disability. They are comparatively common in the age group 20–40 years, with certain occupations (nurses, labourers, farmers, baggage handlers and so on) particularly prone to disc problems. Workers with physically active jobs suffer more from ailments of this nature and their work is more affected than with sedentary workers.

Over-exertion and lower back pain

Over-exertion injuries, especially in the lower back, account for about one-quarter of all reported occupational injuries in the USA. Some industries report that more than half of the total reported injuries are due to over-exertion. Approximately two-thirds of over-exertion injury claims involved lifting loads and about 20 per cent involved pushing and pulling loads.

The British Health and Safety Executive (Health and Safety Executive, 1992) reported that in the UK more than a quarter of the reported industrial injuries in 1990/1 were associated with manual handling — the transporting or supporting of loads by hand or by bodily force. Of these injuries, 45 per cent occurred to the back, 22 to the hands and 13 to the arms. Similar data are reported from the USA (e.g. Marras *et al.*, 1995). According to Krämer (1973), in Germany disc disorders were the cause of 20 per cent of absenteeism and 50 per cent of premature retirements. Clearly, lower back pain is among the most common causes of injury and disability in many industrial populations.

Reasons for back problems

Many victims of back trouble cannot tell how it started. In most cases they cannot point to sudden pain related to a certain action but rather the problem developed slowly until it was strong enough to disable.

The widespread occurrence of lower back pain calls for the efforts of ergonomists to avoid or at least reduce acute overloading and wear and tear on the intervertebral discs. When the 'work seat' was discussed in Chapter 5 we considered the anatomy of intervertebral discs and the nature of disc troubles. Here are a few supplementary remarks.

Disc troubles

The vertebral column, or spine, has the shape of an elongated S: at chest level it has a slight backwards curve, called a kyphosis, and in the lumbar region it is slightly curved forwards, the lumbar lordosis. Another lordosis is in the cervical spine supporting the head. This construction gives the column of vertebrae the elasticity to absorb the shocks of running and jumping.

The loading on the vertebral column increases from the neck downwards and is at its greatest in the five lumbar vertebrae.

Adjacent vertebrae are separated by an intervertebral disc. Degeneration of the discs may simply be a function of ageing or a result of many repeated motions; ageing and wear often come together, and a sudden over-exertion can lead to an acute injury. Disc degeneration first affects the outer layers of the disc, which are normally tough and fibrous. A tissue change is brought about by loss of water so that the fibrous ring becomes brittle and fragile, and loses its strength. At first the degenerative changes merely make the disc flatter, with the risk of damage to the mechanics of the spine or even of displacement of vertebrae. Under these conditions quite small actions, such as lifting a light weight or just one's own body, a slight stumble or similar incidents, may precipitate severe backache.

When degeneration of the disc has progressed further, any sudden compression force may squeeze the gel-like nucleus of the disc and viscous internal fluid out through the ruptured outer ring and so exert pressure either on the spinal cord itself or on the nerves running out from it. This is what happens in the case of a 'slipped disc' or disc herniation. Pressure on the nerves, narrowing of the spaces between vertebrae, pulling and squeezing at the tissues, muscles and ligaments at the spine are the causes of the variety of discomfort, aches, muscular cramps and paralyses including lumbago and sciatica.

Three different approaches can be distinguished among the studies evaluating the risk of back troubles when loads are lifted:

1 The measurement of intervertebral disc pressure.

2 Biomechanical models to predict compression forces on the lumbar spine.

3 The measurement of intra-abdominal pressure.

These three approaches will be summarised below.

7.2 Intervertebral disc pressure

Loading of the intervertebral discs

In Sweden, Nachemson (1974), Nachemson and Elfström (1970) and Anderson and Ortengren (1974), studied the effects body posture, and the handling of loads, on pressures inside the intervertebral discs.

Figure 7.1 shows the results of handling various weights on these pressures in nine people; two of these had back troubles and the other seven were in good health. The figure shows clearly the effect of a bent back on the loading of the discs when a weight was being lifted.

Bending the back, while keeping the knees straight, puts a much greater stress on the discs in the lumbar region than keeping the back as straight as possible and bending the knees.

Lifting technique and disc pressure

Figure 7.2 shows how pressure develops in the discs during the two types of lifting action. This pressure curve shows very clearly how lifting a load with a bent back brings about a sudden and steep increase in internal pressure in the discs and quickly overloads them, especially the fibrous rings.

Scientific studies confirm the everyday experience that persons who have disc troubles are especially liable to sudden and violent pains, and even paralysis. These symptoms are often precipitated by sudden heavy loads placed on the discs, a risk which is increased by working methods involving unskilful manipulation.

The weights of the head and neck, arms and hands, and of the upper trunk all rest on the spinal column because this is the only solid bony structure in the human rump that keeps the rib cage from falling into the pelvis. The heavier the upper body, the larger the force on the spine. An additional load carried in the hands 'loads' the arms, the shoulders and further compresses the spine. Compared with standing still and erect, walking, bending or twisting the body increases the force on the spinal column,

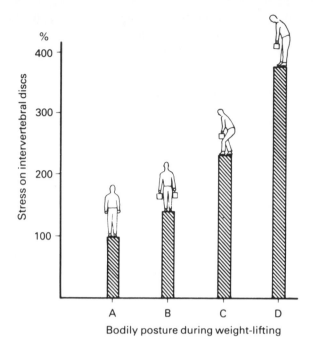

Figure 7.1 The effect of body posture when lifting weights on the intervertebral disc pressure between the third and fourth lumbar vertebrae. *A = upright stance; B = upright stance with 10 kg in each hand; C = lifting a load of 20 kg with knees bent and back straight (correct stance for weight-lifting); D = lifting 20 kg with knees straight and back bent. Pressure on discs during upright stance (A) is taken as 100 per cent. After Nachemson and Elfström (1970).*

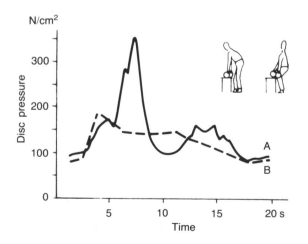

Figure 7.2 Curves of pressure within the intervertebral disc between the third and fourth lumbar vertebrae while lifting a load of 20 kg. *A = back rounded, knees straight; B = back straight, knees bent. After Nachemson and Elfström (1970).*

Table 7.1 Loading of the disc between the third and fourth lumbar vertebrae during various postures and tasks

Posture/activity	N
Standing upright	860
Walking slowly	920
Bending trunk sideways 20°	1140
Rotating trunk about 45°	1140
Bending trunk forwards 30°	1470
Bending trunk forwards 30°, supporting weight of 20 kg	2400
Standing upright holding 20 kg (10 kg in each hand)	1220
Lifting 20 kg with back straight and knees bent	2100
Lifting 20 kg with bent back and knees straight	3270

Source: After Nachemson and Elfstrom (1970).

especially its intervertebral discs. Examples of disc compression forces associated with various body postures are summarised in Table 7.1.

Distribution of loads on discs

When a rounded back causes curvature of the lumbar spine, the loads imposed on the intervertebral discs are not only heavy but asymmetrical, since they are considerably heavier on the front edge than on the back (Figure 7.3). *The resultant stresses on the fibrous ring are certainly bad for it and must be regarded as an important factor in the 'wearing out' of the disc.* Further, it must be assumed that the viscous fluid inside the disc will tend to be

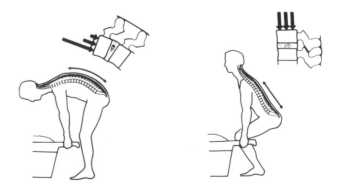

Figure 7.3 How the pressures on the intervertebral discs are distributed when a load is being lifted with bent, and with straight, back. *The round back (left) leads to heavy pressures on the front edge of the disc and increases the risk of rupture. The straight back (right) ensures that the loads on the disc are evenly distributed, thus reducing wear and tear on the fibrous ring of the disc.*

squeezed towards the side with less pressure. If this is on the back side, then there is the danger that this fluid will leak towards the spinal cord. Obviously, there are convincing arguments for keeping the trunk as straight as possible when lifting a heavy load.

7.3 Biomechanical models of the lower back

If a person bends over until the upper part of the body is nearly horizontal, then the leverage effect imposes very heavy pressure on the discs between the lumbar vertebrae.

An average mass of the upper part of the body would be about 45 kg and the length of leverage to the spine is about 0.3 m if one is fairly erect. This results in moment of 132 Nm ($45 \, kg \times 9.81 \, ms^{-2} \times 0.3 \, m$). If a load of 20 kg is held at the same time in the hands, that is about 0.5 m in front of the spine, this adds another 98 Nm for a total of 230 Nm. If one bends strongly forward, the leverage increases, generating a severe bending moment at the lower spine. Figures 7.1, 7.2 and 7.3 illustrate these conditions.

The combined forward-bending moment must be counteracted by muscle action behind the lumbar vertebrae, where only the erector spinae and the latissimus dorsi muscle are available. As seen from the side, they are very close to the spine, about 20 mm behind it. This puts them at great mechanical dis-advantage to counter the forward-bending moment. Dividing the amount of 230 Nm just calculated by the lever arm of 0.02 m shows that the muscles behind the lumbar spine must develop the enormous force of 11 500 N. Adding up all the forces [$11 \, 500 \, N + (45 \, kg + 20 \, kg) \times 9.81 \, ms^2$] results in a total of 12 138 N, all of which is transmitted as compression force through the spinal column. *No wonder that the musculoskeletal structure can buckle under such stress and that discs may herniated.*

Since the late 1800s more and more complex models have been developed to understand and explain the mechanics of the human body. Kroemer *et al*. (1988) describe this in some detail. Tichauer (1968, 1973, 1975, 1976), Chaffin (1969), Frankel and Burnstein (1970) and Kelly (1971) were among the first biomechanicists to develop basic kinematic models. More recently, Chaffin and Andersson (1993), Marras *et al*. (1984, 1989, 1995) and Oezkaya and Nordin (1991) described models to calculate and evaluate the moments and forces acting on and within the human body.

For military applications, biomechanical methods were predominantly employed by researchers at the British Royal Air Force Research Establishment at Farnborough and at the Aerospace Medical Research Laboratories at Wright-Patterson Air Force Base in Ohio, as well as by many other military and civilian institutions in Europe and Asia. (The 1989 book edited by McMillan *et al.* and the 1993 book edited by Peacock and Karwowski provide recent overviews of these developments.)

At several academic institutions (especially at Texas Tech University under the leadership of Professor Ayoub; Professor Chaffin, University of Michigan; Professor Marras, Ohio State University and Professor Laurig, University of Dortmund) models of the human body, in two and three dimensions were developed and tested which indicated the stresses in the body, particularly in the lumbar section of the spinal column which is under so much stress during industrial material handling. One of the major underlying ideas was that of a 'chain of links and articulations' transmitting the forces and moments from the point of application, usually the hands, throughout the body to the point of support, usually the feet. The 'weak' link is often the spinal section of the lower back whose strength limits model capacity.

An important contribution was made in 1984 by Chaffin and Andersson who compiled most of what was known then in the first edition of their book *Occupational Biomechanics*. These authors used the hip moment to predict the expected abdominal pressure and the compression force as well as the shear forces on the spinal disc L5/S1 (fifth lumbar vertebra and os sacrum). The details of this procedure can be looked up in the latest edition of the book; here we shall restrict ourselves to some of the results.

Load handling. First, it must be recognised — as pointed out above — that the distance between the spine and the hands holding the load plays an important role, that is the lumbar spine is greatly affected if the load is moved closer to the torso or further away from it as Figure 7.4 shows.

It is evident that the compression forces increase with the

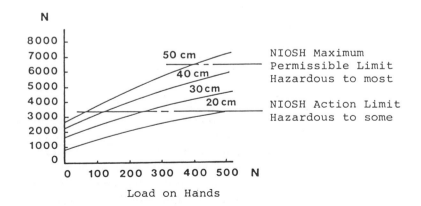

Figure 7.4 Predicted compression forces on L5/S1 of the lumbar spine for varying loads lifted in four different positions from body. *H = Distance from L5/S1 to the hand holding the load. According to Chaffin and Andersson (1984).*

weight and distance of the load: when the lifted load is increased the compression force increases accordingly. Furthermore, the figure reveals that an extension of the load distance is associated with a proportional increase in the compression force. *This confirms the empirical advice to lift a load as close to the trunk as possible.*

Biomechanical modelling also establishes that a squatting posture in order to lift large loads from somewhere around the bent knees involves higher compression forces than lifting the same weight with a stooped lifting posture. In other words, if the load is too bulky to fit between the knees, stooping over the load might be more advantageous than squatting down to lift it. The biomechanical analysis showed that the shear forces, however, were larger in the stoop posture compared with the squat posture. *This suggests that one should keep the trunk in an erect posture whenever possible.*

Recent studies have also dealt with the dynamic aspects of lifting loads, indicating that inertial forces increase the compression forces on the L5/S1 disc during the accelerative phase and that static models underestimate spinal stresses. Ayoub and Mital (1989) and Marras with his colleagues (1995) developed three-dimensional basic dynamic biomechanical models. The measurement of motions and the calculation of reactive forces and moments at joint centres as well as spinal compressive and shear forces under various task variables revealed that lifting loads asymmetrically (involving a rotation of the trunk) or lifting large containers or boxes without handles generated increased spinal strain.

7.4 Intra-abdominal pressure

Load lifting is accompanied by a considerable increase in pressure within the cavities of the abdomen due to the contraction of the abdominal muscles. Intra-abdominal pressure helps to stabilise the spine while we lift loads in the hands. Studies have shown a close correlation between the magnitudes of compression forces acting on the lower spine during load lifting and the magnitudes of intra-abdominal pressure rises.

Davis and Stubbs (1977a,b) measured the intra-abdominal pressure (often simply referred to as IAP) using a capsule containing a pressure-sensitive element and a radio transmitter, which had to be swallowed by the subjects. It was concluded that during a load lifting effort intra-abdominal pressure can be used to give an accurate indication of the spinal stress.

In an epidemiological study the authors observed that occupations in which peak IAPs of 100 mm Hg or more were induced had an increased liability to reportable back injuries. They suggested a 'safe' IAP limit of 90 mm Hg, although values of

150 mm Hg are not uncommon in those who regularly lift weights.

Davis and Stubbs (1977a,b) measured the IAPs with their pressure-sensitive capsules of a number of subjects who had to lift different weights while assuming 36 different positions of the arms and various standing postures. One- and two-handed, frontal and sideward lifts were done in repetitive and occasional activities.

The 90 mm Hg values were incorporated in contour maps (Davis and Stubbs, 1977a) showing suggested limits for lifting forces for 95 per cent of the population. For more details the reader is referred to the publications by Davis and Stubbs (1977a,b).

7.5 Subjective judgements

In everyday life we all continually make judgements about the demands that tasks impose on us. We are able to combine the load sensations throughout our body into one conclusion: this job is 'too hard' but that one is 'easy' for me to do.

Snook and Ciriello (1991) at Liberty Mutual Insurance Company in the USA took advantage of this ability of making subjective judgements about complex strain in our bodies. They let their experimental subjects, all experienced industrial load handlers, determine what combinations of load weights, locations with respect to their bodies, sizes of loads and frequencies of lifting or lowering, pushing or pulling, or carrying they were willing to do over the course of a whole work shift.

Some of the results obtained in these 'psycho-physical' experiments are shown in Tables 7.2, 7.3 and 7.4. These tables show that women, as a group, were willing to exert somewhat smaller efforts than men — a result to be expected according to the discussion earlier in this book about differences in strength between the genders. Of course, groups of workers less accustomed to material handling than Snooks' and Ciriello's subjects, or persons of different body build, are likely to perform less strenuous (or even harder) tasks.

7.6 Recommendations

Acceptable loads for lifting

Just a few decades ago many countries had 'fixed' weights of objects, usually different values for men, women and children, which were believed 'safe' for lifting. Of course, no one limit is suitable for everyone because of different age, fitness, object shape, location, repetitiveness and other circumstances. Furthermore, the actual force exerted on the object, and hence felt in the body, depends not only on the mass of the object but also on the acceleration imparted, as Newton's Second Law

Table 7.2 Maximally acceptable weights (kg) for lifting. Note that this table is only an excerpt from the much more detailed tables for North American workers by Snook and Ciriello (1991) which should be consulted

Width (a)	Distance (b)	Per cent (c)	Floor level to knuckle height								Knuckle height to shoulder height								Shoulder height to overhead reach							
			One lift every								One lift every								One lift every							
			5 s	9	14	1 min	2	5	30	8 h	5 s	9	14	1 min	2	5	30	8 h	5 s	9	14	1 min	2	5	30	8 h
Males																										
34	51	90	9	10	12	16	18	20	20	24	9	12	14	17	17	18	20	22	8	11	13	16	16	17	18	20
		75	12	18	18	23	26	28	29	34	12	16	18	22	23	23	26	29	11	14	17	21	21	22	24	26
		50	17	20	24	31	35	38	39	46	15	20	23	28	29	30	33	36	14	18	21	26	27	28	31	34
Females																										
34	51	90	7	9	9	11	12	12	13	18	8	8	9	10	11	11	12	14	7	7	8	9	10	10	11	12
		75	9	11	12	14	15	15	16	22	9	10	11	12	13	13	14	17	8	8	9	11	11	11	12	14
		50	11	13	14	16	18	18	20	27	10	11	13	14	15	15	17	19	9	9	10	12	13	13	14	17

(a) Handles in front of the operator (cm).
(b) Vertical distance of lifting (cm).
(c) Acceptable to 50, 75 or 90 per cent of industrial workers.

Conversion: 1 kg = 2.2 lb; 1 cm = 0.4 in.

Source: From Snook and Ciriello (1991).

Table 7.3 Maximally acceptable weights (kg) for lowering. Note that this table is only an excerpt from the much more detailed tables for North American workers by Snook and Ciriello (1991) which should be consulted

Width (a)	Distance (b)	Per cent (c)	Knuckle height to floor level								Shoulder height to knuckle height								Overhead reach to shoulder height							
			One lower every								One lower every								One lower every							
			5 s	9 s	14 s	1 min	2 min	5 min	30 min	8 h	5 s	9 s	14 s	1 min	2 min	5 min	30 min	8 h	5 s	9 s	14 s	1 min	2 min	5 min	30 min	8 h
Males																										
34	51	90	10	13	14	17	20	22	22	29	11	13	15	17	20	20	20	24	9	10	12	14	16	16	16	20
		75	14	18	20	25	28	30	32	40	15	18	21	23	27	27	27	33	12	14	17	19	22	22	22	27
		50	19	24	26	33	37	40	42	53	20	23	27	30	35	35	35	43	16	19	22	24	28	28	28	35
Females																										
34	51	90	7	9	9	11	12	13	14	18	8	9	9	10	11	12	12	15	7	8	8	8	10	11	11	13
		75	9	11	11	13	15	16	17	22	9	11	11	12	14	15	15	19	8	9	10	10	12	13	13	16
		50	10	13	14	16	18	19	20	27	11	13	13	14	16	18	18	22	10	11	11	12	14	15	15	19

(a) Handles in front of the operator (cm).
(b) Vertical distance of lowering (cm).
(c) Acceptable to 50, 75 or 90 per cent of industrial workers.

Conversion: 1 kg = 2.2 lb; 1 cm = 0.4 in.

Source: From Snook and Ciriello (1991).

Table 7.4 Maximally acceptable forces (N) for pushing and pulling. Note that this table is only an excerpt from the much more detailed tables for North American workers by Snook and Ciriello (1991) which should be consulted

Height (a)	Per cent (b)	One 2.1 m push every							Height (a)	Per cent (b)	One 30.5 m push every				
		6 s	12	1 m	2	5	30	8 h			1 min	2	5	30	8 h
Initial push forces															
Males 95	90	106	235	255	255	275	275	334	95	90	167	186	216	216	265
	75	275	304	334	334	353	353	432		75	206	235	275	275	343
	50	334	373	422	422	442	442	530		50	265	294	343	343	432
Females 89	90	137	147	167	177	196	206	216	89	90	118	137	147	157	177
	75	167	177	206	216	235	245	265		75	147	157	177	186	206
	50	196	216	245	255	285	294	314		50	177	196	206	226	255
Sustained push forces															
Males 95	90	98	128	159	167	186	186	226	95	90	79	98	118	128	157
	75	137	177	216	216	245	255	304		75	108	128	157	177	206
	50	177	226	275	285	324	334	392		50	147	167	196	226	265
Females 89	90	59	69	88	88	98	108	128	89	90	49	59	59	69	88
	75	79	108	128	128	147	159	186		75	79	88	88	98	128
	50	108	147	177	177	196	206	255		50	98	118	118	128	167

		One 2.1 m pull every									One 30.5 m pull every				
Height (a)	Per cent (b)	6 s	12	1 min	2	5	30	8 h	Height (a)	Per cent (b)	1 min	2	5	30	8 h
Initial pull forces															
Males 95	90	186	216	245	245	265	265	314	95	90	157	177	206	206	255
	75	226	265	304	304	314	324	383		75	196	215	255	255	314
	50	275	314	353	353	383	383	461		50	235	265	304	304	373
Females 89	90	137	157	177	186	206	216	226	89	90	128	137	147	157	177
	75	157	186	206	216	245	255	265		75	147	157	177	186	206
	50	186	226	245	255	285	294	314		50	177	186	206	215	245
Sustained pull forces															
Males 95	90	98	128	157	167	186	196	235	95	90	88	98	118	137	167
	75	128	167	206	216	245	255	294		75	118	128	157	177	206
	50	157	206	255	265	304	314	363		50	137	167	187	215	255
Females 89	90	59	88	98	98	108	118	137	89	90	59	69	69	69	98
	75	79	118	128	128	147	157	186		75	79	88	88	98	128
	50	98	147	157	167	186	196	245		50	98	118	118	128	167

(a) Vertical distance from floor to hands (cm).
(b) Acceptable to 90, 75 or 50 per cent of industrial workers.

Source: From Snook and Ciriello (1991)

(force = mass × acceleration) describes. More detailed guidelines were needed.

This attempt to develop such knowledge raises several questions and problems. First, one must keep in mind that intervertebral disc injuries are in many cases an 'idiopathic disease', that is a degenerative process which is not caused by external factors. It is therefore obvious that maximal loads for lifting will hardly prevent the occurrence of intervertebral disc injuries. Furthermore, intervertebral discs will become less resistant to physical loads and especially sensitive to load-lifting activities with increasing age. These considerations suggest that recommendations for acceptable loads as well as for adequate lifting techniques are worthwhile.

Restrictions to load limits

Pheasant (1986) pointed out that 'Industrial manual handling tasks are characteristically "undesigned" activities, and have something of an extempore quality'. For such occasional lifting tasks it will be difficult to set up standards. For repetitive and continuous load lifting help should be available in the near future through mechanical solutions, such as robots or conveyor equipment. Finally, it should be emphasised that many lifting tasks are associated with turning actions, which impose a rotation or twist upon the spine. Such lifting activities are particularly hazardous and this has seldom been taken into consideration in proposals for maximum permissible loads to be lifted.

The NIOSH guidelines

In 1981 the US National Institute for Occupational Safety and Health (NIOSH) deduced limits for load lifting from various studies concerning epidemiological, physiological, biomechanical and psychophysical aspects. The NIOSH recommendations considered not only the horizontal distance of the load from the body but also the frequency of lifting, the vertical travel distance and height of the load at the beginning of lifting. Under optimal conditions 40 kg (392 N) were considered admissible for 75 per cent of all American women and for 99 per cent of men. The load limit is equivalent to a compression force of 3400 N in the lumbar spine.

In 1991 the NIOSH guidelines were revised. Keeping the assumption of a compression limit of 3.4 kN, better protection was sought, especially for female load handlers. A major drawback of the older recommendations was that they covered only symmetrical two-handed lifting performed directly in front of the body. In fact, most lifting activities on the shop floor involve sideward movements, rotation of the trunk or some other asymmetrical elements. The 1991 guidelines take asymmetric lifting (body twisting) into account and specify the coupling between hands and object. The 1991 recommendations apply to both lifting and lowering loads. The maximum recommended weight is only 23 kg (225 N) even under the most favourable conditions.

Maximal acceptable forces by Davis and Stubbs (1977a,b)

The results of the studies by Davis and Stubbs (1977a,b) were adopted by the UK Ministry of Defence in 1984 for standards for maximum permissible forces for lifting activities. These limits are based on the above-mentioned maximum acceptable intra-abdominal pressure of 90 mm Hg. The acceptable limits cover a wide range of one- and two-handed exertions in standing, sitting and kneeling positions, occasional and frequent lifts for different groups and for both sexes.

A few examples of maximally acceptable forces for lifting loads are reported in Tables 7.5 and 7.6.

Some remarks by Pheasant (1986) relating to Table 7.6 should be emphasised here:

'the restriction of loads to within these (or any other) levels does not guarantee safety. . . There is no such thing as a safe load. An unfit person may injure his or her back (or more accurately trigger an attack of pain in his or her already degenerate spine) by reaching awkwardly to pick up the most trivial loads. A fit person may injure him, or herself handling a very modest load if he or she slips and loses his or her footing. It is not possible to specify a load which guarantees safety'.

These considerations are in accordance with the reflections on the idiopathic character of intervertebral disc diseases (made at

Table 7.5 Maximal acceptable loads for young men while lifting (N). *Frequency not more than once per minute and for a maximum intra-abdominal pressure of 90 mm Hg. For more frequent movements these values have to be reduced by 30 per cent. The N values are rounded off by approximately 2 per cent*

Condition	Grasping distance, expressed as a fraction of arm's length			
	1/4	1/2	3/4	4/4
Standing up				
Two-handed lift, frontal	350	250	150	100
One-handed lift, frontal	300	220	140	100
One-handed lift, sideways	270	200	130	100
Seated				
Two-handed lift, frontal	270	170	120	110
One-handed lift, frontal	350	220	140	100
One-handed lift, sideways	330	210	140	90

Source: After Davis and Stubbs (1977a,b).

Table 7.6 Maximal acceptable loads under various lifting conditions (N). *The loads given should be safe for 95 per cent of the indicated age and sex groups. It is assumed that the activity will be performed in an upright standing position. Two-handed lifts should be carried out in front of the body; if not, loads should be reduced by 20 per cent (according to Pheasant (1986)). The N values are rounded off by approximately 2 per cent*

| | Men | | | | Women | | | |
| | Under 50 | | Over 50 | | Under 50 | | Over 50 | |
Activity	Occa-sional	Fre-quent	Occa-sional	Fre-quent	Occa-sional	Fre-quent	Occa-sional	Fre-quent
Two-handed lift; compact load, close to the body, within preferred range of heights	300	210	240	140	180	130	140	100
One-handed lift; compact load close to the body	200	140	120	80	120	80	70	50

Note: Occasional = Occasional lifts; less than once per minute. Frequency = frequent lifts; more often than once per minute.
Source: All tabulated data are taken from Davis and Stubbs (1977b).

the beginning of this chapter), which modify the importance of maximum permissible load limits.

Practical hints   The following rules are based on general experience as well as on scientific knowledge, for example as described by Kroemer *et al.* (1997) and Mital *et al.* (1993):

1 Seize the load and lift it with a straight back and with bent knees, as shown in Figures 7.3 and 7.5.

2 Get the load as close to the body as possible by grasping the load whenever possible between the knees and by good foot placement, as shown in Figure 7.5.

3 Make sure that your hold on the load is not lower than knee height, A lift starting at knee height can be continued easily to hip or elbow height. Lifts starting at elbow height may be continued to shoulder height; higher levels require much more strength.

4 If the load does not have handles, tie a rope sling around the load and use a harness or hooks.

5 Avoid a rotating or twisting movement of the trunk when lifting or lowering a load.

Figure 7.5 Lift loads as close to the body as possible, with suitable foot placement.

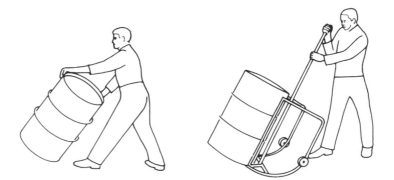

Figure 7.6 Handling casks. *(Left) Tilting and rolling, with the upper part of the body held upright. (Right) Lightening the work by using a trolley.*

6 *Try, wherever possible, to use a mechanical aid such as a trolley, a lifting ramp or similar mechanical aid. An example is shown in Figure 7.6.*

7 *Try to replace lifting and lowering by pushing or pulling. Often a conveyor can be used to make push and pull easy.*

Best protection by lowest load   As discussed in more detail in Chapter 6, *when a heavy load is being carried it is advisable to have it well distributed close to the centre of gravity of the body, as far as possible.* In this way the effort of balancing will be minimised and unnecessary static muscular work avoided.

Often one finds different recommendations in the literature for seemingly the same work task, for example, when perusing either the NIOSH 1981 or 1991 recommendations, the limits proposed by Snook and Ciriello (1991) or the various national and international standards listed by Dickinson (1995). In such a case it is a good idea to accept the lowest load because it will protect best against over-exertion.

Summary

Lifting and other load handling is often associated with over-exertions of the lower back. Trouble can be avoided by proper lifting techniques ('do pull the load close to the body; don't bend or twist') and by work design, especially by keeping the load light. There are many detailed recommendations available to make handling loads easier.

CHAPTER EIGHT

Skilled work

8.1 Acquiring skill

Skilled jobs call for a high degree of:

quick and accurate regulation of muscular contraction

coordination of the activities of the individual muscles

precision of movements

concentration

visual control

In practice, skilled work is mostly a matter for the hands and fingers only. The most important nervous processes that accompany a movement with an element of skill involved are shown in Figure 2.4. It is clear that to perform a delicate movement with speed and precision calls for a whole series of sensory nerve impulses followed by motor directives from the brain.

Learning

During the period in which a skilled operation is being learned we can distinguish between two processes:

1 *Learning the movements.*

2 *Adaptation of the organs involved.*

From a physiological point of view *learning is essentially a matter of imprinting a pattern of the necessary movements upon the medulla of the brain*. To begin with all the movements must be performed consciously but as training progresses the conscious element is gradually reduced. New pathways and junctions are built up in the brain and control of movements is gradually taken over completely by cerebral nerve centres. In simpler terms: *acquiring a skill consists mainly of creating reflex arcs which replace conscious control*.

Learning to write is a good example. First, a child learns one letter after another by consciously copying what the teacher draws. Writing is very laborious and difficult at this stage. Gradually, after months or perhaps years, the different letters, and later on complete words, become 'impressed on the brain as patterns for the necessary hand movements'. Writing has become 'automated', at least as far as finger movements are concerned, and conscious awareness is diverted more and more to finding words and constructing sentences.

Another phenomenon appears during the learning phase: the gradual elimination of all muscular activity that is not essential to the skilled work in hand. The skilled person is relaxed and economical in his or her movements, whereas the novice's work is cramped and tiring. Hence the level of energy consumption for a given task decreases during the training period as non-essential movements are gradually eliminated.

Recommendations for training

The following recommendations will make training easier:

1 *Short training sessions*. To acquire a skill calls for the highest concentration. A tired person easily develops bad habits, which are difficult to unlearn afterwards. It may be said that the higher the level of skill to be acquired, the shorter the training sessions should be. For very delicate skilled work four sessions per day, each of 15–30 min in length, are enough to begin with, gradually becoming longer later on.

2 *Splitting the job into separate operations*. A work study should be made to determine how to divide the job into a number of distinct operations or processes. It is then possible to make tests to determine which parts of the job are the most difficult and to set the expected performance accordingly. It is also possible, and advantageous, to practise the most important operations by themselves, later practising several operations in sequence and finally switching from one to the other and linking them in sequence.

3 *Strict control and good examples*. As we have seen, the learner must avoid acquiring bad habits, so it is important that he or she should be strictly supervised throughout the training. Young people learn a good deal by direct and largely subconscious imitation so they should be given highly skilled instructors to learn from. Older learners depend less on imitation and more on visual aids such as diagrams.

Adaptation of the body

The second process in learning — adaptation of the body organs involved — is a matter of making gradual changes in the muscles and occasionally other organs, such as the heart or the skeleton. Muscular adaptation involves thickening the muscle fibres and thereby increasing the strength of the total muscle. Training for very rapid movement means not only increasing muscle power but concurrently reducing internal friction by getting rid of some of the non-contractile material, such as connective tissue and fat.

8.2 Maximal control of skilled movements

With the objective of increasing the precision and speed of skilled work, many researchers have made detailed studies of several typical movements of the hand and forearm. The most important of their results include the following.

Posture of the arms and shoulders

The height of the working surface affects manual performance and physiological effort. In 1951 Ellis was able to confirm an old empirical rule: the maximal speed of operation for manual jobs in front of the body is achieved by keeping the elbows down near the sides of the trunk and the arms bent at about right angles at their elbows. In a similar situation in practice Tichauer (1968) studied 12 female foodstuff packers to find out how the position of the upper arms affected performance and working metabolism. The results are reproduced in Figure 8.1 and show that the best performance was achieved with the forearms hanging down or slightly abducted sideways, at angles of 8–23° with the vertical.

In another study Tichauer (1975) found that if the arms were held out as much as 45° to the sides, the shoulders took up a

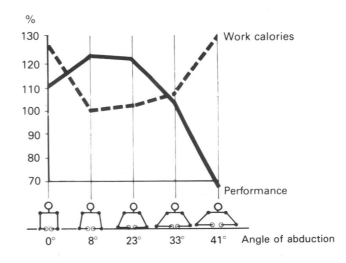

Figure 8.1 Performance and metabolism for various attitudes of the upper arms (angles of abduction) when packing groceries. *After Tichauer* (1968).

balancing posture, causing fatigue in the shoulder muscles. The need to do this may be a result of too high a work area or of too low a seat.

Time and motion

Criteria often used in 'time and motion studies' to discover the optimal type and sequence of movements are the time consumed and the achieved precision; modern ergonomists would say that not enough attention is given to the operator's feelings of monotony and effort, and to signs of fatigue. With this reservation, the most important findings are summarised below.

Time studies show first of all that reaction time (interval between the signal and the beginning of the motor reaction) has a fairly constant value of about 250 ms, which is hardly affected by either the nature of the movement or its distance from the body. The reach and grasp time and distance are not quite linearly related, as Barnes (1936) had already shown. Thus:

Grasping distance (mm)	Time (%)
130	100
260	115
390	125

Brown et al. (1948) and Brown and Slater-Hammel (1949) showed that adjusting movements with the right hand were faster from right to left than from left to right. If the distance was increased from 100 to 400 mm, the time for the movement increased from 0.70 to 0.95 s. The movement is both slower and requires more effort if the upper arm participates in it, as Kroemer discovered in (Kroemer, 1965b).

Precision of movement

McCormick and Sanders (1987) reported some unpublished work by Briggs showing that outward movements were more precise than inward movements and precision fell as the distance of the movement increased. In similar fashion Schmidtke and Stier (1960) showed that the speed of a horizontal movement with the right hand is greatest in a direction 45° to the right, measured from a sagittal plane in front of the body, and slowest at 45° to the left. These studies show that a hand operation can be performed faster and more precisely if most of the movement comes from the forearm.

Optimal working field

Experimental studies of the speed and precision of manual operations are useful when deciding the layout of controls in vehicles of all kinds, including aircraft, but they are also significant for industrial workplaces when the skilled operations involved are complex, often repeated and must be performed with precision.

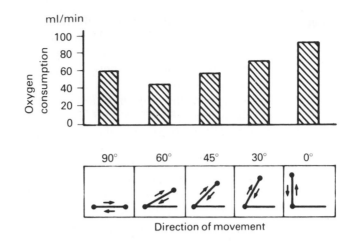

Figure 8.2 Energy expenditure and working field. *The test subjects must move weights of 1 kg repeatedly in the given directions. The direction of 60° is the most efficient provided that the movement does not have to be controlled by eye. After Bouisset and Monod* (1962).

The work of Bouisset and Monod (1962) and Bouisset et al. (1964) is relevant to this problem. These authors made their research subjects shift weights of up to 1 kg and determined the best conditions as far as physical stress was concerned. Oxygen consumption and electrical activity were measured in several muscles of the arm and trunk, both for one-handed and two-handed actions. The weight had to be moved over a distance of 300 mm in different directions, starting from a point immediately in front of the centre of the body, and repeated 24 times/min. The oxygen consumption for various directions is shown in Figure 8.2.

The lowest oxygen consumption corresponds to the direction of about 60° from the front, which agrees with the findings of Kroemer (1965b), McCormick and Sanders (1987), and Schmidtke and Stier (1960). This was also the best angle for two-handed actions, for all weights tested and for all the different frequencies.

Electromyography explained these findings. Electrical activity in various muscles was distinctly less when the arms were held out to the side than when they were held forwards, the latter setting up heavy static loads in the arms, shoulders and trunk. These optimal angles cannot, however, be applied in practice without reserve. An object placed at an angle of 60° to the side is located well only when there is little need to look directly at it at the same time. When forward vision is involved, or when both arms are used, angles of only a few degrees right or left are preferred.

Keyboarding

Typing is skilled work, practised worldwide by millions of operators. Typing speed is usually very high; most operators manage three to five strokes per second. Typing, however, involves not only the control of the muscular activity in fingers, hands and arms but also sensory (usually visual) input, the perception and interpretation of the entering information, and finally the generation of muscle-control signal patterns in the central nervous system where the typing movements for words are 'engraved' as complete movement patterns. That part of perception, interpretation and generation of command patterns can be considered the mental work in typing; it is certainly much more important than the automated process of operating the keys. It is well known that the time required for the mental part is considerably longer than the one for operating the keys. With modern keyboards typing performance is no longer limited by typing speed but mainly by the mental load of the tasks. It is therefore useless to design keyboards with the aim of further accelerating the speed of typing. In fact, increasing the keyboarding speed could lead to biomechanical overloading of the musculo-skeletal system of the body, especially the tendons and their sheaths, as discussed by Putz-Anderson (1988), Kroemer (1972, 1989, 1995) and Kuorinka and Forcier (1995). The problems related to posture of hands, wrists, arms and shoulders could be solved with keyboard design and proper placement. These aspects have been discussed already in Chapter 5.

8.3 Design of tools and equipment

Hand grips

The design of hand grips has a high priority in skilled work. Handles which are not shaped to fit the hand properly, or which do not conform to the biomechanics of manual work, may lead to poor performance and may even be injurious to the operator. The hand and fingers are capable of a wide range of movements and grasping actions, which depend partly on being able to bend the wrist and rotate the forearm. The most important of these many movements are illustrated in Figure 8.3.

The maximum grasping force can be quadrupled by changing over from holding with the fingertips to clasping with the whole hand. The power of the fingers is greatest when the hand is slightly bent upwards (dorsal flexion). In contrast, grasping power, and consequently level of skilled operation, is reduced if the hand is bent downwards or turned to either side.

Tichauer (1975) said that inclining the hand either outwards (ulnar deviation) or inwards (radial deviation) reduces rotational ability by 50 per cent and if the hands are held in such positions every day and frequently this may cause inflammation of tendon sheaths. This possibility was mentioned in Chapter 1, where Figure 1.9 shows an example of using a pair of wire-cutters at

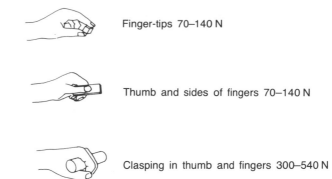

Figure 8.3 Three ways of gripping with the hand. The figures indicate the range of finger pressures. *According to Taylor* (1954).

work. The conclusion from this is that, on biomechanical grounds, the hands should always be kept in line with the forearms as much as possible.

Design of hand grips

The work of Barnes (1949) can be taken as a basis for the design of handles to be gripped by the whole hand. In general, a cylindrical shape is best. It should be at least 100 mm long and its effectiveness increases with thickness up to 30–40 mm.

It is not possible to enumerate all the range of different tools that should be designed to suit the shape of the hand but Figure 8.4 shows a few examples of good design; Fraser (1980) and Kroemer et al. (1994) provide more information.

Hints and equipment for maximum skill

Some of the important conditions for the maximum control of skilled movements have already been discussed earlier in this chapter. The following 10 points summarise ways of making skilled work easier to perform:

1 The working field must be so arranged that manual operations can be performed with the elbows lowered and the forearms at an angle of 85–110°.

2 For very delicate work the working field must be raised up to suit the visual distance, the elbows lowered, head and neck slightly bowed, and the forearm supported (in this context see the seven guidelines in Chapter 3).

3 Skilled operations should not call for much force to be exerted, since heavily loaded muscles are more difficult to control and to coordinate with others. Above all, avoid imposing a static stress at the same time. It is equally bad to have to exercise skill immediately after physical effort. These recommendations are illustrated in Figures 8.5 and 8.6.

4 Concentration on a manual operation is improved if it is not

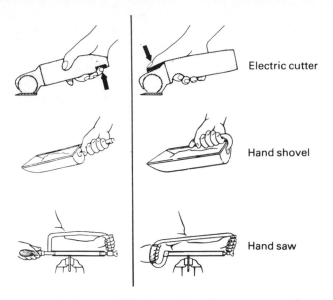

Electric cutter

Hand shovel

Hand saw

Figure 8.4 Handgrips sensibly shaped to the anatomy and functioning of the hand. *Left: Bad solutions. Right: Good design.*

necessary to do other things at the same time. Hence it is helpful to support the work and if possible to have foot pedals with which to clamp and unclamp it, and to switch the machine on and off. Chutes are useful for taking finished work away and delivering fresh items.

5 The arrangement of working materials, parts and controls should be such as to allow the operations to follow rhythmically in a sensible sequence.

6 A free rhythm is better than any kind of imposed tempo, whether it is time controlled, cyclical or on a conveyor belt. A free rhythm consumes less energy (because there are fewer secondary movements), motor control is easier, fatigue is reduced, and monotony and boredom less frequent. It should be mentioned here that too slow a rhythm is bad because the work has to be supported and too fast a rhythm is even worse because of the nervous stress it imposes and the fatigue it engenders. Operators usually find their own rhythm instinctively, to suit their liking.

7 When both hands are used in the work the working field should extend as little as possible to each side, to give the best visual control. The muscular effort should be symmetrical — as nearly as possible equal for the two hands, which should begin and end each movement together.

8 Movements of the forearms and hands are at their most skilful, both in speed and precision, if they take place within an arc of 45–50° to each side. Both for grasping things and for

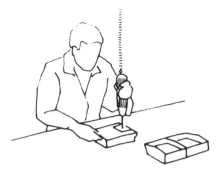

Figure 8.5 During skilled operations the muscles should be kept free from having to exercise force, especially static effort. *The electric screwdriver is suspended from a spring support which reduces manual effort to a minimum.*

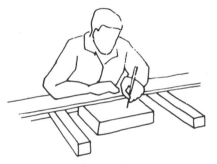

Figure 8.6 A good example of relieving muscular tension during skilled work. *Suitable supports for the forearms make the precision work of the lithographer much easier.*

working, their optimum arc is two-thirds of their maximum reach, i.e. over a radius of 350–450 mm from the tip of the lowered elbow.

9 Horizontal movements are easier to control than vertical ones and circular movements easier than zig-zag ones. Each operation should end in a good position for starting the next one.

10 Handles of controls and tools should be shaped to fit the hand and to operate when the hand is held in line with the forearm.

Summary

Becoming skilful depends on how well one is being trained and on how ergonomically equipment, workstations and work processes are designed.

CHAPTER NINE

Human–machine systems

9.1 Introduction

Closed loop system

A 'human–machine system' means that the human and the machine have a reciprocal relationship with each other. Figure 9.1 shows a simple model of such a system. This is a closed cycle in which the human holds the key position because the decisions rest with him or her.

The pathways of information and their direction are, in principle, the following. The *display* provides information about the progress of production; the operator *perceives* this information and must be able to understand and assess it correctly. On the strength of his or her *interpretation*, and in light of previously acquired knowledge, the human makes a *decision*. The next step is to communicate this decision to the machine by using the *controls*, the settings of which may be displayed by an instrument (e.g. how much water has been mixed in with reagents). The machine then carries out the *production process* as programmed. The cycle is completed when various significant parts of the process, such as temperature or quantities, are displayed for the operator to see.

Roles of machine and human

The technical components of this system (the 'machine') are capable of high speed and precision, and can be very powerful. The human is, in comparison, sluggish and releases only small amounts of energy, yet is much more flexible and adaptable. Human and machine can combine to form a very productive system, provided that their respective qualities are sensibly used.

The control of machines was no great problem until recently when the development of electronics resulted in more elaborate controls and higher output, and the consequent need for accurate interpretation of the information displayed made the operator's

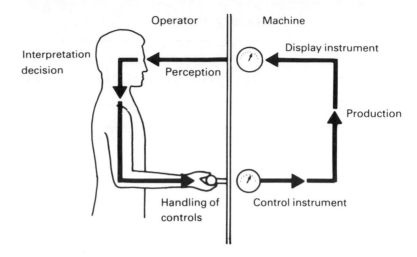

Figure 9.1 The 'human–machine system'.

task both more delicate and more demanding. As a result, the 'human factor' in such systems became increasingly important. In an aircraft the speed of the pilot's or engineer's reaction can be vital; in chemical processes alertness and making correct decisions may alone avert catastrophe. So modern human–machine systems need to be ergonomically sound.

Interfaces

The 'points of interchange' (called interfaces) of information and energy *from human to machine* and *from machine to human* are of paramount importance to the human factors engineer. Two interfaces are of particular interest:

1 The *displays* providing 'feedback' to the human about the status of the machine or the behaviour of the whole system.

2 The *controls* by which the operator inputs 'feed forward' affecting the system.

Below we shall consider these interfaces from an ergonomic point of view.

9.2 Display equipment

A display conveys information to the human sense organs by some appropriate means. In human–machine systems it is usually a matter of visual presentation of dynamic processes, for example about fluctuations of temperature or pressure throughout a chemical production.

Display equipment used commercially mostly falls into three categories:

1 *Digital display in a 'window'.*

2 *A circular scale with moving pointer.*

3 *A fixed marker over a moving scale.*

Each of these displays has its advantages in certain circumstances, some of which are listed below.

Reading-off values

If it is simply a matter of reading off the exact value of some quantity, the digital 'counter' window is best, that is the fastest and most accurate.

If it is necessary to see how a process is going on, to note the amplitude or direction of some change, then the pointer moving over a fixed scale gives that information most easily.

A scale moving by a fixed indicator mark may also be used for these purposes, but suffers from the disadvantage that it may be hard to memorise the previous reading or to assess the extent of movement.

If a process must be set to some particular value (e.g. steam pressure or electrical voltage) it is easier to do this with a moving pointer because a second pointer or a marker can be pre-set and the process accurately controlled as the two pointers come together. If the process covers a wide scale range, then the moving-scale instrument will serve better than a moving-pointer display. Figure 9.2 illustrates the three types of dials and the processes for which they are appropriate.

Fixed or moving scale

While a digital display is best for checking a stable value, in general a fixed scale with a moving pointer is better than the converse arrangement. The moving pointer catches the eye and conveys magnitude and direction of change. Fixed scales are acceptable and

Type of display	Moving pointer	Fixed marker moving scale	Counter
Ease of reading	Acceptable	Acceptable	Very good
Detection of change	Very good	Acceptable	Poor
Setting to a reading: controlling a process	Very good	Acceptable	Acceptable

Figure 9.2 Different ways of displaying information.

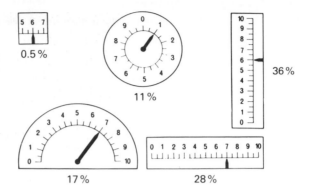

Figure 9.3 How the type of dial affects the precision of reading. *The figures give percentages of reading errors during a display time of 0.12 s. After Sleight (1948).*

cover a wide range of values while displaying only that part of the scale moving against a fixed indicator.

The recent development of electronics, backed by computerisation, has brought various new display techniques, some of which combine aspects discussed above. Books by Cushman and Rosenberg (1991), Kroemer *et al.* (1994) and Woodson *et al.* (1991) provide ergonomic information on electronic diplays.

Reading errors

In the years after World War II it was realised how important the layout of instruments in aircraft and transport vehicles is, in the interests of speed and accuracy of reading, and many studies were carried out to find the best design and arrangement. Figure 9.3 shows results from a study made by Sleight (1948). Each of the scales was shown 17 times to a total of 60 research subjects, for a period of 0.12 s each. The results show significant differences in the frequency of reading errors, the 'open window' with 0.5 per cent error being undoubtedly the best. It appears that with the open window no time is lost in locating the pointer.

Altimeters

An instructive example of the need for sound instrumentation is the design of altimeters, which have not infrequently been the cause of aircraft accidents. The traditional altimeter had three pointers (hands), one reading hundreds of feet, the second thousands and the third tens of thousands. Many studies have shown that both in the laboratory and in flight, reading errors can be reduced by designing the altimeter differently. In 1965 Hill and Chernikof were the first to develop an instrument which for decades was the most trouble free and pleasant to use. It consists of a round dial with a single moving pointer for the 100-ft scale,

Figure 9.4 An aircraft altimeter designed to reduce reading errors. *A weakness is that the pointer sometimes covers up one or other of the counters.*

and two windows, one of which shows the higher ranges and the other the setting for the atmospheric pressure (Figure 9.4).

Matching the display with information requirement

It is important that the instrument gives the operator only the information required, for instance by displaying the smallest unit that the operator is likely to read off. Thus, if there is need to read pressures to the nearest 100 N, the smallest division should be 100 N.

Sometimes the operator does not need a precise reading but just to know a range, say between a lower, safe limit and an upper danger line, or 'cold', 'warm' and 'too hot'. Here a moving pointer is best and the various ranges should be marked by different colours or patterns. Figure 9.5 shows an example of an instrument on which information is kept to a necessary minimum.

Scale graduation

Even more important than the shape of the dial itself is the size of the scale graduations. Since lighting and contrast are not always ideal, and other adverse factors are present in a real workplace, we recommend rather big graduations as follows: if a is the greatest viewing distance to be expected, in mm, then the minimum dimensions of graduations should be as follows:

Height of biggest graduations:	$a/90$
Height of middle graduations:	$a/125$
Height of smallest graduations:	$a/200$
Thickness of graduations:	$a/5000$
Distance between two small graduations:	$a/600$
Distance between two big graduations:	$a/50$

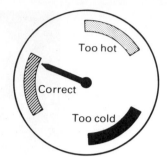

Figure 9.5 A display instrument should convey the required information as simply and unmistakably as possible.

Recommendations for the design of scale graduations

From the above discussion, plus one or two other obvious considerations, the ergonomic design of scale graduations can be summarised as follows:

1 The height, thickness and distance of scale graduations must be such that they can be read off with minimum likelihood of error, even if lighting conditions are not ideal.

2 The information presented should be what is actually wanted: scale divisions should not be smaller than the accuracy required; qualitative information should be simple and unmistakable.

3 Scale graduations should give information that is easy to interpret and to make use of. It is laborious to have to multiply the reading of the instrument by some factor and if this is unavoidable, then the factor should be as simple as possible, say ×10 or ×100.

4 Subdivisions should be $\frac{1}{2}$ or $\frac{1}{5}$: anything else is difficult to read off.

5 Numbers should be confined to major scale graduations and, once again, subdivisions should be $\frac{1}{2}$ or $\frac{1}{5}$.

6 The tip of the pointer should not obscure either the numbers or the graduations and if possible should not be broader than a scale line. It is best if the tip of the pointer comes as close as possible to the scale, without actually touching it. Figure 9.6 shows good and bad pointers.

7 The pointer should be as nearly as possible in the same plane as the graduated scale, to avoid errors of parallax, and the eye must be positioned so that the line of sight is at right angles to the dial and pointer.

Letters and numerals

Western literature contains many studies of the effects of size and shape of letters and numerals on visual perception but these have mostly concerned aircraft. A current view can be summarised as follows.

Bad   Good

Figure 9.6 Bad and good arrangements of numbers and pointer on a dial. *The tip of the pointer should be only as broad as one of the scale lines and it should not obscure the number.*

Black letters on a white background are preferred, in principle, because white characters tend to blur, and a black background may set up relative glare against its lighter surroundings. (This is true both for printed and electronically produced display.) On the other hand, white symbols show up better in poor lighting, especially if the symbols and the pointer are luminous.

Size of characters

The size of letters and numbers, thickness of lines, and their distance apart must all be related to the viewing distance between the eye and the display. The following formula may be used:

Height of letters or numerals in mm = viewing distance in mm divided by 200.

Table 9.1 gives examples.

Capitals and lower case letters are easier to read than letters all of the same size. Most letters and numerals should have the following proportions:

Breadth	$\frac{2}{3}$ of height
Thickness of line	$\frac{1}{6}$ of height
Distance apart of letters	$\frac{1}{5}$ of height
Distance between words and figures	$\frac{2}{3}$ of height

These proportions are illustrated in Figure 9.7.

The recommended dimensions of characters on VDTs will be discussed in Chapter 17.

Table 9.1 Recommended heights of lettering

Distance from eye (mm)	Height of small letters or figures (mm)
Up to 500	2.5
501–900	5.0
901–1800	9.0
1801–3600	18.0
3601–6000	30.0

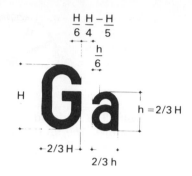

Figure 9.7 Recommended proportions for letters or figures.
*H = height of capitals; h = height of lower-case letters. The
absolute sizes recommended are listed in Table 9.1.*

Controls

**Adequate
controls**

Controls constitute the 'feed forward' second 'interface' between
human and machine. We may distinguish between the following:

1 *Controls which require little manual effort*: push-buttons, toggle
   switches, small hand-levers, rotating and bar knobs; all of which
   can easily be operated with the fingers.

2 *Controls which require muscular effort*: hand-wheels, cranks,
   heavy levers and pedals; which involve the major muscle groups
   of arms or legs.

The right choice and arrangement of controls is essential if
machines and equipment are to be operated correctly.
   The following guidelines should be followed:

1 Controls should take account of the anatomy and functioning
   of the limbs. *Fingers and hands should be used for quick, precise
   movements; arms and feet used for operations requiring force.*

2 Hand-operated controls should be easily reached and grasped,
   between elbow and shoulder height, and be in full view.

3 Distance between controls must take account of human
   anatomy. *Two knobs or switches operated by the fingers should
   not be less than 15 mm apart; controls operated by the whole
   hand need to be 50 mm apart.*

4 Push-buttons, toggle switches and rotating knobs are suitable
   for operations needing little movement or muscular effort, small
   travel and high precision, and for either continuous or stepped
   operation (click-stops).

5 Long-radius levers, cranks, hand-wheels and pedals are suitable
   for operations requiring muscular effort over a long travel and
   comparatively little precision.

There is abundant literature dealing with the ergonomic design and layout of controls. Reviews are given by Cushman and Rosenberg (1991), Kroemer *et al.* (1994), Sanders and McCormick (1993), Schmidtke (1981), Woodson *et al.* (1991) and in military as well as national and ISO standards. Practical recommendations from these sources are summarised below.

Coding

Big machines in industry, agriculture and transport often have control panels with many similar controls and it is important to be able to pull the correct lever or turn the right knob, even without looking at it. According to McFarland (1946) the American airforce in World War II suffered 400 crashes in 22 months because the pilots mistook some other lever for that controlling the wheels.

To avoid confusion *any controls that might be mistaken for each other should be so designed that they can be identified without difficulty*. Identification can be assured by:

1 *Arrangement*. For example, sequence of operation or the difference between vertical and horizontal movement can be used, but only a small number of controls can be identified in this way.

2 *Structure and material* . Figure 9.8 shows knobs of 11 different shapes developed from early experiments by Jenkins (1947). These were the shapes that were least often confused by blindfolded operators. In addition to their shape and size, knobs can be made still more distinctive by their surface texture (smooth, ridged and so on). These characteristics are most helpful if the control must be handled unseen, either in darkness or while attention is being directed elsewhere.

3 *Colour and labelling*. These can be useful, but only in good light and under visual control.

Distance apart

If controls are to be operated freely and correctly, without unintentionally moving adjoining controls, they must be a certain

Figure 9.8 Types of hand grips (knobs) that are easy to distinguish.

Table 9.2   Distance apart of adjoining controls

Control	Method of operation	Distance apart (mm) Minimum	Distance apart (mm) Optimum
Push-button	With one finger	20	50
Toggle switch	With one finger	25	50
Main switch	With one hand	50	100
With both hands		75	125
Handwheel	With both hands	75	125
Rotating knob (round or bar-shaped)	With one hand	25	50
Pedal	Two pedals with same foot	50	100

minimum distance apart. Table 9.2 shows minimal and optimal separations.

Resistance

Controls should offer a certain amount of resistance to operation, so that their activation requires a positive effort and they are unlikely to be inadvertently triggered off. Schmidtke (1974) recommended the following torque resistances:

Single-handed rotation   about 2 Nm
Pressing with one hand   10 to 15 Nm
Pedal pressure   40 to 80 Nm

Higher resistances than these are sometimes desirable to isolate particular controls from the others.

Controls suitable for precision operation with little force include push buttons, toggle switches, hand levers and knobs.

Push-buttons for finger or hand operation

Push-buttons to be operated by the pressure of either finger or hand take up little space and can be made distinctive by colour or other marking. The surface area of the knob must be big enough for the finger or hand to be able to press the knob easily and apply the necessary force without slipping off. Recommended dimensions are:

Diameter   12–15 mm
For an isolated emergency stop   30–40 mm
Travel   3–10 mm
Resistance to operation   2.5–5 N

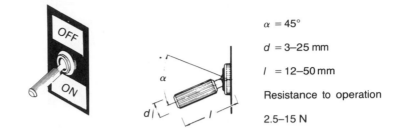

Figure 9.9 Dimensions of toggle switches.

Finger-operated push-buttons should be slightly concave, whereas buttons to be pushed by hand should be mushroom-shaped. Dimensions of the latter should be:

Diameter	60 mm
Travel	10 mm
Resistance to operation	10 N

Toggle switches

Toggle switches are easily seen and reliable in operation. They should preferably have only two positions ('off' and 'on'). Several toggle switches can be placed side by side, provided that each is clearly identified.

The direction of travel should be vertical, and the 'off' and 'on' positions marked above and below (convention whether the 'on' position is up or down varies in different countries). If a toggle switch with three positions is used, there should be at least 40° of travel between two adjoining positions, which should all be clearly identified. Recommended dimensions for toggle switches can be seen in Figure 9.9.

Hand levers

When the toggle is longer than 50 mm, it is called a *hand lever* and allows greater force to be exerted than a mere toggle. The direction of movement should always be either up–down or forward–backward. When a hand lever has several positions, not just 'on' and 'off', each position should have a definite notch. If the lever is capable of continuous fine adjustment, then some suitable form of support should be given to the elbow, forearm or wrist. Hand levers may have different knobs, according to their function:

a finger grip, diameter 20 mm
a grip for the palm, diameter 30–40 mm
a mushroom-shaped grip, diameter 50 mm

Figure 9.10 shows a hand lever suitable for delicate continuous control movements with the fingers. Hand levers which need considerable force to operate them are called *switch levers* and

Figure 9.10 Control knob for precise operation with the fingers. *Lower arm and wrist should be supported by a smooth surface on which they can easily move.*

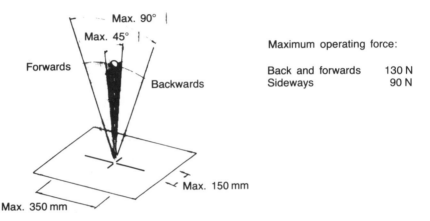

Figure 9.11 Large switch lever requiring moderate force to operate it, with ranges of movement.

come into the category of heavy controls. Suitable dimensions for a switch lever are indicated in Figure 9.11.

Rotating knobs

Rotating knobs may take a variety of shapes: round, arrow-shaped, combinations of knob and handle, or even several concentric knobs on the same spindle.

An important requirement for all types is that *they must fit the hand comfortably and turn easily, and they should be in full view throughout the operation.*

Rotating switches

Figure 9.12 summarises data about rotating switches with click stops. These should have a somewhat higher resistance than knobs for continuous rotation, so that the operator receives a clear tactile signal at each position. A resistance level of 0.15 Nm is recommended. Furthermore, the successive positions should be not less than 15° apart if the switch is kept in view and at least 30° apart if it is operated solely by feel.

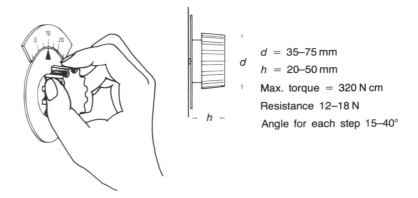

d = 35–75 mm
h = 20–50 mm
Max. torque = 320 N cm
Resistance 12–18 N
Angle for each step 15–40°

Figure 9.12 Knob for step-by-step adjustment (click stops). *The notched margin makes it easier to control.*

Knobs for continuous rotation

Knobs without click stops are suitable for fine and precise regulation over a wide range. An arc of 120° can be turned without shifting the grip and under precise control; to turn further, the grip of the hand can be changed without difficulty. Such a knob can be turned either with the fingers or the whole hand, and it helps if the surface of the knob is slightly grooved or roughened. The following dimensions are recommended:

Diameter for use with two or three fingers	10–30 mm
Diameter for use with the whole hand	35–75 mm
Depth for finger control	15–25 mm
Depth for hand control	30–50 mm
Maximum torque for small knobs	0.8 Nm
Maximum torque for large knobs	3.2 Nm

Pointed bar knobs

Arrow-shaped or pointed bar knobs have the advantage of locating the setting quickly and easily, and should be 25–30 mm, measured along the bar. Figure 9.13 shows such a bar for use with click stops.

Controls which call for muscular strength and long travel, but not a high degree of precision, can be operated by a crank, handwheel or pedal.

Cranks

Geared levers, or cranks, are appropriate for either setting a control or providing continuous adjustment when a wide range of movement (long travel) is necessary. The gearing may be coarse or fine, according to the degree of precision required. The crank can be operated more quickly if the handle can rotate on its own

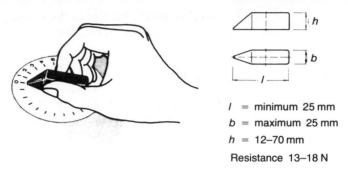

l = minimum 25 mm
b = maximum 25 mm
h = 12–70 mm

Resistance 13–18 N

Figure 9.13 Pointed bar knob for click stops.

axis but a fixed handle is more precise. The following dimensions are recommended:

Length of lever arm for low torque (up to 200 rpm)	60–120 mm
Length of lever arm for high torque (up to 160 rpm)	150–220 mm
Length of lever arm for quick setting	up to 120 mm

According to Van Cott and Kinkade (1972) one can expect the following rotation speeds (rpm) in relation to the crank radius:

mm	rpm
20	270
50	255
120	185
240	140

Hence it appears that the slower the speed of rotation of the crank, the longer it should be, and vice versa.

The relation between crank length and resistance to operation should be:

Crank 120 mm long, resistance 3–4 Nm
Crank 240 mm long, resistance 0.5–2.5 Nm

This range, 120–240 mm, is the best crank length for fine and precise control.

Dimensions of hand grips should be as follows:

Diameter	25–30 mm
Length for one-handed operation	80–120 mm
Length for two-handed operation	190–250 mm

Handwheels

Handwheels are recommended when large forces must be applied because they allow the use of both hands and a relatively long leverage when the turning speed is low.

Notches inside the rim of the wheel give a surer grip and allow force to be applied more efficiently than with a smooth rim.

Pedals

Pedal controls do not often require foot forces of more than 100 N, although more can occur in machinery used in agriculture, earth moving, building construction and occasionally in industry. The brake pedals of motor vehicles also fall into this category, especially if the power assist fails. Pedals are very suitable for this purpose because the human foot is capable of very high force, up to 2000 N, under suitable sitting conditions.

In order to exert strong pedal force, it is necessary to have the following:

a backrest that provides reaction support
an angle at the knee between 140 and 160°
an angle at the ankle between 90 and 100°
the pedal nearly at seat height.

Pressure should be applied with the ball of the foot and the direction of force should be along the line from the point of support on the backrest to the ankle. The following recommendations are made for *pedals with a heavy tread*:

Travel of pedal   50–150 mm
(the smaller the knee angle, the longer the travel)
Minimum resistance to operation   60 N

Pedals which require only light force, such as the accelerator of a car, are operated solely with the foot and their initial resistance is low. It is a help if this type of pedal can be operated with the heel on the floor and the foot resting lightly on the pedal (Kroemer, 1971). The following are recommended dimensions:

Travel of pedal   at most 60 mm
Maximum pedal angle   30°
(the pedal angle = operating angle between the two extreme positions)
Optimum pedal angle   15°
Resistance to operation   30–50 N

Pedals of all types should have a non-slip surface.

As already shown in Table 9.2, the distance apart (clearance) of adjacent pedals should be between 50 and 100 mm. Under special circumstances, e.g. for use with heavy shoes or boots, the pedals must be even wider apart.

Pedals for standing use

When a machine has a multiplicity of controls, pedals are sometimes used to relieve the hands but this is undesirable if the operator stands at work. In any case they should be restricted to operating an on-and-off switch. If pedals cannot be avoided, then they should be of the type recommended in Figure 9.14.

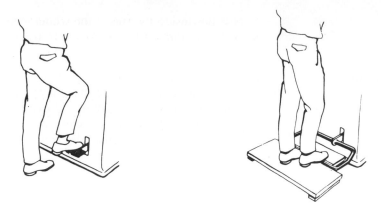

Figure 9.14 Pedals are undesirable for standing work since they set up heavy static loads in the legs. *Left: Bad arrangement, heavy loads on one leg. Right: A better arrangement, a treadle that can be operated with either foot at will.*

9.4 Relationship between controls and displays

Relative speeds of movement

To set up a particular reading on a display, the operator starts by moving the control quickly until the reading is approximately correct, then he or she moves the control slowly to make the precise adjustment. The relative distances of travel of the control lever or knob and the pointer of the instrument are important during this second, precision stage.

A coarse adjustment is easier when the pointer moves more quickly than the control but *for precise adjustment the control should travel faster than the pointer*. The best ratio varies so much in different situations that it is not possible to formulate quantitative rules that would apply generally. Here we shall merely quote a recommendation of Shackel (1974) for precise adjustment using a rotating knob and a pointer moving over a scale. One complete rotation of the knob should make the pointer move through 50–100 mm, allowing a reading tolerance of 0.2–0.4 mm. For a higher reading tolerance of 0.4–2.5 mm the pointer movement should be 100–150 mm per round.

Stereotyped expectations

Even when steering an unfamiliar motor car, we expect that turning the steering wheel clockwise will turn the wheels to the right: no one would expect to turn the wheel to the left in order to steer to the right. Similarly, it is reasonable to expect that when a control knob is turned to the right, the pointer of the corresponding instrument will also move to the right. Such expectations are called *stereotyped*: experience has engraved the corresponding pattern on the brain, as we have already indicated in Chapter 8. *Stereotypes are conditioned reflexes which have become subconscious and 'automatic'.*

Ethnic differences

Chapanis (1975) and Kroemer and co-authors (1994) reminded us that some stereotyped reactions might be influenced by cultural traditions. That is why some arrangements of equipment must be adapted to national conventions including stereotyped reactions. Among many examples are electrical switches, which should always be 'on' in the same direction but this is 'up' in some countries and 'down' in others.

Furthermore, not all stereotypes are equally firmly established and occasionally there are considerable individual deviations. Thus, not all layouts designed for right-handed operation are equally suitable for left-handed people.

There is also the 'rule' that clockwise rotation means an increase in whatever is being controlled but water and gas supplies are often controlled by stopcocks which turn off clockwise. This can lead to problems when one person has to control water/gas and electricity at the same time. In such a case some solution must be sought that involves the least risk of accidents and some degree of conscious control.

Controls and corresponding displays

Below we shall briefly discuss some rules which are valuable for industrial equipment all over the world. An important principle is that controls and instruments which are functionally linked should make corresponding movements that comply with our own stereotypes. Figure 9.15 shows examples of coordination between the direction of movement of controls and instruments with vertical or horizontal scales. Some rules can be summarised as follows:

1 When a control is moved or turned to the right, the pointer must also move right over a round or horizontal scale; on a vertical scale the pointer must move upwards.

2 When a control is moved upwards or forwards, the pointer must move either upwards or to the right.

3 A right-handed or clockwise rotation instinctively suggests an increase, so the display instrument should also record an increase.

4 Hoyos (1974) recommended that a moving scale with a fixed indicator should move to the right when the control is moved to the right *but* the scale values should increase from *right to left*, so that a rotation of scale to the right gives increased readings.

5 When a hand lever is moved upward, or forward, or to the right, the display readings should increase or the equipment should

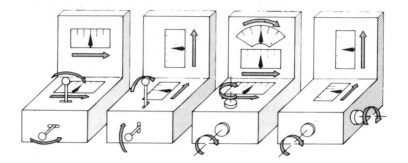

Figure 9.15 Examples of logical relationship between movements of controls and movements of indicators.

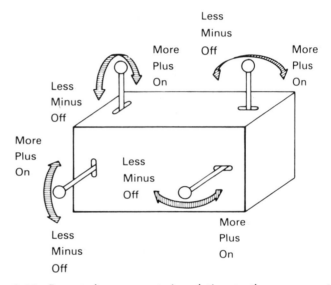

Figure 9.16 Expected movements in relation to the movements of a hand lever. *After Neumann and Timpe* (1970).

be turned 'on'. To reduce the reading, or to switch 'off', it is instinctive to pull the lever toward the body or move it to the left, or downward. These stereotypes affecting a hand lever are illustrated in Figure 9.16.

Control panels

Problems of stereotyped reactions might also come up when designing large control panels. A sensible layout of both controls and display instruments will make supervision easier and reduce the risk of false readings and actions.

Five principles should be considered when designing control panels:

1 As far as possible, display instruments should be located close to the controls which affect them. The controls should be placed either below the display or, if need be, to the right of it.

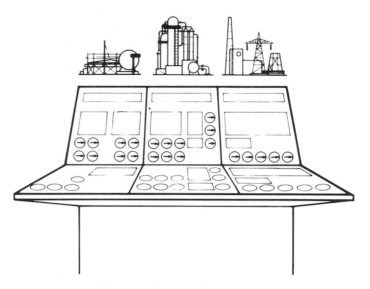

Figure 9.17 Logical layout of a control panel, with controls and display instruments in associated groups. *After Neumann and Timpe* (1970).

2 If it is necessary for controls to be in one panel and display instruments in another, then the two sets should be laid out in the same order and arrangement.

3 Identification labels should be placed *above* the control and identical labels *above* the corresponding display.

4 If several controls are normally operated in sequence, then they and the corresponding displays *should be arranged on the panel in that order, from left to right.*

5 If, on the other hand, the controls on a particular panel are not operated in a regular sequence, then both *they and the corresponding displays should be arranged in functional groups*, in order to give the panel as a whole some degree of orderliness. The grouping can be emphasised by choice of colours, labelling, using control sets that have different sizes and shapes, or simply by arranging members of a group in a row or column. Those controls and displays that are used most often should be directly in front of the operator, the less important ones to each side.

Figure 9.17 shows an example of a logical grouping of controls and displays on a control panel.

These recommendations may seem trivial and pedantic but we must remember that long and often monotonous work can cause boredom and fatigue which lead to reduced alertness and increased likelihood of errors. Under these conditions a logical layout of controls and displays, which takes advantage of stereotyped, 'automated' behaviour, is beneficial, since a weary operator tends to fall back upon his conditioned reflexes.

Summary

Displays and controls as the major interfaces between humans and machines have been the subjects of 'classic' research in human engineering. Hence, comprehensive ergonomics recommendations for their design and use are at hand.

CHAPTER TEN

Mental activity

10.1 Elements of 'brain work'

What constitutes mental activity?

Until a few years ago there was a simple line of demarcation between manual work, performed by blue-collar workers, and brain work which was the domain of white-collar workers, but today this distinction is less clear on two grounds. Some jobs call for a good deal of mental activity without really coming into the category of brain work, for example information processing, supervisory work, taking important decisions on one's own responsibility. Moreover, this sort of work is by no means restricted to white-collar workers but is often delegated to manual operatives. *Hence the expression 'mental activity' as a general term for any job where the incoming information needs to be processed in some way by the brain.* Such activity can be divided into two categories:

1 Brain work in the narrow sense.
2 Information processing as part of the human–machine system.

Brain work

Brain work in the narrow sense is essentially a thought process which calls for creativity to a greater or lesser extent. As a rule the information received must be compared to and combined with knowledge already stored in the brain and committed to memory in its new form. Decisive factors include knowledge, experience, mental agility and the ability to think up and formulate new ideas. Examples include constructing machines, planning production, studying files and extracting the essential facts from them, summarising them, giving instructions and writing reports.

**Information
processing**

Information processing as part of human–machine systems has been discussed in Chapter 9 but nevertheless the essentials will be summarised again. They are:

> perception
> interpretation
> processing of the information transmitted by the sense organs

This 'processing' consists of combining the new information with what is already known, so providing a basis for decision taking.

The mental load at workplaces such as we have discussed is conditioned by the following:

1 *The obligation to maintain a high level of alertness over long periods*.

2 *The need to make decisions* which involve heavy responsibility for the quality of the product and for the safety of people and equipment.

3 Occasional lowering of concentration *by monotony*.

4 *Lack of human contacts*, when one workplace is isolated from any others.

**The barrier that
confronts us**

Researchers in neurophysiology, psychology and other branches of science try very hard to get some degree of insight into the basic processes of mental effort. In the 1988 edition of this book Grandjean used the following description, which he credited to the neurophysiologist Penfield: 'Anyone who studies mental processes is like a person who stands at the foot of a mountain range. He has cut himself a clearing on the lowest foothill, and from there he looks towards the mountain top, since that is his objective, but the summit is obscured in dense cloud'. It is this 'clearing' that we are about to examine.

Mental performances that are important in ergonomics include:

> uptake of information
> memory
> sustained alertness.

10.2 Uptake of information

**Information
theory**

In 1949, Shannon and Weaver made an important contribution to the understanding of how information is taken up. They established a mathematical model representing quantitatively the transfer of information and they devised the term *bit* (binary unit) for the smallest unit of information. The simplest definition of a bit is that it is the quantity of information conveyed by one of two alternative statements. For example, in olden times one flash of light from a watchtower might mean 'enemy approaching from the sea', whereas two flashes would mean 'enemy approaching from the land'. These alternative pieces of information are a bit.

As soon as there are more than two choices, of varying probability, the situation becomes much more complicated. This theory therefore has its limitations when applied to human beings, since the full significance of a stimulus conveying information cannot be interpreted by information theory. This theory is valid only for comparatively simple situations which can be split into units of information and coded signals. It is already useless when applied, for example, to the information being received by the driver of a car.

We need not go further into information theory at this point, since the interested reader needs only to consult Abramson (1963).

Channel capacity theory

Another theory is based upon comparison of information uptake with the capacity of a 'channel' . According to the channel capacity theory the sense organs deliver a certain quantity of information to the input end of the channel and what comes out at the other end depends upon the capacity of the channel.

If the input is small, all is transmitted through the channel but if the input rises it soon reaches a threshold value beyond which the output from the channel is no longer a linear function of the input. This threshold is called the 'channel capacity' and can be determined experimentally for a variety of different sorts of visual and acoustic information.

Human beings have a large channel capacity for information communicated to them by the spoken word. Thus it has been calculated that a vocabulary of 2500 words requires a channel capacity of 34–42 bits/s. This capacity is very modest when compared with the channel capacity of a telephone cable, which can handle up to 50 000 bits/s.

In everyday life the incoming information is much greater than the 'channel capacity' of the central nervous system so that a considerable 'reduction process' must be carried out. It is estimated that this results in the number of bits shown in Table 10.1 being characteristic of different parts of the system.

These approximate figures make it clear that only a minute fraction of the information available is consciously absorbed and processed by the brain. The brain selects this small fraction by some kind of filtration process but we know very little about its details.

Table 10.1

Process	Information stream in bits/s
Registration in sense organ	1 000 000 000
At nerve junctions	3 000 000
Conscious awareness	16
Lasting impression	0.7

10.3 Memory

Information store

Memory is the process of storing incoming information in the brain, often only a selected portion of it, after it has been processed. How this selection is carried out is not known. We do know, however, that the process is subject to the emotions of the moment and we must further assume that information to be stored must have some relevance to what is already there. Each person determines what is subjectively relevant and what is not.

Two kinds of memory can be distinguished:

1 *Short-term or recent memory*.

2 *Long-term memory*.

Short-term memory comprises immediate recollection of instantaneous happenings, up to the remembrance of events which occurred a few minutes or an hour or two ago. Recall of events months or years after they occurred is from long-term memory.

Items of information which become part of the memory leave 'traces', or *engrams*, in certain areas of the brain. Information thus stored can be recalled at will, although, regrettably, not always as fully as one could wish.

Short-term memory

A model for short-term memory assumes that the information received leaves a 'trace', which continues to circulate as a stimulus inside a network of neurones and which, by a kind of feedback, can be recalled into the conscious sphere at any moment.

This reservoir of short-term memory is expendable. This is what happens during 'retrograde amnesia' following some mechanical or emotional disturbance to the brain, whereby the memory of events immediately before the shock is obliterated. Periods of hours or even weeks can be lost from memory in this way. We must assume, therefore, that there is a period during which memories are being consolidated, or 'engraved', in the brain and that during this period they are vulnerable to destruction. Afterwards they become more stable and surprisingly resistant engrams, which constitute long-term memory.

The role of the limbic system in memory

Experiments on animals, as well as clinical observations on humans, have shown that parts of the *limbic system* of the brain, especially that in the hippocampus, play an important role in the engraving process of short-term memory.

Before we discuss this, let us take a general look at the anatomy of the brain, which may be considered in three sections:

1 The *fore brain*, including the cerebral cortex, where, among other functions, memory, learning, consciousness, perception and all the thinking processes are located.

2 The *mid-brain* lies beneath the fore brain and links it with the medulla. This is where autonomic functions are localised: such basic sensations and reactions as hunger, thirst, anger, defensiveness and flight, as well as vegetative control of the internal organs. Among its important constituents are the thalamus and hypothalamus.

3 The *hind brain* forms the link with the spinal cord and includes the cerebellum. Here are located such vital functions as respiration, control of heart and blood circulation, as well as hiccuping, coughing, swallowing and vomiting. Nearly all the nerves from the brain connect with the medulla, which has a dense and complex network of synapses, the reticular formation.

Figure 10.1 shows a diagram of these parts of the brain.

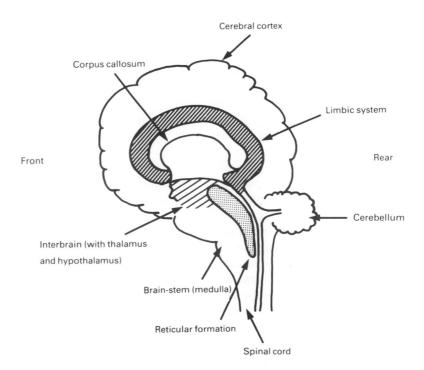

Figure 10.1   General plan of the layout of the brain.

The limbic system is an elongated structure lying beneath the cerebral hemispheres (the two parts of the cerebral cortex) and forming a link with the deep parts of the brain (mid-brain). It consists in part of cortical tissue and in part of nervous structures belonging to the underlying nerve centres. The limbic system is believed to play an important part in generating emotions and the physical excitement that goes with them. In addition, this system is involved in the direction of day/night rhythms, appetite, sexual behaviour, motivation and excitements such as rage and fear.

If the lateral part of the limbic system (the hippocampus) is destroyed it results in serious deficiencies in short-term memory, although long-term memory remains intact. We still do not know the exact location of short-term memory.

The reticular formation is an activation system essential to the processes of learning and memory, since it is only during waking hours that memories can be stored away. The more active the brain, the more learning and memorising go on, and doubtless the higher the level of motivation.

Long-term memory

Long-term memory is very stable and resistant to both brain disturbance and electric shock. This fact has led to the conclusion that long-term memory must depend upon some form of intra-molecular storage of stimuli that is, to changes in the chemical substratum of the nerve cells. Many experiments suggest that this may be a durable engraving in the ribonucleic acid (RNA).

Fascinating experiments with flatworms (planarians) have shown that acquired behaviour is not lost when these worms are cut up. If a worm is trained to react in a particular way to a certain stimulus and is then cut into two, both head and tail regenerate complete worms and both retain the acquired behaviour. This led to the hypothesis that acquired behaviour is based upon changes in the RNA. It is well known that the RNA provides the template for the synthesis of proteins during cell division; hence it is assumed that the cells of each cut half are replicated into the regenerated half complete with their 'marked RNA'. This hypothesis is confirmed by a further experiment: if either half is treated with ribonuclease, an enzyme that destroys RNA, then the acquired behaviour is lost and the regenerated worm is no longer conditioned.

Many other experiments suggest that long-term memory is engraved upon the RNA, including several studies of learning in mammals, which were accompanied by an increase in the RNA in the brain. In one study, reported by Müller-Limmroth (1973), rats were trained to use only the left forepaw to reach for food, and the corresponding centre in the brain showed an increase in RNA content. Information theory also gives interesting insights into memory processes. The number of nerve cells in the brain that are concerned with memory storage may be estimated at 10 billion (10^{10}) and, assuming that all these cells are actively involved, *the storage capacity of the human memory may be as great as $10^8–10^{15}$ bits*. This is an inconceivably large storage capacity.

A problem for human beings is to be able to 'recall' the stored information. Everyone knows the difficulties that older people have and how sadly we say 'I can't call it to mind any more'. So far science can do nothing about this problem.

10.4 Sustained alertness (vigilance)

Some jobs, in industry and transport such as driving or flying, call for sustained alertness which is especially demanding mentally. Before we consider the actual problems of remaining alert, it is reasonable first to discuss some of the activities that are often used as indicators of the level of mental efficiency.

Reaction time and response time

Psychologists and ergonomists have particularly concerned themselves with *speed of reaction*: psychologists because a study of reaction times gives them an insight into mental problems and ergonomists because reaction time can often be used as a way of assessing the ability to perform mental tasks. *Reaction time* means the interval between the appearance of a signal and the required response. Wargo (1967) divided this time into these parts:

conversion into a nerve impulse in the sense organ	1–38 ms
transmission along a nerve to the cerebral cortex	2–100 ms
central processing of signal	70–300 ms
transmission along a nerve to musculature	10–20 ms

The total reaction time cannot be much shorter then about 100 ms under the most favourable conditions; in reality it takes usually several hundred milliseconds to react to a stimulus. Furthermore, the ensuing motion, such as of the foot from the gas to the brake pedal in a car or truck, takes additional time. The sum of reaction time and motion time is called *response time*.

The considerable range in each of these times must be attributed to the wide variety of types and quantities of stimuli: they may trigger sensations of seeing, hearing, touch, taste, smell, electricity or pain. The transmission times to and from the brain are variable as well, depending on the length and properties of nerves involved. A substantial part of the reaction time is taken by the mental processing of the signal in the brain. Thus, the ergonomist trying to minimise response time would select the stimulus with fastest recognition, make mental processing and deciding as easy and snappy as possible, and employ the smallest, quickest and most accurate body motions — compare this with the brake design in even today's automobiles.

Simple reaction time

A simple reaction time is one that involves a single and expected signal which is answered by a simple and practised motor reaction. The time for such a reaction can be as short as 0.15 s. Swink (1966) reported average *simple reaction times* under various conditions:

light signal	0.24 s
siren	0.22 s
electrical shock on skin	0.21 s
light signal + siren	0.20 s
all three at once	0.18 s

Choice reaction time

If a variety of signals must each be answered with a different reaction, or if one response must be the selected from several possible responses upon the same stimulus, we speak of *selective or choice reaction*. For example, a signal may be green, red or yellow, and each must be answered by pressing a different key, or a threatening traffic situation can avoided by either braking or lane changing. These times for complex reactions are substantially longer than for simple reactions because the incoming signal must be processed intricately by the brain and the number of possible choices increases the decision-making time.

A summary by Damon *et al.* (1966) suggests the following relationship between the number of possible choices and the consequent reaction time.

Number of answers	1	2	3	4	5	6	7	8	9	10
Approximate reaction time in 1/100s of a second	20	35	40	45	50	55	60	60	65	65

Thus, Hilgendorf (1966) assumed, to a first approximation, that selective reaction time increases about linearly with the number of bits of information, up to a threshold of about 10 bits.

Anticipation

As mentioned before, both types of reaction time are shortened if the signals are anticipated. This is always the case in experimental work, whereas in everyday life most reactions are to unexpected stimuli. This was demonstrated by Warrick *et al.* (1965) who gave shorthand typists a knob to be pressed whenever they heard a buzzing tone. This signal was sent out no more than once or twice a week, over a period of six months, in the middle of the typists' ordinary work. Reaction times under these conditions averaged 0.6 s, in place of the normal 0.2 s.

Movement time

So far no allowance has been made for the time taken up by the movement of the control, which can be especially important when controlling a vehicle since it may be at least 0.3 s. Response time, the total time elapsed from the signal until the response has been executed, can easily reach a full second. This can be of grave consequence to the driver of a vehicle who does not keep sufficient distance from the vehicle in front; if that vehicle suddenly brakes, there is not enough space (time) for the slow braking response. The slowness of response reaches an extreme meaning for the pilot of a supersonic aircraft: if the plane is flying at 1800 km/h it will travel a distance of 300 m during a response time of only 0.6 s.

Limits of mental load

It has long been known that thinking and other mental processes become less effective as time goes on. It is common experience that the longer one reads, the harder it becomes to take in the information and the more often it is necessary to read a passage again before the words and sentences make sense. Who has not suffered the wandering of attention that occurs during a long and boring speech?

Bills' blocking theory

These simple, everyday observations were made as early as 1931 when Bills demonstrated that people cannot concentrate on a mental activity without breaks. In psychological experiments he found that interruptions occur at frequent intervals in the processing of information, which Bills called 'blockings' or 'blocks'. The duration of a block extended to at least twice the length of the average processing time. Bills saw these *blocks as a kind of autonomic regulatory mechanism which has the effect of allowing the level of mental effort to be as high as possible for as long as possible*.

Long-continued mental effort brings about more frequent and longer blocks, which can be seen as a fatigue symptom. Broadbent (1958a) studied fatigue symptoms of this kind in an experiment during which the test subjects had to remain continuously alert for the appearance of a weak optical signal of short duration, which was given only infrequently (15 times/h). Broadbent found that various distractions (noise, heat, sleep deprivation) made blocks appear more quickly and more signals were missed. The author compared blocking with the closing reflex of the eyelids, which is an interruption of continuous perception of light and which also becomes more frequent and lasts longer as fatigue increases.

**Variability of
heart rate**

*The variability of the heart rate has been used as an indicator of
mental stress.* In practice heart rate is not regular from one beat
to the next but constantly varies between speeding up and slowing
down. The physiological term for this variation is *sinus arrhythmia.* It is linked to breathing. With each inspiration the heart rate
rises, to slow down again during the following expiration. This
arrhythmia seems to be directed, in the first instance, by the
autonomic nervous system, with the vagus nerve playing an
important role as 'pacemaker'. Several authors, among them
Kalsbeck (1971) and O'Hanlon (1971), agree in finding that the
heart rate is less variable under either physical or mental stress.
O'Hanlon found that when his research subjects relaxed their
concentration, their heart rate became more variable and he
suggested that this rise in variability might be used as an indicator
of the level of concentration. It may be said, in simple terms
that:

> *a fall in variability of heart rate may be a sign of increasing
> concentration
> a rise in variability may accompany any fall in concentration.*

**Sustained
vigilance**

*Sustained concentration is the ability to maintain a given level of
alertness over a long period of time.* 'Vigilance' is another term
for this ability, which was found to be of special importance during
World War II when it was noticed that the frequency with which
look-outs noticed U-boats on the radar screen diminished with the
length of the period on watch. Half of all occurrences were
reported during the first 30 minutes on watch; during subsequent
30-minute periods the reported sightings fell to 23 per cent, then
to 16 per cent and finally to 10 per cent. Obviously, alertness
became less the longer the time on watch. This wartime experience
led to many studies of alertness, which came to be called 'vigilance
research'.

Missed signals

In an early investigation, which has become a classic, Mackworth
(1950) set up a number of research subjects in a quiet situation
and made them watch an electric clock. Each rotation of the hand
occupied one minute and was divided into 100 steps of 1/100
minute each. Occasionally the hand jumped two steps at once and
the observers were required to notice this and respond to it. The
experiment lasted two hours and during each 30-minute period
there were 12 critical signals (double jumps of the hand) at
irregular intervals. The results are shown in Figure 10.2.

These results confirm the experience of the radar observers
detecting submarines: more signals were missed as time went on.
Following this pioneering work by Mackworth there have been
hundreds of other experiments to find out how the level of

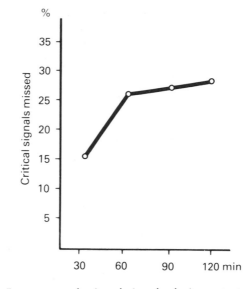

Figure 10.2 Frequency of missed signals during a test of vigilance of 25 sailors lasting 2 h. The results are expressed as a percentage of those during the preceding 30-min period. *After Mackworth* (1950).

alertness is affected by different kinds of signals, critical and non-critical, by their number and arrangement, and by the duration of the experiment. Leplat (1968) rightly criticised vigilance research, which dissipated itself in the study of too many variables using questions that had too little relevance to practical conditions. In spite of these limitations it is possible, with hindsight, to pick out a number of results that have a practical value and to formulate certain laws. Reviews of the literature were published by Broadbent (1958a,b), Leplat (1968), Schmidtke (1973), Mackworth (1969) and Davis and Tune (1970); for engineering applications, the newer compilations by Boff et al (1986) and Boff and Lincoln (1988) are indispensable.

Signal frequency and performance

Among all these results there is one that is significant for our later discussion of the problem of boredom in Chapter 13. Several investigations have shown unmistakably that the frequency with which signals are noticed rises when there are more such signals per unit time. Schmidtke (1973) has shown that this rise continues up to an optimal frequency of 100–300 critical signals per hour. If this limit is appreciably exceeded, then the observational performance falls off again. These results are set out in Figure 10.3.

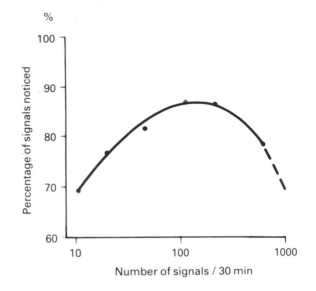

Figure 10.3   Relation between signal frequency and actual observations. *After Schmidtke* (1973).

Schmidtke concluded that the relation between the frequency per unit time of the critical signals presented and the observational performance has the shape of an inverted 'U'. Hence we may assume that too few signals leave the research subjects under-stressed and conversely that excessively frequent signals make too heavy demands on them.

Jerison and Pickett (1964) have observed in addition that the frequency of irrelevant signals also affects performance; the more there are, the poorer is alertness.

Results from vigilance research

The most important results to date from vigilance research may be summarised as follows:

1 Sustained alertness (measured by the number of signals noticed) decreases the longer the duty period. The decline begins to be evident, as a rule, after the first 30 minutes.

2 Within certain limits the observational performance is *relatively improved* if:

the signals are more frequent
the signals are stronger
the subject is informed about his or her own per-formance
the signals are more distinct in shape or contrast.

3 Performance is *worse* if:
intervals between signals vary a great deal;
the research subject has previously been under physical stress or aroused from sleep;

the research subject has performed under unfavourable conditions of noise, temperature, humidity and so on.

4 Many results of vigilance tests are strikingly similar to analogous research into reaction times. Both show a close relationship with the strength of stimulus, the interval between signals and the current state of information available to the subject.

Theories

Psychologists have not missed the opportunity to construct theories about vigilance. Thus, we have at least 10 different theories to explain the varied observations made during concentration experiments. Three theories of special interest may be descibed briefly, as follows:

1 *Mackworth's internal inhibition theory* (1969) invokes our knowledge of conditioned reflexes and explains the decline in performance as a consequence of inhibition because of the absence of 'rewards' and 'incentives'.

2 *The 'arousal' or activation theory* explains the fluctuations in concentration in terms of the level of activity of the cerebral cortex, as we know it from neurophysiology. According to this theory, a lack of external stimuli reduces the activity of the reticular formation, which in turn induces changes in the cerebral cortex, leading to increased drowsiness and a lower level of concentration (see Chapter 11). Numerous experiments which have clarified different aspects of vigilance can be readily explained by the effect of the activation system on thc functional state of the brain and hence on observational efficiency.

3 *Broadbent's filter theory* (1958) is based upon the concept of 'channel capacity' whereby a filtering mechanism allows only signals with certain characteristics to pass through in the stream of information. During the course of a task requiring alertness the filter becomes less and less discriminating and allows irrelevant signals to pass. These overload the channel, exceeding its capacity and crowding out some of the relevant signals.

None of these classic theories is able to explain satisfactorily all outcomes of the numerous experiments. Today it seems doubtful whether sustained concentration can be considered an isolated process, complete in itself. Perhaps that is why no one theory can fully explain it.

From a physiological point of view there is support for the belief that *sustained concentration depends on the functional state, i.e. the level of activity, of the cerebral cortex*. This is a dynamic equilibrium controlled by a variety of influences and by stimuli which excite or dampen activity. No doubt the activation system of the reticular formation also plays a decisive role but this is certainly not the only efficient factor. There are good reasons for assuming that other neurophysiological processes affect the level of activity and hence the degree of sustained concentration. Among these may be *habituation* (becoming accustomed to unimportant stimuli), *adaptation* (decline in intensity of stimuli transmitted by the sense organs) and finally the part played by *the limbic system of the brain in motivation and emotional reactions*. The processes of habituation and adaptation are discussed in more detail in Chapter 13.

The obvious conclusion is that the arousal (activation) theory best fits the observed facts of sustained concentration. If we consider all the many processes such as habituation, adaptation and the influence of the limbic system there are few experimental results in vigilance research that cannot be accounted for by neurophysiological processes in the brain. Still, to say it is all a neurophysiological phenomenon is hardly the same as formulating a theory.

Summary

Mental activities rely on afferent information supply and on the use of short- and long-term memory to make decisions. Proper ergonomic design of the work system avoids mental overloads, including missing or false interpretation of signals, and facilitates correct and fast actions.

CHAPTER ELEVEN

Fatigue

Terminology

'Fatigue' is a state that is familiar to all of us in everyday life. The term usually denotes a loss of efficiency and a disinclination for effort but it is not a single, definite state. It does not become clearer if we define it more closely as physical fatigue, mental fatigue and so on.

The term 'fatigue' has been used in so many different senses that its applications have become almost chaotic. A reasonable distinction is the common division into *muscular fatigue* and *general fatigue*.

The former is a painful phenomenon which arises in the overstressed muscles and is localised there. General fatigue, in contrast, is a diffused sensation which is accompanied by feelings of indolence and disinclination for any kind of activity. These two forms of fatigue arise from different physiological processes and must be discussed separately.

11.1 Muscular fatigue

External symptoms

Figure 11.1 illustrates the external signs of muscular fatigue as they appear in an experiment with an isolated muscle from a frog. The muscle is stimulated electrically, causing it to contract and perform physical work by lifting a weight. After several seconds it is seen that:

> the height of lift decreases
> both contraction and relaxation become slower
> the latency (interval between stimulation and beginning of the contraction) becomes longer.

Essentially the same result can be obtained using mammalian muscle. The performance of the muscle falls off with increasing strain until the stimulus no longer produces a response.

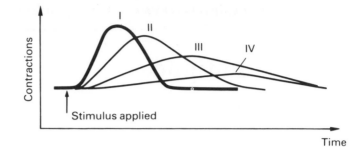

Figure 11.1 Physical manifestations of fatigue in an isolated muscle from a frog's leg. *(I) Contraction and relaxation of a fresh muscle. (II) The same, after moderate stress. (III) The same after heavy stress. (IV) The same after most severe stress.*

Human beings show this process, whether the nerve or muscle is stimulated electrically or the research subject makes voluntary and rhythmical contractions of a muscle over a period.

This phenomenon of reduced performance of a muscle after stress is called 'muscular fatigue' in physiology and is characterised not only by reduced power but also by slower movement. Herein lies the explanation of the impaired coordination and increased liability to errors and accidents that accompany muscular fatigue.

Biochemical changes

We know that during muscular contraction chemical processes occur which, among other things, provide the energy necessary for mechanical effort. After contraction, while the muscle is relaxed and resting, the energy reserves are replenished. Both energy-releasing breakdown and energy-restoring synthesis are going on in a working muscle. If the demand for energy exceeds the powers of regeneration, the metabolic balance is upset, resulting in a loss of muscular performance capability.

After a muscle has been heavily stressed, its energy reserves (sugar and phosphorus compounds) are depleted while waste products multiply; the most important of these are lactic acid and carbon dioxide. The muscular tissue becomes more acidic.

Electro-physiological phenomena

There are statements in the literature to the fact that even after a muscle has been exhausted by repeated voluntary contractions it still responds to an electrical stimulus applied to the skin, suggesting that this form of fatigue is a phenomenon of the central nervous system (the brain).

This interpretation cannot, however, be confirmed in every case. Many physiologists have made the opposing observation: when the muscle itself is in an exhausted state it does not contract any more, even though further motor nerve impulses sent by the brain are visible on the electromyogram. It seems that fatigue has now become a peripheral phenomenon, affecting the muscle fibres, as Scherrer (1967) suggested.

We must presume that one group of researchers were looking for the first sign of fatigue, whereas the others were studying muscles that were already in a state of exhaustion.

Electro-myograms of fatigued muscle

Comparing all the experimental results, we are led to the assumption that the central nervous system acts as a compensatory mechanism during the early stages of fatigue. Thus, many studies with the electromyograph have shown that when a muscle is repeatedly stimulated its electrical activity increases, even though its contractions remain at the same level or decline. This must mean that more and more of its individual fibres are being stimulated into action (recruitment of motor units). The electromyogram in Figure 11.2 demonstrates this increasing electrical activity as fatigue increases.

These electromyographical phenomena have also been observed under practical conditions. For example, Yllo (1962) recorded after 60–80 minute periods of perforating punchcards, increased electrical activity in the muscles of forearms and shoulders. It is reasonable to conclude that, at this stage of muscular fatigue, it still can be compensated for by increased activity of the muscular control centres.

Another indicator of muscle fatigue seems to be the decrease in the frequency of the discharges of the muscular control centres. In fact, during fatiguing static contraction both an amplitude increase and a frequency decrease of the EMG activity are observed. In a rested muscle the mean frequency of the myoelectric signal may be twice as high as that of a fatigued muscle.

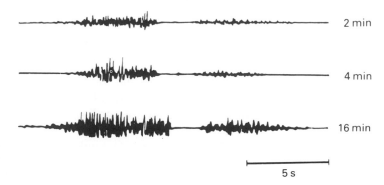

Figure 11.2   Three sections from the electromyogram of the extensor muscle of the upper arm after it has been fatigued by a long series of contractions of equal strength. *From top down, the EMG after 2, 4 and 16 min work. After Scherrer* (1967).

If we now turn to a state of exhaustion, which is certainly located in the muscles themselves, we find that this is accompanied by a reduction in muscular strength. Initially this can be partly compensated for by increasing discharges of the motor neurones.

11.2 General fatigue

A sensation of weariness

A major fatigue symptom is a general sensation of weariness. We feel ourselves inhibited and our activities are impaired, if not actually crippled. We have no desire for either physical or mental effort; we feel heavy, drowsy, tired.

A feeling of weariness is not unpleasant if we are able to rest but it is distressing if we cannot allow ourselves to relax. It has long been realised, as a matter of simple observation, that weariness, like thirst, hunger and similar sensations, is one of nature's protective devices. Weariness discourages us from overstraining ourselves and allows time for recuperative processes to take place.

Different kinds of fatigue

In addition to purely muscular fatigue, other types of fatigue can be distinguished:

1 Eye fatigue: arising from overly straining the visual system.
2 General bodily fatigue: physical overloading of the entire organism.
3 Mental fatigue: induced by mental or intellectual work.
4 Nervous fatigue: caused by overstressing one part of the psychomotor system, as in skilled, often repetitive, work.
5 Chronic fatigue: an accumulation of long-term effects.
6 Circadian fatigue: part of the day–night rhythm and initiating a period of sleep.

This classification of types of fatigue is based partly on the cause and partly on the way in which the fatigue manifests itself, with the obvious corollary that the two should be linked as cause and effect. This should be particularly true for the different sensations of fatigue, which vary according to source. Grandjean believed, however, that certain regulatory processes in the brain are common to fatigue of all kinds and considered these in more detail, as follows.

Functional states

At any moment the human organism is in one particular functional state, somewhere between the extremes of sleep on the one hand

and a state of alarm on the other. Within this range there are a number of states, as shown in the following summary:

Deep sleep
Light sleep, drowsy
Weary, hardly awake
Relaxed, resting
Fresh, alert
Very alert, stimulated
In state of alarm

Seen in this context, fatigue is a functional state which in one direction grades into sleep, and in the opposite direction grades into a relaxed, restful condition.

The electro-encephalogram

Before we discuss the neurophysiological basis of this functional state it would be sensible to take a look at the most important method by which fatigue can be studied: the *electroencephalogram (EEG)*, which records the electrical activity of the brain. When studying human subjects the electrodes are usually applied to the skin of the head, where they detect and register the waves of electrical potential in the cerebral cortex. The resulting EEG makes it possible to study the varying amplitudes and frequencies of these waves.

In greatly simplified form, the most important features recorded in an electroencephalogram are as follows:

1 The *alpha rhythms* comprise waves in the frequency band 8–12 Hz. Alpha waves are present during waking hours and are blocked by sensory impulses, so that a high alpha wave component indicates a relaxed condition and a reduced readiness to react to stimuli. A lower alpha component, coupled with a higher beta component, indicates a more alert state.

2 The *theta rhythms* (4–7 Hz) are slow, long-period waves, which replace the alpha waves when we go to sleep. Sleep is characterised by a whole series of other phenomena, which have been discussed by Horne (1988) and with respect to shiftwork by Kroemer *et al.* (1994).

3 The *delta rhythms* are also slow waves, of less than 4 Hz, which are present only during sleep.

4 *Desynchronisation*, producing beta rhythms of 14–30 Hz, is an irregular electrical activity of very low amplitude. It occurs after the receipt of a sensory stimulus and is the expression of an interruption of the synchronised activities of neurones which make up the alpha rhythm. *Desynchronisation is the sign of a state of increased alertness and it is also known as an arousal reaction*. It may be mentioned here that adrenalin (a stimulant derived from the adrenal medulla), which is released into the

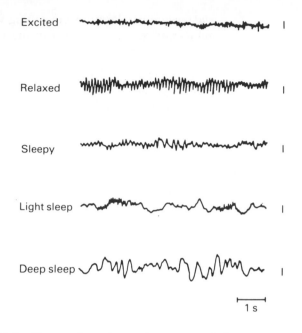

Excited

Relaxed

Sleepy

Light sleep

Deep sleep

1 s

Figure 11.3 Five sections from electroencephalograms, characteristic of various functional states. *The vertical lines indicate the scale for 1 µV. After Jasper (1974).*

blood as a reaction to stress, acts by inducing desynchronisation.

5 *Evoked potentials* are single fluctuations of potential which are induced by isolated sensory stimuli. They are located in that area of the cerebral cortex where the ascending nerve tract from that particular sense organ terminates (this has made it possible to plot a 'map' of the cortex showing which area serves which sense organ).

Figure 11.3 shows five extracts from electroencephalograms, each characteristic of a different functional state of the body.

What has been said up to now gives the impression that for every functional state that we can describe by terms such as 'weary' or 'lively' there should be a distinctive pattern on the EMG. Unfortunately, this is not the case. The patterns shown in Figure 11.3 indicate no more than a very rough correspondence between the encephalogram trace and the underlying physiological state.

The reticular activating system

There is one very important nervous structure which controls to a large extent the functioning of the brain and, consequently, the whole organism: this is the reticular formation of the medulla which can increase or decrease the sensitivity of the cerebral cortex (Figure 10.1). It is in the cerebral cortex that conscious awareness is localised, including the powers of perception, subjective feeling, reflection and willpower. Increasing the sensitivity of the cortex will therefore bring all the conscious functions to a higher level of alertness.

Thus it appears that the reticular formation controls the degree of alertness, including attention, and readiness for action. Its level of activity is very low during deep sleep, increases when sleep becomes shallow and rises steeply on awakening. The higher the level of reticular activity, the higher the level of alertness, culminating in a state of alarm.

These reticular structures, which control the sensitivity of the conscious functions, are called the *ascending reticular activating system*, whereas those structures that increase the readiness of the skeletal musculature are the *descending* activating system. In what follows we shall be concerned solely with the arousal activating system.

The activating structures of the reticular formation do not, however, act on their own initiative. They must be activated by afferent nerve stimuli, which come mainly from *two sources, the conscious sphere of the cerebral cortex and the sense organs*. What significance do these two have?

Feedback control   *Nervous tracts coming from the cerebral cortex* carry impulses from the conscious sphere into the reticular activating system. Such impulses arise, for example, when an idea, or something noticed outside, seems ominous and calls for increased alertness. A closed circuit is then set up; the reticular activating system arouses the cerebral cortex and alerts conscious perception. If this results in any significant signals being received, these stimuli send impulses back along the nerve tracts from the cortex to the reticular activating system. The result is a feedback system, analogous to that in many electronic devices.

Afferent sensory system   *The other source of stimuli to the reticular activating system is the stream of afferent stimuli from the sense organs*; nerve fibres branch off from all the afferent nerve tracts and pass to the reticular activating system and this sensory inflow has its effect upon the level of reticular activity. Every strong impulse that comes from the ear, the eye or nerves conveying pain can raise the level of reticular activity in a flash. The importance of this is obvious. Signals coming into the body from the outside world are passed over to the reticular activating system and make this more active; this alerts the cerebral cortex and so ensures that the brain is ready to notice and act on what is happening outside the body. *This close linkage between the reticular activating system and the afferent sensory system is an essential prerequisite for conscious reaction to the external world.*

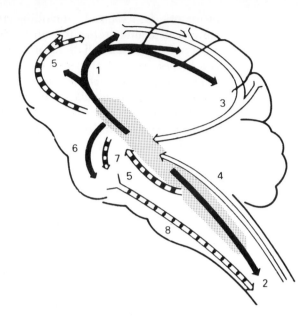

Figure 11.4 The activating and inhibitory systems in the brain.
*Dotted area = reticular formation; 1 = ascending reticular
activating system; 2 = descending activating system; 3 = pathways
from the cerebral cortex; 4 = incoming sensory pathways;
5 = inhibitory (damping) system; 6 and 7 = links with vegetative
(autonomic) centres; 8 = vegetative tracks leading to the
autonomic nervous system of the internal organs.*

For example, suppose that a loud noise is heard. Any sudden
noise stimulates the reticular activating system by way of the
afferent sensory stream and thereby increases the alertness of the
cerebral cortex. There the noise is perceived and interpreted,
perhaps resulting in precautionary action against possible danger.
In this case the reticular activating system is operating as a
distributor and amplifier of signals from the outside world and
makes sure that the organism is alerted as much as is necessary
for the preservation of its life.

Figure 11.4 shows diagrammatically the flow of stimuli from the
cortex and the sense organs to the reticular activating system, as
well as the presumed pathways in the cortex itself.

**Limbic system
and level of
activity**

The reticular activating system is not the only nervous organ that
affects the state of readiness of the cerebral cortex and, through
this, of the entire body. Yet another of the numerous functions
of the limbic system must be mentioned: the part it plays in
excitement, emotion and motivation. Figure 10.1 shows the
anatomy of the limbic system and briefly enumerates its functions.
The general level of alertness is greatly dependent upon the limbic

centres for circadian rhythm (day–night periodicity), fear, rage and calmness, as well as on motivation.

Since the limbic system is involved in the generation of emotional states and the build-up of motivation, it seems obvious that *the activity level of the cerebral cortex, and through it of the entire organism, will be influenced by what goes on in the limbic system*. Links between the flow of afferent sensory impulses and the reticular activating system lead to the further assumption that the two systems have a reciprocal effect on each other.

Inhibiting systems

These two activating systems are not alone. There are other structures which clearly work in opposition to them. In fact many physiologists have conducted experiments showing the existence of nervous elements both in the inter-brain and in the medulla which have an *inhibitory or damping effect on the cerebral cortex*. These inhibiting pathways are shown in Figure 11.4. It is generally accepted that such inhibiting structures can finally induce sleep but there is certainly no proof yet that they are related to states of fatigue.

Relations to vegetative functions

We know that there are close links between the autonomic (vegetative) nervous system, which controls the activities of the internal organs, and the activating and inhibitory systems. In fact, any increase in stimulation of the reticular activating system is accompanied by a whole series of changes in the internal organs, which include:

increase in heart rate
rise in blood pressure
more sugar released by the liver
increased metabolism.

Ergotropic setting

This increased sensitivity spreads from the activating system to all parts of the body — brain, limbs and internal organs — until the entire organism is braced up for a period of high-energy consumption, whether this is for work, for fighting, for flight or whatever. Hess (1948) coined the term 'ergotropic setting' for this process.

Trophotropic setting

By an analogous process, increased activity of the inhibitory system lowers the heart rate and blood pressure, cuts back respiration and metabolism, and relaxes the muscles, while the digestive system works more vigorously to assimilate more energy. Hess' term 'trophotropic setting' means that this process accelerates the recuperative functions by assimilating more food and replacing lost energy.

Thus we see that the nervous structures in the inter-brain and brain stem (medulla) play a key role in the regulation and coordination of functional states of the body. *The brain as well as the internal organs are directed in a logical manner by these nerve centres*, as an example will show

A fire alarm sounds in a factory. The acoustic signal raises the level of activity of the reticular formation in all the persons within the danger zone. It also activates the cerebral cortex to a greater awareness of its surroundings, while alerting the internal organs to be ready for a greater consumption of energy and physical performance.

Humoral control

Humoral control means regulation by means of chemical substances which circulate in the body fluids. We have seen that the reticular activating system is not autonomous but is dependent upon stimulation from the cerebral cortex as well as from afferent sensory signals. In addition to these purely nervous mechanisms there are also the humoral effects which have a part to play in tuning-up the activating system as well as regulating the sensitivity of the limbic system.

Adrenalin

We have already mentioned desynchronisation, the effect produced by the performance hormones *adrenalin* and *noradrenalin* which raise and lower the degree of alertness and activation level of the body. Bonvallet and co-workers (1954) postulated that every time the activation level is raised by stimuli from outside the body there is a release of adrenalin, which in turn stimulates activity of the reticular formation. The increased activity triggered by outside stimuli would soon fade away if this hormone did not intervene to maintain it. In this theory, the reticular activating system is responsible for short-term reactions, while effects of longer duration depend upon the action of adrenalin.

We can say, therefore, that *the level of activity of the reticular activating system depends upon*:

1 *The inflow of sensory stimuli.*

2 *Stimulation of the cerebral cortex.*

3 *Level of adrenalin.*

Adrenalin and noradrenalin are also considered 'stress hormones' since their secretion into the blood is greatly increased in stress situations. This will be discussed in Chapter 12.

Our concept can be summarised in different terms, as follows. The level of readiness to act lies somewhere between the two extremes of sleep and the highest state of alarm. *Control mechanisms in the medulla and the inter-brain regulate the*

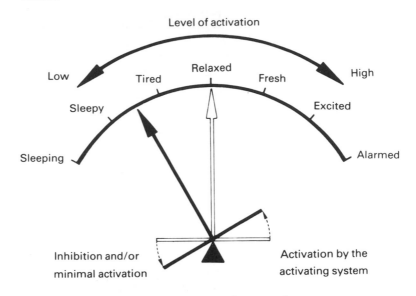

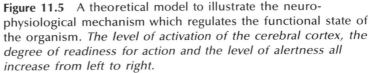

Figure 11.5 A theoretical model to illustrate the neuro-physiological mechanism which regulates the functional state of the organism. *The level of activation of the cerebral cortex, the degree of readiness for action and the level of alertness all increase from left to right.*

readiness level and adjust it to meet the momentary demands of the organism. If external influences are dominant, then the activating systems prevail; the person feels keyed up, even in a state of alarm, and is ready for action both physically and mentally. If, however, inhibiting influences from inside the body predominate, then the damping system prevails; the person feels sluggish, drowsy and lethargic.

Figure 11.5 compares the neurophysiological model with a balance scale in which the activating system is regulating the general functional state of the central nervous system.

11.3 Fatigue in industrial practice

Causes of general fatigue

We know from everyday experience that fatigue has many different causes, with the most important illustrated in Figure 11.6. The degree of fatigue is an aggregate of all the different stresses of the day. Visualise this as a barrel partly filled with water. The recuperative rest periods are the outflow from the barrel. To make sure that the barrel does not overflow we must ensure that inflow and outflow are of the same order of magnitude. In other words, to maintain health and efficiency the recuperative processes must cancel out the stresses. Recuperation takes place mainly during night-time sleep but free periods during the day and all kinds of pauses during work also make their contributions.

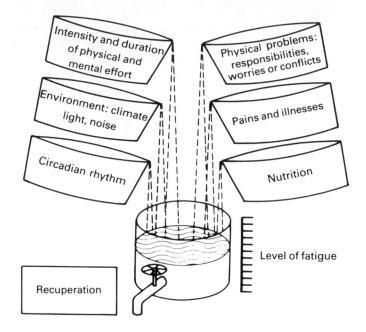

Figure 11.6 Theoretical diagram of the combined effect of everyday causes of fatigue and the recuperation necessary to offset them. *The total stresses must be balanced by the total recuperation within the 24-h cycle.*

Strain and recuperation must balance over the 24-hour cycle so that neither is carried over to the next day. If rest is unavoidably postponed until the following evening, this can be done only at the expense of well-being and efficiency.

Symptoms of fatigue

Fatigue symptoms are both subjective and objective, the most important being:

1 Subjective feelings of weariness, somnolence, faintness and distaste for work.

2 Sluggish thinking.

3 Reduced alertness.

4 Poor and slow perception.

5 Unwillingness to work.

6 Decline in both physical and mental performance.

Chronic (or clinical) fatigue

Some of the fatigue states arising from industrial practice are of a chronic nature. These are conditions that are brought about not by a *single* instance of overstrain but by stresses which recur over days or even longer periods. Since conditions such as these are usually also accompanied by signs of ill health, this may correctly be called *clinical or chronic fatigue*.

Under these conditions the symptoms occur not only during the period of stress or immediately afterwards but are latent almost all the time. The feeling of tiredness is often present on waking up in the morning, before work has begun. This form of fatigue is often accompanied by feelings of distaste for work, which have an emotional origin. People so fatigued often show the following symptoms:

1 Increased psychic instability (quarrelsomeness and associated behaviour).
2 Fits of depression (baseless worries).
3 General weakening of drive and unwillingness to work.
4 Increased likelihood of illness.

These ailments are mostly vague and come under the heading of psychosomatic disorders. This term is applied to functional disturbances of the internal organs or the circulation which are judged to be external manifestations of psychological conflicts and difficulties. Some of the more common symptoms are:

    headaches
    giddiness
    loss of sleep
    irregular heart beat
    sudden sweating fits
    loss of appetite
    digestive troubles (stomach pains, diarrhoea, constipation).

More ailments mean more absences from work, especially short absences, indicating that the cause of the absenteeism is the need for more rest.

People who have psychological problems and difficulties easily fall into a state of chronic fatigue and it is often difficult to disentangle their mental from their physical problems. In practice, cause and effect are hard to distinguish in cases of clinical fatigue. The cause may be dislike of the occupation, the immediate task or the workplace, or conversely these may be the cause of maladjustment to work or surroundings.

11.4 Measuring fatigue

Why measure fatigue?

The science of ergonomics is just as interested in the quantitative measurement of fatigue as is industry itself. What is the relationship between fatigue, work output and the level of stress? The reaction of the human body to different stresses can be measured to develop ways of improving work and making it less laborious.

The question asked by industry is often simply whether the working conditions make excessive demands on the operatives or whether the stresses involved are physiologically acceptable.

We are measuring only 'indicators' of fatigue

Discussion of measuring methods is subject to one serious limitation: *to date there is no way of directly measuring the extent of the fatigue itself.* There is no absolute measure of fatigue, comparable to that of energy consumption which can be expressed in such simple units as kilojoules. All the experimental work carried out so far *has merely measured certain manifestations or 'indicators' of fatigue.*

Methods of measurement

Currently used methods fall into six groups:

1 *Quality and quantity of work performed.*
2 *Recording of subjective perceptions of fatigue.*
3 *Electroencephalography (EEG).*
4 *Measuring frequency of flicker-fusion of eyes.*
5 *Psychomotor tests.*
6 *Mental tests.*

Measurements such as these are frequently taken *before, during* and *after* the task is performed, and the extent of fatigue is deduced from these. As a rule the result has only relative significance since it gives a value to be compared with that of a fresh subject or at least with that of a 'control' person who is not under stress. Even today we have no way of measuring fatigue in absolute terms.

Correlation with subjective feelings

More recently it has been the practice to study a combination of several indicators to make interpretation of the results more reliable. It is particularly important that subjective feelings of fatigue should also be considered. *A measurement of physical factors needs to be backed up by subjective perceptions before it can be correctly assessed as indicating a state of fatigue.* The six indicators listed above will now be briefly discussed.

Quality and quantity of output

The quality and quantity of output is sometimes used as an indirect way of measuring industrial fatigue. Quantity of output can be expressed as number of items processed, time taken per item or conversely as the number of operations performed per unit time. Fatigue and rate of production are certainly interrelated to some extent but the latter cannot be used as a direct measure of the former because there are many other factors to be considered: production targets, social factors and psychological attitudes to the work.

Sometimes fatigue needs to be considered in relation to the *quality* of the output (bad workmanship, faulty products, outright rejects) or to the frequency of accidents, once again with the reservation that fatigue is not the only causal factor.

Subjective feelings

Special questionnaires have been used to *assess subjective feelings*. One of these is the so-called bi-polar questionnaire, where the subject is required to place a mark between the opposed items according to his or her feelings — the location between the extreme statements provides a scale for the strength of the feeling. An example of such a bi-polar questionnaire shows the following opposing items:

Fresh	—————— Weary
Sleepy	—————— Wide awake
Vigorous	—————— Exhausted
Weak	—————— Strong
Energetic	—————— Apathetic
Dull, indifferent	—————— Ready for action
Interested	—————— Bored
Attentive	—————— Absent-minded

A simpler procedure makes the person choose between one of the two statements: this is called a forced choice. There are also many more complicated questionnaires and scaling procedures described in the psychological literature.

The electroen-cephalograph

The electroencephalograph is particularly suitable for standardised research in the laboratory, where variations in the trace in the sense of increasing synchronisation (increase of alpha and theta rhythms, reduction of beta waves) are interpreted as indicating states of weariness and sleepiness (See Figure 11.3).

The techniques of detecting and recording have been improved recently so that the electroencephalograph can now be used successfully to monitor sedentary activities, such as driving a vehicle (O'Hanlon *et al.*, 1975; Zeier and Bättig, 1977).

Flicker-fusion frequency of the eye

Based on the original work of Rey and Rey (1965), the *flicker-fusion frequency of the eye* has been used as an indicator of the degree of fatigue. The procedure, modified by Gierer *et al*. (1981) is as follows. The research subject is exposed to a flickering lamp and the frequency of flickering is increased until the flickers appear to fuse into a continuous light. The frequency at which this occurs is called the subjective flicker-fusion frequency. The source of light should have an area that subtends an angle of 1–2° at the eye and should be so placed that it does not call for any optical accommodation.

Lowering of the flicker-fusion frequency

It has been observed that reductions in the flicker-fusion frequency of 0.5–6 Hz take place after mental stress, as well as under various industrial stresses. Yet a survey of the literature shows that not every kind of stress brings about such a reduction. Experience to date can be summarised to a first approximation as follows:

1 *A distinct lowering* of flicker-fusion frequency can be expected during unbroken high-level mental stress. Examples are doing mental arithmetic, working as a telephonist, piloting an aircraft and carrying out demanding visual work.

   We shall see later (Chapter 13) that dull, repetitive and monotonous situations produce a distinct lowering of the flicker-fusion frequency.

2 *Either little lowering or none at all* result from work that requires only moderate mental effort and which allows the comparative freedom of action or involves physical effort. Examples are office work, sorting jobs and repetitive work at a moderate level.

During several studies the reduction in flicker-fusion frequency was accompanied by parallel changes in other signs of fatigue, notably by increased feelings of weariness and sleepiness. For example, in an experiment by Weber *et al.* (1973) eight subjects in each of three tests were given a dose of 5 mg of Diazepam (Valium) to produce 'pharmacological fatigue'. In a control experiment the same subjects were given a placebo (a similar tablet without the drug). Measurements were taken of both the flicker-fusion frequency and the subjective feelings of fatigue, the latter by using a bi-polar questionnaire.

The results, set out in Figure 11.7, show that after the administration of Valium there was a distinct lowering of the flicker-fusion frequency, amounting to about 2 Hz on average. Simultaneous records of subjective feelings by the bi-polar questionnaires showed that in 10 out of 15 of the opposing items of feelings there was a significant shift in the direction indicating increased fatigue.

This result shows a good correlation between flicker-fusion frequency and subjective feelings of fatigue. The statistics show that the same subjects who showed a marked lowering of flicker-fusion frequency also exhibited considerable subjective fatigue.

In spite of these and other similar observations we can still say nothing of general application about the relationship between flicker-fusion frequency and subjective feelings of fatigue, since the individual correlations recorded here have not been confirmed by other experiments.

Nevertheless, these experiments collectively have encouraged most authors to interpret a *lowering of the flicker-fusion frequency as a sign of fatigue*.

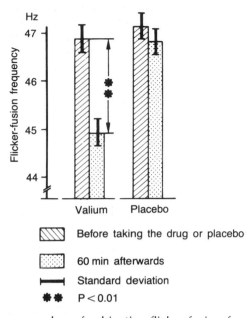

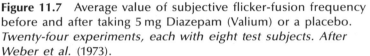

Figure 11.7 Average value of subjective flicker-fusion frequency before and after taking 5 mg Diazepam (Valium) or a placebo. *Twenty-four experiments, each with eight test subjects. After Weber et al.* (1973).

In recent years the subjective flicker-fusion frequency has been less used in fatigue studies. The main reason for this might be some controversial results, the impossibility of obtaining a quantitative measure of fatigue and the rare correlations with other symptoms of fatigue.

Psychomotor tests

Psychomotor tests measure functions that involve perception, interpretation and motor reactions. The following are tests which are used very often:

    simple and selective reaction times
    tests involving touching or pricking squares in a grid
    tests of skill
    driving tests under simulated conditions;
    typing
    tachistoscopic tests to measure performance
    involving perception.

Reservations

In tests like these it is also assumed that a decrease in performance can be taken as a sign of a state of fatigue. Since, however, ability to perform a psychomotor test is dependent on other factors such as, for example, motivation, it is sometimes doubtful whether a general state of fatigue is really the main cause for decreased performance.

A further disadvantage of psychomotor tests arises from the fact that often the test itself makes heavy demands on the subject, thereby raising the level of excitability. In view of what we have said previously, it is very likely that such tests will cause some kind of cerebral activity, which may at least temporarily mask any possible signs of fatigue.

Performance of mental tests

Performance of mental tests often involves:

arithmetic problems
tests of concentration (e.g. crossing-out tests)
estimation tests (e.g. estimation of time intervals)
memory tests.

The same reservation must be made as for psychomotor tests: the test itself may excite the interest of the person being examined and so cancel out any signs of fatigue. Other disturbing factors are the effects of training and experience, and, if the test is protracted, fatigue brought on by the test itself.

Fatigue has been investigated in many field studies carried out under industrial conditions, in traffic and in schools. As a rule, their significance is limited to a particular problem in a particular setting and almost nothing can be deduced from them that is of wider application or would lead to generalisation about the relationship between stress and fatigue.

Nevertheless, a few indications about some of the field studies shall be given here; for more details the reader is referred to the literature.

Traffic fatigue

For a long time fatigue was a problem among telephone operators, and has been the subject of studies by Grandjean (1959, 1971) and Grandjean et al (1966, 1970) who also performed investigations on employees of postal and railway services and air traffic controllers.

A particular importance attaches to studies of fatigue in traffic because it is reasonable to suppose that fatigue is an important contributory factor in mistakes and accidents. As early as 1936 Ryan and Warner concluded from an extensive study on truck drivers that long periods of driving led to a reduced ability to discriminate between certain sensory impressions and a loss of efficiency in some motor functions. Several authors have shown unmistakably that about 4 h of continuous driving is enough to bring on a distinct lowering of the level of alertness and thereby increase the risk of accidents. Most studies done until the mid-1970s on fatigue in road traffic have been described in detail by Lecret (1976).

Investigations among bus drivers and air traffic controllers showed interesting parallels. Both occupations call for sustained vigilance; in both cases the first signs of reduced efficiency appear after about 4 h and this becomes very marked after 7 or 8 h. *This decline in vigilance is a symptom of a fatigue state, which shows itself in both groups as*:

subjective fatigue
fall in flicker-fusion frequency
decrease in psychomotor efficiency
decrease in driving precision
more irregular heart beat
fall in heart rate
rise in alpha waves in the EEG.

It is difficult to deny the obvious assumption that all these symptoms are expressions of a decline in the level of activation (arousal) of the central nervous system.

One final conclusion is inevitable. *Tasks that demand sustained vigilance must be planned with working periods and rest periods so that the risk of accidents is not increased through fatigue of the operators.* The research work detailed above shows that these conditions are often not fulfilled at the present time.

Summary

There are various types of 'fatigue' ranging from specific muscular to general phenomena. Long hours of strenuous work as well as night work are well-known causes of fatigue.

Occupational stress

12.1 What is stress?

**The original
definition of
stress by Selye**

The term 'stress' was introduced by the Selye in the 1930s into
the fields of psychology and medicine. *He defined stress as the
reaction of the organism to a threatening or oppressing situation*.
He distinguished between the 'stressor' as the external cause and
'stress' as the reaction of the human body. (This choice of terms
was unfortunate: engineers consider stress the cause and 'strain'
the result. The confusing use of terms has contributed much to
the muddled popular complaint of 'feeling stressed'.)

**Physiological
reactions**

Selye (1978) had discovered that stress was essentially a result
of chain reactions of neuroendocrine mechanisms, beginning with
an excitation in the brain stem, followed by an increased secretion
of hormones from the adrenal gland, especially of *adrenalin* and
noradrenalin, known as 'performance hormones' since they keep
the whole organism in a state of heightened alertness. These
hormones, also called *catecholamines*, can be found in urine and
this provides a way of determining stress. These hormones have
already been mentioned in Chapter 11 in connection with the role
of the activating system, mainly located in the reticular formation.
It was said there that an increase in stimulation of the reticular
formation is accompanied by an increase in heart rate and blood
pressure as well as by an increase in sugar level and metabolism.
This reaction is called the 'ergotropic setting' and is essentially
identical to the basic mechanisms of the stress reaction. It reflects
an intensified readiness to defend life, including fighting, fleeing
or other physical achievements. Selye (1978) also observed
that this emotional state, resulting from the feeling of being
threatened, was responsible for the adverse effects of stress. In
fact, long-lasting or recurrent stress situations can be detrimental
to health by inducing functional troubles, particularly in the
gastro-intestinal or the cardiovascular systems. These effects are

psychosomatic disturbances which, in the long run, can turn into organic illnesses.

Health troubles

The most common forms of stress diseases are probably gastro-intestinal disorders, which can lead to gastric or duodenal ulcers. Selye explained the adverse stress effects on health as a maladaptation of the organism to stress.

Is stress always harmful?

It is obvious that stress is part and parcel of our life; it is a necessary condition for all living creatures to react to threatening situations in an appropriate way. A life without stressors and stress would not only be unnatural but also boring. Stress cannot be divorced from life, just as birth, death, food and love are inseparable. However, if a person feels subjectively overloaded, then he or she is in *distress*.

Paracelsus, a physician of the early sixteenth century, said that *the dose determines whether a compound is toxic or not.* (*Dosis sola facit venenum.*) The same is true for stress: the amount determines whether stress will have adverse effects on health or whether it will increase human ability to cope with life. Where the borders between normal physiological and pathological stress can be drawn is still an open question. Only one thing is certain: this border varies from one individual to another. One person can bear a great amount of stress all his or her life; another suffers immensely and will sooner or later be overwhelmed by it.

The more the term 'stress' was used, the more it became a myth. Eventually the word was used for nearly every kind of pressure on people. In the last two decades, however, psychologists and social scientists (for example, McGrath (1976), Lazarus (1977), Harrison (1978), Caplan *et al.* (1980) and Cox (1985)) have done detailed research on the phenomenon of stress and formed a much clearer concept of it, particularly with respect to occupational stress.

Occupational stress

The emotional state (or mood) which results from a discrepancy between the level of demand and the person's ability to cope defines occupational stress. It is thus a subjective phenomenon and exists in people's recognition of their inability to cope with the demands of the work situation.

A stressful situation can become a negative emotional experience which may be associated with unpleasant feelings of anxiety, tension, depression, anger, fatigue, lack of vigour and confusion. These moods are often studied with specially designed questionnaires such as Profile of Mood States (POMS).

Person–environment fit

Research on occupational stressors has come up with the concept of the *person–environment fit. The basic assumption is that the degree of fit between the characteristics of a person and the environment can determine the well-being and performance of workers.* Environment is used here in its largest sense and includes the social as well as the physical environment. Some authors distinguish the fit between the person's needs and their satisfaction through the job environment, others refer to the fit between the demands of the job environment and the relevant worker's ability to meet those demands.

Stressors in the work environment

Surveys as well as theoretical considerations suggest that the following conditions may become stressors in work environments:

1 *Job control* is the person's participation in determining their own work routine, including control over temporal aspects and supervising work processes. Several studies suggest that lack of control may produce emotional and physiological stress.

2 *Social support* means assistance through supervisors and peers. Social support seems to reduce the adverse effects of stressors while lack of support increases the load of stressors.

3 *Job distress* is mainly related to job content and workload. It is the perceived excessive stress in job and career and often leads to *job dissatisfaction.*

4 *Task and performance demands* are characterised by the workload, including demands upon attention. Deadlines may be major stressors.

5 *Job security* today refers mainly to the likelihood of further employment or, conversely, the threat of unemployment. Many office and shopworkers worry about being made redundant. The recognition of the availability of similar or alternative employment and future needs for the person's professional skills is important.

6 *Responsibility* for the lives and the well-being of other people may be a heavy mental burden. It seems that jobs with great responsibility are associated with an increased proneness to peptic ulcers and high blood pressure. Responsibility in itself is perhaps not the key stressor. The crucial question is rather whether the amount of responsibility exceeds one's resources.

7 *Physical environmental problems* include noise, poor lighting, unpleasant indoor or outdoor climate and small, enclosed or crowded offices.

8 *Complexity is* defined as the number of different demands involved in a job. Repetitive and monotonous work is often characterised by a lack of complexity, which seems to be an important predictor of job dissatisfaction. On the other hand,

complexity that is too high can arouse feelings of incompetence and lead to emotional stress.

Any individual may experience a number of other stressors and this list could be easily extended but the stressors mentioned above are those which are often considered by social scientists when preparing their questionnaires to evaluate people's experience with occupational stress.

12.2 The measurement of stress

Cox (1985) wrote that stress, as an individual psychological state, has to do with the way a person sees and then experiences the environment. Because of the nature of the beast, there can be no direct physiological measures of stress. The measurement of stress at work must focus on the individual's psychological state. A first step is thus to ask the person about individual emotional experiences or mood in relation to the situation at work. This means using state-dependent subjective data.

Questionnaire surveys

Today almost all field studies on occupational stress are based on extensive questionnaire surveys on working conditions, potential stressors, workers' health and well-being, job satisfaction and states of mood. Many authors use scales that have become standardised and widely-used instruments for which normative data are available.

A popular method of evaluating psychological response criteria is the use of mood checklists. These procedures serve to gauge the worker's feelings. The checklist of Mackay et al. (1978), for instance, distinguishes between stress and arousal. To give an example, after a prolonged and monotonous repetitive task significant increases in self-reported stress were found together with significant decreases in self-reported arousal.

Another approach is that of psychosocial questionnaires which evaluate perceptions and feelings about the job situation, including job satisfaction, perception of workload, work pace, career opportunities, supervisory style and organisational environment. One of the most often cited and widely employed questionnaires was described by Caplan et al. (1980); it was used to measure various psychosocial aspects of 23 different occupations. About a decade later, Carayon (1993) found in a questionnaire study on 170 employees in a midwestern US government agency that, compared with expectations, job control was not the primary determinant of stress but job demands and concerns about future and career were the leading causes.

Several scientists combined the use of questionnaires with measurements of physiological parameters of stress such as the excretion of catecholamines in the urine, while also assessing heart rate and blood pressure. These measurements are interesting correlate to the questionnaire survey procedures; they will be discussed in Chapter 13.

12.3 Stress among VDT operators

General experience

Anecdotal reports as well as general experience indicate that the introduction of VDTs into offices often creates psychological problems. In some cases the new technology imposes a performance increase and therefore a greater workload. A bank example illustrates this: without a computer, about 30 payment transfers were settled per employee per working day but with the aid of a VDT the same employee handled about 300 transfers per day. On the other hand, some new VDT jobs became more and more repetitive and monotonous, especially data-entry jobs.

Some people worry

Computerisation can be worrisome for some employees; they are afraid of the new technology, automation and unemployment. This rather complex and only vaguely recognisable situation sometimes gives rise to a general negative attitude towards the new VDT job.

Some people have fun

However, a contrary reaction has often observed that some employees are proud to be involved in the new information technology and look forward to interacting with a computer. Jobs that require creative participation from the operator are perceived as interesting. Many managers observed that clerical employees showed some resistance to word-processing procedures in the beginning but after a few weeks clearly preferred the new secretarial work with computers to their former work conditions.

In general, it seems that many psychological problems were more acute when VDTs were first introduced and that they are becoming less pressing as time goes by and computers are simply a normal work tool.

The Swedish study on stress at VDTs

One of the first surveys focusing on stress and job satisfaction at VDT workplaces was conducted in 1977 by Johansson and Aronsson (1980) on 95 employees in a large insurance company.

The questionnaire survey revealed rather positive attitudes to the job, but a certain anxiety characterised opinions about computerisation: the 'data-entry' group had slightly higher catecholamine levels than the control group. The most striking result was observed during a temporary breakdown of the computer system: adrenalin, blood pressure and heart rate were raised, while at the same time the subjects felt more irritated, tired, rushed and bored. The authors concluded that *computer breakdowns are an important cause of mental strain for persons with extensive VDT work*. In fact, an interruption meant that the VDT operators were condemned to idleness while their own work piled up, which presumably increased the next day's workload. The authors believe that stress at VDTs can be counteracted partly at the technical and organisational levels by reducing the duration and frequency of breakdowns, shortening the response times in the system and redistributing unavoidable but monotonous data-entry work.

Many other studies have been carried out in the 1980s on alleged stress among VDT operators. They are discussed in detail in Grandjean's 1987 book *Ergonomics of Computerized Offices*. Only a summary of a few selected surveys shall be given here.

The French survey

The ergonomists Elias and Cail (1983) observed an increased incidence of gastro-intestinal symptoms, anxiety, irritation and sleep disturbances among a French 'data-acquisition' group.

The American studies

Two field studies were carried out in the USA by Smith *et al.* (1980, 1981). The studies found that VDT operators as well as the control subjects were subjected to a large number of psychosocial stressors. In general, the VDT operators reported more psychosocial stress than the control groups. The authors concluded that the job content may be an important factor for increased occupational stress and health complaints. The results suggested that the use of VDTs is not the only factor contributing to operator stress but that job content also plays an important role in this matter. Carayon (1993) reported that concerns about one's career and professional future were the main determiners of stress in American office workers.

Sauter of the National Institute for Occupational Safety and Health (1984) and Sauter et al (1983) also conducted a large survey, focusing on job-attitudinal, affective and somatic manifestations of stress among VDT office workers. None of the well-being indices relating to job stressors and moods disclosed a strong indication of increased strain for the VDT group. The VDT users had a higher incidence of unfavourable working conditions and rated their working place environment less pleasant and their chairs less comfortable than the control subjects. The authors summarised their findings as follows: 'Other than a tenuous indication of increased eye strain and reduced psychological disturbances among VDT users, the two groups were largely undifferentiated on job-attitudinal, affective and somatic manifestations of stress'.

Most of the psychosocial surveys on stress among VDT operators have been criticised, mainly because the designs of the studies have not allowed conclusiveness. In spite of some methodological deficiencies it is possible to draw the following tentative conclusions:

Computer work can be satisfying

1 *Generally speaking, clerical VDT operators as a group do not show symptoms of excessive stress.*

2 *There is one important exception to the above: some VDT operators, engaged in very fragmented, repetitive and monotonous jobs, such as data-entry or data acquisition, experience strong psychosocial stressors, report low job satisfaction and indicate a high frequency of mood changes for the worse, as well as gastro-intestinal or other psychosomatic troubles.*

3 *The fact that other studies including repetitive VDT jobs did not reveal more psychosocial stressors or symptoms of stress than control groups, leads to the conclusion that it is not the work with the VDT as such but the poor work structure of some specific repetitive jobs which is responsible for observed adverse effects.*

4 *Apart from repetitive and monotonous jobs, VDT operators are, on the whole, satisfied with their work and consider the computer an efficient tool.*

Summary

The psychologic term 'stress' indicates a mismatch between demands, such as imposed on the job, and the individual's capabilities. Some amount of stress increases aspiration and motivation and leads to improved capabilities to meet the demands; conversely, underuse of a person's capabilities often leads to boredom and dissatisfaction. If the demands exceed the individual's ability to cope, 'distress' develops.

CHAPTER THIRTEEN

Boredom

Definition of boredom

A monotonous environment is one that is lacking in stimuli. *The reaction of an individual to monotony is called boredom. Boredom is a complex mental state characterised by symptoms of decreased activation of higher nervous centres, with concomitant feelings of weariness, lethargy and diminished alertness.*

Boring situations are common in industry, travel and commerce. They can be found, for example, at a control desk if there are too few signals to which the operator needs to respond. An engine driver can be in a similar situation if signals are too far apart. An example of a monotonous job is being in charge of a stamping press and having to carry out exactly the same operation 10–30 times per minute, for hours, days and years on end. Occupations such as this are *repetitive* as well as monotonous and boring.

Psychologists, as well as a few physiologists, have concerned themselves with the problem of boredom. The psychologists have mainly described the external causes of boredom and the behaviour of persons suffering from it. The physiologists have concerned themselves more with the nervous mechanisms of boredom and related these to the measurable indicators of this condition. Although boredom is a *single* condition, it will be considered from these two distinct points of view.

13.1 Causes

External causes

Experience shows that the following circumstances give rise to feelings of boredom:

1 *Prolonged repetitive work that is not very difficult yet does not allow the operator to think about other things entirely.*

2 *Prolonged, monotonous supervisory work that calls for continuous vigilance.*

The decisive factor in these situations is obviously that there are not enough matters that call for action.

Observations in industry have shown that certain conditions make boredom more likely. Examples are a very brief cycle of operations and few opportunities for bodily movements. Others are dimly lit or warm workrooms and solitary working without contact with fellow workers.

Personal factors enhancing boredom

Personal factors have a considerable impact on the incidence of boredom or, put another way, *on the ability to withstand boredom*. Proneness to boredom is higher for the following people.

1 People in a state of fatigue.

2 Not-adapted night workers.

3 People with low motivation and little interest.

4 People with a high level of education, knowledge and ability.

5 Keen people who are eager for a demanding job.

Conversely, the following are very resistant to boredom

1 People who are fresh and alert.

2 People who are still learning (for example, a learner driver has no time to be bored).

3 People who are content with the job because it suits their abilities.

Extroversion

Grandjean suggested in 1988 that extrovert people are very susceptible to boredom. The opinion often expressed that women are more resistant to boredom than men is scientifically questionable; equally, the supposed relationship between intelligence and susceptibility to boredom is still disputed.

Satiation

Grandjean also indicated that one may distinguish between boredom itself and its emotional manifestations, which may be called satiation. This means a state of irritation and aversion to activity which is provoking boredom. A person feels that he or she has 'had enough'. This is a state of actual conflict between a feeling of duty to work and the desire to have done with it, which puts the person involved under increasing internal tension.

Job satisfaction

A decline in work satisfaction can be looked upon as a precursor of mental satiation. Several studies have shown that in practice work satisfaction is lower when monotonous, repetitive work is

Table 13.1 Assessment of work and index of job satisfaction by 557 workers in two car factories with differently organised work

Work	Factory	Production line (motorised)	Assembly line (non-motorised)	Free assembly
'Interesting'	A	35%	56%	67%
	B	34%	57%	94%
'Boring'	A	54%	42%	39%
	B	55%	37%	42%
Index of job	A	0.53	0.92	0.96
satisfaction	B	0.57	1.00	1.17

Source: After Wyatt and Marriott (1956).

required than with jobs that allow freedom of action. As an example of such a survey the results of Wyatt and Marriott (1956) may be recalled. These authors questioned 340 workers in one car factory (Factory A) and 217 in another (Factory B) about their attitudes to their work. The authors compiled an 'Index of Satisfaction' from certain replies, and from self-assessments carried out by the workers themselves. Table 13.1 summarises the most important results.

The workers at the two factories agreed to rate working on the motorised production line as rarely 'interesting' and more often 'boring' than work at either the non-motorised assembly line or the free assembly. The index of job satisfaction was comparable for both factories. Wyatt and Marriott (1956) also compared this index with the length of shift of a selection of workers and found that the shorter the shift, the greater the job satisfaction.

13.2 The physiology of boredom

We have seen already that situations with few stimuli, or lacking in variety, induce a state of boredom, recognisable by weariness and somnolence as well as by a decline in alertness.

Neurophysio-logical basis

This state of affairs is not difficult to explain in neurophysiological terms: *when stimuli are few the stream of sensory impulses dries up, bringing about a reduction in the level of activation of the cerebrum and thereby of the functional state of the body as a whole.*

In addition to the reduced sensory inflow during quiet conditions there are two other physiological processes that should be noted because they too are responsible for the decline in the level of stimulation, particularly in situations where the existing stimuli vary very little. These are *adaptation* and *habituation*.

Adaptation

Most sense organs have the peculiarity that *under a prolonged, steady stimulus the discharge from the receptor organ declines*. Obviously one function of this is to protect the central nervous system against prolonged overloading with impulses from the peripheral sense organs. *The term adaptation indicates that the stream of sensory impulses is adapted to the needs of the organism.*

In principle, all sensory organs have this power of adaptation, even though they differ in the extent and speed of it. Adaptation is particularly well developed in the sensitivity of the skin to pressure (we soon get used to wearing a wrist-watch), the stretch receptors of the muscles and the photoreceptors of the eyes.

Adaptation is not confined to the receptors of the peripheral sense organs but also occurs in the synapses joining one nerve fibre to another.

What significance has adaptation in the problem of boredom? The sense organs and synapses adapt to external circumstances in such a way that they respond mainly to changes in stimuli and are relatively insensitive to a sustained level. Under uniform stimulation, therefore, the activating structures in the brain do not pass on any stimulus to those organs (reticular and limbic activating systems) responsible for the general level of activation of the body.

Habituation

Habituation can be regarded as adaptation on a higher level, which leads to a reduction in brain activation by repetitive stimuli and operates not peripherally but in the zone between the cerebral cortex and the limbic and reticular activating systems. The following example illustrates this. If a note of a regular pitch is sounded close to a sleeping cat, the cat will wake up the first time. If the same note sounds at regular intervals the effect on the cat will gradually diminish. If, however, the pitch of the note is changed, its original waking effect will be restored, whereas a note of the original pitch will still fail to wake the cat. This experiment shows that identical stimuli lose their effect with repetition, provided that the stimulus is meaningless, that is it has no significance in the life of the animal. *The essential nature of habituation is the elimination of reactions to meaningless stimuli.*

Habituation is a protecting filter

The mechanism of habituation may be compared to a filter which does not let stimuli pass that are meaningless under the existing circumstances, allowing only those that have a certain relevance to pass.

The biological significance of habituation is the same as that of adaptation: the protection of the cerebral cortex (and thereby the entire organism) against being inundated with irrelevant alerting or alarm stimuli. Without habituation, the organism would need to maintain itself constantly in a state of maximum alertness.

It is an obvious assumption that the process of habituation also has a part to play in monotonous situations, when it nullifies the effect of irrelevant and repetitive events.

It appears from these considerations that adaptation and habituation are neurophysiological mechanisms which are to be taken as indicating the existence of monotonous conditions. *Situations in industry and in transport which give rise to the phenomena of adaptation and habituation certainly involve an increased risk of monotony and boredom.*

A summary of the neuro-physiology of boredom

The physiological aspects of boredom may be summarised as follows. Situations which are characterised by a low level of stimulation, or by a regular repetition of identical stimuli, or just by making few mental or physical demands upon the operator, lead to a diminution in the flow of afferent sensory impulses, as well as to a lower level of stimulation of the conscious spheres of the brain. The consequent fall in activation level of the reticular and limbic systems reveals itself in a reduced reactivity of the entire organism. It is hardly possible to distinguish between fatigue and boredom on physiological grounds since both of these states are characterised by a reduction in the level of cerebral activation. Nevertheless, differences do exist, as will be shown later in the evidence of results from certain experimental work on boredom.

Medico-biological aspects of boredom

Until a few decades ago the science of work physiology was mainly interested in finding out how to relieve the worker of excessive physical load. *Increasing mechanisation and automation, as well as the tendency to divide up the work into as many simple operations as possible (Taylorism), has now led, in many occupations, to a new problem: insufficient demands on physical and mental capacities.* Unused physical and mental capacities characterise a state which we call 'underload'.

Nearly all the organs of the human body have the important biological characteristic of being able to respond to demand by stepping up their performance. This is true not only of the muscles, heart and lungs but also of the brain. Human development from childhood onwards is heavily dependent upon this ability to adapt to the stresses of life.

Conversely, if an organ is not exercised, it atrophies. A good example is the wasting of muscle which becomes distinctly noticeable only a few weeks after a fracture of a limb. Cessation of development, followed by decline, takes place on a mental level as well as a physical one. It is known from experiments on animals that the brain becomes better developed, functionally, morphologically and biochemically, when the animal is subjected to various mental demands and stresses than when it is allowed to grow up in a quiet situation with few external stimuli.

From these considerations it is evident that underload, such as a person experiences from monotonous, repetitive work, is basically unhealthy from a medico-biological point of view.

The relationship between stress and biological reactions can be broadly summarised as follows:

1 *Underload leads to atrophy.*

2 *The right amount of load leads to healthy development.*

3 *Overload wears out the body.*

Boredom and adrenalin

An interesting contribution to a better understanding of the different aspects of monotonous work was made by several Swedish studies such as Levi (1975) and Frankenhäuser (1974). They analysed the catecholamine excretion in the urine and found that the most diverse physical and emotional stress situations led to a measurable increase in the adrenalin excreted in the urine, which was interpreted as a mobilisation of the performance reserves of the body. One study by Frankenhäuser *et al.* (1971) is particularly relevant to the problem of boredom. Their experiment on mental under- and overload yielded the following results:

1 *Overload*, created by a long-lasting serial reaction time test, produced an increased flow of adrenalin (about 9.5 ng/min).

2 *Moderate load*, in the form of reading a newspaper, gave only a small increase in adrenalin excretion (about 4 ng/min).

3 *Underload*, as a consequence of a uniform, repetitive operation, also produced a higher flow of adrenalin, amounting to about 5.7 ng/min and so falling between the levels of 'overload' and 'moderate load'.

The authors concluded that adrenalin production is increased not only when acting under pressure, against the clock and with a high inflow of information but also in conditions that are monotonous and lacking in stimulation. This shows that the physiological reaction is produced by the mental and emotional stress, rather than by the physical effort as such.

A field study by Johansson *et al.* (1976) also produced interesting results. A group of sawmill workers whose work was repetitive and at the same time responsible, secreted much more adrenalin than other groups of workers. They also exhibited a higher incidence of psychosomatic ailments and more absenteeism. The authors concluded that the combination of monotonous, repetitive work with a higher level of mental stress called for a continuous mobilisation of biochemical reserves, which, in the long term, adversely affected the general health of the workers.

13.3 Field studies and laboratory experiments

Laboratory experiments have the advantage that they are conducted under controlled conditions and therefore it is usually possible to distinguish clearly between 'cause' and 'effect'. They have, however, the drawback that the working conditions studied are usually simulated and do not compare exactly with either industrial practice or everyday traffic. This reservation is particularly necessary when studying boredom. Even if it is possible to reproduce the physical working conditions very closely, the important psychological factors, e.g. the profit motive or the role of social contacts, can be simulated only roughly in the laboratory. On the other hand, it is usually possible to carry out field studies under everyday working conditions although many factors are not under the investigator's control and they may affect the study outcome. Some results may be difficult to assess or may be accepted only with reservations.

The following account concerns two field studies into the effects of monotonous work on operators.

On the production line in factories

Haider (1963) studied boredom among 337 female workers in various factories; 207 worked on a moving production line, while the other 130 worked individually. He assessed the subjective feelings of these workers by using a self-assessment card with 12 pairs of contrasting states. Comparison of the two groups gave, on average, the following differences: the production line workers were more 'tense', 'bored' and 'dreamy' than those who worked at their own speed. Haider concluded that one may expect to see 'satiation phenomena', with increasing tension, restlessness, lack of incentive and a declining performance of the boring work when the job consists of a long succession of simple, repetitive acts. Indeed, the percentage of discontented and tense workers on the production line was as high as 20–25%.

Checking bottles

Saito *et al.* (1972) carried out a similar investigation in the food industry. The workers studied were engaged in the visual control of bottles and their contents moved so quickly that the job was rated as arduous, even though monotonous and repetitive. It was found that after a short time at work the number of rejected bottles became distinctly lower, a fact which the authors considered to indicate diminishing alertness. Concurrently a distinct fall in the flicker-fusion frequency could be recorded, accompanied by such subjective symptoms as increasing fatigue, sleepiness, headache and a sense of time 'dragging'. The fall in flicker-fusion frequency was greatest at times when the operative was observed to talk less to the others and tended to doze or even to fall asleep. (See Chapter 11 for more information about the flicker-fusion test.)

The progression of all effects was as follows:

First hour: no change.

Second to fourth hours: marked impairments.

Final hour before lunch break: workers both felt and performed better.

Experiments on boredom

The following are a few selections from the very large number of experiments that have been carried out to investigate boredom and its effects on people at work. First, we must mention the many vigilance studies already discussed in Chapter 10. In practice, nearly all these involved a monotonous task which nevertheless called for a constant state of alertness and they have overwhelmingly shown that prolonged concentration on a monotonous task results in a steady decline in alertness. In the same chapter it was also pointed out that vigilance is dependent upon the functional state of the brain, the level of cerebral activation.

In order to induce a state of boredom Hashimoto (1969) used simulated driving tests. One test series was easy, the second series much more difficult. The easy condition produced a distinct lowering of the flicker-fusion frequency and an increase in alpha rhythms on the electroencephalogram, as a result of boredom. Under the more difficult condition boredom was less evident.

Experiments with simulated repetitive tasks

Whereas the experiments described above were predominantly concerned with simulated traffic conditions, Martin and Weber (1976) and Baschera and Grandjean (1979) elected to produce the state of boredom by means of a very uniform and repetitive task. Over a period of several hours, their test subjects were required to pick up nails singly, count them and place a specified number into a series of envelopes. This task fulfils many of the conditions conducive to boredom: it is extremely repetitive, undemanding and does not require much alertness, yet it does not leave the mind entirely free for daydreaming.

In all these nail-counting experiments the work caused a lowering of the flicker-fusion frequency. Simultaneous checks, either with a questionnaire or a bi-polar self-assessment chart, showed changes in the subjective well-being — either less efficient and less willing manipulation of the nails or a displacement of the bi-polar charts in the direction of increased sleepiness, fatigue, inattention and boredom.

Figure 13.1 reproduces some of Martin and Weber's work in which the results of the nail-counting experiment were compared with a study of more stimulating conditions, which required the research subjects to carry out a variety of psychomotor tests, between which they listened to music of their own choice.

As the diagram shows, the monotonous counting of nails quickly provoked a distinct fall in the flicker-fusion frequency, which

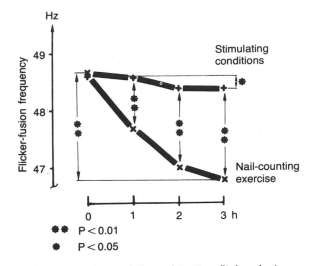

Figure 13.1 Average values of the subjective flicker-fusion frequency of the eyes of 25 test subjects in a nail-counting exercise, as compared with a stimulating situation. *After Martin and Weber* (1976).

reached an average value of 1.7 Hz after 3 h. In contrast, the more stimulating conditions of the second experiment produced only an insignificant fall in flicker-fusion frequency in the same length of time. Similar results were obtained from questioning the test subjects about how they felt. The nail-counting experiment produced an increased sense of strain (through loss of manual dexterity) and a loss of motivation (disinclination for the task), whereas almost all the effects of the more stimulating conditions were beneficial.

Such research shows that *a monotonous, repetitive job can quickly lead to boredom, as shown by a fall in flicker-fusion frequency and changes in subjective feelings*.

Both in driving tests (Hashimoto, 1969) and field studies among drivers of motor vehicles (O'Hanlon, 1971; Harris *et al.*, 1972) boredom could be avoided or at least postponed either by stimulating conditions or by making the task more difficult. This state of affairs leads one to ask whether repetitive jobs could be made less boring simply by making them more difficult. To test this Baschera and Grandjean (1979) extended the nail-counting test by introducing a moderately difficult and a very difficult mental task. The test subjects were required to perform the following three tasks, each lasting 3 h:

Repetitive tasks with varying mental demands

1 The nail-counting test as described above: *low mental demand.*

2 Selection of nails of different colours and with different ring markings in such a way as to fill an envelope with the correct

number, colour and ring marking. We rated this exercise as making only a *moderate mental demand*.

3 Selection of nine nails, with a prescribed combination of colours and markings, and arrangement of them into a board in an ascending order of rank according to their ring markings. This was the exercise that made *heavy mental demands*.

Figure 13.2 shows the average fall in subjective flicker-fusion frequency of the 18 research subjects who were tested under the three sets of conditions.

Underload and overload

The outstanding result of these three series of tests is the distinct relation between the flicker-fusion frequency and mental stress: the low stress level is accompanied by the greatest fall in flicker-fusion frequency; the high level of mental stress equally provokes a fall in this frequency, though it is somewhat smaller than before. On the other hand, the moderate mental stress leaves the flicker-fusion frequency almost unaltered. We can draw a theoretical U-curve through the tops of the three columns in the block diagram, which suggests that both understress and a high level of overstress cause a reduction in the level of cerebral activity. Between these two extremes lies a zone of moderate mental stress which has no adverse effect on the functional state of the central nervous system. The analysis of the simultaneously filled in questionnaires related to subjective feelings revealed some inter-

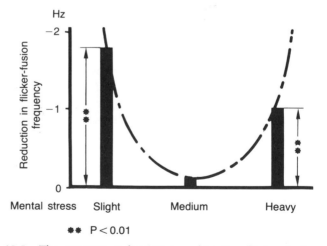

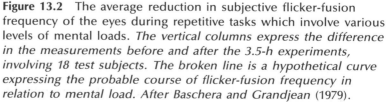

Figure 13.2 The average reduction in subjective flicker-fusion frequency of the eyes during repetitive tasks which involve various levels of mental loads. *The vertical columns express the difference in the measurements before and after the 3.5-h experiments, involving 18 test subjects. The broken line is a hypothetical curve expressing the probable course of flicker-fusion frequency in relation to mental load. After Baschera and Grandjean (1979).*

esting results. Several feelings, e.g. weariness and sleepiness, show a parallel with flicker-fusion frequency. They were increased both for low and high levels of mental stress. In contrast, 'boredom' was predominantly associated with low level of stress. This shows that it is obviously possible to differentiate between the effects of under- and overstress by asking the appropriate questions. It is doubtful, however, whether it is yet possible to make this distinction by physiological methods.

These results are in accordance with those of Frankenhäuser *et al.* (1974) and Frankenhäuser (1971) mentioned above, showing an increased adrenalin secretion not only under work pressure but also in monotonous conditions.

Summary

Boredom is a reaction to a situation in which there are too few stimuli and is characterised by decreased activation of the central nervous system. However, different persons react quite differently to monotonous prolonged tasks.

CHAPTER FOURTEEN

Job design to avoid monotonous tasks

Taylorism and boredom

For decades, critical concerns have been raised by social scientists over the Tayloristic principle of splitting a job into a large number of identical tasks which are repeated over and over. Work organised on this principle is characterised by short cycles per piece and few demands on the operative. *The result of such a specialisation of the job is that individual freedom of action is severely curtailed, mental and physical abilities lie fallow and the potential of the worker is wasted.*

14.1 The fragmented work organisation

Several surveys favour the hypothesis that there is a link between the quality of one's working life and that of one's life in general. Early comments on these social and ethical aspects of job satisfaction can be found in the work of Friedmann (1959) and in the 1972 proceedings of the International Conference on Enhancing the Quality of Working Life.

Some escape into daydreams and like it

Some studies revealed that there are individuals who like their repetitive and monotonous work. It seems that some people are able to escape with their thoughts into a world of daydreams and that they appreciate working conditions which permit them to do so and do not want a job that is more varied or challenging. On the other hand, personnel managers report that it is becoming increasingly difficult to find workers to do monotonous and repetitive jobs.

Different attitudes really exist. For some workers, male or female, continuous working on a production line must actually be more relaxing than free assembly since it allows them to express their personalities better by conversation, thinking or daydream-

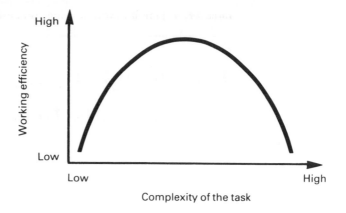

Figure 14.1 Conjectured relationship between level of complexity of a job and working efficiency. *The level of complexity is deduced from the number and variability of the operations involved.*

ing. For other workers, however, monotonous work on a production line seems meaningless because it does not provide them with opportunities to develop their personalities by exercising their brain power at work.

Whatever the individual preference may be, all social scientists and occupational psychologists agree that a job that considers a person's potential and inclinations will be carried out with interest, satisfaction and good motivation. Conversely, it is obvious that an undemanding job, which does not develop the potential of the person will prove to be boring and lacking in motivation for most. At the other extreme, a job that requires more from the worker than he or she is capable of will be overtaxing. Hence *work should be so planned that it is matched to the capabilities of the individual operative, without asking either too little or too much.*

This basic recommendation embodies the idea that the efficient performance of a complex or difficult task is at its optimum in the range between being under- and overdemanding. This concept appears in Figure 14.1 as an inverted U-shape.

Similarly Blum and Naylor (1968) concluded that the level of frustration in relation to the level of complexity of the task was represented by a U-shaped curve and that frustration was least when the demands of the job were most closely matched to the capabilities of the worker.

Summary of the consequences of extreme fragmentation of work

Table 14.1 summarises the most important critical objections, culled from various branches of science, against extreme Tayloristic fragmentation of human tasks into simple, monotonous, repetitive elements.

Table 14.1 Simple, monotonous, repetitive jobs from the viewpoint of various sciences

As seen by	Probable consequences
Physician	Atrophy of mental and physical powers
Work physiologist	Boredom; risk of errors and accidents
Occupational psychologist	Increasing discontent with the job
Social scientist	Human potentialities not fully realised
Industrial engineer	Increased absenteeism; increasing difficulty in finding personnel to do the job

14.2 Principles of job design

The various consequences that may follow from repetitive work have led in recent years to the development of different ways of organising and restructuring assembly work and similar serial jobs.

Aims of restructuring job design

The main objective of these efforts is to give the operator more freedom of action in the following two ways:

1 *Reduction of boredom*, with its concomitant feelings of fatigue and satiation.

2 *Making the work more worthwhile* by providing a meaningful job and one which allows the operator to develop his or her full potential.

Basic to both these improvements is the assumption, already indicated above, that there will be a reduction in absenteeism, turnover of the workforce and social stress, and that the new conditions will attract more workers. Hence in the long run higher productivity should result.

These desirable new forms of work organisation involve various improvements, ranging from variety of work, through various ways of broadening the scope of the job, to job enrichment by giving the worker more information, more responsibility, more participation in decision making and more control of the work process.

Increased variety of work

A first step to improve repetitive working conditions is the attempt to increase the variety of work. This is a scheme where each individual worker is entrusted with different jobs at different workplaces, which he or she carries out in rotation. Such attempts have been made in industry: for instance, the rotation of workers among jobs with frequent changes of assembling operations. The best solutions were judged to be those which at the same time introduced a certain group autonomy, allowing the workers to control the making of their own product.

An example may be taken from the assembly of electronic calculators. The complete assembly is deployed round a workbench. There are eight places, but only six operatives so that there are always two vacant seats. The resulting accumulation of components forces the operatives to change their seats frequently. It is an essential feature of this system that each person must be trained to work at any of the eight places and so the complete assembly is a function of the entire group.

However, one point must be emphasised: if variety of work merely means moving to and fro between jobs that are equally monotonous or repetitive, the risk of boredom may be slightly reduced but the desirable matching of the difficulty of the job to the capabilities of the worker is not being achieved. Adding yet another monotonous, repetitive job is not going to lead to job enrichment.

Broadening and enriching the job

For these reasons a special importance attaches to types of organisation which strive to enrich the work by broadening its scope, thus helping to develop the personalities and self-realisations of the employees. In such organisations the tasks are planned so that a worker moves to a succession of different jobs, each of which makes different demands on the person's abilities. Additional responsibilities, such as quality control or the installation and maintenance of machinery, make an important contribution to the enrichment of the job.

An actual observation in a factory making electrical apparatus may serve as an example. A certain piece of equipment was originally assembled on an assembly line by six successive operations performed by six workers. In the new plan, one worker carried out all six operations, being solely responsible for the quality of the entire assembly.

Autonomous working groups

A further step towards more worker participation is the organisation of autonomous working groups. The workers employed for each production unit are organised in a group and the planning and organisation of the work as well as the control of the end-product are delegated to them. The autonomous working group, therefore, has planning and control functions.

Here is a Dutch example reported by den Hertog and Kerkhoff (1974). A firm making television sets introduced more attractive forms of work organisation, by stages, as it became a problem to find workers. At first the sets were assembled on a long assembly line and the 120 workers pushed their sets along once a minute, at a signal. The first stage of improvement was to have a new assembly line with 104 working places, divided into five groups. At the end of each group there was a barrier and quality check. It was found that with this arrangement assembly times were shortened and quality improved. Next, the experiment tried allowing more time before the sets moved on as well as a certain amount of job switching. The final step was to introduce *autonomous working groups*. The groups were reduced in size to seven people and the range of each person's tasks broadened so that intervals between moves rose from 4 to 20 min. The working groups undertook many responsibilities which previously had fallen upon the foreman or overseer. The groups performed their own quality control and were self-administering. A certain amount of interchange within the group made them more flexible.

The authors reported the results to include the following:

more positive approach to the work
greater cooperation among members of a group
creation of a certain critical attitude to production levels

less absenteeism
less waiting time
fewer 'passengers'
more working areas and new machines
more consultation
higher wages.

The groups were a nuisance to the management in one sense since autonomous groups acted as a controlling force on the management.

Taken as a whole, the groups produced a television set more cheaply than did their predecessors.

All efforts to broaden and enrich people's work must be regarded as experiments that are by no means completed, nor can their results as yet be fully evaluated. There are reports on successful attempts but there are also cases where such projects had to be given up because of resistance on the part of employees, unions or managers. In fact, the search for new ways of restructuring monotonous, repetitive and meaningless work is still going on.

Social contacts

This catalogue of organisational improvements may be concluded by emphasising the importance of social contacts in the workplace. *The opportunity to talk to one's fellow workers is an effective way of avoiding boredom.*

Conversely, social isolation brings monotony and increases the tendency to become bored with work.

Sitting along a straight assembly line is bad: it is much better if the line follows a semi-circle or is sinuous. Any arrangement is good as long as it brings several workers within conversation distance of each other.

Other ways of reducing the incidence of boredom include:

more frequent short breaks
opportunity to move about during these stops
a stimulating layout of the surroundings, making use of light, colour and music.

Job design for supervisory work

The same principles apply to supervisory work and to driving a vehicle. An essentially dull situation, lacking in stimuli, must be enlivened just enough so that it is interesting but not overstressing.

In both types of job the problem is often to recognise a critical situation at the right moment. The necessity to remain alert, which is often associated with a high level of responsibility, may result in boredom and increases the risk of mistakes and accidents. Hence, preventive measures concentrate on making sure that relevant items of information are correctly assimilated.

To summarise, the following arrangements can be recommended for supervisory jobs at control panels, machines, projection screens and similar places of work:

1 Alarm signals and similar safety limits must be clear and decisive. A combination of a light signal and a noise (buzzer, gong or siren) is particularly effective.

2 If a series of signals must be noticed without missing any of them, there should be between 100 and 300 of them per hour.

3 The operator must be fresh and must avoid getting tired beforehand. Night-shift workers are particularly liable to boredom (and fatigue) until they have become night-adapted.

4 The surroundings should be brightly lit. Music is helpful in the right circumstances. Room temperatures should vary only within comfortable limits.

5 A change of work must be arranged as soon as boredom causes dangerous lapses of alertness. In extreme situations it may be necessary to consider hourly or even half-hourly changes.

6 Short and frequent pauses help to avert boredom and improve alertness.

7 In certain particularly critical situations it may be necessary to employ two people to keep a look-out together, e.g. in the driver's cab of a locomotive.

And what about VDT jobs?

As shown above, attempts have been made in industry to avoid adverse effects of repetitive, monotonous jobs by means of alternative ways of organising production. Johansson (1984) wrote that unless countermeasures are taken *"there is a risk that computer technology creates highly repetitive tasks which require little skill, allow little social interaction and generate the type of negative consequences associated with mechanised mass production."* Data-entry work is a case in point.

Many authors agree with these words of warning and point out that some clerical jobs had little content before the era of computers and that VDT jobs were often fragmented and simplified versions of traditional clerical activities. This is certainly true for some very simple data-entry or data-acquisition tasks, whereas many other computerised jobs are characterised by a high degree of complexity, judged interesting and challenging by operators.

Job redesign is necessary for data entry or data acquisition

Attempts to improve job design are mainly needed for the highly repetitive and monotonous data-entry and data-acquisition jobs.

No projects or results of restructuring fragmented and repetitive VDT jobs with the aim of improving job design have been published yet. For that reason claims in connection with job design are mainly based on general considerations of the relationships between working conditions and job satisfaction.

Broadening data-entry tasks can be rather difficult unless one changes the job structure from the narrow traditional data 'entry' to 'processing' or even 'control'. Some banks have improved the situation by creating 'mixed jobs', alternating pure data entry with payment transfers and other more demanding tasks. Furthermore, break rooms have been provided where operators can take their breaks together. Other banks have engaged only part-time employees for pure data-entry jobs, which, however, has been less successful than 'mixing activities'.

Job control

Job control is characterised by the power over and mastery of task and work environment. One distinguishes between two types of control: *instrumental control* is at the task level, e.g. over work pace; *conceptual control* is at a higher level, for example regarding company policies or decision latitude. Having job control means *participation*, lack of job control seems to be an important social stressor. Performance feedback is a vital part of the worker's control of the work process. It allows for corrective actions toward better performance; improved performance seems to enhance job satisfaction. The importance and meaningfulness of work often appear low to an employee if his or her work is just a fragment of a larger task. The person is likely to lose interest especially if the job is small and simple and repetitious. If fragmentation is necessary, the importance of the person's contribution to the end-product should be made clear — yet of course, instead of breaking the job into small meaningless pieces, it is better to entrust a large portion to the person to generate the feeling of importance and personal contribution.

VDT work should be meaningful

The meaning or content of work may be low in some VDT tasks. As work is fragmented, it is also simplified. Thus workers fail to identify with their jobs and lose interest in the product of their work. They should know and feel that their contribution is important, which heightens their satisfaction and self-esteem. However, fears about one's professional future, including one's job security, can become more important to the worker than job control, as Carayon (1993) reported.

Social contacts should be facilitated

One of the obvious drawbacks of some VDT work is the low chance for social contacts, particularly with colleagues. This can lead to an isolation of individual operators, much more than traditional clerical activities. Therefore it is advisable to enhance and encourage social interaction during non-task periods; this is an argument in favour of interrupting work to meet colleagues in break rooms conveniently located near the workplaces.

Careful introduction of computers

For traditional non-computer offices, a well-planned careful introduction of VDTs is an important measure to prevent hostility towards office automation. A good transition policy should include proper information and clear instructions, adapted to the worker's capabilities. It is certainly insufficient to leave the employee alone with a manual that explains how the system works. Classroom teaching, followed by practical application, and on-the-spot help should be given by a well-qualified expert able to coach the trainees. A good training programme will increase acceptance and reduce psychological fears since well-trained operators will come to consider themselves an important investment.

Some words of caution

Some restrictions must be made, though, when discussing the above-mentioned principles of job design for VDT work. It must not be overlooked that the main problems of job design refer to some repetitive, monotonous and meaningless data-entry or data-acquisition jobs. The proportion of these among the total number of VDT activities is not known but a figure of between 15 and 25 per cent is a reasonable assumption. Another restriction concerns the above-mentioned fact that not every employee dislikes repetitive jobs. Salvendy (1984) reported that 10 per cent of the labour force in the USA do not like work of any type; the remaining workers are evenly split between those who prefer to work in enriched jobs and are more satisfied and productive in them and those who prefer to do simplified jobs in which they are more satisfied and productive.

Avoiding physical overstrain

Designing keys for lighter operation (see Chapter 5), getting more people to do keyboarding and making many of them do more keying than ever before has brought about 'epidemics' of repetitive trauma disorders. Though well known to occur in pianists, typists, meat-cutters and in many other occupations, disorders of the connective tissues (especially tendons) appeared in very large numbers in Australia in the early 1980s and then in North America. Many factors are involved to cause, aggravate or precipitate but overuse of body tissues due to the repetitiveness of use of computer keys or mice is without doubt the culprit. Repetitiveness is one of the characteristics of monotonous tasks. This is another example of the intricate connections that often exist between job conditions and well-being, both physical and psychological, of the human. *Boredom and fatigue often come together*.

Summary

Personal well-being and work efficiency can be greatly reduced by over-specialisation and job fragmentation. Often it is better to broaden and enrich jobs, provide responsibility, allow control over one's work and facilitate social contacts.

CHAPTER FIFTEEN

Working hours and eating habits

15.1 Daily and weekly working time

Daily working time

Many studies have shown that changes in the length of the working day may result in either higher or lower output. In the fourth edition of this book, Grandjean (1988) reported that at one factory it was found that shortening the working day from $8\frac{3}{4}$ to 8 h raised the output by between 3 and 10 per cent, with predominantly manual work showing a bigger increase than machine operation.

Figure 15.1 sets out the results of an old British survey (Vernon, 1921). These results confirmed the general experience that shortening a working day can result in a higher hourly output; the work is finished off more quickly, with fewer voluntary rest pauses. This change in working rhythm generally takes place within a few days, although occasionally it may be several months before the effect can be seen.

Conversely, making the working day longer may cause the tempo of the work to slow down and the hourly output to fall. The schematic relationship between a day's work and a day's total output has been sketched by Lehmann (1962) as shown in Figure 15.2. From this it can be established that as a rule the total daily output does not increase in proportion to the daily working time according to curve A in the figure. Most often it behaves like curves B and C. In many cases of medium or heavier physical work it has even been found that increasing the daily hours beyond 10 results in a fall in total output because the slowing down of the work done per hour, resulting from fatigue, more than offsets the gain expected from the longer work shift.

A longer working day or overtime work are common in wartime and boom periods but the results are often disappointing. Because of the relationship between hours and output just mentioned,

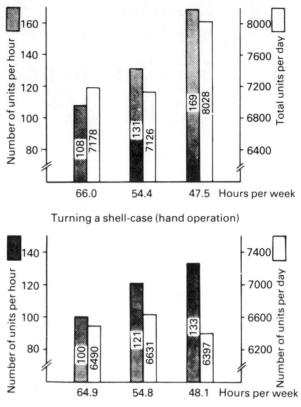

Figure 15.1 Working time and output. *(Top) For predominantly hand work, shorter hours (either weekly or daily) bring about an increase in both hourly and overall daily output. (Bottom) For a job that is mostly operating a machine, shorter hours, while they increase hourly production, have little effect on the overall daily output.*

productivity does not increase as much as is desired and expected; it even may fall if people become 'overworked', physically or mentally.

These observations point to the conclusion that most workers tend to maintain a particular daily output and if the working day is varied will to some extent adjust their working rhythm to compensate. This is only true, however, for work that is not linked to the speed of a machine or process. Workers on a conveyor belt, and others who must fit their work into an externally given rhythm, cannot do much to compensate for changes in working times. This can be seen in Figure 15.2.

Of course, the extent to which working speed is varied to compensate for variations in working hours is also affected by pay incentives and other changes in motivation.

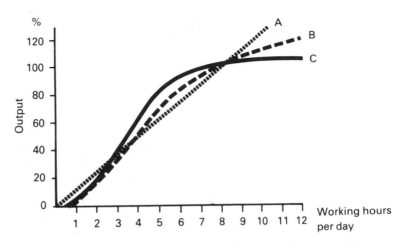

Figure 15.2 Relation between working hours and output. *100 per cent = output from an 8-h day. A = curve of working hours and output; B = output in relation to working hours for moderately strenuous work; C = same for heavy manual work. After Lehmann (1962).*

Effects on sickness rates

Many observations have provided evidence that excessive over-time not only reduces the output per hour but is also accompanied by a characteristic increase in absences for sickness and accidents. A working time of, say, 8 h per day, which makes the operator moderately but not seriously fatigued, cannot be increased to 9 h or more without negative effects. These include a perceptible reduction in the rate of working and a significant increase in nervous symptoms of fatigue, often resulting in more illnesses and accidents.

Our physiological knowledge and present-day experience point to the conclusion that a working day of 8 h cannot be exceeded without detriment if the work is heavy. Modern firms, organised according to the principles of industrial science, usually arrange most of their work sensibly, whether the demands on their workers are heavy or only moderate. *An extended working day is tolerable in jobs where the nature of the work provides plenty of rest pauses.*

Historical review of weekly working time

The worktime of 'western' employees is usually agreed in number of hours per week, which have become markedly less during the last 100 years.

In Switzerland, the first federal Factory Act was accepted in a plebiscite in 1877 and stipulated 65 h per week (11 each weekday and 10 on Saturday). An amendment in 1914 shortened this to a 48 h week.

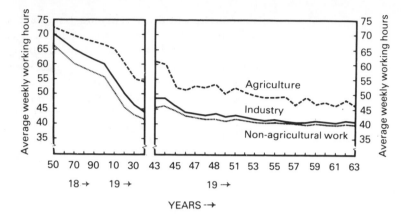

Figure 15.3   The trend of weekly working hours in the US from 1850 to 1963. *After Northrup* (1965).

Figure 15.3 shows the historical development of weekly working hours in the USA. It is evident from the figure that the working week in the USA has tended to shorten almost continuously since 1850. Today the 40-h week or less is common, not only in the USA but in most industrial countries. In Germany, the average work week consisted of about 38 h in 1995 and was at 32 h in the metal and electronics industry in 1997.

The five-day week

Today the five-day week is common everywhere and it presents little problem to combine this with a 40-h week. Nevertheless, the experience of introducing the five-day week is of some general interest. Occasionally factories found that the changeover from a six- to a five-day week led to less absenteeism. Experience shows that the workforce in general, and especially women, prefer the five-day week, mainly on social grounds. These social factors, combined with increased opportunities for rest and relaxation, are mostly responsible for the reduced absenteeism.

The four-day week

A four-day week has come under discussion in recent years. About 600 American firms, and recently some German and French ones, are said to have had favourable experience of this (Maric, 1977). The total time worked in four days is usually 40 or fewer hours in the USA and Europe. In Europe particularly there is a trend toward a four-day week of some 30 h which is already practised in some industries and offices. Advantages put forward are three days free at the weekend, less traffic to and from work, and the possibility of increasing employment by taking on more workers which is especially important during conditions of widespread unemployment.

A reduction in the weekly working days to four or even three will, in most cases, be accompanied by a considerable increase in daily working hours. Some medical experts as well as ergonomists consider such overly long work shifts as possibly damaging to health. This criticism is certainly valid. It must be possible for a person to recuperate within each 24-h period, not to go on exhausting oneself for four days and hoping to recover over the following three rest days. Continuously working shifts of 9 or 10 h often leads to excessive fatigue and increased absenteeism through sickness — unless only a few physical efforts are required which are diverse and interesting: firefighters' shifts are often 12 h, but most of the time they are on stand-by.

Attitudes and opinions change. In 1995, Germany had an average working week of less than 40 h, often compressed into a four-day period, but the results of a public opinion survey in Germany in 1971 yielded the following figures:

Four-day week of 10 h daily   preferred by 46 per cent
Five-day week of 8 h daily   preferred by 47 per cent
Uncertain   7 per cent

In the 1988 edition of this book the four-day, 40-h week was summarily rejected on medical and physiological grounds. This is still true for 'exhaustive' work but theoretical considerations as well as experiences reported in the meantime (compiled by Costa, 1996; Kogi, 1996; Kroemer et al., 1994, 1997) indicate that 'compressed work weeks' can be suitable and are accepted, even preferred by persons with jobs such as nursing, clerical work, supply operations or supervising of automated processes.

Flexible working hours

Flexible working hours are a fairly recent form of work organisation which has found many adherents. It is characterised by a fixed working period, called the block or core time, and a flexible period at each end of this, the lengths of which are at the discretion of the individual. All employees must be present during the core time and a certain number of hours must be worked per day, week or month as a minimum.

Reviews of experiences in various countries can be found in the publications of the International Labour Office (e.g. Maric, 1977; Kogi, 1991) and of Kroemer et al. (1994). On the whole it can be said that *flextime* is popular with many employees and employers.

A valid objection is that not all kinds of work are suitable for this flexible hours arrangement and that in nearly every set-up there is at least one group of employees who must be excluded such as providers of certain services, e.g. receptionists, people at the information desk, telephonists, assembly workers organised in groups, police or air traffic controllers.

Looking over the very extensive literature on flexible working hours, one gets the distinct impression that the advantages greatly exceed the drawbacks and that both on industrial and social grounds this new way of regulating working hours, where suitable, is a distinct step forward.

15.2 Rest pauses

Biological importance

Every function of the human body can be seen as a rhythmic balance between energy consumption and energy replacement or, more simply, between work and rest. This dual process is an integral part of the operation of muscles, of the heart and, if we take all the biological functions into account, of the organism as a whole. Rest pauses are therefore indispensable as a physiological requirement if performance and efficiency are to be maintained.

Military commanders have always known that a marching column should be halted once an hour because the time lost will be more than compensated for by a better performance by the men at the end of the march. Rest pauses are essential, not only during manual work but equally during work that taxes the nervous system, whether by requiring manual dexterity or by the need to monitor a great many incoming sensory signals.

Different kinds of rest pause

Work studies have shown that people take rest pauses from work of various kinds and under varying circumstances. Four types can be distinguished:

1 Spontaneous pauses.

2 Disguised pauses.

3 Pauses as part of the nature of the work.

4 Pauses prescribed by management.

Spontaneous pauses are the obvious pauses that workers take on their own initiative to interrupt the flow of work in order to take a break. These are usually not very long but may be frequent if the job is strenuous; in fact, many short pauses have greater recuperative value than a few long ones.

Disguised pauses are times when the worker temporarily takes up some activity other than the main job. Usually the disguising activity is a shorter, easier routine task which allows the worker to relax from the main job. Most jobs offer many opportunities for such disguised pauses, for example cleaning some part of the machine, tidying the workbench, sitting down more comfortably or even leaving the workplace on the pretext of consulting a workmate or the supervisor. Such disguised pauses are justified from a physiological and psychological point of view since nobody can do either manual or mental work continuously without interruption.

The problem with disguising pauses is often that they do not provide sufficient relaxation because another activity is performed. For full effect, pauses should be taken openly.

Work-conditioned pauses are all those interruptions that arise either from the operation of the machine or the organisation of the work; for example, waiting for the machine to complete a phase of its operation, for a tool to cool down, for a piece of equipment to warm up, for a component, a machine or a tool to be repaired. Waiting periods are especially common in the service industries, for example waiting for customers or orders. On a conveyor belt the length of work-conditioned pauses depends on belt movement and dexterity of the operative. The faster he or she works, the longer the wait for the next piece to come along. Since a worker's speed and dexterity decline with age, younger workers have long pauses whereas older workers often have to work almost continuously to keep up. Hence on conveyor belts the older operatives, as well as the less skilled, often have to work hastily and may be overstressed.

Prescribed pauses are breaks in the work that are laid down by the management. Examples are the midday break and pauses in the morning and afternoon for snacks (tea/coffee breaks).

Interrelationship between pauses

The four types of rest pause are to some extent interrelated, as was shown in 1954 by Graf with the help of a time study. In particular, Graf found that the introduction of a few prescribed pauses led to fewer spontaneous and disguised pauses being taken. Figure 15.4 shows the results of a time study in the electrical industry where a highly skilled job on piece rate was performed.

Graf showed that during tiring skilled work done standing up both disguised and spontaneous pauses increased progressively during the course of an 8 h study. There were about three times as many pauses during the last three hours as during the first five, showing that the need for breaks in the work increased as the operator became more fatigued. To meet this need there was an increase in both disguised and spontaneous rest pauses.

In general, it can be said that all the different types of rest pauses (disguised, spontaneous, prescribed and work-conditioned) should amount to 15 per cent of the working time. Often a ratio of 20–30 per cent is allowed and this is certainly necessary in some jobs.

Rest pauses and output

Many investigations into the effect of rest pauses on production have been reported in the literature and in general their results agree with those of research into working time and production.

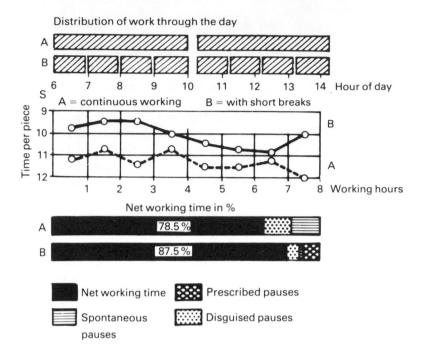

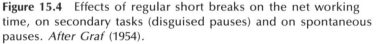

Figure 15.4 Effects of regular short breaks on the net working time, on secondary tasks (disguised pauses) and on spontaneous pauses. *After Graf* (1954).

Introducing organised rest pauses often actually speeds up the work and this compensates for the time lost during the prescribed pauses; it also reduces disguised and spontaneous pauses.

The hourly output of fatiguing work usually declines towards the end of the morning shift, and even more towards evening, as the rate of working slows down. Various studies have shown that if prescribed pauses are introduced the appearance of fatigue symptoms is postponed and the loss of production through fatigue is less.

On the whole rest pauses tend to increase output rather than to decrease it. Ergonomics attributes these effects to the avoidance of excessive fatigue or to the periodic relief of fatigue symptoms by an interval of relaxation.

Rest pauses in heavy work

For heavy work obligatory pauses should be written down and evenly distributed throughout the eight working hours of the shift. If the pauses are only optional, workers tend to work continuously and save up all the permitted rest time until the end so they are able to leave work earlier. This leads to overstress, particularly among older workers. The rest pauses must be arranged in such a way that the total energy expenditure of 20 000 kJ per working day is not exceeded.

Rest pauses in moderately heavy work

For all other jobs in manufacturing, assembly and other industries, even in offices, a prudent recommendation is a rest pause of 10–15 min in the morning and the same again during the afternoon, in addition to a longer midday break. These pauses serve the following purposes:

Prevent fatigue
Allow opportunities for refreshment
Allow time for social contacts.

If the job is demanding, physically and/or mentally, it is unthinkable not to have some rest pauses, which one values as much for social as for medical reasons.

Problem of time-linked jobs

Time-linked work, such as on an assembly line, poses a special problem. Many studies in the laboratory, as well as in actual work sites of all kinds, have shown that pauses of 3–5 min every hour reduce fatigue and improve concentration. They are especially necessary when the job is repetitive, time-driven and calls for constant alertness.

Under training

Rest pauses have a dramatic effect on the learning of a skilled operation, as we have already seen in Chapter 8. If training is interrupted frequently for short periods of relaxation, a new skill is acquired much more quickly than if training is continuous, as Rohmert *et al.* (1971) described.

Rest pauses during training do more than just prevent fatigue: during rest pauses a trainee will look ahead and understand the process so that it becomes easier to acquire the automatic skills required. When a skilled operation is involved, the rest pauses provide additional periods for mental training.

Time schedules for computer jobs

Opinions about the impact of working hours on the adverse effects of VDT work are controversial. Some unions and a few scientists have claimed that the number of working hours each day spent at computers needs to be reduced from the usual 8 h. However, *neither change nor reduction in working time should be considered until the display, workstation, task and work environment fulfil ergonomic recommendations.* It would be nonsense to reduce the working time because of a badly designed workplace. There are good reasons to believe that the work at ergonomically well-designed VDT workstations is not more strenuous than other office jobs.

However, there are some very repetitive and speedy computer jobs, often of the data-entry kind. Such jobs may warrant special consideration but, as a rule, only with the aim of eliminating or facilitating them. Providing more pauses, or shortening the working time, does not solve the problem: the job remains unsuitable, only its negative effects on the operator are averted, reduced or postponed.

Recommenda-
tions

To summarise, the following arrangements for rest periods are recommended:

1 *When heavy work is done, or when working in great heat, rest pauses should be arranged often so that the acceptable maximum demands on the worker's circulatory and metabolic system are not exceeded (see Chapter 6).*

2 *For jobs demanding moderate physical or mental effort, there should be organised breaks of 10 to 15 min about halfway through the work periods before and after the longer (about 30 min) mid-shift rest period.*

3 *A job making heavy mental demands, especially if it is timed work with little embedded rest, should have several short pauses of a few minutes' length in addition to the organised breaks described under (2) above.*

4 *When learning a skill or serving an apprenticeship, many pauses should be the rule, varying in rate and duration to suit the difficulty of the task.*

5 *Supervisors should encourage their staff to make, and everyone at work should take, as many spontaneous breaks as needed to maintain attention, concentration, endurance and well-being. They should be open rather than disguised, frequent and of short duration. Don't wait for fatigue, avoid it!*

15.3 Nutrition at work

Comparison with
the automobile
engine

A combustion engine requires three essentials for smooth running:

1 *Petrol as a source of energy.*

2 *Lubrication to protect the moving parts.*

3 *Cooling water.*

Similarly the 'human engine' requires:

1 *Foodstuffs (sugar, protein, fat) as sources of energy.*

2 *Protective materials (vitamins, mineral salts, iron, iodine, unsaturated fatty acids, etc.) as 'lubricants'.*

3 *Liquids for cooling purposes.*

Figure 6.1 demonstrates how chemical energy in the form of

nutrients is taken in by the body and converted into heat and mechanical energy. This is a process comparable to that taking place in the engine of a car. A car can be driven only as long as the petrol lasts; a human being can go on working only as long as the food provides chemical energy. The more manual work done, the greater the demand for energy, which can be met only by increasing the intake of food.

Food requirement and occupation

The energy content of foodstuffs can be measured and is expressed in kilojoules (kJ). (The kilocalorie (kcal) has also been commonly used in the past; 1 kcal = 4.2 kJ.) The same unit is used for the energy consumption of the human body, which increases the more physically active is a person's occupation. The average daily requirements of energy for men and women in various occupations were summarised in Chapter 6.

In recent years the overall pattern of the working population has changed in regard to physical labour. In every industrial country the proportion of workers with sedentary jobs has greatly increased; we can expect this to account for about 70 per cent of employed persons. Conversely, the proportion of manual workers has fallen. A German example is given in Table 15.1 and this can be extrapolated into the future.

In general terms present-day adults can be divided into two categories according to their energy requirement in their occupation:

1 Sedentary workers and all female operatives: energy requirement 8400–12 500 kJ per day.

2 Heavy workers whose average daily requirement is 12 500–17 000 kJ (ignoring the few whose exceptionally severe work calls for 17 000–21 000 kJ per day).

Table 15.2 summarises a proposal for distributing the requirement among the five meal times customary in Europe and North America.

To illustrate the energy content of some of the most important foodstuffs, Table 15.3 gives the amounts of each which must be eaten to yield 100 kcal or 420 kJ.

Needs of 'sedentary workers'

For 'sedentary workers', as well as for the great majority of female occupations, it is broadly true that *the quantity of food should be restrained in favour of high quality*.

Sedentary workers would be well advised to cut down on energy-rich and highly refined foodstuffs, and give preference to natural foods containing protective elements such as vitamins, minerals and trace elements, i.e. vegetables, salads, raw fruit, milk, brown bread and liver.

Table 15.1 Distribution of employed persons, including housewives, in Germany, in percentages

Work demands	1882	1925	1950	1975
Light, sedentary work	21	24	58	70
Moderately heavy work	39	39	21	23
Heavy to severe work	40	37	21	7
Workforce in millions	16.9	32.0	32.2	39.8

Source: After Wirths (1976).

Table 15.2 Recommended distribution of daily food intake, in kJ

Meals	Sedentary workers	Manual workers
Breakfast	1200–1700	2500–2900
Morning break	100–200	600–1000
Midday meal	3300–3700	3700–4200
Afternoon break	100–200	600–1000
Evening meal	5200–5900	5900–7000
Total	9900–11 700	13 300–16 100

Under normal circumstances a person takes in just enough food to supply the energy he or she needs, regulated by feelings of hunger. Disturbances of this energy balance are fairly common among sedentary workers, who have a tendency to eat more than they need for their everyday life. Such people are often visibly overweight.

Needs of manual workers

Manual workers have quite different problems. They need a diet that is energy-rich, but not bulky, and so tend to prefer food that has protein and fat. Carbohydrate foods are more bulky because they contain more indigestible roughage. They include all kinds of sugars (sweet and non-sweet), which make up the greater part of flour, potatoes, pastas and of course all sweet foods.

If a worker wants to take in 15 000 kJ in the form of potatoes, he or she will have to eat 5 kg of them. Such a diet would be too bulky and would overload the digestive organs. The recommended course is for manual workers to increase their intake of proteins and fats to approximately double the normal value. For a man weighing 70 kg this means about 200 g of protein per day and about 50 g of fat to get about 5700 kJ, with the remainder made up from carbohydrates. Of course, individual diets depend on needs, preferences and availability.

Table 15.3 Energy equivalents of some important foodstuffs. *The quantity of each that must be eaten to yield 100 kcal (420 kJ) of energy is given*

Foodstuffs	Quantity (grams/litres
Green vegetables	670 g
Turnips and swedes	400 g
Skimmed milk	0.3 l
Full-rich milk	0.2 l
Potatoes	150 g
Hens' eggs	60 g
Jam	50 g
Meat	50 g
Cheese	45 g
Bread	4–45 g
Legumes and pastas	30–40 g
Sugar	25 g
Butter or margarine	12 g

Muscular work requires increased amounts of phosphates. The energy-rich western diet of manual workers usually provides enough protective vitamins and minerals.

The importance of proteins and fats

Since protein of animal origin is more valuable than plant protein for body-building and muscular strength, half of the intake should come from meat, eggs and milk. A 70 kg person could find the necessary 50 g of animal protein in 1.5 liters of milk, 300 g of meat or seven eggs.

Fats are the foodstuffs that are richest in energy: 100 g of fat contains as much as about 300 g of bread or even 1 kg of potatoes. A further advantage of fat for physical workers is that it remains in the digestive organs longer and so postpones the onset of hunger; on the other hand, unsaturated fats are blamed for a variety of ailments.

Generally speaking, too much fat is eaten in many 'developed' countries. Heart and circulation ailments can be partly traced to excessive fat in the food. The risk is smaller for manual workers than for 'paper pushers', who are strongly advised to avoid greasy foods. The origin of the fats is also of some significance to health: one half of the fats should be plant oils, such as from olives, and the other half animal fats. Milk and milk products should be the preferred forms of animal fats since they contain many vitamins and minerals.

Overall working time and nutrition

Old 'continental' working hours, with a long 2-h break at midday, allowed the worker to go home to take midday dinner with the family and still have time for a rest. Today nearly all industries and most services have adopted the short lunch break and an earlier release in the afternoon.

A shorter midday break usually means eating in a company canteen or a nearby restaurant. A meal away from home is more expensive and there is less time for eating, and workers instinctively feel that it would be unwise to go back to work immediately after a large meal; hence a change in eating habits, transferring the main meal to the evening and making the midday meal little more than a large snack. This has the advantage that the afternoon's work is less affected by digestive problems.

From a physiological point of view it should also be noted that *a midday break of 45–60 min is usually enough for relaxation, provided that there are also rest pauses of 10–15 min, both in the morning and afternoon, for relaxation and eating a snack.*

Food intake and biological rhythm

In a Swedish gas works the reading errors of several inspectors were tabulated daily over a period of 19 years. Figure 15.5 shows these reading errors, according to the time of day at which they occurred. It is clear, among other things, that they reached a maximum immediately after the midday break, that is while digestion was still going on.

This observation confirms a biological law by which loading the digestive system dampens the state of readiness of the entire organism. This effect of large meals has been known for a long time and finds expression in the familiar proverb: 'A full stomach makes a lazy student'.

There are several old studies showing the effects of the distribution of meal times. A classic study was carried out by Haggard and Greenberg (1935) who observed that small snacks every 2 h kept the blood sugar and efficiency at a higher level throughout the working day. These results were confirmed in a shoe factory where three meals plus two snacks were associated with a higher productivity than conditions with fewer meals and snacks. These results, as well as more recent studies, lead to the conclusion *that the intake of food five times daily (such as the 'western' arrangement of three meals and two snacks) is good for both health and efficiency, and is to be recommended, provided it does not lead to excessive energy intake.* This recommendation applies particularly for continuous work with a short midday break since then there is need for extra periods to relax and eat. Figure 15.6, based on current scientific knowledge and general experience, shows a 'theoretical' curve of eagerness for work, both during traditional working hours and during a working day that is planned according to physiologically sound principles.

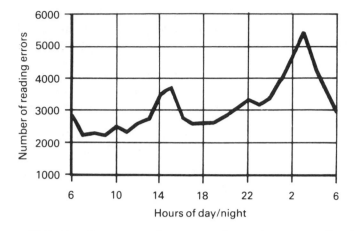

Figure 15.5 Reading errors of gas meter inspectors, according to the time of day. *Evaluation of 175 000 records, with a total of 75 000 errors, between 1912 and 1931. Besides the distinct peak after the midday meal, mentioned in the text, there is a much higher peak in the early hours of the morning, when readings are taken in the dark. After Bjerner et al.* (1955).

Intake of fluids

A person needs not only food, to provide energy, but also water, to maintain a correct water balance. The average requirement is about 35 g water for every kilogram of body weight per 24 h (2–2.5 litres/day).

The water that is drunk and that which is obtained from the food eaten is continuously excreted through the kidneys and the sweat glands. The liquid excreted is not pure water but a fluid rich in waste materials (urea, sodium chloride and many other metabolic end-products). Although foodstuffs have a high fluid content which is released during digestion (e.g. meat 70–80 per cent; bread 45 per cent; fruit 85 per cent; potatoes 80 per cent; pastas 20 per cent), we still need more liquid. The amount varies among individuals but is generally about 0.5–1 litres, rising to 2 litres and more on a hot summer day.

The amount of water we drink is governed by feelings of thirst, which in turn depend mainly on the concentration of salts in the blood. An increase in salt concentration makes a person more thirsty.

In summer, much of the water is lost by sweating. Since sweating is essential to maintaining body temperature, like the cooling of an engine, the consequent water loss must be replaced in summer, in tropical countries and in hot jobs in industry. This is best done by drinking water or easily assimilated liquids such as tea, coffee or soft drinks. The loss of salt during heavy sweating is usually offset by salt contained in the food; salt tablets are seldom required.

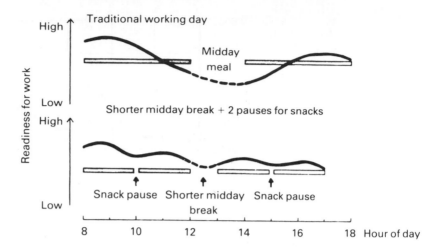

Figure 15.6 Working time, eating and readiness for work. *The dotted blocks cover the working periods and the curves indicate the conjectural rise and fall of readiness for work. During the working day there should be two breaks for snacks, in addition to a short midday mealtime (below) when the curve does not fall as low as it does with a larger midday break (above).*

Snack summary

Beverages and snacks between meals are important both for sedentary and manual workers to meet a significant part of their daily fluid requirements, as well as of additional energy-producing food, according to their bodily needs. Table 15.4 shows the approximate energy content of between-meal snacks but people vary greatly in their eating habits and hence in their choice of snacks.

The range of choice lies between a refreshing drink that is low in kJ and, at the opposite extreme, a heavy snack equivalent to some 1600 kJ, with many intermediates. A few suggested snacks between meals are listed in Table 15.4, the choice depending primarily on the fluid and energy requirements of the person.

Effects of snacks on teeth

An important aspect of between-meal snacks is the concern with dental health because the link between daily sugar intake and the incidence of dental caries is undeniable. The stickiness and physical consistency of sugar foods, especially pastries, nutty crunch, and many kinds of sweet biscuits, chocolate, bananas and crystallised fruit is bad for the teeth. On the other hand, bread and fruit are rarely harmful in spite of their starchiness. From a dental point of view the following are recommended items for between-meal snacks: *apples, nuts, fresh fruit, water or skimmed milk, bread and butter, cheese, yoghurt, sausage and meat.*

A drink with few kJ (soup, tea or coffee) and no solid food is recommended for desk workers, whereas manual workers need

Table 15.4 Suggestions for between-meal snacks and their energy content

Type of snack	Energy content in kJ
1 cup of water	0
1 cup of soup	40–60
1 cup of tea with two lumps of sugar	150
1 cup of coffee with milk and two lumps of sugar	155
1 cup of apple juice	270
1 cup of milk or yoghurt	275
1 cup of ovaltine in milk	540
Bread (50 g)	500
Bread with fruit	1000
Bread with cheese	1250
Bread and sausage	1250

an energy-rich drink such as fruit juice, milk or yoghurt, supplemented by bread and cheese, meat or fruit.

Coffee, tea and soft drinks with caffeine are especially popular drinks as snacks because they have an immediate stimulating effect, though this is slight and does not last very long. If a worker feels the need for frequent stimulants of this kind, provided that they are not drunk to excess, there is no objection on medical grounds. A certain amount of stimulation is a good thing when the work is monotonous.

Summary

We are 'trapped' by the traditional western division into seven-day work weeks with two days off for the weekend and the work day divided into 8-h shifts. There are many jobs that should be shorter than 8 h because they are so demanding and there are other jobs that could last longer. Depending on the actual conditions, proper regimens of work and rest, and of supply of food and beverage, can be recommended.

CHAPTER SIXTEEN

Night work and shift work

More and more shift work

In recent times all industrial countries have turned increasingly to continuous production. This is why shift working is no longer a fringe problem but of increasing importance. The main reasons for going to continuous work are economic ones. Many processes are said to be feasible or profitable only when used 24 h a day.

16.1 Day- and night-time sleep

The problem

The human organism is naturally in its ergotropic phase (geared to performance) in the daytime and in its trophotropic phase (occupied with recuperation and replacement of energy) during the night. Hence the night worker approaches work not in a mood for performance but in the relaxed phase of the daily cycle. Herein lies the essential physiological and psychological problem of night work. Another aspect is the burden it puts on family life and social isolation. Ergonomics is therefore faced with the problem of planning work schedules in such a way that shift work does as little harm as possible to health and to social life.

Comprehensive surveys of the problem of night work and shift work may be found in the publications by Costa (1996), Costa *et al.* (1989), Kogi (1996), Folkard and Monk (1985), Horne (1988), Rutenfranz and Knauth (1976), Tepas and Mahan (1989), and, summarised for engineers and managers, by Kroemer *et al.* (1994, 1997).

The circadian rhythm

The various bodily functions of both humans and animals fluctuate in a 24-h cycle, called the diurnal or circadian rhythm (*diurnus = daily, circa dies* = approximately a day).

Even if the normal influences of day and night are excluded, for example in the Arctic or in a closed room with unchanging

Table 16.1   Circadian bodily functions

Body temperature
Heart rate
Blood pressure
Respiratory volume
Adrenalin production
Excretion of 17-keto-steroids
Mental abilities
Flicker-fusion frequency of eyes
Release of hormones into the
bloodstream
Melatonine production

---

artificial lighting, a kind of internal clock comes into play, the so-called endogenous rhythm. This varies in different individuals but usually operates a cycle of between 22 and 25 h.

Under normal conditions endogenous circadian rhythms are synchronised into a 24-h cycle by various 'time-keepers':

changes from light to dark and vice versa
social contacts
work and its associated events
knowledge of clock time.

The bodily functions that are most markedly circadian are sleep, readiness for work, as well as many of the autonomic vegetative processes such as metabolism, bodily temperature, heart rate, blood pressure and hormone release. Table 16.1 lists a few characteristic day/night changes.

Effect of circadian rhythms

The bodily functions listed above show these diurnal trends throughout the 24 h but they do not all reach their maxima and minima at the same time; there are some distinct phase differences among them. However, taken as a whole they confirm the rules mentioned above, ie.:

1 *During daytime all organs and functions are ready for action (ergotropic phase).*

2 *At night most of these are dampened and the organism is occupied with recuperation and renewal of its energy reserves (trophotropic phase).*

Normal sleep

The most important function that is geared to circadian rhythm is sleep. While it is still not possible to say just what is the specific

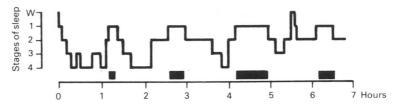

Figure 16.1 The cyclic course of night sleep. *W = is awake; black bars = periods of rapid eye-movements (REM).*

function of sleep, certainly having sufficient undisturbed sleep is a prerequisite for health, well-being and efficiency.

An adult human requires 6 to 8 h sleep per night, although there are considerable individual variations. Some people must have 10 h sleep if they are to be fresh and alert, others need only 5 h, or even less. It was said of Thomas Alva Edison that he needed only about 3 h and that he himself dismissed even these 3 h as merely a bad habit.

Length of sleep is mainly a matter of age. A newborn child needs 15 to 17 h daily during its first six months, whereas many ageing people sleep less and less, and often in broken periods.

The quality of sleep is not uniform but cyclical, and has various stages of different depth (Figure 16.1). The following stages may be distinguished by events observed in the *electroencephalogram* (EEG):

Stage 1. The EEG shows low amplitudes, with many theta waves. This is the stage of going to sleep. The stage lasts 1–7 min.

Stage 2. EEG shows low amplitudes. Besides the theta waves there are also the so-called 'sleep spindles', strong peaks between 12 and 14 Hz, following in quick succession. Stage 2 is a condition of light sleep and its total duration is about 50 per cent of total sleeping time.

Stage 3. EEG shows deeper sleep with increased amplitudes and a decrease in frequencies, up to 50 per cent of the waves being below 2 Hz. Many delta rhythms, interspersed with sleep spindles.

Stage 4. More than 50 per cent of the waves in the EEG are below 2 Hz. Maximum synchronisation and deepest phase of sleep.

Electro-oculograms (EOGs) are used to record activities of the muscles moving the eyes. Presence or absence of *rapid eye movement* (REM) is also used to describe sleep conditions. Frequent salvoes of quick movements of the eyes often occur in sleep Stage 5, with alpha and beta EEG waves and dreams especially common. Despite the activities of the eye muscles, the REM sleep is characterised by maximum relaxation of the other muscles and a great resistance to being awakened; hence REM is also known as the 'paradoxical sleep stage'.

Quality of sleep

Although little is yet known about the significance of the conditions of sleep, it appears that *stages 3, 4 and 5 have particular recuperative properties*. These stages seem to determine the quality of one's sleep.

As mentioned above, cyclical changes take place during sleep, with about four descents into deep sleep linked by intervening shallow periods. Figure 16.1 shows the regular course of an ordinary night's sleep.

Daytime sleep of night workers

For a long time researchers and practioners have recorded frequent cases of disturbed daytime sleep among night workers – discussed further in Chapter 19. Part of this disturbed sleep must be attributed to noise, which is usually greater in a residential area during the day than at night, but many night workers say in addition that they feel a certain restlessness during the day and their daytime sleep is not refreshing enough.

Length and quality of daytime sleep

EEG studies of length and quality of sleep among night-shift workers are of interest. Figure 16.2 shows the results of a study by Lille (1967) who analysed in detail the daytime sleep of 15 regular night workers.

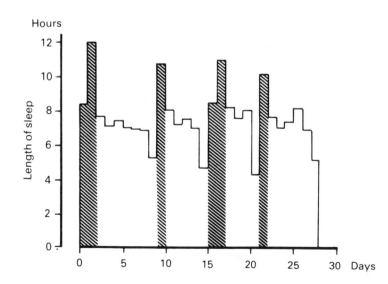

Figure 16.2 Length of daytime sleep of night-shift workers. *White columns = total of daytime sleep; shaded columns = night sleep on rest days. Average of 15 workers. After Lille (1967).*

It appeared that daytime sleep was distinctly shorter than the night sleep the workers took on their rest day. The average length of sleep in the daytime was 6 h, whereas on the rest day the average varied between 8 and 12 h, with longer sleep on the second of the two rest days than on the first. Lille concluded that the night-shift worker accumulated a 'sleep debt' which was 'paid back' on the two rest days. Evidently a single day's rest was not enough for this purpose.

Daytime sleep reflected in the EEG

Detailed analysis of EEGs showed the quality, as well as the duration, of daytime sleep was impaired in night workers, as evidenced by a greater number of periods of light sleep and more body movements. Comparison between sleepers in noisy surroundings and a soundproof room showed that the disturbance was not caused by noise but was an integral feature of daytime sleep.

To summarise, all these studies show that *sleep following a night shift is often curtailed and of too little restorative value*.

Capacity for work at night

Both mental and physical working capacity show a characteristic circadian rhythm. As an ergonomic example, the reading errors of the Swedish gas inspectors may be mentioned once again. Figure 15.5 shows that psychophysiological readiness for work is at a maximum in the morning and in the second half of the afternoon, whereas it is poor immediately after the midday break and declines even more at night.

One more example may be given from early 1950s. Prokop and Prokop (1955) asked approximately 500 truck drivers at what times of the 24-h day they had fallen asleep at the wheel at least once. The responses for each of the hours are shown in Figure 16.3.

The statements about falling asleep show a clear daily cycle, with one peak in early afternoon and an even more pronounced peak during the night. These examples, to which others could be added, show that readiness for action is high during the daytime and low at night. These results reflect the rule that was formulated in Chapter 15: *the human organism is performance orientated during the daytime and ready for rest at night*.

Productivity and frequency of accidents

These facts led to the assumption that night work would be conducive not only to lower output but also to more frequent accidents. Several authors have recorded accident statistics, compiled by Costa (1996). Yet the facts do not clearly support the hypothesis: in some cases, the accident rate at night seems scarcely altered, or even reduced. Perhaps this contradiction between theory and practice reflects the conditions surrounding the night worker, such as fewer disturbances from other people,

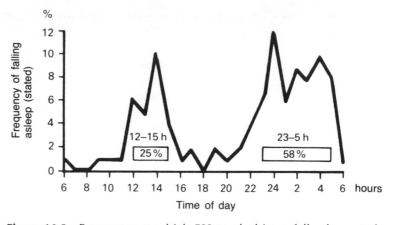

Figure 16.3 Frequency at which 500 truck drivers fell asleep at the wheel, in relation to the time of day. *After Prokop and Prokop* (1955).

higher wages, different kind of work and so on, compared with the circumstances of day work. Moreover, at least a portion of night workers have chosen to do well-paid night work or decided to continue doing it.

Reversal of circadian rhythm

It has already been pointed out that the circadian rhythm is affected by a variety of time-markers. (*Zeitgeber* is the German technical term.) In addition to real clocks, light and darkness they include meals and all the other regular habits. The circadian rhythm can be 'reset' by new time markers; this is commonly experienced when a person travels to another time zone by flying, say, from France to Australia. When doing night work regularly the rhythm is permanently altered, even reversed by the primary Zeitgeber 'work'.

For the shift worker who returns to the common day shift after just a few days (for example at the end of the work week of four or five days) this reversal is not complete; the diurnal curves of physiological functions have become flattened, yet the maxima are not really changed in position along the 24-h time scale. *Biological circadian rhythms show the first signs of readjustment after several night shifts but the reversal is not usually complete even after several weeks.*

16.2 Night work and health

When shift work increased after World War II the matter of circadian rhythm adjustment (or lack thereof) and the fear of negative health outcomes became an important topic in in-

Table 16.2 Sickness rates among shift workers in Norwegian factories. *Studies made between 1948 and 1959 by Thiis-Evensen (1955) and Aanonsen (1964). The percentages quoted relate to the total number of workers in each group studied*

Ailments	Thiis-Evensen		Aanonsen		
	Day work	Night work	Day work	Night work	Former night workers
Stomach troubles	10.8	35.0	7.5	6.0	19.0
Ulcers	7.7	13.4	6.6	10.0	32.5
Intestinal disorders	9.0	30.0	11.6	10.2	10.6
Nervous disorders	25.0	64.0	13.0	10.0	32.5
Heart trouble	–	–	2.6	1.1	0.8

dustrial medicine. It was analysed in detail in several large field studies.

Sickness rate

The first large surveys came from Scandinavia. In Norway Thiis-Evensen (1958) and Aanonsen (1964) studied sickness rates among 6000 and 1100 workers, respectively (Table 16.2).

This-Evensen's survey showed that shift workers had significantly more digestive ailments and nervous disorders, and Aanonsen's work revealed an interesting corollary. Among the day workers investigated there were many who had abandoned shift work either on health grounds or because they did not like it. This group of former night workers, who had chosen to 'get out', showed a distinct increase in digestive and nervous troubles.

'Positive choice' of night workers

This discovery proved that comparisons with so-called normal groups must be carried out with caution. Night-shift workers should be regarded as being a 'positive selection' of particularly tough workers: about 2 out of 10 people who try out working during the night give up. This might also account for the contradictory results of surveys of sickness rates among the 'surviving' shift workers and day workers.

In spite of the difficulty of finding a good comparison group, an increased illness rate has been observed during the past 20 years among 'active' as well as former night-shift workers. Costa (1996) listed as possible health impacts of night and shift work:

- disturbances of sleeping habits
- disturbances of eating habits
- gastro-intestinal disorders

- neuro-psychic functions
- cardiovascular functions.

Grandjean (1988) was concerned that night-shift workers might misuse drugs by taking stimulants during the night work and sleeping tablets during the day. He stated *chronic fatigue and unhealthy eating habits* as primary reasons for their increased liability to nervous disorders and ailments of the stomach and intestines .

Occupational sickness among night workers

Symptoms of occupational sickness among night workers are, in addition to *chronic fatigue*, the following:

weariness, even after a period of sleep
mental irritability
moods of depression
general loss of vitality and disinclination to work.

The state of chronic fatigue is accompanied by an increased liability to psychosomatic disorders, which in night workers commonly takes the following forms:

- *loss of appetite*
- *disturbance of sleep*
- *digestive troubles*

Chronic fatigue of night-shift workers, combined with unhealthy eating habits, probably is the main cause of their increased digestive troubles.

The causes

What, then, are the actual causes of occupational health problems among night workers? The answer to this question lies in what we have already said about the circadian rhythm and disturbance arising from the change from day work to the night shift. A conflict is generated in the body of a night worker by 'desynchronisation' of internal rhythms: the enforced 'working' cycle is opposed to the natural 'light–dark' and the 'social contact' cycles. None of these mechanisms seems to be fully dominant so that the functional unity of the body is lost and the harmonisation between the separate biorhythms is impaired.

Symptoms

Since a complete adjustment to night work does not take place quickly enough, the worker's body control system is only partly switched over to 'working' at night and 'sleeping and resting' by day. The result is insufficient sleep, both in quantity and quality, with inadequate recuperation, resulting in chronic fatigue with its associated symptoms.

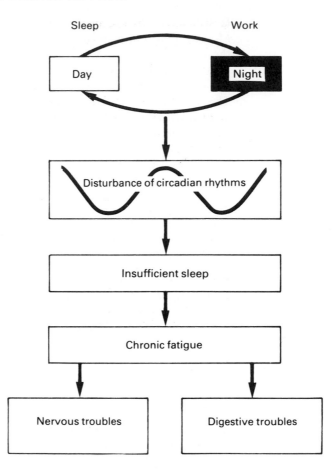

Figure 16.4 Diagram illustrating causes and symptoms of occupational ailments among shift workers who periodically work at night.

Since at the same time eating habits are unhealthy (inappropriate meals at unfamiliar times), psychosomatic symptoms tend to show themselves mainly in digestive disorders.

The nature of occupation sickness of night workers, with its causes and symptoms, is set out diagrammatically in Figure 16.4.

Individual susceptibility

These ailments do not afflict every worker in the same way and even if they show the same symptoms the extent of the disorders varies very much from one person to another. *It is broadly true that about two-thirds of shift workers suffer some degree of ill health and about one-quarter sooner or later abandon shift work because of health problems or the inability to adjust socially.*

Effect of age

Resistance to the special stresses of night work declines with age. Most burdensome is the need to change over to being active during the sluggish period of the night. The older worker is less adaptable and tires more easily. On the other hand many older workers do not require as much sleep as they did when they were younger. The sleep of the elderly is easily disturbed. *Hence, some older night workers suffer both from greater stresses and fewer opportunities to recuperate from them.* In fact, many surveys have shown that shift workers in age groups over 40 are distinctly more prone to disturbed sleep and complain of ill health. Having reviewed the literature, Haermae (1996) recommends:

- *Working time arrangements should consider older workers' personal preferences.*
- *Many older workers prefer to start work earlier than younger shift workers and dislike night shifts.*
- *Continuous night work should be voluntary.*
- *Regular health checks should be done after the age of 40 years.*

Specific risks for women

Costa (1996) states that shift work, especially night work, may have specific adverse effects on women's health. This may be related to their periodical hormonal body functions and to additional domestic activities, particularly for those who have children. Costa found evidence for more frequent pertubations of the menstrual cycle and of more menstrual pain as well as more frequent abortions and lowered rates of pregnancies and deliveries. Female night workers with children have shorter and more frequently interrupted day sleep and suffer from more cumulative tiredness than men and women without children.

Social aspects of shift work

Since 'social well-being' is closely related to physical health, we must now briefly consider the social effects of shift work.

In the forefront of these is the disruption to family life, interference with social contacts among friends and fewer opportunities to participate in group activities.

Many investigations show that most of the people questioned give priority to complaints about fewer meal times with their families.

Free time

Another frequent complaint concerns disruption of the social life outside the family. Active participation in group activities, whether in sport or politics or 'prime time' TV programmes, is so limited that the night worker often feels excluded from society altogether. There are similar impediments to the cultivation of friendships, especially if there are not many other shift workers living nearby. This state of affairs also dictates what shift workers can do in their free time so that they often pursue solitary hobbies. Some authors talk in this context of a tendency for shift workers to feel on 'the fringe of society' or even to be in 'social isolation'.

Opinion polls of shift workers

Opinion polls have shown that many people have two opposing opinions about shift work. On the one hand they are opposed because of impediments to health and social life, while on the other they see certain advantages in it, such as more pay or more freedom to plan their leisure. On the whole, however, the drawbacks predominate.

Three work shifts

Dividing the 24-h day into work sets of 8 h results in the common three shifts. Each has its advantages and drawbacks.

The day shift, typically from 8 to 16 h, complies with the regular day/night rhythm of the body and the current set-up of 'Euro-American' lifestyle. All family, communal and leisure activities are possible, either in the afternoon or evening. However, a shift which begins very early (e.g. 6 a.m.) is tiring because night sleep is cut short.

The evening shift, typically from 16 to 24 h, is particularly bad for social life. On the other hand, sleep is good after this shift and there are opportunities for family life and leisure activities especially in the early afternoon. Thus, people who can socially adjust to this schedule usually have few health problems.

The night shift is bad from all angles. Family life is often limited to taking the evening meal together and all social activities must be geared toward the working hours to follow. Leisure activities are usually possible only in the second half of the afternoon. Sleeping habits vary: some night workers interrupt their daytime sleep for a midday meal and then lie down again afterwards; others sleep through until early afternoon. All sleeping must be attempted during the noisy daytime. Whatever physiological adjustment to night work is achieved during the work week is partially lost during the free weekend.

16.3 The organisation of shift work

Distribution of shifts during the day

With the three-shift system the day is divided into three equal periods of 8 h each. A common European system was this:

Early shift   0600–1400 hours
Late day-shift 1400–2200 hours
Night-shift   2200–0600 hours

There are, however, many variants. In the USA, for example, 8–16–24 h is commonly worked and this arrangement seems to have advantages, both physiological and social. Each shift allows the family at least one meal together and also provides times for enough sleep for persons working the early and late day shifts.

A few firms work a system of two shifts of 12 h each but a 12-h working day cannot be recommended from the standpoint of either industrial medicine or ergonomics (see Chapter 15). At most, exceptions might be made for undemanding jobs, with long built-in pauses. When shifts are as long as this, each shift, whether day or night, is followed by two rest days and many workers like this.

Rotation of shifts

In Europe periodic rotation of shifts is the rule but in the USA it is not uncommon to work the same shift all the year round. Mott and co-authors (1965) and Kroemer *et al.* (1994) see certain social advantages in this arrangement but in the long run continuous night work in not acceptable, either on social or medical grounds, to at least 2 out of 10 people who try it.

Shift rotation cycle

Until about 1960 many experts had been of the opinion that the intervals between shift rotation should be as long as possible. Recommendations for rotation every three or four weeks were based on the idea that people need several days to change their biological rhythm and adaptation to the new shift can take place only if several weeks are allowed. Today we know that this interpretation is misleading because the work-free weekend partly negates the just initiated adjustment. Even after several weeks adaptation is not complete, especially with regard to sleep, one of the most important bodily functions. The daytime sleep of many workers on the night shift remains inadequate, both quantitatively and qualitatively, for a long time.

Thus, in opposition to proponents of long-cycle rotation, *the latest recommendation is that rotation of shifts should be short term*.

Obviously, there are many implicit assumptions made in either of these opinions, such as current 'western' life styles, the presence of or need for work-free weekends, the unavailability of suitable sleeping arrangements during the day and that all persons prefer social interactions during the usual hours of the day, especially the late afternoon and the evening. Deviations from such norms can lead to very different choices of shift work.

Criteria for shift rotation

As a start, it may be helpful to consider what criteria apply to shift systems.

The following are the most important requirements for the worker:

1 Loss of sleep should be as little as possible, so as to minimise fatigue.

2 There should be as much time as possible for family life and other social contacts.

One shift plan to meet these requirements has single, isolated night shifts, each followed immediately by a full 24-h rest. Figure 16.5 shows a shift plan which meets most of the requirements.

From this it can be seen that over a period of four weeks there is only one set of three consecutive night shifts. All the other night shifts are scattered singly and each is followed immediately by a rest day. A very good feature of this plan is the distribution of free shifts, which, throughout the year, include 13 complete weekends, Saturday to Monday inclusive.

Two plans widely used in the UK are the 2–2–2 system (the so-called 'metropolitan rota') and a 2–2–3 system that is called the 'continental rota'. Both of them are short rotations. These are shown in Figures 16.6 and 16.7.

It can be seen that in one system the free days follow two nights' work and the other system follow three nights' work. The 2–2–2 system is the slightly less favourable because a free weekend (Saturday/Sunday) comes only once in eight weeks. The 2–2–3 system is more advantageous in this respect because a free weekend occurs every four weeks.

Short-term rotations are made more difficult because they sometimes bring production to a halt at weekends.

16.4 Recommendations

Shift work that includes night shifts is socially burdensome and often leads to health disorders which can rightly be classified as occupational. Rutenfranz and Knauth (1976) wrote in this context: '. . . *from the standpoint of medical safeguards in industry, continuous production is permissible only where it is unquestionably essential to the manufacturing process. Its introduction simply to increase profits is to be deplored.*'

Since there is no way of planning shift work that covers the 24-h day and that significantly reduces the occupational risk, it should be introduced only with the greatest hesitation. During the last decade or so, much research has been performed on human functions related to sleep, loss of sleep and 'tiredness'. Knauth (1996) as well as Kroemer *et al.* (1994) tried to compile and consolidate the new information but acknowledge that, in the end, it is very difficult to formulate simple and general recommendations when rotation to work during the night is necessary in addition to day and evening work.

If night-shift work is unavoidable, then the following recommendations should be considered:

Mon	Tue	Wed	Thur	Fri	Sat	Sun
N	—	D	E	N	—	—
—	D	E	N	—	D	D
D	E	N	—	D	E	E
E	N	—	D	E	N	N

Weekend patterns and frequency per year			
Saturday	Sunday	Monday	Frequency per year
—	—	—	13
D	D	D	13
E	E	E	13
N	N	N	13

Figure 16.5 (Top) An example of a shift rota in which the night shifts are widely scattered. (Bottom) Summary of free shifts (rest periods) over the year. *D = day shift; E = evening shift; N = night shift.*

1 *Night-shift workers should not be engaged when they are below 25 years old or over 50.*

2 *Workers should not be employed on night work if they have a tendency towards ailments of the stomach and intestine, are emotionally unstable, prone to psychosomatic symptoms or to sleeplessness.*

Week 1	M T W Th F S Su	D D E E N N —	Week 5	M T W Th F S Su	N N — — D D E
Week 2	M T W Th F S Su	— D D E E N N	Week 6	M T W Th F S Su	E N N — — D D
Week 3	M T W Th F S Su	— — D D E E N	Week 7	M T W Th F S Su	E E N N — — D
Week 4	M T W Th F S Su	N — — D D E E	Week 8	M T W Th F S Su	D E E N N — —

Figure 16.6 The 2–2–2 shift system ('metropolitan rota'). *D = day shift; E = evening shift; N = night shift.*

3 *The usual three-shift system, changing over at 6–14–22 h, would be better altered to 7–15–23 or 8–16–24 h.*

4 *Short-term rotations are better than long-term ones.*

5 *Continuous night work without rotation should be avoided.*

6 *A good shift rotation either calls for scattered single nights at work or else the 2–2–2 or 2–2–3 rotation.*

7 *Whether one, two or three nights are worked in a row, they should be followed immediately by at least 24 h of rest.*

8 *Forward rotation is preferred.*

9 *Any shift plan should include some weekends with at least two consecutive rest days.*

10 *Every shift should include one longer break for a hot meal, to ensure adequate nourishment.*

Week 1	M T W Th F S Su	D D E E N N N	Week 3	M T W Th F S Su	N N — — D D D
Week 2	M T W Th F S Su	— — D D E E E	Week 4	M T W Th F S Su	E E N N — — —

Figure 16.7   The 2–2–3 shift system ('continental rota'). *D = day shift; E = evening shift; N = night shift.*

Summary

Briefly stated, the human body and mind are meant to sleep at night and be active during daylight. Some workers can adapt to working continually during the night but even among these certain health disorders are prominent. If continuous or rotating work during the night hours must be done, then certain means to make such a work regimen tolerable should be followed.

CHAPTER SEVENTEEN

Vision

**Visual
perception**

The eyes, very important receptor organs for the human, sense energy from the outside world in the form of light waves and convert these into a form of energy that is meaningful for a living organism – into nerve impulses. It is only through the integration of the retinal impulses by the brain that we have visual perception. Perception in itself does not give a precise image of the world outside: our impressions are a subjective modification of what is reported by the eye. Here are two examples:

A particular colour seems darker when it is seen against a bright background than against a darker one.
A straight line appears distorted against a background of curved or radiating lines.

Individual variations in the interpretation of sense data may be critical in certain situations: people differ in experience, attitude and preconceived ideas. People differ greatly in the intensity with which they react to sensory data.

17.1 The visual system

**Control
mechanisms**

The successive complex stages of 'seeing' can be explained simply as follows. Light rays from an object pass through the pupil aperture, the lens and the interior of the eyeball (which is filled with vitreous humour) and converge on the retina where specific sensors (cones and rods) are stimulated. Here the light energy (see Chapter 18) is converted into the bioelectric energy of a nerve stimulus which then passes along the fibres of the optic nerve to the brain. At a first series of intermediate nerve cells – called neurones – new impulses are generated which branch off to the centres which control the eyes, regulating the width of the pupil,

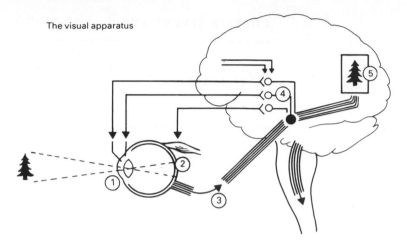

The visual apparatus

Figure 17.1   Diagram of the visual system. *1 = cornea and lens; 2 = light received on the retina; 3 = transmission of optical information along the optic nerve to the brain; 4 = synapses and feedback to the eye; 5 = visual perception of the external world in the conscious sphere of the brain.*

the curvature of the lens and the movements of the eyeball. These control mechanisms keep the eyes continuously directed at the object and this is automatic, not under conscious control. At the same time the original sensory impulses travel further into the brain and after various filtering processes end up in the cerebral cortex, the seat of consciousness. Here all the signals coming from the eye are integrated into a picture of the external world. Here, too, new impulses arise which are responsible for coherent thought, decisions, feelings and reactions. These processes of the visual system are shown in Figure 17.1.

The essential processes of vision are nervous functions of the brain; the eye is merely a receptor organ for light rays. The complete visual system controls about 90 per cent of all our activities in everyday life; it is especially important in a great many jobs. If the numerous nervous functions that are under stress during seeing are considered, it is not surprising that the eyes are sometimes an important source of fatigue.

The eye

The eye has many elements in common with a photographic camera: the pupil, with its variable aperture, the transparent cornea and the adjustable lens represent the optics of the camera. Cornea and lens together refract the incoming rays of light and bring them to a focus on the retina which corresponds to the light-sensitive film. The principal parts of the human eye are shown in Figure 17.2.

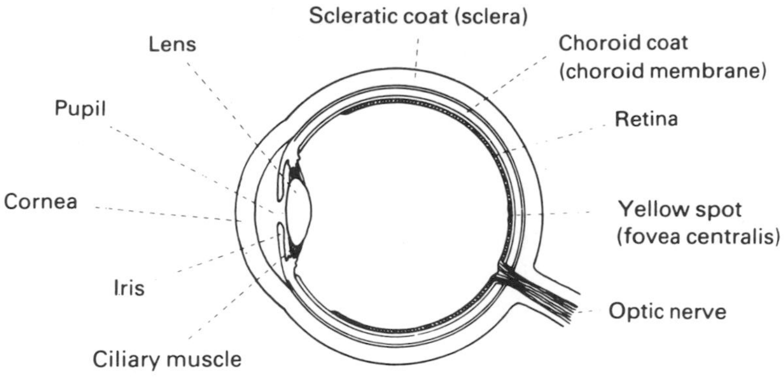

Figure 17.2 Diagrammatic horizontal section through the right eye.

The retina

The actual receptor organs are the visual cells embedded in the retina, consisting of 'cones' for colour vision in bright light and of highly sensitive 'rods' for vision in dim light. The visual cells convert light energy by photochemical reactions into nervous impulses which are then transmitted along the fibres of the optic nerve.

The fovea

The human eye contains about 130 million rods and 7 million cones, each of which is approximately 0.01 mm long and 0.001 mm thick. On the posterior surface of the eye, a few degrees on either side of the optical axis (directly across from the pupil), is the retinal pit, or *fovea centralis*, characterised by a thinner covering than the surrounding area. The thin covering allows the light rays to pass directly to the visual cells, which, in the fovea, consist entirely of cones, here at their maximum density of about 10 000 cones per mm^2. Each foveal cone has its own fibre connecting it to the optic nerve. For these reasons the fovea has the highest resolving power of any part of the retina, up to about 12 s of arc.

Since vision is most acute in the area of the fovea, it is instinctive to look at an object closely by turning the eye until the image falls onto this spot of the retina, which is called the area of central vision. Any object that is to be seen clearly must be brought to this part of the retina, which covers a visual angle of only 1°.

Rods and cones

Outside the foveal area there are considerably fewer cones and one nerve fibre serves several rods and cones. Here the rods are distinctly more abundant than cones and they become more numerous as they are located farther from the fovea. Cones detect fine differences in either colour or shape but need high illumination for this. Rods are more sensitive even in dim light but conceive only shades of grey between black and white. They are the most important light-detecting organs in poor visibility and at night.

The sharp picture

Only objects focused on the fovea are seen clearly, while other images become progressively less distinct and blurred as the focus distance from the fovea increases. Normally the eye moves about rapidly so that each part of the visual field falls on the fovea in turn, allowing the brain to build up a picture of the whole surroundings.

The visual field

The visual field is that part of one's surroundings that is taken in by the eyes when both eyes and head are held still. Only objects within a small cone of 1° apex are focused sharply. If the eyes are kept still when reading, only a few letters can be focused. More details about the physiology of reading will be discussed later in this chapter.

As shown in Figure 17.3, the visual field can be divided as follows:

Area of sharp vision   viewing angle 1°
Middle field:viewing   angle 40°
Outer field:viewing   angle 40–70°

Objects in the middle field are not seen clearly but strong contrasts and movements are noticed: alertness is maintained by quickly shifting the gaze from one object to another. The outer field is bounded by the forehead, nose and cheeks; objects in this area are hardly noticed unless they move. With rods more numerous away from the fovea, in dim light objects can be noticed at the periphery of the field of view but not when one tries to focus on them.

17.2 Accommodation

Accommodation means the ability of the eye to bring into 'sharp focus' objects at varying distances from infinity down to the nearest point of distinct vision, called the 'near point'. If we hold up a finger in front of the eye, the finger can be sharply focused, leaving the background blurred, or the background can be sharply focused, leaving the finger indistinct. This demonstrates the phenomenon of accommodation.

An object is seen clearly only when refraction through the cornea and lens produces a tiny but sharp image on the retina, the three components forming an optical system. Focusing on near

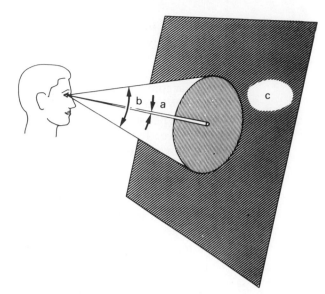

Figure 17.3   Diagram of the visual field. *a = zone of sharp vision, angle of view of 1°; b = middle field: vision unsharp, angle of view from 1° to 40°; c = outer field: movements perceptible, angle of view from 41° to approximately 70°.*

objects is achieved by changing the curvature of the lens, by contraction of *the muscles of accommodation, the ciliary muscles.*

Distant objects   When the ciliary muscles are relaxed, the refraction of cornea and lens is such that parallel rays from distant objects are focused onto the retina. Therefore, when attention is allowed to wander over distant objects, the eyes are focused on 'infinity' and the ciliary muscles remain relaxed.

Resting accommodation   For a long time it was assumed that accommodation focused on infinity was also the resting position of the eye. Several studies, however, revealed that in the dark the resting position corresponds to focusing distances lying somewhere between the near point and infinity, for most people around 1 m (Heuer and Owens, 1989; Jaschinski-Kruza, 1991), This distance is individually different and seems to move gradually towards 'infinity' as we get older.

Near vision   Without accommodation, the image of an object close to the eye would fall behind the retina, resulting in a blurred impression. To avoid this the ciliary muscle increases the curvature of the lens so that the image is focused on the retina. When we look at objects in the near field of vision, the lens is continuously adapting its focal length so that sharp images are always projected onto the retina. Thus, to maintain focus on near targets the ciliary muscles must continuously exert contracting forces.

The accommodated lens does not hold still but is in constant motion. When viewing a target the lens oscillates at a rate of about four times per second. Even when reading a book the lens remains quite active. It seems that these movements of the lens and the perception of blur are important for the automatic regulation of accommodation.

After viewing a near object for some time the lens may not immediately return to its relaxed position. This condition, referred to as 'temporary myopia', may remain for several minutes.

The key to comfortable near viewing is accommodation which keeps the image well focused on the retina.

The near point

As already mentioned the shortest distance at which an object can be brought into sharp focus is called the *near point* and the furthest away is the *far point*. The nearer an object upon which we must focus, the greater is the effort of the ciliary muscle. The near point is a measure of the power of the ciliary muscle and of the elasticity of the lens. *Thus, every individual has his or her own near point.*

The near point moves further away as the ciliary muscle becomes tired after a long spell of close work. Many experiments have shown that prolonged reading under unsuitable conditions is associated with an increase in the near point distance, a phenomenon considered one symptom of *visual fatigue*.

Age and accommodation

Age has a profound effect on our powers of accommodation because the lens gradually loses its elasticity. As a result the near point gradually recedes, whereas the far point usually remains unchanged or becomes slightly shorter.

The average distance of the near point at various ages is reported in Table 17.1.

Presbyopia

When the near point has receded beyond about 250 mm close vision becomes gradually more strenuous, a condition called *presbyopia*. It is ususally caused by loss of elasticity of the lens due to age. This inhibits the lens from changing its curvature. The correction for presbyopia is to wear glasses.

Presbyopia is a frequent cause of visual discomfort while doing close work, as quite a few people find out when they start to do more computer work. The uncomfortable sensation is due to the increased muscle effort which is needed to compensate for the loss of lens pliability. This additional muscular activity might be one

Table 17.1 Average near point distances at different ages

Age (years)	Near point (mm)
16	80
32	120
44	250
50	500
60	1000

of the reasons for *visual fatigue*. A rule of thumb is that no more than two-thirds of the available accommodation power should be used to maintain a comfortable focus.

Speed and accuracy of accommodation

The level of illumination (discussed in more detail in Chapter 18) is a critical factor in accommodation. When the lighting is poor the far point moves nearer and the near point recedes, while both speed and precision of accommodation are reduced. The better the luminance contrast of visual targets against the background, such as letters in a printed text, the faster, easier and more precise the accommodation.

The speed and the precision of accommodation decrease with age. According to Krueger and Hessen (1982) these two functions show a marked decrease from about the age of 40.

17.3 The aperture of the pupil

The 'diaphragm' of the eye

Two different muscles control pupil aperture: one constricting and the other widening the pupil size. This part of the eye is called the *iris*. Its function can be compared to that of the diaphragm in a camera which is used to avoid under- or over-exposure. The pupil aperture is under reflex control to adapt the amount of light to the needs of the retina. When light levels increase the iris constricts and the pupil size is reduced. When light levels decrease the iris opens, making the pupil larger. For any given lighting condition however, the pupil is in continual adjustment motion, much like the accommodated lens of the eye.

Speed of pupil reaction

The adjustment of the aperture of the pupil takes a measurable time which may vary from a few tenths of a second to several seconds. Fry and King (1975) demonstrated that when stimuli producing a significant change in pupil size are presented at a slightly higher rate, about 3 Hz, than the pupil can respond to, the pupil reaction is slowed and discomfort is produced. In fact, if the level of lighting changes frequently and strongly, there is a danger of over-exposure of the retina since the reaction time of the pupil is relatively slow.

Brightness and pupil size

Pupil size reflects to a large extent the brightness of the visual field. It seems that the central vision is of greater importance for the regulation of the pupil size than the outer areas of the retina. During daylight the aperture may have a diameter of 3 to 5 mm, increasing at night to more than 8 mm.

Other regulating factors

The aperture of the pupil is also affected by two other factors:

1 The pupil contracts when near objects are focused and opens when the lens is relaxed.
2 The pupil reacts to emotional states, dilating under strong emotions such as alarm, joy, pain or intense mental concentration. The pupil narrows with fatigue and sleepiness.

Pupil size and acuity

Under normal conditions, however, the general level of lighting is the dominant regulating factor of pupil size.

When the pupil becomes smaller, the refractive errors of the lens are reduced and this improves visual acuity. One of the reasons that higher levels of lighting increase visual acuity is the effect of light on reducing the pupil size. Here too it is possible to make a comparison with the camera: a small aperture of the diaphragm will increase the depth of field and generate a sharper image.

17.4 Adaptation of the retina

If we look into the headlights of a car at night we are dazzled but the same headlights do not bother us in daylight. If we walk from daylight into a darkened cinema where the film has already started we can see very little at first but after about 5 or 10 min the surroundings of the room gradually become visible. These are everyday examples of how the sensitivity of the retina is continuously adapted to the prevailing light conditions. In fact, this sensitivity is many times higher in darkness than in daylight.

The process, called *adaptation*, takes place in the retina through photochemical and nervous regulation. Thanks to this facility we can see almost as well in moonlight as in the brightest sunlight, even though the levels of illumination differ by more than 100 000 times.

Adaptation to darkness

Adaptation to darkness or brightness takes time. Darkness adaptation is very quick in the first 5 min, becoming progressively slower afterwards. Eighty per cent adaptation takes about 25 min and full adaptation takes as much as 1 h. Hence, sufficient time must be allowed for adaptation to darkness; at least 30 min is needed to acquire good night vision.

Adaptation to light

Adaptation to lightness is much quicker than to darkness. The sensitivity of the retina can be reduced by powers of 10 in a few tenths of a second yet full light adaptation takes several minutes.

The abrupt reduction in sensitivity during light adaptation involves the entire retina. Whenever the image of a bright surface (a window, a light source or a bright reflection) falls onto any part of the retina, sensitivity is reduced all over, including the fovea. This phenomenon, most important for precision work or for reading tasks, is illustrated in Figure 17.4.

Partial adaptation

If the visual field contains a dark (or a bright) area, adaptation will occur in the corresponding part of the retina. This adaptation appears in one part of the retina and is called local or *partial adaptation*. As mentioned before, this partial adaptation spreads over the whole retina, including the fovea. Such partial adaptation therefore changes the sensitivity of the retina and affects vision.

Furthermore, adaptation of one eye has some corresponding effect on the other, a fact that may be significant at workplaces where only one eye is employed.

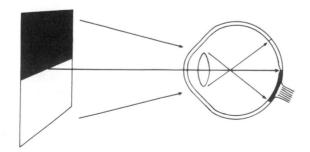

A bright area reduces the sensitivity of the whole retina

Figure 17.4 Effects of light and dark surfaces on the retina. *A light patch reduces the sensitivity of the entire retina and thereby reduces the visual acuity in the fovea. This form of disturbance is called relative glare.*

**Ergonomic
principles**

Two general ergonomic principles can be deduced from this knowledge:

1 *All important surfaces within the visual field should be of the same order of brightness.*

2 *The general level of illumination should not fluctuate rapidly because pupil reactions as well as retinal adaptation is a relatively slow process.*

**Glare in
physiological
terms**

Physiologically speaking, *glare is a gross overloading of the adaptation processes of the eye, brought about by overexposure of the retina to light.* Three types of glare may be distinguished:

1 *Relative glare*, caused by excessive brightness contrasts between different parts of the visual field.

2 *Absolute glare*, when a source of light is so bright (e.g. the sun) that the eye cannot possibly adapt to it.

3 *Adaptive glare*, a temporary effect during the period of light adaptation, e.g. coming out of a dark room into bright daylight outside. This phenomenon is also called 'transient adaptation'.

Practical hints

In this context the following hints are important for the layout of workplaces:

1 The effect of relative glare is greater the nearer the source is to the optical axis and the larger its area.

2 A bright light above the line of sight is less disturbing than one below or to either side.

3 The disturbance is greater in a dim room than a bright one since the retina is then at its most sensitive.

17.5 Eye movements

Tremor

The eyeball has several external muscles which direct the eye to the point of interest. It continuously makes small movements which keep the retinal image in slight motion. Without this continuous tremor the perceived image would fade away. This is like placing your hand lightly on a rough surface and feeling the roughness only as long as the fingers move back and forth.

In general, eye movements are very precise and fast. An eye movement of 10° may be accomplished in about 40 ms.

Vergence

For good vision the *movements of convergence and divergence* are of special importance. Binocular vision requires the lines of gaze, the optical axes, of the two eyes to meet ('converge') on the object being looked at, so that the image falls on the corresponding parts of the retina in each eye.

When viewing a near object, the visual axes are turned inward distinctly. If the gaze is shifted to another object further away, the angle between the two lines of gaze of the eyes must be opened until the optical axes again cross at the object.

This movement is brought about by activity of the outer eye muscles; it is a very delicate adjustment upon which distance perception depends. This specific sensitivity is gradually developed in infancy until we learn by experience to estimate distance mainly from the angle of convergence of our two eyes. In monocular vision distances must be guessed from the apparent size of objects, from foreshortening by perspective and from other visual cues.

The incredible number of eye movements

The number of eye movements required when reading a book may be as many as 10 000 coordinated eye movements per hour. Walking a rocky trail in the mountains demands even more from the eye muscles. When the head is in motion, as in walking, the external eye muscles are in constant activity to adjust the position of the eyes in order to maintain steady fixation points. That is why objects viewed by an observer, even when walking or sitting in a car, appear stable.

If the coordination of external eye muscles is disturbed, the phenomenon of double images will appear. This can be easily demonstrated by slightly touching one eyeball with a finger. In case of excessive fatigue transitory double images can cause annoying sensations.

17.6 Visual capacities

The various functions of the eye are not usually pushed to the limits of their performance capabilities in everyday life but may sometimes occur in industry or busy traffic conditions. The most important visual capacities are:

visual acuity
contrast sensitivity
speed of perception

Visual acuity

Visual acuity is the ability to detect small details and to discriminate small objects. This includes perception of two lines or points with minimal intervals as distinct or apprehending the form and shape of signs or discerning the finest details of an object. By and large, visual acuity is the resolving capacity of the eye. The ability to resolve one minute of arc-wide separation between two signs is often considered as 'normal' acuity. In this case the minimum distance between two points in the image on the retina is 5×10^{-6} m. However, under adequate lighting conditions a person with good vision should be able to resolve an interval of about half that size. Measurement of visual acuity commonly uses standardised black stimuli (such as Landolt rings or Snellen or Sloan letters) on a white background.

Influences on visual acuity

Visual acuity is related to illumination and to the nature of observed objects or signs as follows:

1 Visual acuity increases with the level of illumination, reaching a maximum at illumination levels above 1000 lx. (For a definition of illumination, see Chapter 18).

2 Visual acuity increases with the contrast between the test symbol and its immediate background, and with the sharpness of signs or characters.

3 Visual acuity is greater for dark symbols on a light background than for the reverse. (Light background decreases the pupil size and reduces refractive errors.)

4 Visual acuity decreases with age. This is shown in Figure 17.5.

Contrast sensitivity

Sensitivity to contrast is the ability of the eye to perceive a small difference in luminance. (For the exact meaning of lighting terms see Chapter 18.) Contrast sensitivity allows us to appreciate grades of shading and nuances of brightness, all of which may be decisive for the perception of shape and form. Contrast sensitivity is probably more important in everyday life than visual acuity, especially for many jobs of inspection and product control.

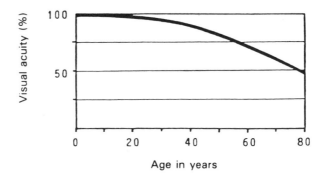

Figure 17.5 Decrease of visual acuity with age. *According to Krueger and Müller-Limmroth* (1979).

To measure contrast sensitivity a procedure is used in which the luminance of a standardised target is compared with its surroundings.

Influences on contrast sensitivity

Contrast sensitivity follows these rules:

1 It is greater for large areas than for small ones.

2 It is greater when boundaries are sharp and decreases when the change is gradual or indefinite.

3 It increases with the surrounding luminance and is greatest within the range of $70 \, cd/m^2$ and more than $1000 \, cd/m^2$ (see Chapter 18).

4 Within this luminance range the just-perceived contrast corresponds to about 2 per cent of the surrounding luminance, i.e. the background must be at least 2 per cent lighter or darker than the target.

5 It is greater when the outer parts of the visual field are darker than the centre and weaker in the reverse contrast.

Figure 17.6 shows results of experiments carried out as early as 1937 by Luckiesh and Moss. It appears that raising the illumination level from approximately 10 lx to 1000 lx increases visual acuity from 100 to 170 per cent and contrast sensitivity up to 450 per cent. At the same time the investigators recorded a decrease in muscular tension (measured as the continuous pressure of a finger on a key) and in the rate of blinking the eyelids. This was interpreted as a reduction in nervous tension as a result of better lighting.

Speed of perception

The speed of perception is defined as the time interval that elapses between the appearance of a visual signal and its conscious perception in the brain. Speed of perception is commonly measured by the technique of tachistoscopy. In this procedure a series of words is presented to the test subjects for a short time. The minimum display time required for correct perception is measured and used as a parameter. Speed of perception measured by such a procedure is, of course, mainly a function of neural and mental mechanisms in the brain.

Speed of perception increases with improved lighting as well as with higher luminance contrast between an object (or sign) and its surroundings. That means that lighting, visual acuity, contrast sensitivity and speed of perception are closely connected to each other.

Speed of perception can be vital in transport. We need only think of an airliner flying at the speed of sound and how much can happen during a perception time of 1.2 s. Speed of perception is also an important factor in reading.

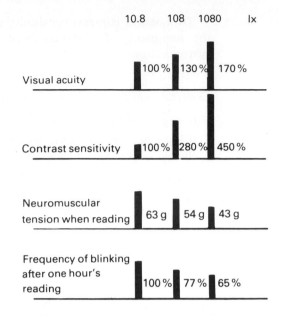

Figure 17.6   Effect of light intensity on visual acuity, contrast sensitivity, nervous strain and frequency of blinking the eyelids. *After Luckiesh and Moss (1937).*

17.7 **Physiology of reading**

Saccades

There is a distinction between reading, which is taking in information, and search, which is locating needed information. In both activities the eyes move along a line in quick jumps rather than smoothly. These jumps are called *saccades*. They are so fast that no useful information can be picked up during their occurrence. Between the jumps the eyes are steady and fix on the target which is projected through the eye. Only in the fovea and in the adjacent area is detailed vision sufficiently accurate for the recognition of normal print.

Three forms of reading saccades are of importance: the *rightward reading saccades*, the *correction saccades* and the *leftward line saccades*.

The *rightward reading saccades* along a line cover in each jump an area of about 8 ± 4 letters. Occasionally small leftward saccades may occur, the so-called *correction saccades*. The *line saccades* start just before the end of a line is reached and jump to the beginning of the next line.

Bouma (1980) studied eye saccades and eye fixations in reading subjects. Figure 17.7 shows the succession of eye saccades and eye fixations when reading a Dutch text. All types of saccades may be different for different texts and different subjects.

Figure 17.7 Saccades and fixations of eyes during silent reading of a Dutch text. *Three different types of saccades are indicated: reading saccades (circles), correction saccades (triangles) and line saccades (squares). Numbers indicate the order of fixations within each line. According to Bouma* (1980).

Character recognition

Bouma observed that the eye pauses between saccades mostly between 120 and 300 ms. During these pauses characters are recognised in foveal and near-foveal vision. For rapid and good recognition it is important that characters are *acceptable, identifiable and distinctive.*

Acceptability is the degree to which characters correspond to the 'internal model' a reader has of them. This is the fundamental process of reading.

Identifiability requires letter details which must be designed clearly.

Distinctiveness means that each character has such a specific design that no confusion can occur. The extension of descending letters (such as p and q) and of ascending letters (b or d) can be important for good distinctiveness.

Visual reading field

The fovea and the adjacent area on the retina pick up visual information from a rather small portion of the total print surface, the so-called *visual reading field*. The visual reading field is larger for words than for numbers because sufficient word knowledge renders the recognition of the full word possible at the sight of merely a few letters. In comparison, numerals allow only a small field when each number must be recognised separately during a single glance.

When reading a text the eyes make about four fixations per second. In well-printed texts the visual reading field can easily be as wide as 20 letters, about 8 to the left of fixation and 12 to the right. The visual reading fields overlap, that is to say that words within the visual reading field may appear at least twice.

According to Dubois-Poulsen (1967) the following time fractions are about normal:

Gaze fixation between saccades 0.07–0.3 s
Rightward reading saccades   0.03 s
Line saccades   0.12 s

Line saccades

Correct line saccades require sufficiently large distances. The lines above and below the reading line will interfere with parafoveal word recognition unless line distances are sufficiently wide. If they are too narrow the visual reading field becomes restricted so that less information can be picked up during a single eye pause. *Thus a wide visual reading field calls for sufficient interline distance.* According to Bouma (1980) the visual reading area around fixation, which is free from interferences by the two adjacent lines of print, decreases by shortening the inter-line distance. If the reading field covers 15 letters the inter-line distance must be equal to about five times the height of lower-case characters; if the reading field is restricted to seven letters the inter-line distance must still be equal to two lower-case characters. Bouma recommended a minimum admissible inter-line distance of about $\frac{1}{30}$ of the line length (in this book it is $\frac{1}{24}$).

As a consequence, inter-line distance should increase with line length. For VDTs it seems advantageous to use screens oriented vertically since such a screen design would require shorter lines and smaller inter-line distances.

Contrast and colour

According to Timmers (1978) parafoveal word recognition is critically dependent on character contrast. The lower the contrast, the narrower is the visual reading field and the lower, therefore, the readability. Similar effects were observed on VDTs with coloured letters. Engel (1980) showed that coloured letters and digits can only be read when quite close to the fixation, although colour itself may well be discernible far away from the fixation. *This indicates that colour is a useful aid for visual search but actual reading takes place in a restricted visual reading field.*

If a reader is familiar with the significance of colours, then colours will help to locate the required information quickly but the recognition of a word or symbol itself depends on the legibility of characters and not on their colour.

17.8 Visual strain

Excessive eyestrain can have two main effects: tiring the eyes and adding to general fatigue.

Visual fatigue

Visual fatigue comprises all those symptoms that arise after excessive stress on any of the functions of the eye. Among the most important of these are straining the ciliary muscle of accommodation by looking too closely at very small objects and the effects of strong local contrasts on the retina. Visual fatigue manifests itself as:

> painful irritation ('burning') often accompanied by lachrymation (tearing, 'watering'), reddening of the eye and conjunctivitis
> double vision
> headaches
> reduced powers of accommodation and convergence
> reduced visual acuity, sensitivity to contrast and speed of perception.

These symptoms are brought about in particular by strenuous fine work, reading poorly printed texts or low-quality computer images, inadequate lighting, exposure to flickering light or optical aberrations of the viewer's eyes. Elderly people are, of course, more prone to visual fatigue.

Obviously, all types of visual work can contribute to the general fatigue discussed earlier since every job that calls for more rapid and precise eye movements will make heavier demands on perception, concentration and motor control. So whenever the eyes are overstressed for long periods the symptoms of eyestrain (sore eyes and headaches) will be added to those of general fatigue.

The effects of visual fatigue on a person's occupation may include:

> *loss of productivity*
> *lowering of quality*
> *more mistakes*
> *increased accident rate*
> *visual complaints.*

Accident rate

Grandjean (1988) referred to a report by the American National Safety Council in which experts reckoned that bad lighting was the cause of 5 per cent of all industrial accidents and, together with the optical fatigue it engendered, contributed to as much as 20 per cent of them.

The experience of an American heavy industry (Allis Chalmers) in the early 1950s may be mentioned as an example. After the level of illumination of an assembly line had been increased to 200 lx, there was a fall of 32 per cent in the accident rate. As a further step the walls and ceilings were painted in light colours, to reduce contrast and provide a more uniform illumination, and the accident rate fell by another 16.5 per cent. Similar surveys in the UK and France showed drastic reductions in accident rates when the lighting conditions were improved, especially in shipyards, foundries, large assembly lines and engineering shops.

Lighting and productivity

There are also many reports of increased productivity after the lighting was improved. These increases are partly a direct effect (through more rapid visual assessment of the work) and partly indirect (through the reduction in fatigue). McCormick and Sanders (1987) provided a table summarising the results of 15 industrial studies, all of which showed increases in output, ranging from 4 to 35 per cent, after increasing the level of illumination. The original level had been very low, however, less than 100 lx. However, the authors voiced reservations because of the existence of other uncontrollable factors that were present in such situations.

An interesting survey in a US cotton-spinning factory showed a stepwise improvement in productivity when the general illumination level was increased. When the illumination was raised from 170 to 340 lx, the production rose by about 5 per cent, while simultaneously the amount of rejected product was sharply reduced. As a result, the total costs fell by 24.5 per cent. This result encouraged the management to increase the illumination still further up to 750 lx, whereupon production rose to 10.5 per cent above the original level and the reduction in wastage brought costs down by almost 40 per cent.

Similar results were obtained in the UK, France, Germany and other countries, often showing increases in productivity, reduction in rejected products and fewer accidents as the level of illumination was increased.

Visual strain of VDT operators

The expansion of computers in recent decades has been accompanied by complaints from many VDT operators about visual strain. Systematic research has been carried out in various countries. Grandjean's 1987 book *Ergonomics in Computerized Offices* provided an overview of the findings up to the mid-1980s. Most studies disclosed an increased incidence of visual discomfort together with the above-mentioned symptoms of visual fatigue. A few studies, however, did not confirm these results since the frequency of complaints among VDT operators did not significantly exceed that of control groups. The controversial results might be explained to some extent by the choice of control groups. Complaints of visual discomfort might be frequent in control groups engaged in strenuous office work but seldom occur among control groups occupied in traditional office work.

Correlates of visual discomfort

Some studies revealed significant relationships between the photometric characteristics of computer displays and symptoms of visual discomfort. Thus, screen flicker, excessive luminance contrast ratio between screen and environment, reflected glare on the screen and poor readability were related to an increased incidence of visual complaints. These findings lead to the assumption that sharpness of characters, luminance contrasts, stability, character flicker, screen reflections and the geometric design of characters might decrease the legibility and produce occasional visual fatigue. Bräuninger *et al.* (1984) measured photometric characteristics for a great number of VDT makes and models and found many of inferior designs.

We shall restrict ourselves here to summarising the main ergonomic recommendations.

Recommended luminances and contrasts

Displays should have dark characters on a light background. A luminance contrast ratio between background and characters of 1:6 is sufficient for good readability.

No flicker

The display must be free of perceived flicker for all operators. As a general rule refresh rates of 80 to 100 Hz with a phosphor decay time of approximately 10 ms are recommended.

Character sharpness

The characters should show sharp edges; no blurred border zone should be perceived. If the blurred border zone is less than 0.3 mm, characters appear to have sharp borders.

Poor sharpness is often due to an insufficient focusing device, a character luminance that is adjusted too high or unsuitable anti-reflective devices.

Character stability

The electronic control of the electron beam must ensure good character stability. Neither drifts nor jitter should be perceived by the operators.

Reflections on screen surfaces

Reflections on the screen meeting the operator's eye should be eliminated. This is best done by 'turning off' the sources, such as luminaires, bright windows, even white clothing of the operator, or by turning the screen surface so that no visible reflections appear.

If these actions fail, reflected glare on the screen surface should be reduced. All anti-reflective technologies available on the market today have serious drawbacks. Some are associated with a decrease of sharpness and an excessive dark screen background, others are easily soiled. If efficiency is weighed against drawbacks, quarter-wave coatings and etching-roughening procedures are to be preferred.

Size of characters and fonts

The range for appropriate character sizes on computer displays is 16–25 min of visual angle. This means that 3 mm is a suitable height for characters at a viewing distance of 500 mm and 4.3 mm at 700 mm. The following sizes are recommended:

Height of capital letters	3–4.3 mm
Width of characters	75 per cent of height
Distance between characters	25 per cent of height
Space between lines	100–150 per cent of height

The spaces between pixels should not be visible. Thus, a dot matrix of 7×9 offers better legibility than one of 5×5.

A number of fonts are available that are easily read on the computer screen. They consist of character designs and arrangements which are simple, clear and without 'decorations'; examples of good typefaces are Helvetica, New York and others where 'function determines form'.

Dark versus light characters

VDTs with dark characters on a light screen background offer the following advantages: reading conditions similar to printed texts, low contrast ratios to the visual environment and less disturbing reflections on the screen. The main drawback is an increased risk of flicker. Thus a refresh rate of 90 Hz and a phosphor with a decay time of approximately 10 ms to reach the 10 per cent luminance level are recommended.

Summary

The human visual system is well understood. Its limitations are known and can often be enhanced, such as by corrective lenses or by well-designed visual objects such as printed text or computer displays.

CHAPTER EIGHTEEN

Ergonomic principles of lighting

18.1 Light measurement and light sources

In order to understand what follows later it is worth defining the main terms employed in the field of lighting. (Colours are discussed in Chapter 21.)

Illumination

Illumination (also called illuminace) is the amount of light falling on to a surface. The light may come from the sun, lamps in a room or any other light source. The unit of measurement is the *lux*, defined as

> 1 lux (lx) = 1 lumen (lm) per square metre, the lumen being the unit of luminous flux.

A formerly used unit in the English-speaking world was the *footcandle* (ft c). 1 lux is approximately 0.1 footcandle.

The human eye responds to a very wide range of illumination levels, from a few lux in a darkened room to hundreds of thousands of lux outside under the midday sun. Illumination levels in the open vary between 2000 and 100 000 lx during the day, whereas at night artificial light levels between 50 and 500 lx are normal.

Luminance

Luminance is the amount of light reflected or emitted from a surface. The unit of measurement is the candela per m^2 (cd/m^2).

In the USA the terms *millilambert* (mL) and *footlambert* (ft L) are still used to measure luminance. One millilambert is the amount of light emitted from a surface at the rate of 0.001 lm/cm^2. A footlambert is the amount of brightness of an ideally reflecting surface illuminated by one footcandle.

The light that we see on the surfaces of walls, furniture and other objects depends on the absorptive or, conversely, reflective property of the surface (see below). The luminance of lamps, on the other hand, is an exact measure of the light they emit.

A few examples illustrate the approximate luminance of some common objects in an office with an illumination of 300 lx:

Window surface	1000–4000 cd/m^2
White paper lying on a table	70–80 cd/m^2
Table surface	40–60 cd/m^2
Bright enclosure of a VDT	70 cd/m^2
Dark enclosure of a VDT	4 cd/m^2

A fluorescent lamp of 65 watt power intake typically has a luminance of 10 000 cd/m^2.

Reflectance

Various surfaces absorb different amounts of the incident light: a dark suface absorbs more and therefore reflects less than a light surface. This can be measured and compared by the ratio between reflected and incident amounts of light. *It is usually expressed as reflectance, the percentage of reflected to incident light.* With the luminance in cd/m^2 and the illuminance in lx the formula is as follows:

$$\text{Reflectance } (\%) = \frac{\text{Luminance}}{\text{Illuminance}} \times \pi \times 100$$

A simple example is as follows. If a table surface has a reflectance of 70 per cent and the incident light has an illumination of 400 lx, the luminance of the table will then be 70 per cent of $400/\pi = 89$ cd/m^2.

Brightness and Dimness

Sensory stimuli that are the same physically are perceived by different people in different ways. This applies to sounds, touch and colours. Regarding lighting, the so-called psychophysical correlates are 'bright' for high amounts of light falling on the retina and 'dim' for little incident light.

Direct and indirect lighting

Among the various lighting technologies one can distinguish between direct and indirect lighting.

Direct(ional) lighting means that a surface is illuminated by light rays that come straight from a source. For example, a lamp may shine 90 per cent of its light towards a desk surface in the form of a cone of the light. Such a 'task' light can generate high local luminance and cast hard shadows behind objects in the path of the light. Excessive luminance tends to produce glare that is especially difficult to tolerate if there are 'dim' shadows adjacent to 'bright' patches, which generates 'relative' glare as discussed in Chapter 17.

Direct lighting systems can be recommended in two cases: either where the general illumination is high enough to avoid 'relative' glare or when one has to accept such lighting contrast to allow sufficient luminance at a specific part of the workplace. At computer workstations such lighting is used when the general illumination is insufficient for reading source documents with poor legibility. (See below for more information on VDT lighting.)

Indirect lighting systems throw 90 per cent or more of the light flux onto the ceiling and walls which reflect it back into the room. For energy efficiency this requires the ceiling and walls to be light-coloured. Indirect lighting generates diffuse (non-directed) light and casts practically no shadows. In general, it can provide a high level of illumination with a low risk of glare but in offices with VDTs the bright ceilings and walls can produce reflections on the screens and cause relative glare.

A combination of direct and indirect lighting is widely used. Often the luminaires have a translucent shade and about 40–50 per cent of the light radiates to the ceiling and walls while the rest is thrown directly downwards. This type of lighting casts only moderate shadows with soft edges. The whole room, including furniture and shelves placed at the walls, is fairly evenly lighted.

Opalescent globes and similar free radiants shine light equally in all directions and throw slight to moderate shadows. Because they are of high emissive luminance, they can cause glare and so they should not be used in workrooms. They are suitable for storerooms, corridors, entrance halls, vestibules, lavatories and so on.

Light sources

Current electric room lighting sources are mainly of two kinds: *candescent (common filament) bulbs and incandescent (such as fluorescent or other gas-filled) luminaires.*

Filament lamps

The light of filament lamps is relatively rich in red and yellow rays. In energy terms, they are very inefficient because about half the electrical energy input is converted into heat, which can become a temperature problem in a workplace. On the other hand, their warm glow creates a pleasant atmosphere.

Incandescent lighting

Incandescent lighting is produced by passing electricity through a gas (argon or neon) or through a metal (such as mercury) vapour. This procedure converts electricity into light much more efficiently than a heated filament. The inside of the tube may be

covered with a fluorescent substance which converts the ultraviolet
rays of the discharge into visible light, the colour of which can
be controlled by the chemical composition of the fluorescent
material. Fluorescent tubes have a series of advantages:

> *High output of light and long life*
> *Low luminance*, when adequately shielded
> *Ability to match the light to daylight* or at least to a pleasant
> and slightly coloured light.

Fluorescent luminaires have serious drawbacks. Since they
operate from alternating current, fluorescent tubes produce a
flickering light at a frequency of 100 Hz in Europe and 120 Hz in
the USA. *This is above the normal flicker-fusion frequency, the
so-called critical fusion frequency of the human eye* (mentioned in
more detail in Chapter 11) but it can become noticeable as a
stroboscopic effect on moving objects. Furthermore, old or
defective tubes usually develop a slow visible flicker.

Visible flicker

*Visible flicker has adverse effects on the eye mainly because of the
repetitive over-exposure of the retina. Flickering light is extremely
annoying and causes visual discomfort.*

When fluorescent lighting was first introduced on a large scale
in European offices, a series of complaints about irritated eyes
and eyestrain was reported. On the assumption that the oscillating
character of fluorescent light was the cause of visual discomfort,
phase-shifted equipment was developed which produced an almost
constant light. Complaints seem to have stopped in offices where
phase-shifted fluorescent tubes have been installed.

A study by Collins (1956) revealed another interesting aspect
of fluorescent tubes. On some models of fluorescent tubes Collins
recorded small 50 cycle/s fluctuations superimposed on the main
100 Hz cycle. These subharmonic 50 Hz oscillations come from a
partial rectifying action in the discharge due to asymmetrical
emissions by the electrodes. Small amounts of subharmonics were
found to be perceptible by subjects and Collins assumed that such
tubes are sufficiently common to account for the complaints which
had arisen with fluorescent lighting.

**Phase-shifted
fluorescent tubes**

*Offices should never be lit with single fluorescent tubes but always
with two or more phase-shifted tubes inside one luminaire.*

Figure 18.1 shows recordings of the luminance oscillation of
fluorescent tubes. It illustrates the effects of phase-shifting which
generates an almost constant luminance.

It is obvious that appropriate equipment will avoid the
disadvantages of fluorescent tubes so that their undisputed
advantages can be fully utilised.

3 fluorescent tubes:

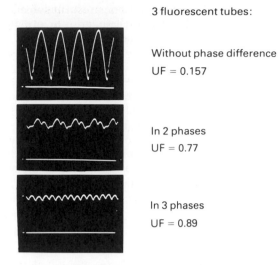

Without phase difference
UF = 0.157

In 2 phases
UF = 0.77

In 3 phases
UF = 0.89

Figure 18.1 Effects of different arrangements of fluorescent tubes on the uniformity of light. *The curves express the invisible flicker, as registered by a photoelectric cell, with the horizontal lines marking the zero level of light intensity. UF (uniformity) = min/max light intensity.*

18.2 Physiological requirements of artificial lighting

For visual comfort and good optical performance the following conditions should be met:

> *suitable level of luminance*
> *spatial balance of surface luminances*
> *temporal uniformity of lighting*
> *avoidance of glare with appropriate lights.*

The physiological requirements under these four headings are just as valid for artificial light as for natural daylight but since the practical problems are somewhat different, the requirements for artificial light will be considered first.

Luminance and illumination levels

Decades ago illumination levels of 50 to 100 lx were generally recommended for workshops and offices. Since then the figures have increased steadily and today levels between 500 and 2000 lx are quite common. The general attitude towards lighting has been 'the more the better'. This does not necessarily hold true, however, especially not for offices.

First, it is not the level of illumination that counts; in fact what we perceive and what helps us to 'see' is luminance.

If 1972 recommendations are compared, it is obvious that the

Table 18.1 Comparison between German (DIN 5035, 1972) and US (Illuminating Engineering Sciety, 1972) recommendations for intensity of illumination, each in lx

	DIN	IES
Rough assembly work	250	320
Precise assembly work	1000	5400
Very delicate assembly work	1500	10800
Rough work on toolmaking machine	250	540
Fine work on toolmaking machine	500	5400
Very precise work on toolmaking machine	1000	10800
Technical drawing	1000	2200
Book-keeping; office work	500	1600

American Illuminating Engineering Society (IES) prescribes significantly higher levels than the German DIN standard. A few examples are given in Table 18.1.

Drawbacks of too high illumination levels

A 1971 study by Nemecek and Grandjean in open-plan offices showed that a very high level of illumination is often unsuitable in practice. Levels above 1000 lx increase the risk of troublesome reflections, deep shadows and excessive contrasts. In the study 23 per cent of 519 employees reported that they were disturbed by either reflections.

Another interesting observation in the same study was the significantly higher incidence of eye troubles in offices with illumination levels above 1000 lx. All employees preferred illumination levels between 400 and 850 lx.

Obviously it would be going too far to interpret these results to the effect that there is a direct causal relationship between illumination level and eye troubles, but there are good reasons to believe that brightly lit open-plan offices can create glaring reflections, deep shadows and relative glare, and that these possibly contribute to the eye troubles recorded. However, these results conflict with several studies carried out in well-lit test rooms where the preferred illumination levels were 1000–4000 lx. The carefully designed reflectances of the surroundings may account for the contradictory results.

The values given in Table 18.2 may be recommended as a basis for comparison of workrooms for different purposes.

If a strong light is necessary this is best achieved by the use of task ('spot') lights, but these should always be used in conjunction with a good general illumination, to avoid creating too much contrast. Guidelines might be as follows:

Table 18.2 Examples of suitable lighting levels in work rooms

Type of work	Examples	Recommended illumination (lx)
General	Storeroom	80–170
Moderately precise	Packing; despatch;	200–250
	Simple assembly; winding thick wire onto spools; work on carpenter's bench; turning; boring; milling; locksmith's work	250–300
Fine work	Reading; writing; book-keeping; laboratory technician; assembly of fine equipment; winding fine wire; woodworking by machine; fine work on toolmaking gig	500–700
Very fine to precision work	Technical drawing; colour proofing; adjusting and testing electrical equipment; assembling delicate electronics; watchmaking; invisible mending	1000–2000

Task light(s)	*General illumination*
500 lx	150 lx
1000 lx	300 lx

Specifications for illumination levels can be no more than general guidelines and other circumstances must be taken into account in any particular situation. For example:

the reflectance (colour and material) of the working materials and of the surroundings
the extent of the difference from natural lighting
whether it is necessary to use artificial lighting during the daytime
the age of the people concerned.

Spatial balance of surface luminances

The distribution of luminances of large surfaces in the visual environment is of crucial importance for both visual comfort and visibility. In general, the higher the ratio of change or difference in luminance levels, the greater the loss in comfort and visibility.

How to express luminance contrast

Although there are many ways to define relative luminances, the most common procedure is simply to specify the ratio of two luminances that exist on side-by-side surfaces that have a distinct border between them. Accordingly luminance *contrast* (C) is calculated from the following formula:

$$C = (L_{max} - L_{min}) \cdot (L_{max})^{-1}$$

In the special case of an object (such as a printed letter or a character on a computer display screen) on a background field:

$$C = (L_0 - L_B) . (L_B)^{-1}$$

where L_0 = luminance of the target and L_B = luminance of the background.

Age

In 1968 the Blackwells determined the contrast necessary to satisfy people of different ages. When the age group 20–25 was taken as unity (1), then for older people the multiplication factors were approximately:

40 year olds   1.2
50 year olds   1.6
65 year olds   2.7

Sharp contrasts ('glare')

Sharp luminance contrasts between large surfaces located in the visual environment reduce visual comfort and visibility, but the degree of acceptable contrast ratios depends on the specific circumstances. Many factors are involved, such as age of the viewer, size of the source of glare, its distance from the viewer's line of sight and the intensity of the general illumination in the room. Furthermore, results of experiments also vary depending on whether quantitative visual performance or subjective visual discomfort is measured.

The early study by Guth

In 1958 Guth observed a decrease in contrast sensitivity and an increase in the eye blinking rate in subjects when the centre area of the visual field was five times brighter or darker than the adjacent area. Figure 18.2 shows the results of this study.

According to these experiments relative contrast ratios of 1:5 in the middle of the visual field significantly impair the efficiency of the eye as well as visual comfort. If the adjacent areas are lighter than the centre area the disturbances seem to be more strongly felt than in the opposite case.

Two types of 'glare' at work

There are two types of glare that can make visual work difficult or impossible. *Direct* glare occurs when the eyes look directly into a light source (the sun, headlights of an oncoming car, task light at work). *Indirect* glare is reflected from a surface into the eyes (the headlights of a following car reflected in the rear-view mirror, a task light or a bright window pane reflected in the computer screen). Both kinds can be avoided by proper ergonomic measures.

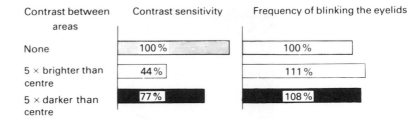

Figure 18.2 Physiological effects of contrasting areas in the middle of the visual field. *The contrast is measured between the 15 per cent of the visual field in the centre and the area immediately adjacent to this. After Guth (1958).*

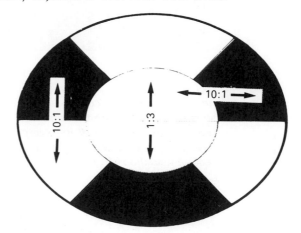

Figure 18.3 Acceptable contrasts between luminances of various parts of the visual field. *Within the middle field, 3:1; within the outer field 10:1; between middle and outer field 10:1.*

General rules

The following general rules for avoiding glare are widely accepted:

1 All the objects and major surfaces in the visual field should appear to be about equally bright.

2 Surfaces in the middle of the visual field should not have a luminance contrast of more than 3:1 (Figure 18.3).

3 Contrast between the middle field and the rim of the visual field should not exceed 10:1 (Figure 18.3).

4 The working field should be brightest in the middle and darker towards the edges.

5 Excessive contrast is more troublesome if it occurs at the sides of and below the visual field than at the top of the field.

6 Light sources should not contrast with their background by more than 20:1.

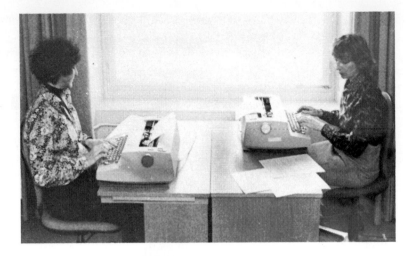

Figure 18.4 Correct and incorrect arrangement of the workplace. *(Left) The secretary, when reading, has a bright window surface in her visual field; the contrast between the window and other surfaces is more than the ratio 10:1. (Right) The bright window is not in the visual field of the secretary; the luminance contrasts have been decreased and the ratio corresponds to the recommendations in Figure 18.3.*

7 The maximum permissible range of luminous contrast within the entire room is 40:1.

In everyday practice these guidelines are often neglected. Contrast problems in the visual environment which can be avoided easily include:

- dazzling white walls contrasting with dark floorings, dark furniture or black office machines

- reflecting tabletops

- black typewriters on bright underlays

- polished machine parts

- bright windows contrasting with computer displays.

The choice of colour and material is of great importance in the design of walls, furniture and larger objects in a room because of their varying reflectance. The following reflectances are recommended:

Ceiling	80–90 per cent
Walls	40–60 per cent
Furniture	25–45 per cent
Machines and equipment	30–50 per cent
Flooring	20–40 per cent.

Windows

Windows should be equipped either with adjustable blinds or with translucent curtains so that excessive contrast can be avoided on sunny days. Both direct and indirect glare can be averted if workplaces are at right angles to the window, as shown in Figure 18.4. This applies equally to schoolrooms, meeting-rooms, conference halls, libraries and so on.

Some workplaces where delicate visual work is being done illustrate this rule. The bench is often placed across the window so that the bright daylight comes from the front. To avoid glare, the operator must bend his or her head so far forward that it is almost horizontal, held on top of the work. Hence, frontal lighting is often a cause of bad posture of the neck and body.

Designers' imagination

Some designers come up with innovative and individual ideas in an attempt to design attractive furniture for offices. They visualise pitch-black office machines on a bright table or dark furniture neighbouring bright walls. Such designers do not care about ergonomic principles or balanced surface luminances.

The instructions for designers of a VDT workstation can therefore be summed up as follows. *Select colours of similar brightness for the different surfaces, renounce eye-catching effects with black and white contrasts, avoid reflecting materials and give preference to matt (roughened) surface treatments including colour paints.*

Specific recommendations for computer workstations are given at the end of this chapter.

Temporal uniformity of lighting

Even more disturbing than static contrast in the visual field are surfaces whose brightness fluctuates regularly. This can occur if the work requires the operator to glance alternately at a light and a darker object, if bright and dim objects pass by on a conveyor belt, if moving parts of a machine are reflective or if a light source flickers.

As we have already seen, the pupil and the retina of the eye can cope with changes in luminance only after a certain delay so that fluctuating brightness leaves the eyes either under- or over-exposed for much of the time. Hence, such lighting conditions are particularly disturbing. Research has shown that if two luminance levels in the ratio 1:5 fluctuate rhythmically, visual performance is reduced as much as if the level of illumination had been lowered from 1000 lx to 30 lx.

To avoid fluctuating levels of brightness as far as possible:

1 *Cover moving machinery with an appropriate housing.*

2 *Equalise brightness and colour along the main axes of sight.*

3 *Take the precautions mentioned earlier to avoid flickering light sources.*

18.3 Appropriate arrangement of lights

Avoid glare with appropriate lights

Inadequate lights or lighting arrangements can be sources of glare which make viewing difficult and uncomfortable. *Avoidance of glare inside a room is one of the most important ergonomic considerations when designing offices.*

Figure 18.5 sets out the results of classical research by Luckiesh and Moss (1937). Their test subjects carried out a visual task in which a light source of 100 W was moved closer to the optical axis step by step. Visual performance was gradually impaired.

A bad example

Figure 18.6 shows a very unsatisfactory arrangement of lights. Opalescent light globes are being used in a drawing office, where they often come into the visual fields of the draughtsmen. The lamps' images are also reflected in the polished floor covering. The result is glaring contrasts, far exceeding the recommended maximum of 10:1.

The following recommendations should be considered in order to arrive at a good arrangement of lights and appropriate overall distribution of light.

1 *No source of light should appear in the visual field of any worker during working operations.*

2 *All lights should be provided with shades or glare shields to prevent the luminace of the light source from exceeding 200 cd/m².*

3 *The line from eye to light source must increase at an angle of more than 30° above the horizon* (Figure 18.7). If a smaller angle cannot be avoided, e.g. in large workrooms, then the lamp must be effectively shaded.

4 *Fluorescent tubes should be aligned at right angles to the line of sight.*

5 *It is generally better to use more lamps, each of lower power, than a few high-powered lamps.*

6 *To avoid glare from reflection, imaginary lines connecting the workplace and lamps should not coincide with any of the directions in which the operator normally needs to look.*

7 *No reflection giving a contrast greater than 10:1 should be within the visual field (Figure 18.8).*

8 *The use of reflective colours and materials on machines, apparatus, table tops, switch panels and so on should be avoided.*

18.4 Lighting for fine work

Fine and delicate work

Very precise work, say from the size of normal print down to fractions of a millimetre, needs special lighting to supplement

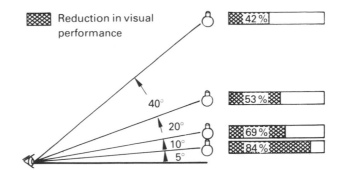

Figure 18.5 Effect of glare on visual performance. *The hatched blocks indicate the reduction in visual performance as a percentage of the normal performance when there is no glare present. Visual performance becomes worse the closer the light source is to the optical axis. After Luckiesh and Moss (1937).*

Figure 18.6 Unsuitable lighting in a drawing office. *The opalescent globes are sources of much glare. The dark floor contrasts too strongly with the white working surfaces (relative glare) and throws back strong reflections of the lamps.*

the general illumination. Examples of such work include the following:

1 Colour testing in a chemical works, paper factory and the textile industry.
2 Delicate assembly work, adjusting and testing electronic equipment, making watches and clocks and precision engineering.
3 Grinding, etching, polishing and engraving glass.
4 Weaving, sewing, knitting, colour printing, 'invisible' mending and quality testing in the textile industry.

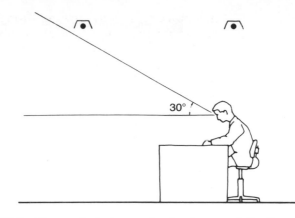

Figure 18.7   The angle between the horizon and the line eye-to-overhead-lamp should be more than 30°.

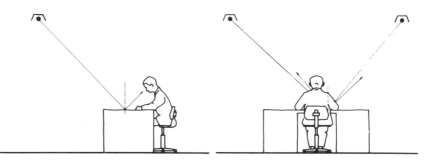

Figure 18.8   (Left) Poor placing of a single lamp. *Its reflection is projected into the eyes of the operator, creating glare. (Right) the reflections of two lamps placed to the sides are not in the line of sight, so reflected glare is avoided.*

Very small objects may need to be magnified and lenses, magnifying glasses and other optical aids should be provided.

Requirements for good vision

The following considerations are important in this context:

1 The level of illumination at the workplace.
2 The distribution of bright surfaces within the visual field.
3 The size of the objects to be handled.
4 How much light is reflected from these objects.
5 The contrast between objects and surroundings/background/shadows.
6 How much time is available for seeing whatever is necessary.
7 The age of the person concerned.

As already mentioned, fine and delicate work requires illumination levels between 1000 and 10 000 lx. Such high levels of illumination are usually demanded by operators who need to concentrate on tiny objects, often by creating strong contrast and shadows to be able to discern contours, small details or exact locations.

The general principle is that work demanding high visual acuity (recognition of the tiniest shapes or objects) and contrast sensitivity (e.g. checking colour or pattern in textiles, reading X-ray pictures) calls for a high level of illumination, often with directed light. The values given in Tables 18.1 and 18.2 provide some general guidelines.

Lights that are too bright can be damaging

Occasionally, however, the lighting can be too bright and harmful. Reflections from shiny surfaces may impair vision. Furthermore, fine structures in materials or surface irregularities in sheets of metal or glass, for example, may actually be seen more easily with sharply directed but moderate lighting than if over-illuminated.

Contrasts in fine work

In contrast to large areas, very small objects are usually best seen with strong contrasts. Dark markings or objects on a light background are often easier to see than bright objects on a dark background. For this reason, when working with very small objects it is generally better not to have the work lit from the side but rather from the front because the back of the object is in deep shadow and the object stands out against a bright, possibly reflective surface.

Frontal lighting

The incidence of the light and the casting of shadows can make a considerable difference to the recognition of objects and the interpretation of their surface structure.

Very diffused light, without shadows, makes everything look flat and featureless, whereas lighting that casts shadows makes things more obvious and discernable.

However, strongly directed light can be accompanied by the problem of disabling glare. This leads to the conclusion that for very fine work neither a completely diffused light nor one fully directional is entirely suitable. For example, when checking metal parts for uneven areas and spots of rust, the visual task is easier with a lamp that is half diffused than with one that gives strongly directed light. Figure 18.9 shows this workplace schematically.

Different kinds and arrangements of light for very fine work should always be tested with experienced workers. Many kinds of precise work pose special visibility problems which cannot be solved by stereotyped methods.

General recommendations for fine work

In spite of this reservation we can formulate general principles that are especially valid for precise assembly work or delicate mechanical tasks

1 Use frontal lighting.

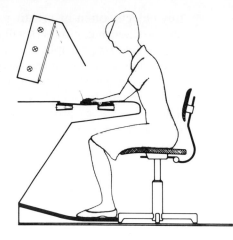

Figure 18.9 Diagram of lighting at a workplace where metal parts passing on a conveyor belt are visually inspected. *The light comes from three fluorescent tubes, out of phase, giving a large light source, with a diffuser in front of it and a shade to protect the eyes from direct light. The frontal lighting enables defective parts to be quickly recognised.*

2 Screen the lamps from being directly visible.

3 Lamps should have ribbed or frosted glass generating partly diffused light.

4 The diffusing surface should be broad and deep to make the illumination of the workbench as uniform as possible.

5 The light source should emit from a large area.

6 Phase-shifted fluorescent tubes are preferable to filament lamps since the latter give off more heat.

18.5 Lighting in computerised offices

Illumination levels for VDT workstations

The general recommendations for illumination levels are not valid for offices with computer workstations. A VDT operator who is alternately looking at a dim screen and a bright source document is exposed to great luminance contrasts. We will see on the following pages that the contrast ratio between screen and source document should not exceed 1:10, which implies that the illumination level on source documents should be kept low. On the other hand, however, the reading task requires that the source document be well illuminated. This conflicting situation calls for a compromise. Hence, it is not surprising that the assessment of the optimal illumination level is a controversial matter.

Preferred illumination levels

In computerised offices one often finds that single fluorescent tubes have been removed or switched off by operators. When questioned, the operators cannot give a specific reason for this but claim that a lower illumination level suits them better.

Some early research has been done in the field of preferred lighting conditions at VDT workstations. Shahnavaz (1982) carried out a field study in a Swedish telephone information centre. The operators could adjust the level of illumination on the working desk. The preferred mean illumination levels on the telephone catalogue were 322 lx during the day and 241 lx for night shifts with similar levels on the desk and keyboard.

A study in Germany carried out by Benz *et al.* (1983) revealed that 40 per cent of the VDT operators preferred levels between 200 and 400 lx, whereas 45 per cent had levels between 400 and 600 lx.

During a study of 38 CAD (computer-aided design) workstations van der Heiden et al. (1984) noticed that a number of lights had been switched off, reducing the mean illumination levels to around 120 lx.

Recommended illumination levels

These early studies and many others that followed made it clear that ambient illumination in traditional offices is often not suitable if computers are introduced. The illumination in computer offices should be reduced to a level compatible with luminous contrast suitable for the work at VDTs. Such levels are in the order of 200 lx but this generally causes the office to appear dimly lit. Furthermore, 200 lx is in most cases inadequate for reading hard-copy documents.

It must be emphasised that it is not prudent to recommend merely one figure since the working conditions might differ from one job to another. For instance, Läubli *et al.* (1981) observed in a field study that operators working on data-entry tasks tended towards higher illumination levels than those engaged in conversational tasks.

General experience as well as several field studies lead to the recommendations given in Table 18.3.

Surface luminances at VDT workstations

The surfaces in the visual field and visual surroundings of a computer operator are the screen, frame and enclosure of display, desk, keyboard, source documents and other elements of the immediate environment, such as walls, windows, ceiling and furniture.

The contrasts of surface luminances at older VDT workstations which commonly had light characters on a dark background were often excessive, similar to the contrast between a dark screen background and a light paper document.

Figure 18.10 illustrates an example of a very badly arranged workstation; unfortunately this condition is not an exception.

3:1 or 10:1?

According to the general guidelines the contrast ratio between the dark background of a screen and a well-illuminated source document should not exceed 1:3. It is obvious that this recommendation cannot be realised at the majority of those VDT workstations that have light characters on a dark background. Haubner and Kokoschka (1983) observed no decrease in performance up to luminance contrasts of 1:20 between screen and source document. Rupp (1981) pointed out that the contrast between bright characters and source document might be more important than the contrast between screen background and source document.

A solution: dark characters on a light background

The contrast problem is solved, at least rendered much less severe, if an operator uses computer screens that show dark symbols on a light background instead of the light-on-dark arrangements of the late 1970s. This so-called 'reverse video' has the additional advantage of being less prone to causing reflected glare because the light screen acts less like a mirror than the dark screen.

Although not all problems of spatial and temporal differences of surface luminances in the visual environment of VDT operators have been solved, it is certainly realistic and reasonable to make the following propositions. *The luminance contrast between the screen and the source document should be between a ratio of 1:3 and 1:10. The same luminance contrast range applies to dark characters on a light monitor screen, with ratios of 1:5 to 1:7 preferred.*

These recommendations are depicted in Figure 18.11 in terms

Table 18.3 Recommended illumination levels at computer workstations. *The lx figures refer to measures taken on a horizontal plane*

Working conditions	Illumination level (lx)
Conversational tasks with well-printed source documents	300
Conversational tasks with reduced readability of source documents	400–500
Data-entry tasks	500–700

Figure 18.10 Excessive luminance contrasts in the visual environment of a VDT operator. *The numbers in the circles indicate the measured luminances expressed in cd/m². Screen and source document have a contrast ratio of 1:50, screen and window of 1:450.*

of reflectance values. The illustration shows one of the older dark-screen computer displays; with a light screen reflectances can be more easily avoided.

Contrasts with bright screens

Most modern terminals have dark characters on bright backgrounds ranging between 50 and 100 cd/m². It is obvious that such bright screens do not raise the problem of excessive contrast ratios with source documents or other bright surfaces in the visual environment, an undoubted merit.

Reflections on screen surfaces

A screen surface made of glass reflects about 4 per cent of the incident light; this is sufficient to reflect clear images of the office surroundings such as lights, the keyboard or the image of the operator, particularly if he or she is wearing light clothes.

First, the object is reflected from the surface of the screen; it produces a mirror-like image. Then the object is reflected from the phosphor layer, producing a veiled and diffuse reflection. Figure 18.12 shows a common example of a reflected window on a dark screen.

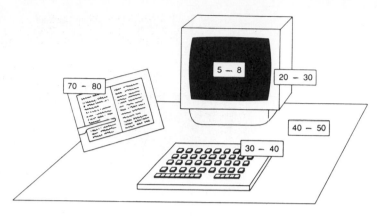

Figure 18.11   Recommended reflectances for a VDT workstation with a dark screen and a light workstation environment. *Reflectance is the percentage of reflected light related to the luminous flux falling onto the surface concerned.*

... produce glare or annoyance

Bright reflections can be a source of disabling glare; image reflections are at least annoying. They interfere with focusing mechanisms; the eye is forced to focus alternately on text and reflected image. They also interfere with contrast adjustment and sensitivity of the visual system. (The functioning of human sight was discussed in Chapter 17.) Thus reflections in the computer screen are a source of visual fatigue and distraction. Bright reflections on the screen are often the principal complaint of operators.

Positioning of computer workstations

The most effective ergonomic measure is the adequate positioning of the screen with respect to lights, windows and other light surfaces. Many other human factors procedures, such as adjustment of screen angle and use of anti-reflective devices on the screen surface are discussed in more specialised publications including Grandjean's 1987 book *Ergonomics in Computerized Offices* or in Kroemer *et al.* (1994) *Ergonomics* book.

If the light source is behind the back of the VDT operator it can easily be reflected on the screen and cause reflected glare. If it is in front it can cause direct glare. These conditions are illustrated in Figure 18.13.

... with respect to light fixtures

Light fixtures directly above the operator can veil the displayed characters with blurred reflections generated in the phosphor layer. Thus it is preferable to install the light fixtures parallel to and on either side of the medial plane of the operator.

Figure 18.12 The reflected image of a window behind the back of the operator is superimposed on the screen text and puts a veil of light over the display.

... and with respect to windows

In offices, windows play the role of huge lights: a window in front of an operator disturbs through direct glare; when behind it produces reflected glare. For this reason the VDT workstation should be placed at right angles to the window. In offices with only one or two parallel window walls this is an efficient protective measure. Figure 18.14 shows such an arrangement.

Cover windows

In offices with two or more window walls some form of window covering should be used. Windows should be covered at night also because the reflections of interior office lights may cause glare. Two types of window coverings can be useful.

Louvers or mini-blinds. Horizontal as well as vertical louvers can be used. Their purpose is to occlude the window on a bright day or to absorb light from indoor sources at night.

Curtains. These are especially efficient if they are also used to control heat flux into or out of the window. Preference should be given to material of low reflectance, say of 50 per cent. Curtains can only be completely opened or closed while blinds can be partially closed.

Finally, there is the possibility of placing intermediate screens between the VDT workstation and bright windows. Such a screen should not have a higher reflectance than about 50 per cent.

Appropriate light fixtures

The best light fixtures for offices with VDTs are not the same as those for traditional offices. Fixtures which provide a great deal of mainly horizontally directed light should be avoided since such

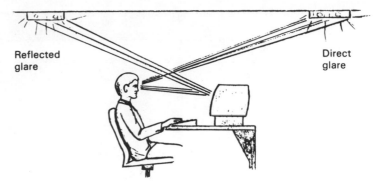

Figure 18.13 Light sources behind the operator create a risk of reflected glare; lights in front of the operator cause direct glare.

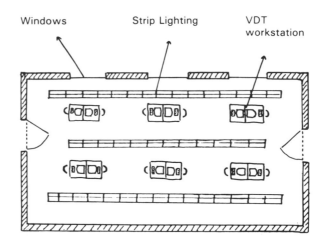

Figure 18.14 *Plan of an office layout with one window wall. Computer workstations should be arranged at right angles to the window. .*

light illuminates the vertical screen and generates reflections on it. It is advisable to use fixtures which provide a confined, primarily downward (or upward – see below) directed flow of light, such as louvers, curved mirrors or prismatic pattern shields. The luminous flux angle should not exceed 45° from the vertical. Suitable and well-arranged light fixtures are shown in Figure 18.15. Such fixtures cause neither direct nor reflected glare since screen and keyboard are in a shadow area.

Indirect lighting   Some lighting engineers suggest suspending luminaires below the ceiling, thus permitting the fixtures to direct the greater part of the light upwards to the ceiling. Standard lamps emitting the full light up towards the ceiling and the upper part of the walls are also used in some offices. These lighting systems may produce a

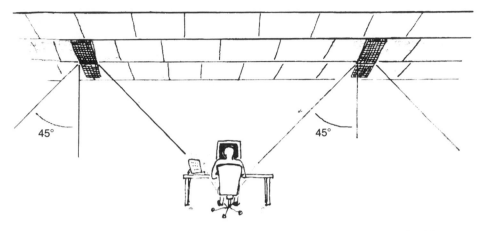

Figure 18.15 Ceiling lighting with a prismatic pattern shield generating light flux confined to a cone with an angle of 45° to the vertical.

pleasant aesthetic effect but have the drawback of bright ceilings and walls, which, in turn, may cause unwanted reflections on the screen.

Task lighting   If there is a low general illumination level, supplementary lighting may be provided for certain visual tasks such as reading source documents. It is important that such task lighting be confined to the arca of the targets and directed so that it does not shine into the eyes of people working in the dim room. Spot lighting can be very efficient but for the avoidance of either direct or indirect glare task light fixtures must be carefully selected and arranged.

Summary

Providing good lighting is a theoretically well- understood task that can be technically well done although it does require some effort on the part of the engineer.

CHAPTER NINETEEN

Noise and vibration

Sound perception

The physiological processes of perception of sound are essentially the same as those already discussed for visual perception. In this case the inner ear (instead of the eye) provides the 'interface' at which soundwaves are converted into nerve impulses along the auditory nerve. *The actual perception of sound is the integration and interpretation of these sensory impulses in the brain, more precisely in the auditory cortex.*

19.1 Perception of sound

Perception of sound is not just a faithful reproduction of the whole band of frequencies 'played' in the brain. This fact is especially important in people's reaction to noise, which varies greatly from person to person. What is noise to you may be music to someone else. Another example of varying perception is the assessment of loudness in relation to the frequency of a sound. A low-pitched sound seems less loud than a shrill one, even though the energy content may be the same.

Sounds

Any sudden mechanical movement sets up fluctuations in the air pressure which spread out as waves, just like waves when water is stirred. As long as these variations of pressure occur with a regular frequency and intensity, the human ear reacts to them as sounds. *The extent of pressure variation determines the sound pressure and this determines the intensity of the sensation.*

Sound pressure is subjectively perceived as *loudness*.

The frequency of a tone is the number of pressure fluctuations or vibrations per second, expressed in Hertz (Hz). Most sounds contain a mixture of tones of different frequencies; if high frequencies predominate, we regard it as a high-pitched sound, and vice versa.

Frequency is subjectively perceived as *pitch*.

The decibel

The physical unit of sound pressure is the micropascal (μPa). The weakest sound that a healthy human ear can detect is about 20 μPa. This pressure wave of 20 μPa is so low that it causes the membrane in the inner ear to deflect it by less than the diameter of a single atom. However, the ear can also tolerate sound pressures up to more than 1 million times higher. The range of hearing encompasses every sound from the gentle murmur of a small creek to the scream of a jet engine.

To accommodate such a wide range in a practical scale, a logarithmic unit, the decibel (dB), was introduced.

The decibel scale uses the hearing threshold of 20 μPa as a reference pressure. Each time the sound pressure in μPa is multiplied by 10, 20 dB are added to the decibel level, so that 200 μPa corresponds to 20 dB. One dB is the smallest change the ear can distinguish; a 6 dB increase is a doubling of the sound pressure level, although a 10 dB increase is required to make the sound twice as loud.

Sound pressures are recorded logarithmically by using the sound pressure (SPL), according to the formula:

$$SPL_{dB} = 20 \log_{10} \frac{P_x}{P_o}$$

where SPL_{dB} = sound pressure level in dB p_x = sound pressure in μPa and p_0 = lowest sound pressure that humans can detect, internationally fixed at 20 μPa.

Pitch and loudness

As already mentioned, the loudness of a sound depends a good deal on its pitch. The young human ear is sensitive to sounds in the frequency range from 16 to 20 000 Hz, a span of nearly nine octaves. Sounds below 16 Hz (infrasonic) are perceived as vibrations; if above 20 000 Hz (ultrasonic) we humans cannot hear them but dogs and other animals do. *Low-pitched sounds seem much less loud than high-pitched ones.* This is shown clearly by the curve of auditory threshold for different frequencies, the lowest curve in Figure 19.1. It shows that the greatest sensitivity of human hearing lies in the range 2000–5000 Hz. It is of some interest to note that much of human speech is between 300 and 700 Hz: most vowels are below 1000 Hz but consonants may be up to 10 000 Hz, especially if they are sibilant.

Curves of equal loudness

As long ago as 1933 Fletcher and Munson plotted curves of equal loudness in relation to sound pressure and frequency. For this purpose they worked from a base of 1000 Hz. At tones of higher

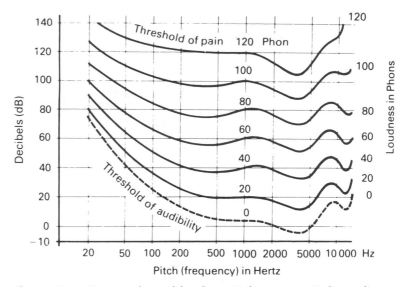

Figure 19.1   Curves of equal loudness ('phon curves') depending on sound level (in decibels) and frequency (in Hertz). *The lowest curve indicates the threshold of audibility, the least sound level that can be perceived.* After Robinson and Dadson (1957).

and lower frequencies they determined the sound pressures necessary to give their research subjects the impression of equal loudness. In this way they obtained curves of equal subjective loudness. Robinson and Dadson (1957) carried these studies further and their results (incorporated in an ISO standard since 1957) are displayed in Figure 19.1. Note that the 'phon' values are the same as the sound pressure in dB at the reference frequency of 1000 Hz.

These curves of equal loudness ('phon curves') are valid only for pure tones; they no longer agree with subjective impressions of loudness if the sound includes many different frequencies. Since nearly all noises, and many signal sounds, are a mixture of frequencies, the use of phon values as a measure of loudness has become obsolete. Nevertheless, these curves of equal loudness still retain their value for assessing the effects of different frequency ranges on the human ear.

Weighted noise level

Today the so-called weighted sound levels have come into use as measures of loudness. Weighted sound level is essentially a process of filtering out the sound energy in the lowest and highest frequencies, at which, according to the curves of equal loudness, sensitivity is least. Hence, sound pressure has little significance in these frequency ranges. The term 'weighted sound (or noise) level' has a similar derivation. Figure 19.2 sets out three weighted curves in dB, (A), (B) and (C), that are in current use.

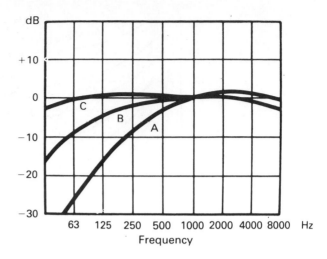

Figure 19.2 Curves showing the relationship between dB(A), dB(B) and dB(C). *They show how much of the sound energy is filtered out in each frequency range.*

The curve of weighted sound level in dB(A) is most often used because many psychological studies have shown that *sound levels measured in dB(A) give a fair assessment of subjective perception of sound (or noise).*

Other units of measurement which are useful in particular circumstances are discussed later.

Hearing organs

A sensation of hearing is produced when sound waves pass through the external auditory passage of the ear canal, then through the middle ear, finally into the inner ear, where the energy from sound pressure is converted into nervous impulses. These travel along the auditory nerve to the brain, where the sound is 'heard'.

The most important components of the hearing organs are shown in Figure 19.3. Sound waves cause the eardrum (tympanic membrane) to vibrate and these vibrations are transmitted by the auditory bones (ossicles: hammer, anvil, stirrup) to the membrane closing the 'oval window' of the inner ear. Here fluids transmit the vibrations along the cochlea and back to the membrane closing the 'round window'. The so-called basilar membrane divides the cochlea lengthwise into two chambers and contain the *organs of Corti* with their sound-sensitive cells. These are the organs which convert the pressure waves into nervous impulses. Each cell is sensitive to a particular range of frequencies and passes its stimulus to a single nerve fibre, which conveys it to the brain.

The basilar membrane has a length of about 30 mm. The sense cells on the cochlea near the entrance (adjoining the oval window) respond to high frequencies whereas those at the apex of the

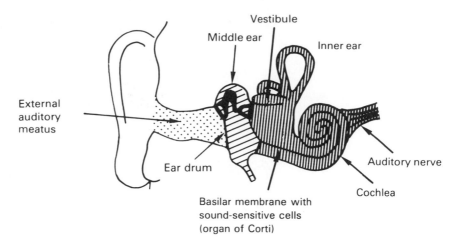

Figure 19.3 The anatomy of the ear. *The shaded parts belong to the inner ear, in which the cochlea detects the sound whereas the curved tubes overlying it (the vestibule) perceive acceleration and balance.*

cochlea are stimulated by low-frequency sounds. Movements of the membrane of the oval window create a series of waves along the cochlea and the distance from the window to the crest of each wave (wavelength) is a function of its frequency. High notes set up short waves, with their crests near the beginning of the cochlea; low notes create longer waves, with their crests nearer the apex of the cochlea. Hence the point at which the basilar membrane receives maximum pressure depends on the frequency of the sound.

This phenomenon may be compared with the breaking of waves in surf; short waves (high frequencies) 'break early', at the beginning of the cochlea, whereas longer waves (low frequencies) travel farther before 'breaking'. In this way the inner ear carries out a sort of 'frequency analysis' of the incoming sound waves and transmits information to the brain of all the frequency components making up the sound. The cerebral cortex integrates these components again so that we are conscious only of the sound as a whole and not of its constituent parts.

Auditory pathways

The nervous impulses generated by the sound waves travel along the auditory nerve to the brain, entering into the brainstem or medulla, and passing through two synapses or nerve connections to the auditory sphere of the cerebral cortex. Here the separate nerve impulses are localised, as if the cochlea of the ear were spread out over the cerebral cortex. This is where the brain finally integrates these impulses into an impression of the sound. Figure 19.4 shows these auditory pathways schematically.

It should be emphasised again that conscious hearing is a

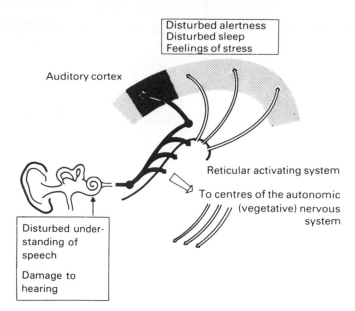

Figure 19.4 The auditory pathways (solid arrows) and the accessory tracks (open arrows) which set up secondary effects of noise.

phenomenon of the brain, more precisely of the cerebral cortex. The inner ear and the auditory pathways are no more than the transmitting mechanism, the 'interface' between the atmospheric sound and the conscious perception of the brain.

Side-effects

Figure 19.4 also shows that between the two synapses, mentioned above, nerve fibres lead out to the reticular activating system. From here fibres run to the entire conscious sphere of the cerebral cortex so that excitation of the system by incoming acoustic signals may induce alarm throughout the consciousness of the individual, disturbing sleep, reducing concentration when awake and producing other distressing symptoms. This kind of 'alarm call' has a biologically important role in alerting a person, providing an opportunity to interpret the sound and react suitably to it. Here is an example. A pedestrian is strolling dreamily along a beautiful country road and the noise of an approaching motor car grows steadily louder. At first the pedestrian remains unconscious of this but when the noise has reached a certain level the reticular activating system is excited and sends a signal to the brain. The pedestrian is immediately alerted so that he or she can:

1 Take conscious notice of the sound.

2 Interpret it.

3 Get out of the way of the car.

Hearing is an alarm system

It is clear that hearing has two principal functions:

1 To convey specific information as *a basis for communication between individuals*. This function is highly developed in humans.

2 As an alarm system by activating secondary pathways leading to the brain *it plays an essential part in waking, increased alertness and finally alarm*.

The alarm function of the sense of hearing may be used to advantage in planning transport and in industry. It is essential to recognise dangerous situations quickly, whether supervising a control panel or machinery, driving a locomotive or flying an airplane. To this end a suitable combination of acoustic signals and visual aids is needed. Often the acoustic signals serve to 'alert' the brain and the visual aids convey the necessary information.

19.2 Noise

Definition

The simplest definition is that *noise is any unwanted sound*. In practice we call it simply 'sound' when it is not unpleasant and 'noise' when it annoys us at that moment. This definition is particularly apt when applied to noise at work. Often, but not always, noise is "loud", ie of high sound pressure level.

Measurements of noise load

The *noise load* can be measured in physical units, taking account of all the acoustical factors over a given time. Various pieces of research work have shown that the *noise level* is not the only factor operating but that the frequency with which the noise occurs and other quantities contribute to the total noise load.

These studies have led to the development of units of measurement which combine various components of the noise load into one quantity so that it is then possible to characterize the noise load at a particular point by a single figure.

Two such units are important in assessing noise problems at the workplace:

1 *The equivalent level of sustained noise (continuous sound level).*

2 *The summated frequency level.*

The equivalent noise level

The equivalent level of sustained noise (L_{eq}) expresses the average level of sound energy during a given period of time (the energy level). This quantity is an integration of all the sound levels which vary during this time and so compares the disturbing effect of the fluctuating noises with a continuous noise of steady intensity.

The summated
frequency level

The summated frequency level is measured with a sound level indicator and a frequency counter, operating over a given time. Commonly used units of sound measurement include:

1 L_{50} (*average noise level*). '$L_{50} = 60$ dB' means that the level of 60 dB was reached or exceeded during 50 per cent of the relevant time.

2 L_1 (*peak noise level*). '$L_1 = 70$ dB' means that the level of 70 dB was reached or exceeded for 1 per cent of the time.

These two levels, L_{50} and L_1, are related to the equivalent noise level by the following approximation:

$$L_{eq} = L_{50} + 0.43 \ (L_1 - L_{50}) \simeq \frac{L_{50} + L_1}{2}$$

Sources of noise

Disturbing noise may be either *external*, coming from outside the building, or *internal*, generated within the building itself. The most important sources of external noise are traffic, industry and neighbours.

The most important sources of internal noise in factories are machines, engines, compressed air, milling machines, stamping machines, looms, sawmills and any other pieces of noisy machinery. In addition, the office has its own internal noise which comes from telephones, typewriters and computer keyboards, printers and people walking and talking.

Street noise

External noise is a disturbing factor in offices, conference rooms, schools and in the home. Table 19.1 lists some of the noise levels to be expected in premises alongside a road.

Industrial noise

Internal noise in factories is highly variable. It may be continuous or intermittent, and may take the form of banging, clattering, rattling or whistling. Table 19.2 shows some of the peak levels that may be attained.

To assess the extent of the danger to hearing from such noise we need to consider the average level of L_{eq} during the 8-h shift. According to ISO TC43 (1971), the measurement should be in dB(A) and the equivalent level of sustained noise (L_{eq}) calculated for intermittent noise.

Table 19.3 gives a few examples of L_{eq} for an 8-h day.

Noise in offices

Concentrated mental work or jobs at which understanding of speech is important, are 'noise-sensitive' occupations and even if the noise level is comparatively low it can be disturbing.

Noise levels to be expected in offices are listed in Table 19.4.

Table 19.1 Noise levels from street traffic, given as L_{eq} in dB(A). *Measured behind a closed window: if this is opened, the noise level in the room is only 5 to 10 dB(A) less than that recorded outside*

Traffic density	L_{eq} in dB(A)	
	Day	Night
Heavy (main road with through-traffic)	65–75	55–65
Moderate	60–65	50–55
Light (local street)	50–55	40–45

Table 19.2 Peak levels of industrial noise, in dB(A)

Source	Noise level, SPL, in dB(A)
Rifle-shot; engine test bench	130
Pneumatic bore-hammer	120
Pneumatic chisel	115–120
Rocking sieve; chain-saw; compressed air rivetter; electric cutter; compressed air hammer	105–115
Milling or weaving machine; crosscut saw; stamping machine; boiler-house; weaving shed	100–105
Electric motor; rotary press; wire-drawer; sawmill; composing room; bottle-filling machine	90–95
Toolmaking machine (running light)	80
Jet engine	120 and lighter

The peak noise level L_1 lies in the zone between 56 and 65 dB(A), which is about 4–8 dB(A) above the average noise level L_{50}.

Nemecek and Grandjean (1971) studied noise problems in 15 large offices in 1970, at a time before VDTs had been introduced. The results showed equivalent noise levels, L_{eq}, of 52–62 dB(A) with peak noise levels, L_1, of 56–65 dB(A). It was concluded that large open offices are rather noisy.

Noise emission of office machines

The advent of VDTs brought a number of additional machines into the office: printers, plotters, calculators and so on. Table 19.5 presents the noise emissions of some modern office machines.

Matrix and daisy-wheel printers are torture!

Matrix and daisy-wheel printers produce high noise emissions. Their peak levels, which are highly repetitive, interfere strongly with speech communication and are very annoying, particularly to those who do not benefit from their use. There are some simple

Table 19.3 Equivalent noise level L_{eq} in dB(A) at various work places calculated over an 8-h shift.

Department or machine	L_{eq} in dB(A)
In a flexible-tube factory	
At yarn-spinning machine	95
In the spinning room	90
At the weaving machine	95
In the weaving shed	95
In a soft-drinks plant	
At the mixing plant	95
At the washing check-point	100
At the automatic sealing machine	100
At a can-filling machine	90

Table 19.4 Noise levels usual in offices

Type of office	Average L_{eq} in dB(A)
Very quiet, small offices and drawing offices	40–45
Large, quiet offices	46–52
Large, noisy offices	53–60

means to reduce noise: hoods deaden the noise emission by about 20 dB(A); unfortunately, many employees dislike having to open and shut the hood each time, except when a great amount has to be printed for a long period of time. Where several printers are being used, one should consider locating them in a separate room. The panels which are used to separate units in large open offices reduce the noise emissions only by a few decibels; they do not suffice to dampen noise sources exceeding 70 dB.

Ink-jet and laser printers are nearly noiseless machines and are appreciated as a great improvement over their noisy predecessors.

The computer itself is noiseless but not its cooling fan and keyboard

The computer itself is silent but many makes have built-in electric fans to cool the unit from the heat produced by the CRT (cathode ray tube) in its monitor. As long as the sound level of the fans is below 40 dB(A), it will hardly be noticed but when it exceeds 50 dB(A), it becomes irritating. Personal computers, located in a quiet closed space, frequently create this noise problem.

Modern electronic typewriters as well as the keyboards attached to VDTs should not produce excessive noise but many do

Table 19.5 Approximate noise emissions of office machines. Measurement position at about the head level of operators

Machine	Noise emission in dB(A)
Matrix and daisy-wheel printers	
Basic noise	73–75
Peak levels	80–82
Matrix printer with hood	
Peak levels	61–62
Ink-jet printer	
Basic noise	57–59
Peak levels	60–62
Laser printer	No measurable noise
Cooling fan of VDT	30–60
Old typewriter	approx. 70
Modern electronic typewriters	approx. 60
Two electronic typewriters face to face	68–73
Copying machines	55–70

and are then a source of complaints. Modern copying machines, or fax machines, fall into the same category of noise makers.

19.3 Damage to hearing through noise

Hearing losses

Strong and repeated stimulation by intense sound can lead to loss of hearing, which is only temporary at first but after being 'deafened' repeatedly some permanent damage may occur. This is called *noise-induced hearing loss (NIHL)* and is usually brought about by slow but progressive degeneration of the sound-sensitive cells of the inner ear. The more intense the noise, and the more often it is repeated, the greater the damage to hearing. Noise consisting of predominantly high frequencies is more harmful than low-frequency noise. Intermittent noise, such as hammering, is more harmful than continuous noise and a single very loud noise – for example a shot or an explosion – can damage the cells on the cochlea immediately.

Sensitivity to noise varies greatly from one person to another. Some who are particularly sensitive may suffer permanent deafness after only a few months, whereas less sensitive people may not show the first symptoms until after many years' exposure. NIHL usually starts at frequencies above 4000 Hz and extends gradually to the lower frequencies. At first the person is unaware of it and only gradually notices loss of hearing when it begins to involve the lower frequencies. Noise deafness is progressive and commonly combines with the hearing loss that comes with natural ageing: NIHL is often mistaken for the early onset of the latter. In most industrial countries, noise deafness is one of the occupational hazards of working life.

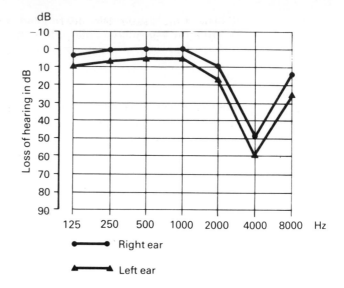

Figure 19.5   Pure-tone audiogram showing impairment of hearing by noise. *The zero line is the normal threshold of hearing. The loss of 50 to 60 dB at 4000 Hz is characteristic of hearing damage by noise.*

Audiometry

A person's ability to hear or the extent of loss of hearing is measured by means of so-called pure tone audiometry, which determines the threshold of hearing of pure tones of various frequencies. The result is recorded on an audiogram, which shows in dB how much the threshold of hearing has been raised for each frequency. Such an audiogram of impairment of hearing by sound is shown in Figure 19.5. It shows a reduction of hearing capability around 4000 Hz, apparently caused by exposure to damaging sound at that frequency.

Temporary loss of hearing

After exposure to intense sound one often experiences a temporary reduction in hearing, already mentioned above. This phenomenon is characterised by the *hearing returning to normal* after the *temporary threshold shift*. There is a close relationship between temporary and *permanent threshold shifts*, with the latter properly called 'deafness' which must be specified as to affected frequency ranges:

1 Sound up to 80 or 90 dB causes slight shifts of 8 or 10 dB only in the threshold of hearing, but if the intensity increases to 100 dB, the threshold goes up much higher.

2 The temporary shift in the threshold of audibility is proportional to the duration of the noise. For example, a 100 dB sound lasting 10 min produces a shift of 16 dB and after 100 minutes, one of 32 dB.

3 The time taken for hearing to return to normal is also proportional to the intensity and duration of the preceding sound exposure. The time of restitution is about 10 per cent longer than the duration of the noise.

Age-related hearing loss

The permanent threshold of hearing rises progressively with age, with the loss of hearing greatest in the higher ranges of frequency and more pronounced in men than in women. Taking a frequency of 3000 Hz as standard, the average loss of hearing to be expected at various ages is as follows:

50 years   10 dB
60 years   25 dB
70 years   35 dB

The audiogram of age-related hearing loss differs from that due to exposure to noise (Figure 19.5) in that the loss of hearing increases progressively as the frequency is raised so that the highest frequency still audible shows the greatest shift in its threshold. The dip in the curve at 4000 Hz, characteristic for noise exposure in the west, is not evident in cases of age deafness.

Older workers often show the combined effects of age deafness and noise deafness, and it may be difficult to distinguish between the two.

The risk of hearing loss

From the evidence of many comparisons between exposure to noise and the frequency of impaired hearing, it is now possible to estimate the risk of hearing damage in noisy factories. The International Standards Organisation (ISO) sets out comprehensive publications about this risk in relation to age, duration of exposure and the intensity of the noise. A simplified extract from ISO TC43 (1971) is given in Table 19.6.

The table data show that the risk of damage increases both with sound intensity and duration of exposure, *the damaging intensities being those above approximately 90 dB(A)*.

Factory workers are often exposed to noise that varies widely and it has been shown that interruptions, or periods of relative quiet, reduce the risk of damage to hearing.

To assess the extent of such risk, the equivalent noise level over the 8-h working day must be calculated. The relation between length of exposure and intensity of sound to create the same degree of risk is as follows:

Hours	*dB(A)*
8	90
6	92
3	97
1.5	102
0.5	110

Table 19.6   The risk to hearing as a percentage of the workforce. *The percentages will be increased by several units with age*

L_{eq} in dB(A)	Length of exposure in years		
	5 (%)	10 (%)	20 (%)
80	0	0	0
90	4	10	16
100	12	29	42
110	26	55	78

To avoid noise damage, Grandjean (1988) *proposed as a limiting value the L_{eq} for an 8-h day not to exceed 85 dB(A)*. Most of the limits in use today approximate to this, although they are often more precisely defined and formulated by national and regional laws and regulations.

19.4 Physiological and psychological effects of noise

Figure 19.4 showed the auditory tracts and how they are linked with the activating and alarm-sensitive structures of the brain. Earlier in this chapter, primary and secondary effects of sound were discussed. Noise may result in:

> *impaired alertness*
> *disturbance of sleep*
> *annoyance.*

Noise can affect the autonomic centres and produce so-called *vegetative effects* in the internal organs. Finally, a special problem of noise is that it makes it *difficult to understand what people say*. The ergonomic aspect of these various problems will now be discussed.

Understanding speech

We all know from our own experience that the sensitivity of the ear to one particular sound – say the voice of a colleague – becomes less and less as the ambient noise increases. The ability to pick out one particular sound from the rest depends on its auditory threshold, which rises linearly with the sound intensity up to 80 dB. Where human speech is concerned, however, it is not enough to hear the pure tones. The 'message' – the content and meaning – must not only be 'heard' but 'understood' and to do this requires a very special discriminating ability in the ear. A critical factor is correct hearing of the consonants which are 'softer' sounds than the vowels because they are spoken with less energy and are therefore more easily masked by ambient noise.

To test the intelligibility of speech, one can simply use the understanding of *syllables* as the criterion. A 'standard' speaker utters a succession of 'standardised' meaningless syllables and the proportion of correctly understood syllables is recorded. It has been shown that a considerable understanding of sentences and their meaning is possible without understanding all the separate syllables. Figure 19.6 shows the ratio between syllable comprehension (S) and sentence comprehension (W).

The graph shows that with a syllable comprehension of only 20 per cent (S = 0.2) it is still possible to understand nearly 80 per cent of the sentences; if half of the syllables are understood, then about 95 per cent of the sentences are intelligible.

Speech comprehension in a workroom depends largely on the loudness of the speaker's voice and the level of background noise. Figure 19.7 shows the relationship between syllable comprehension (S), ambient noise level (N) and voice level (P) solid lines. In addition, the dotted lines join together points at which there are equal differences between sound pressure of noise and sound pressure of speech: $P - N = 20, 10, 0$ and $-10\,dB$ respectively. It is apparent from this graph that a syllable comprehension of 40–56 per cent is possible as long as $P - N = 10\,dB$. According to Figure 19.6, this implies 93–97 per cent comprehension of sentences or sense. *Experience shows that this level of comprehension is good enough for most factories and offices, and therefore*

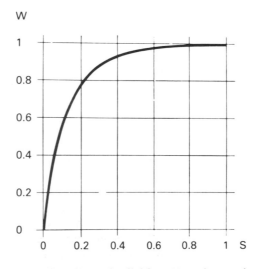

Figure 19.6 Comprehension of syllables (S) and complete sentences or sense (W). *Each is plotted as a decimal fraction of the total number of syllables or sentences offered for comprehension.*

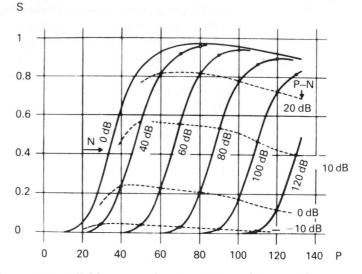

Figure 19.7   Syllable comprehension (S) in relation to the sound-pressure (P) of speech and the noise level (N) in the room. *The dotted lines connect all those points at which P–N is 20, 10, 0 or −10 dB.*

speech comprehension is considered to be unimpaired as long as the background noise level is at least 10 dB below the level of the speaking voice.

If, however, we are concerned with an exchange of verbal information, on a subject that is unfamiliar, with difficult new words, then a higher level of syllable intelligibility is necessary. It has been demonstrated that in these circumstances syllable comprehension must be as high as 80 per cent and this requires a difference of 20 dB between the pressure levels of voice and background noise (Figure 19.7).

There are other more complex tests to assess the intelligibility of voice communication. They include measurement and calculations leading to the Articulation Index (AI) or the Speech Interference Level (SIL) or the Preferred Noise Criteria (PNC). More information on these techniques can be obtained from specialised publications, such as listed in the Kroemer *et al.* (1994) book *Ergonomics*.

Sound pressure of speech in noise

The normal speaking voice, indoors, at a distance of 1 m, operates at the following sound pressure levels in dB:

Quiet conversation	60–65
Speaker at a conference	65–75
Delivery of a lecture	70–80
Loud shouting	80–85

Assuming voice communications over a distance of about 1 m at 65 to 70 dB which must be understood clearly and without strain, the background noise level must not exceed 55 to 60 dB. If the verbal communication is more complex and difficult to comprehend, for instance because it contains strange words or unfamiliar names, then the background noise must not exceed 45 to 50 dB.

If offices or workrooms with these requirements are situated next to roads with moderate or heavy traffic, then a background noise of 60 dB may be exceeded, especially in summer when windows are open. In these circumstances noise levels of 70–75 dB are often reached in offices. One desirable side-effect of air conditioning is that it allows the windows to be kept closed, shutting out external noise.

Effects on performance

Exposure to noise has little effect on manual work whereas we all know from experience that mental concentration, thought and reflection are more difficult in a noisy environment than in a quiet one. Many everyday examples show that noise impairs performance and output of such tasks. It is interesting, though, that what seems to be a truism in everyday life is only partly confirmed by either experiments or field studies. Research into the effects of noise on either mental or psychomotor performance has given contradictory results: *noise may even improve performance but usually it makes it worse.*

Improving performance

Noise can be positively stimulating in the right circumstances. Performance *may* be improved if the work is boring; perhaps this could even apply to mental performance and concentration in situations where there are many other distractions which could be masked by one dominant noise.

Reducing performance

However, these examples of improved performance must be set against more pieces of research that have shown performance to be adversely affected by noise. For example, Broadbent (1957, 1958b) and Jerison (1959) have observed a decline in performance during exacting tests of sustained vigilance.

NASA (1989) has compiled the effects of different kinds of noises (by frequency, duration and origin) on human performance. Kroemer *et al.* (1994) compressed that listing into a table which is reprinted here as Table 19.7.

To summarise, we can say of the negative effects of noise in performance the following:

1 *Noise often interferes with complex mental activities, as well as certain kinds of performance that make heavy demands on skill and on the interpretation of information.*

Table 19.7 Effects of noise on human performance and temporary threshold shift (TTS) of hearing. *With permission from Kroemer et al.* (1994)

Conditions of exposure

SPL (dB)	Spectrum	Duration	Performance effects
155		8 h; 100 impulses	TTS 2 min after exposure
120	Broadband		Reduced ability to balance on a thin rail
110	Machinery noise	8 h	Chronic fatigue
105	Aircraft engine noise		Visual acuity, steroscopic acuity, near-point accommodation, all reduced
100	Speech		Overloading of hearing due to load speech
90	Broadband	Continuous	Vigilance decrement; altered thought processes; interference with mental work
90	Broadband		Performance degradation in multiple-choice, serial-reaction tasks
85	$\frac{1}{3}$-Octave at 16 kHz	Continuous	Fatigue, nausea, headache
75	Background noise in spacecraft	10–30 days	Degraded astronaut performance
70	4000 Hz		TTS 2 min after exposure

Source: Adapted from NASA (1989).

2 *Noise can make it more difficult to learn certain kinds of dexterity.*

3 *Discontinuous or unexpected high levels of noise (over 90 dB as a rule) can impair mental performance.*

Field studies   The effects of noise have occasionally been studied under industrial conditions. Thus Wisner (1967a) reported the following experiences:

1 In one machine shop a reduction of about 25 dB in the noise level led to 50 per cent fewer rejected pieces.

2 In an assembly shop, a reduction of the noise level by 20 dB raised production by about 30 per cent.

3 In a typing pool, a reduction of noise level by about 25 dB was accompanied by a 30 per cent reduction in typing errors.

Such results must always be interpreted with caution, however. Wisner himself hints that creating a new (less noisy) situation in the factory may have a psychological effect on the workers likely to improve performance. (This has been called the *Hawthorne Effect* in the psychological literature.)

Effects of noise in offices

An interesting example of this was revealed during the study of Nemecek and Grandjean (1971) in 15 open-plan offices. The noise level was measured in each of the offices and at the same time 411 office workers were questioned about their experiences and opinions. About 8000 measurements of noise levels gave average values (L_{50}) between 38 and 57 dB(A), and peak noise levels (L_1) between 49 and 65 dB(A). The distribution of replies to the question 'What kinds of noise disturb you?' is shown graphically in Figure 19.8.

It is clear that 'talking' or 'conversation' is the noise problem most often complained of. Many of those questioned added that it was not the loudness of the conversation that disturbed them, but its content.

These indications were confirmed by the results of correlations calculated between noise levels and the incidence of rating noise as 'very disturbing'. Indeed, no correlation could be found between noise levels and disturbance. Does this mean that the disturbance due to office noise is almost independent of the measured noise levels?

Information content of talk is distracting

These results led to the suspicion that the *conversations of other people may be distracting not so much through their sheer loudness as through their information content.*

Conversations of other people, which are usually not very loud, can be masked by the general background noise caused by office staff, rustling paper, walking, typing, office machines and the

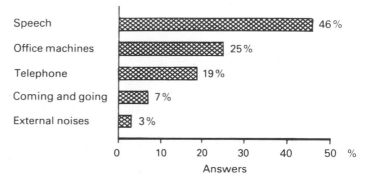

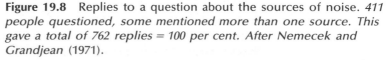

Figure 19.8 Replies to a question about the sources of noise. *411 people questioned, some mentioned more than one source. This gave a total of 762 replies = 100 per cent. After Nemecek and Grandjean (1971).*

room ventilation. One may conclude that a certain average noise level, which covers up the conversations of others, would be appreciated by many office employees. This is why in some large open offices fixtures have been installed that produce a regular, constant background noise, called 'sound conditioning' or 'white noise', intended to mask conversations.

Physiological stress from noise

Many physiological studies have shown that exposure to noise produces:

> raised blood pressure
> accelerated heart rate
> contracted blood vessels of the skin
> increased metabolism
> slowed digestion
> increased muscular tension.

These reactions are symptomatic of a mental 'state of alarm' generated by increased stimulation of the autonomic nervous system. This is actually a defensive mechanism which prepares the whole body for facing possible danger, by being ready for fight, flight or defence. It should not be forgotten that throughout the animal kingdom the sense of hearing is primarily an alarm system and this basic function still remains even in the human organism.

Waking effects

It is essential for the maintenance of good health that the stresses of the day should alternate with the restorative powers of sleep (see Chapter 16).

The acoustic sense is the most effective awakener from sleep. When the eyelids are closed, optical stimuli are largely excluded whereas the sense of hearing is only slightly muffled during sleep. It still retains its primary function as an alarm system.

Experience shows that familiar sounds are less likely to awaken a sleeper than unfamiliar ones. People who live close to a railway are not awakened by passing trains whereas an anxious mother awakes if her child coughs or breathes strangely. Obviously the human brain can 'tune in' to certain sounds and react to these by awakening, while ignoring others, but has little defence against totally unpredictable sounds when neither their nature nor their timing can be foreseen. Such noises have a powerful waking effect.

Noise may either waken people completely or rouse them into a 'twilight' sleep. This is particularly likely if the noise is repeated. Since light sleep is not as restful as deep sleep, it is undesirable and so intermittent noises may be said to impair the quality of sleep even if the subject does not wake up completely.

Noise and sleep   The effects of noise on the duration and quality of sleep under experimental conditions has been studied in detail with the aid of the electroencephalograph. Studies up to now have shown that a noisy environment:

> seriously curtails the total time asleep
> cuts down the amount of deep sleep
> increases the time spent awake, or in light sleep
> increases the number of waking reactions
> prolongs the time to fall asleep.

Annoyance   Everyday experience teaches us that many noises have emotional effects on people. Sounds arouse feelings and sensations, and these are strongly subjective. They are reckoned among the psychological effects of noise. Not all noises or sounds are burdensome. Natural sounds, such as the rustling of leaves or the murmur of a stream, are pleasant. Certain noises (such as 'white' or 'pink' depending on their frequency spectra) may also be agreeable when they mask other sounds that are not pleasant. On the other hand there are many kinds of noise, or noisy situations, which people regard as being subjectively unpleasant and burdensome. The nature and extent of the burden depend on a number of subjective and objective factors, the most important of which are as follows:

1 The louder the noise, and the more high frequencies it contains, the more people are affected by it.

2 Unfamiliar and intermittent noises are more troublesome than familiar or continuous sounds.

3 A decisive factor is a person's previous experience of the particular noise involved. A noise which often disturbs one's sleep, which excites anxiety or which interferes with what one is doing is particularly burdensome.

4 A person's attitude to the source of the noise is often specially important. Motorcyclists, workers, children or musicians are not disturbed by the noise generated by their own activities whereas a bystander or someone not participating is disturbed to an extent which depends on how much they dislike either the sounds being produced, the situation in which they are generated or the person causing them.

5 The extent of the disturbance by noise often depends upon what the person affected is doing and what time of day it is. At home, a person is less disturbed by continual traffic noise and sounds coming from the neighbours all day long, than someone during a short midday break who is wanting to rest and relax. The rustling of papers is disturbing during a lecture whereas out in the street it would pass unnoticed.

The adverse, even angry reactions generated by noise are very important because they are so widespread. They must be regarded together with direct health threats as the decisive reasons for developing techniques for combating noise, and generating administrative and legal rules.

Becoming accustomed to noise

It is still not clear how far people can become accustomed to noise. Experience shows that there may be some degree of adaptation in certain circumstances, yet in others there is either no adaptation at all or even the converse, an increasing sensitivity to noise. These phenomena all depend on so many objective external circumstances, and so many internal subjective factors, that it is not possible to generalise.

Noise and health

The recuperative processes after a day's work that are essential to health take place during night sleep, pauses of all kinds, interruptions of work and a person's leisure time.

If the irritating effects of noise on the autonomic nervous system are not confined to working hours but extend into the hours of rest and sleep, this is then likely to upset the balance between stress and recuperation. Noise will then become a causative factor for chronic fatigue, with all its ill effects on well-being, efficiency and the incidence of ailments (see Chapter 11).

According to the definition of the World Health Organisation (WHO) 'Health is a state of physical and mental well-being'. Accepting this definition, we must reckon among the hazards to health not only noise-induced hearing loss, whether temporary or permanent, but also the noise-caused disturbances to sleep, delayed recuperation and the daily repetition of the psychological and emotional burdens of noise.

19.5 Protection against noise

Guidelines for office noise

Based on general experience, as well as on studies, the guidelines reported in Table 19.8 can be proposed for office noise. It is recommended that the noise in offices with more than 5 to 10 occupants should not be much higher than the given ranges – in fact, it should not be much lower either because a level of 'background noise' helps to mask sounds that by themselves could be bothersome. For offices with one or two people the recommended values can be considered to be the upper desirable limit but lower noise levels would certainly be more favourable. Some experience indicates that the desirable background noise level is naturally achieved in large offices of at least $1000 \, m^2$ containing more than 80 people.

The recommended range of equivalent noise levels L_{eq} of

Table 19.8 Guidelines for noise levels in large open offices. *The office sound levels should principally be neither lower nor higher than the given ranges*

Noise measurement	Desirable range in dB(A)
Equivalent noise level, L_{eq}	54–59
Mean noise level, L_{50}	
(approx. background noise)	50–55
Peak noise levels, L_1	60–65

54–59 dB(A) will to some extent mask conversations and telephone calls of others, while speech communication between two employees remains undisturbed. This was confirmed by the results of a survey by Cakir *et al.* (1983) who found the lowest prevalence of noise disturbances in computerised large open offices with a background noise between 48 and 55 dB(A). Below and above this range the incidence of complaints was clearly higher.

Noise in factories

Noise in factories and elsewhere can be countered in the following ways:

1 *Planning for 'no noise'.*

2 *Reduction of noise at source.*

3 *Interference with noise propagation.*

4 *Personal sound protection.*

Planning for 'no noise'

Not to generate noise is the one fundamentally successful way of battling noise. The most important step is planning to avoid noise. This is primarily the engineering task of selecting such machines and technologies that do not produce unacceptable sounds. This applies to offices, where noisy printers, fans, or keyboards should be eliminated from consideration in the planning stage. This principle of avoiding noise generation also applies to assembly shops where no hand tools should be used (such as many pneumatic cutters, mechanical rivetters or wire wrappers) that produce a racket. This approach applies also to manufacturing where machinery and processes should be selected that by design and nature are quiet instead of noisy: engines must be muffled, clanking dies eliminated, and hammering and rivetting replaced by joining and glueing. Of course, avoiding noise being generated applies to traffic, where mufflers on trucks, buses, cars and motorcycles is required, just as are noise-abating take-off techniques for airplanes from airport runways.

The most effective and rational way to deal with existing noise is to tackle it at its source. To *'engineer out' unnecessary sound* is an important engineering task. But certain jobs, and pieces of machinery, are noisy by nature because heavy, hard surfaces clash together. Sometimes this noise can be reduced by replacing the hard material with something softer, like rubber or felt. For the same reason transport vehicles are quieter with rubber tyres than with metal wheels. All forms of transmission should be examined to see if they are still in good condition since old and worn components create unnecessary noise. Belt drives, using rubber, leather or fabric belts, are quieter than a series of cog-wheels. Toothed belts engaging with toothed wheels are the quietest of all, especially if the toothed wheels are made of some synthetic material.

Tackling noise propagation

The next most important technological step in the battle against noise propagation lies in the choice of building materials and in planning the subdivisions of the building. *Noise protection should be high on the architect's and civil engineer's task list.*

The noise level decreases with the squared distance from its source so it is advantageous to locate offices and any other places where mental work is carried on as far away as possible from the source of noise, for example traffic or machinery. If the factory itself is noisy, the noisy sections also should be as far away as possible from places where work is carried on that calls for concentration and skill; intervening rooms, used for packing and storing, will act as 'buffers'.

When considering the division between two rooms, consideration must be made of the sound-dampening effect of walls, doors, windows and hatches. Table 19.9 gives some examples of these.

Noise radiated from vibrating plates can be reduced by stiffening them, loading them with weights, making them curved or using non-resonant materials. Moving machinery and motors in operation not only send out sound waves but also set the structure of the building in vibration. Such vibrations, and the secondary noises that they set up through resonances, can be disturbing throughout the whole building. For this reason very heavy machines should be rigidly set in concrete or iron mountings. If necessary they can be mounted in special concrete troughs with intervening layers of sound-insulating material. Such sound-insulating layers can be of spring steel, rubber, felt or cork, according to the weight of the offending machine.

Enclosing the source of noise

An effective way of reducing noise propagation is to enclose the source. A housing of suitable material may reduce the radiated noise by 20 or 30 dB. The inside wall of such a housing should be lined with sound-absorbing material, while the wall itself

Table 19.9 Sound-deadening effect of various building items

Item	Attenuating effect in dB(A)	Remarks
Normal single doors	21–29	Speech clearly understandable*
Normal double doors	30–39	Loud speech still understandable*
Heavy special doors	40–46	Loud speech still audible*
Window, single glazing	20–24	
Window, double glazing	24–28	
Double glazing with felt packing	30–34	
Dividing wall, 6–12 cm brick	37–42	
Dividing wall, 25–38 cm brick	50–55	
Double wall, 2 × 12 cm brick	60–65	

*On the other side of the doors

should be as heavy and airtight as possible. The housing should enclose the source of noise, with as few gaps as possible. Of course, some apertures are usually necessary for the passage of wiring or for operating the machine; such apertures impair the soundproofing of the machine and it can be stated as a general guideline that their total area should not exceed 10 per cent of the area of the housing. Doors or sliding panels may be provided to allow the machine to be controlled and serviced.

Sound absorption in a room

When all the possibilities of sound dampening at source and by enclosure have been exhausted, it is sometimes possible, in the right circumstances, to give further protection by covering the walls and ceilings with sound-absorbing material. *Sound-absorbing panels (acoustic tiles) absorb part of the sound and so reduce reflection back (echo effect) into the room.*

Acoustic tiles have been used mainly in offices of more than 50 m^2 in area. Sound reductions of 5 dB, or even up to 10 dB, can be expected in such places. On the other hand, the effects of installing acoustic tiles in noisy workshops, machine rooms and factories are not often successful. It must be remembered that when someone works close to a source of noise he or she is mainly affected by the sound radiated directly from the source, while reflected sound is of little importance. Covering the walls and ceiling with acoustic tiles will be of some benefit only if someone is working several metres from the source of the noise.

Personal hearing protectors

If planning and technical means have failed to reduce the noise to a safe level of below $L_{eq} = 85$ dB(A) and if someone must work in that noisy environment, then as a last resort there remains only personal hearing protection.

The following conventional passive protection devices are worth considering:

> *earplugs inserted into the ear canal*
> *ear canal caps to block the outer ear passages*
> *earmuffs covering the entire external ear*
> *soundproof helmets enclosing the head including the ears.*

Furthermore, 'active noise reduction' technology is now available, as discussed below.

Earplugs

A simple plug of cotton-wool or wax inserted deeply in each ear canal is an old device that was often used successfully; today many fabricated plugs of a synthetic material are on the market. *Properly used, earplugs can reduce the noise level inside the ear by up to 30 dB.* All types of earplugs have the drawback that they need to be inserted carefully and correctly, and that they must stay in that position if they are to give sufficient protection.

Protective caps (earmuffs)

Caps which enclose the whole of the external ear (earmuffs) give good protection and, if correctly fitted, can reduce noise levels by about 40 dB in the frequency range of 1 to 8 kHz, as Robinson and Casali (1995) demonstrated. The insulating pad must fit closely round the ear, on to the skull, so that it does not exert any pressure on the outer ear structures. Figure 19.9 shows the effects of using an earmuff. The graph shows that the noise of a circular saw, with frequencies between 500 and 8000 Hz, is reduced at the eardrum by about 40 dB.

Problems with hearing protectors

Unfortunately many workers object to any kind of hearing protectors. Their main ground for complaint is 'acoustic isolation'; the workers fear they might miss some vital information from the surroundings. This is usually not true in environments above 85 dB(A) and workers quickly learn to hear each other and the important sounds of the work environment. The wearing of ear protectors should be encouraged in the interests of the workers themselves.

A general guideline is to wear earplugs for noise levels 85–100 dB(A) and earmuffs (perhaps in combination with plugs) for noise levels above 100 dB(A).

However, below 80 dB(A), or if one is already hearing-impaired, communication abilities may be reduced with conventional hearing protectors (Casali and Berger, 1996). These

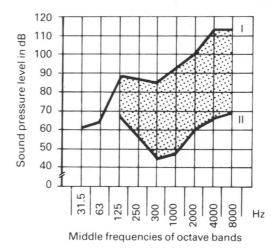

Figure 19.9 Protective effect of an earmuff in the vicinity of a circular saw cutting light metal. *(I) = noise spectrum at the unprotected ear. (II) = noise spectrum inside the earmuff.*

'passive' protectors cannot differentiate between frequencies or sound energies and hence do not improve the critical signal-to-noise ratio. In fact, they usually attenuate high frequencies more than low frequencies, which reduces phoneme sound power and thus impairs the perception of consonants which are especially important in speech.

'Active' hearing devices

New earplugs, caps, muffs or helmets can incorporate electronic components that sense sound in the ambient environment or sound that has penetrated the protector. To reduce such sound, an exactly opposite (180° phase shifted) sound of the same amplitude is generated by the device which cancels the sound arriving at the ear by destructive interference. Current technology works especially well in low-frequency noises below 120 dB, as Casali and Berger (1996) reported.

Furthermore, specific sound (such as of speech or warning signals) can be selectively amplified, thus boosting particularly important acoustic information. Another possibility is to play messages – or music – into the ears of a person wearing such a device, making it 'active' both in reducing unwanted noise and increasing wanted sound.

Medical tests

Noise protection in its widest sense includes medical (audiometric) surveillance measures that are required for noisy places of employment in many countries. The purpose of periodic audiometric tests is to detect noise damage at its onset, to supply a basis for devising protective measures against noise and for the introduction of personal protection into factories.

19.6 Vibrations

Oscillations of body parts'

Vibrations are oscillations of mass about a fixed point. In the human body they are produced by either regular or irregular periodic movements of a tool or vehicle, or other mechanism in contact with a human, which displaces the body from its resting position. Sound is a vibration that affects our hearing cells. 'Mechanical vibrations' are of concern since they cause changes in the position of body limbs and important organs.

If the human body were on a rigid bulk of mass in translation all parts would undergo the same motion in rotation and different body parts would move at different angular displacements. However, the body is not rigid and hence different body segments oscillate in different ways. The nature of vibrations and their effects on people have been written about in detail by Dupuis (1974), Griffin (1990), Guignard (1985), Kroemer *et al.* (1994), Kuorinka and Forcier (1995), Putz-Anderson (1988) and Wasserman (1987). The International Standards Organisation has tried to put the information into ISO Standards 2631 and 7962.

A little physics

The following seven physical facts are important for the understanding of what follows:

1 *Point of application to the body.* Three points at which vibrations enter the body are significant ergonomically: the buttocks and possibly the feet (when driving or riding in a vehicle) and the hands (when operating vibrating tools or a machine).

2 *Direction of application.* The direction of oscillation is important. For the main body the direction mostly lies in the vertical plane (head to foot). For the hand and arm, it is often approximately perpendicular to the line through the hand and arm.

3 *Frequency of oscillations.* The extent of the biomechanical and often pathological effects of vibrations is strongly dependent on the frequency. Particularly important frequencies are those which fall into the range of natural frequencies of the body and so cause resonance (see 6). Often a low and high range of frequencies are distinguished. Vibrations of motor vehicles belong to the low range, those of motor-driven tools to the high range of frequencies.

4 *Acceleration of oscillations.* Within the frequency range that is physiologically important, the acceleration of the oscillations is usually taken as a measure of the vibrational load. A commonly used reference unit is the acceleration due to gravity $g = 9.8 \text{ m/s}^2$.

5 *Duration of effect.* The effect of vibrations depends greatly on their duration. Their ill effects increase very rapidly as time goes on.

6 *Resonance*. Every mechanical system which possesses the elementary properties of mass and elasticity is capable of being set in oscillation. Each system has its own natural frequency at which it vibrates after stimulation. The nearer the frequency of the inducing force comes to this, the greater will be the amplitude of the forced vibrations. When the amplitude of the forced vibrations exceeds that of the inducing force the system is said to be in resonance.

7 *Damping*. The oscillations of any system are subject to damping, which reduces their amplitude. Thus, for example, when we are standing up any vertical vibrations transmitted at the feet are quickly dampened in the legs. Frequencies above 30 Hz are particularly well dampened by the tissues of the body; thus for an inducing frequency of 35 Hz the amplitude of oscillation is reduced to $\frac{1}{2}$ in the hands, to $\frac{1}{3}$ in the elbows and to $\frac{1}{10}$ in the shoulders.

Oscillatory characteristics of the human body

Above frequencies of about 2 Hz the human body does not vibrate as a single mass with one natural frequency; rather it reacts to induced vibration as a set of linked masses. Studies have shown that the natural frequencies are different in different parts. For example, the body of a sitting person reacts to vertical vibrations as follows:

3–4 Hz:	strong resonance in the cervical vertebrae
3–6 Hz:	resonance on the stomach
4 Hz:	peak of resonance in the lumbar vertebrae
4–5 Hz:	resonance in the hands (difficult to make aiming motions)
4–6 Hz:	resonance in the heart
5 Hz:	very strong resonance in the shoulder girdle (up to twofold increase in displacement)
5–20 Hz:	resonance in the larynx (voice changes)
5–30 Hz:	resonance in the head
10–18 Hz:	resonance in the bladder (urge to urinate)
20–70 Hz:	resonance in the eyeballs (difficult to see)
100–200 Hz:	resonance in the lower jaw

In recent decades many studies have been carried out on the effects of linear and rotational vibrations induced in different directions and at various points of application, with many kinds of forces, displacements and accelerations. With the human body being such a complex biomechanical system, the results are diverse and complicated – Griffin (1990) has described this in great detail and with a good measure of humour. Here it can only be said that the natural frequencies of the smaller components of the body, such as groups of muscles, eyes and so on, lie in the higher frequency ranges. Hence the operation of machines with frequencies above 30 Hz are likely to set up resonances in the fingers, hands and arms. On the other hand the damping effect of bodily tissues is greater for these higher frequencies and tends to confine them to the vicinity of the point of application.

Vibrations in the workplace

Up to the present, vibration experienced at work has been measured mainly in construction machinery, tractors, trucks and cars. The studies on various motor vehicles revealed that *the acceleration of vertical oscillations lies between 0.5 and 5 m/s²*, with the highest values being recorded for earth-moving machines and tractors. Operating motor-powered tools involves high levels of vibration in the hands and wrists, some examples of which are summarised in Table 19.10.

Physiological effects

Vibration seriously affects visual perception and psychomotor performance and the musculature, with circulatory, respiratory and nervous systems also reacting, often to a lesser extent.

Vibration seems to generate muscle reflexes which have a protective function, causing the extended muscle to shorten. The reflex activity of the muscles also explains the often-observed increase in energy consumption, heart rate and respiratory rate when a person is exposed to strong vibrations. These vibrational effects on metabolism, circulation and respiration are small and have little significance. However, reflex closure of sphincter muscles around blood vessels can can severely reduce blood circulation in the afflicted body segment, as discussed below under "dead fingers".

Visual powers

The adverse effect of vibration on eyesight is most important because it impairs the efficiency of drivers of tractors, trucks, constructional machines and other vehicles, and increases the risk of accidents. Visual acuity is poorer and the image in the visual field becomes blurred and unsteady.

Visual powers are not affected by vibrations of less than 2 Hz. Measurable optical aberrations appear from 4 Hz upwards and are greatest in the range 10–30 Hz. With a vibration of 50 Hz and an oscillatory acceleration of 2 m/s², visual acuity is reduced by half, according to Guignard (1985).

Skill

Strong vibration impairs performance in various psychomotor tests. Simplifying somewhat, we may say that *vibration can impair visual perception, the mental processing of information and the performance of skilled motor tasks*.

Driving tests

The following psychophysiological effects of vibration are particularly evident in all kinds of simulated driving tests:

Table 19.10 Vibration in motorised hand tools. *Effective accelera-tion = root mean square of the accelerations at various amplitudes. These figures apply to oscillations along the direction of the arm; oscillations perpendicular to this are often higher*

Type of tool	Effective acceleration Ground (m/s^2)		
	On fingers	On wrist	
Power saw	17.5	1.1	120
Soil borer	21.0	3.5	110
Pneumatic compass saw	—	9.9	—
Two-wheeled cultivator	3	2.8	82

Source: After Dupuis (1974).

1 Over the range 2–16 Hz (especially around 4 Hz) driving efficiency is impaired and the effects increase with increasing acceleration of the oscillations.

2 Driving errors increase when the seat is subjected to accelera-tions of the order of 0.5 m/s^2.

3 When the accelerations reach 2.5 m/s^2, the number of errors becomes so great that such vibrations must be rated as positively dangerous.

This consensus of physiological effects of vibration point to the conclusion that *mechanical oscillations can reduce efficiency and in many situations may lead to the risk of errors and accidents.*

Vibration as a nuisance

Vibration is subjectively felt as an imposition and a burden, impressions ranging from a minor annoyance to an unbearable nuisance. The extent of nuisance depends in the first instance on the inducing frequency, the rate of acceleration of the oscillations and on the length of time they continue. The source of the nuisance lies in the physiological effects and in the resonances set up in various parts of the body. Figure 19.10 shows the results of an investigation by Chaney (1964) on seated test subjects; curves are given for subjective feelings of equal intensity in relation to frequency and acceleration of oscillation.

From these and many other results the following conclusions may be drawn for vertical oscillations applied to seated people:

1 *The most intense subjective sensitivity lies in the frequency range 4–8 Hz.*

2 *The average threshold of 'very severe' intensity comes at an acceleration of 1 g (about 10 m/s^2).*

3 *At accelerations of 1.5 g (about 15 m/s^2) the vibrations became dangerous and intolerable.*

Similar tests on standing subjects showed that they felt equal subjective intensity at higher vibration loads to before, because of the damping effects in the legs, to which we referred earlier. The threshold of 'very severe' intensity came 0.2–0.3 g higher than for seated subjects.

Health complaints

The complaints about effects on well-being and health suffered in addition to the annoyance of vibration vary greatly, but some of them are frequency-dependent. The following are the most common complaints:

1 Interference with breathing, especially severe at vibrations of 1–4 Hz.
2 Pains in the chest and abdomen, muscular reactions, rattling of the jaws and severe discomfort, chiefly at 4–10 Hz.
3 Backache, particularly at 8–12 Hz.
4 Muscular tension, headaches, eyestrain, pains in the throat, disturbance of speech, irritation in intestines and bladder, at frequencies of 10–20 Hz.

In addition, we may mention sea and travel sickness, with nausea and vomiting, brought about by low oscillations of 0.2–0.7 Hz, with the greatest effect at 0.3 Hz.

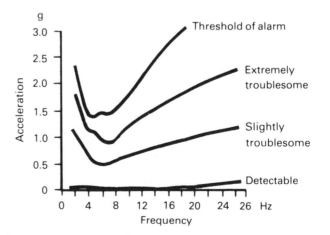

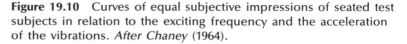

Figure 19.10 Curves of equal subjective impressions of seated test subjects in relation to the exciting frequency and the acceleration of the vibrations. *After Chaney* (1964).

Damage to health

Exposure to vibration at one's place of work can, if repeated daily, lead to morbid changes in the organs affected. The effects are different in the parts of the body most commonly subjected to vibration. Vertical oscillations experienced when either standing or sitting caused by vibration from underneath (e.g. in a vehicle) can cause degenerative changes in the spine whereas the vibrations of power tools and repeated manipulations affect mainly the hands and arms.

Spinal ailments

Tractor drivers in various countries have been recorded as suffering an accumulation of disc troubles and arthritic complaints in the spine, as well as an above-average incidence of intestinal ailments, prostate troubles and haemorrhoids.

The cumulative appearance of spinal damage among workers who have been subjected to a high level of vertical oscillation leads one to suppose that heavy and prolonged vibration causes excessive wear and tear on the intervertebral discs and the joints.

Hand and arm troubles

Workers who use power tools for years on end (e.g. chain-saws or pneumatic hammers) may suffer various ailments of the hands and arms, with the frequency of vibration a decisive factor.

Arthritis

Tools with a frequency vibration below 40 Hz, for example a heavy pneumatic hammer, can cause degenerative symptoms in the bones, joints and tendons of the hands and arms, leading to arthritis in the wrist, elbow and occasionally in the shoulder.

Atrophy

The effects on the bones may lead to atrophy, which in rare cases may involve so much loss of calcium that the risk of fracture is substantially increased. In a few countries these possible consequences of using a pneumatic hammer are classified as an occupational disease.

'Dead fingers'

Power tools with frequencies between 40 and 300 Hz usually have a very small amplitude of oscillation (0.2–5 mm) and their vibrations are quickly dampened in the tissues. Such vibrations may have ill-effects on the blood vessels and nerves of the hands, resulting in one or more of the fingers going 'dead'. Usually the middle finger is most affected, becoming white or bluish, cold and numb. After a little while the finger turns pink again and is painful. The cause is a cramp-like condition of the blood vessels, known as Raynaud's disease.

These symptoms have also been observed among miners using a pneumatic drill with a higher frequency of vibration as well as among forestry workers using power saws with frequencies between 50 and 200 Hz.

'Dead fingers' usually make their appearance, at the earliest, six months after the beginning of the work using the vibrating tool and cold is an important factor in the onset of this condition. Raynaud's disease is more common in northern countries than in warmer latitudes. It must be assumed that cold makes the blood vessels more sensitive to vibration and more liable to vascular cramps.

Users of power tools with even higher frequencies have also been known to suffer from ailments affecting the circulation and loss of sensation. One example of such tools is polishing machines operating at 300–1000 Hz, which cause painful swellings and loss of sensation in the hands. These symptoms often do not disappear when the work is finished.

Repetitive strain injuries (RSI)

Since the 1930s typists, then operators of keyboards on card punches, check-out cash registers and, since the 1970s, computer operators, have been complaining about ailments in the hands and wrists that are similar to those induced by vibrating hand tools. The frequent operation of keys may act like vibrations induced at the fingertips, with similar biomechanical effects. Kroemer (1989) and Kroemer *et al.* (1994), Kuorinka and Forcier (1995) and Putz-Anderson (1988) discussed these cumulative trauma in much detail.

Limiting values

Attention has been given in many countries to setting limits for human exposure to vibrations. In general it may be concluded that vibration becomes intolerable in the following circumstances:

Below 2 Hz	at accelerations of 3–4 g
Between 4 and 14 Hz	at accelerations of 1.2–3.2 g
Above 14 Hz	at accelerations of 5–9 g.

ISO Standards 2631 and 7962 are attempts at providing standarised information. Their guidelines for the assessment of vertical oscillations are shown in simplified form in Figure 19.11. The ISO Standard distinguishes between three criteria and envisages three levels of limiting values:

1 *The criterion of comfort* (reduced comfort boundary). This applies mostly to vehicles and to the automobile industry.

2 *The criterion of the maintenance of efficiency* (fatigue-decreased proficiency boundary). A decisive factor in this criterion is working efficiency (proficiency). It applies to tractors, construction machinery and heavy vehicles. These limiting values are shown in Figure 19.11.

3 *The criterion of safety* (exposure limits). Protection against damage to health is the criterion here.

Starting from the 'efficiency' criterion (Figure 19.11), the other two can be derived as follows:

The criterion of comfort can be derived by dividing the acceleration by 3.15.

The criterion of safety can be derived by multiplying the acceleration by 2.

Recommendations

Vehicles

From the ergonomics viewpoint, tractors, heavy vehicles and construction machinery, with frequencies most often between 2 and 5 Hz and operating for an 8-h day, require a limit of oscillational acceleration of 0.3–0.45 m/s^2. These limits are often exceeded but technically they can be achieved by jointly engineering the suspension of the vehicle's axles and the driver's and passengers' seats.

Hand tools

The same applies in principle to hand-operated tools. Damping elements within the tool itself, and even more between the tool and the hand-grip, can reduce vibration considerably. The grip itself can be dampened by making it of flexible material. Further improvement can be obtained by wearing thick gloves and avoiding working in very cold conditions.

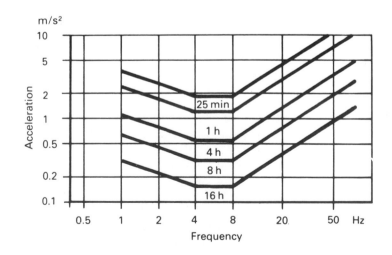

Figure 19.11 Proficiency-based limitations for vertical vibrations. *'Fatigue-decreased proficiency boundary' for exposure times ranging from 25 min to 16 h. For the calculation of the threshold values 'safety' and 'comfort', see text. Note exponential scales. After ISO 2631 (1974).*

*'Vibration
erasure'*

With high-speed sensors, computers and vibrators available, some high frequency but low amplitude/low mass vibrations can be eliminated by counteracting them immediately by the exact opposite vibration, similar to "active noise cancellation" mentioned earlier. This virtually 'erases' the effects of the original vibration.

Summary

Mechanical and acoustic vibrations follow the same physical laws but we either feel or hear them. Mechanical vibrations are mostly a nuisance or even a health threat; noise can have the same effects, but suitable sound is a source of information, even of pleasure.

CHAPTER TWENTY

Indoor climate

Climate components

The term 'climate' applies to physical conditions of the environment in which we live and carry out work. Its principal components are:

air temperature
temperature of surrounding surfaces
air humidity
air movement
air quality.

Measurement techniques of these physical variables have been described in detail by Mairiaux and Malchaire (1995) and Olesen and Madsen (1995).

20.1 Thermoregulation in humans

Body temperature

The temperature of the human body is not, as often assumed, uniform throughout. A constant temperature, which fluctuates a little around 37 °C, is found only in the interior of the brain, heart and abdominal organs. This is the so-called core temperature. A constant core temperature is a prerequisite for the normal functioning of the most important vital functions and wide or prolonged deviations are incompatible with the life of a warm-blooded animal. In contrast, the temperatures in the muscles, limbs and above all in the skin show great variations. This is the so-called shell temperature. It varies according to the body's need to either conserve heat or to dissipate it.

Cool skin in the cold

The underlying rule is that energy always flows from the warmer to the colder. When the surrounding air is cool and if the body wants to prevent heat outflow, blood circulation to the skin is much reduced, making the skin look pale and lowering its temperature so there is little heat loss.

Hot skin in the heat

If the body must lose heat, then the skin is well supplied with warm blood to have a steep temperature gradient to the outside. In warm surroundings the task is usually to dissipate heat; thus, the skin will be kept as hot as possible by bringing 'hot' blood to just below its surface. This helps especially to evaporate sweat, the body's most important means to dissipate heat.

This capacity for controlling the heat exchange with the variable environment by skin tempearature adaptation allows the human body to tolerate a temporary heat deficit which may be in the magnitude of 1000 kJ for the whole body. The muscles too, show considerable fluctuations in temperature, being several degrees warmer during strenuous effort than when they are at rest.

Control processes

The control mechanisms throughout the body, which are necessary to maintain a constant core temperature, are shown diagrammatically in Figure 20.1. The heat control centre is located in the brainstem. It directs the thermal regulation of the body, comparable to a room thermostat.

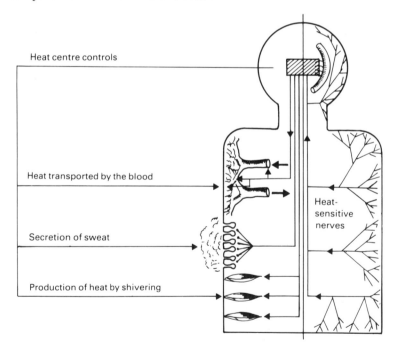

Heat centre controls

Heat transported by the blood

Heat-sensitive nerves

Secretion of sweat

Production of heat by shivering

Figure 20.1 Diagram of the physiological control of the heat balance of the body. *The heat centre located in the brain stem regulates the flow of blood through the capillaries of the skin, as well as the secretion of sweat. These two mechanisms, between them, adjust the heat balance of the body according to external and internal conditions.*

The nerve cells of the heat control centre receive information about temperatures throughout the body. The control centre in turn sends out the impulses that are necessary to direct and control the regulatory mechanisms, mostly of the circulatory system, which keeps the core temperature constant. In this way the heat production of the body, its diffusion by the circulatory system and the heat loss by secretion of sweat in the skin are controlled, thus enabling the process of thermoregulation to be carried on.

Heat transport by the blood

The most important item in thermoregulation is the heat transport function of the blood. The blood picks up heat, especially at the capillaries, from warm tissues and dissipates it to cooler tissues. In this way the blood can transport heat from the interior of the body to areas of skin that are cooled by the colder environment; conversely, if warmth from the outside is needed, heat can be transported from the skin into the interior, for example when we want to be warmed by the sun or a fire. *The key to the control of body temperature is blood circulation, particularly to and from the skin.*

Secretion of sweat

The second regulatory mechanism directed by the heat control centre is the secretion of sweat onto the skin. This, too, is under nervous control. Sweating is a very efficient way to lose body energy to the outside because about 2.4 kJ are needed to evaporate 1 cm^3, all taken from the body's reservoir of excessive heat.

Shivering

The third regulatory mechanism is to raise the rate at which heat is produced by the body, a process that is set in motion whenever the body is subjected to excessive cooling. This occurs by an increase in metabolic heat production, especially by activities of the muscles. We all know the rapid muscular movements called 'shivering'.

Heat exchange

As explained in Chapter 1, the body converts chemical energy into mechanical energy and heat. The body uses this heat to maintain a constant core temperature and dissipates any excess heat into its surroundings.

There is thus a constant exchange of heat between the body and its surroundings which is regulated by physiological control following the ordinary laws of physics. The latter involve four different processes:

1 Conduction

2 Convection

3 Evaporation

4 Radiation.

**Conduction of
heat**

Heat exchange by conduction depends first and foremost on the conductivity of objects and materials in contact with the skin. Anyone who touches different material in the same room can discover this. For example, first grasp a piece of metal, then one of wood. The metal feels colder because it conducts heat away from the body; the wood feels warmer because its conductivity is less. Conductivity of heat is of practical importance in the choice of floorings, furniture and the parts of machinery that have to be touched (control handles and so on). Hence, to avoid loosing body heat, *workplaces should have well-insulated floors (e.g. cork, linoleum or wood), while benchtops, machine parts (particularly handles and controls) and tools should be protected with felt, leather, wood or other material of low thermal conductivity.*

Convection

If the heat exchange is with gas (air) or with fluid (water), we call the process convection instead of conduction. Heat exchange by convection depends primarily on the difference in temperature between the skin and the surrounding air, and on the extent of air movement. Under normal circumstances convection accounts for about 30 per cent of the total heat exchange of the body.

**Evaporation of
sweat**

Loss of heat by sweating occurs continually because some sweat on the skin always evaporates, consuming heat. As discussed above, this evaporation is especially important, and clearly felt, in a hot environment. Under normal conditions each person evaporates about one litre per day (insensible perspiration) and dissipates about one-quarter of the total daily loss of heat.

If, however, the temperature of the surroundings exceeds the limits of comfort, then the hot skin begins reflex sweating, with a sharp increase in the rate of loss of heat.

The extent of heat loss by evaporation depends on the area of skin where sweat can evaporate and on the difference in water vapour pressure between the air next to the skin and that further away. Thus, *the relative humidity of the surrounding air is an important factor, as is air movement.* Air movement replaces humid (and heated) air layers from the skin and thus increases the gradient of water vapour pressure and of temperature, and cools the skin by convection.

Radiation of heat

Warm bodies radiate electromagnetic waves of relatively long wavelengths, which are absorbed by other bodies (objects and surfaces) and converted into heat. This is called infrared radiation or radiant heat. It does not depend on any material medium for its transmission, in contrast to conducted or converted heat. Such heat exchange by radiation takes place between the human body and its surroundings (walls, inanimate objects, other people) in both directions all the time. In contrast to conduction or convection, heat radiation is hardly affected by the temperature, humidity or movement of the air; it depends mainly on the temperature difference between the skin and adjacent surfaces. In temperate countries the surrounding objects and surfaces are usually cooler than the human skin and so the human body loses a considerable amount of radiant heat in the course of a day. Conversely, much heat can be gained by radiation from the sun, a melting furnace or an open fire.

Loss of heat by radiation is not noticeable as long as the amount is not excessive but it may become uncomfortable when someone is standing close to a cold wall or a large window, even though the air temperature is high enough. In such circumstances the loss of heat may be considerable because the decisive factor is not the air temperature but the temperature difference between the skin and the adjacent cold surface.

The amount of radiant heat loss in a day by a fully clothed person varies greatly, according to circumstances. Excluding high summer, the average daily loss in temperate climates is 40 to 60 per cent of the total heat dissipated from the body.

Overall heat balance

Figure 20.2 shows three of the principal ways in which the human body exchanges heat with its surroundings.

To summarise, it can be said that the following four physical factors are decisive:

1 Air temperature for exchange of heat by convection.

2 Air movement for convection.

3 Temperatures of adjacent surfaces (walls, ceiling, floor, machinery) for exchange of heat by radiation.

4 Relative humidity of the air for the loss of heat by evaporation of sweat.

At the environmental temperatures of walls or of the air above 25 °C the clothed human body is hardly able to lose heat by either convection or radiation, and sweating is the only compensatory mechanism left. Hence the dissipation of heat by the evaporation of sweat rises steeply the hotter the environment.

20.2 Comfort

Physiological basis of comfort

One hardly notices the internal climate of a room as long as it is 'normal' but the more it deviates from a comfortable standard, the more it attracts attention.

The sensation of discomfort can increase from mere annoyance to pain, according to the extent to which the heat balance is disturbed. Discomfort is a practical biological device in all

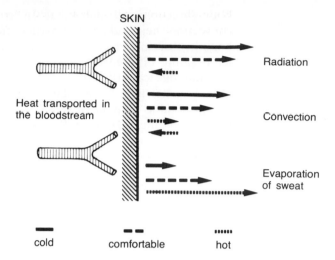

Figure 20.2   Diagram of heat exchange between the human body and its surroundings. *The length of the arrows gives a rough indication of the heat transferred by each of the three processes under different conditions.*

warm-blooded animals, which stimulates them to take the necessary steps to restore a correct heat balance. An animal can only react by seeking out another place that is neither too hot nor too cold, and it can either decrease or increase its metabolism by changing its muscle activity level, but man has the use of clothing as well as being able to modify the environment by technological means.

Side-effects of discomfort

Discomfort brings about functional changes that may affect the entire body. Overheating leads to weariness and sleepiness, reduced physical performance and increased liability to errors. Decreased activity causes the body to produce less heat internally.

Conversely, over-cooling induces restlessness, which in turn reduces alertness and concentration, particularly on mental tasks. In this case stimulation of the body into greater activity causes it to produce more internal heat.

Thus the maintenance of a comfortable climate indoors is essential for well-being and performance at maximal efficiency.

Temperature zones in physiological terms

If a test subject is placed in a climatic chamber and exposed to different temperatures, a range can be found in which the heat exchanges of his or her body are in a state of balance. This is called the *zone of vasomotor regulation*, or the *comfort zone*, because within this range the heat balance is maintained chiefly by regulating the flow of blood between parts of the body. For a clothed and resting person in winter this zone lies between 20 and 23 °C.

If there is a slight excess of (internal or external) heat above this comfort level, this is dealt with by warming the peripheral parts of the body and by increasing perspiration. This is the *zone of evaporative control*. If, however, the heat continues to increase and exceeds a certain level, *the limit of tolerance*, the core temperature rises quickly and steeply, and in quite a short time can lead to death by heat stroke.

Temperatures below the zone of vasomotor regulation are characterised by a negative heat balance for the body since more heat is being lost than is being generated internally. This is the *zone of body cooling*. At first the cooling is confined to the peripheral parts of the body, which can tolerate a heat deficit for a time but extreme hypothermia leads to death.

The heat balance in these three zones is shown diagrammatically in Figure 20.3.

Comfortable ranges of temperature

If test subjects are asked to say when they feel really comfortable, the range is a comparatively narrow one, covering only 2 or 3 °C. Obviously people feel comfortable only when the vasomotor regulation system is not heavily stressed, that is when the blood circulation to the skin is subject to no more than normal fluctuations. On the other hand, either a negative or a positive heat balance (i.e. either a deficit or an accumulation of heat in the body shell) is felt as discomfort.

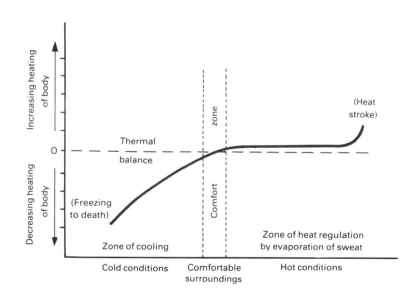

Figure 20.3 Heat balance of the body between exposure to extremes of heat and cold.

The temperature range within which a person feels comfortable is very variable. It depends firstly on the amount of clothing being worn and then on how much physical effort is being performed. Other factors such as food, time of year, time of day, body size, age, gender and habits are important as well.

Four climatic factors and comfort

People's impressions of comfort are influenced by the same four climatic factors that determine the heat exchange, so we may repeat them here:

> *air temperature*
> *the temperatures of adjacent surfaces*
> *air humidity*
> *air movements.*

Each factor contributes to a climate 'balance' and researchers have tried to find an 'objective' unit of measurement that would consider them all. An example is the so-called 'wind-chill temperature' which attempts to use the rate of cooling of naked body surfaces as an index of discomfort but this approach works only in certain conditions of a cold environment. Thus, researchers have come to rely more and more on the subjective impressions of test subjects as a measure of the degree of comfort or discomfort within a thermal environment, resulting in the concept of *effective temperatures*.

Effective temperatures

Houghten and Yaglou (1923) were the first to investigate combinations of temperature and humidity of the air that would result in the same effective temperature. The test subjects were placed in a climatic chamber in still air, at a given temperature, and with 100 per cent relative humidity. They were required to note their impressions of the temperature and to remember them. The relative humidity was then reduced and the temperature varied until the test subjects felt the same sensation of warmth as before. This is the perceived *effective temperature*. These investigations were followed by others in which the effects of temperatures of adjacent surfaces and air movements on the effective temperature were analysed. The investigations led to the formulation of 'indices of comfort' and 'zones of comfort', which could be read off from nomograms in relation to three or four of the climatic factors listed above. We may refer to the experiments of Fanger (1972) who compiled an equation of comfort from ratios among the four climatic factors, the amount of clothing and the extent of physical activity. As expected, this equation is very complex. Fanger set out the most important results in 28 diagrams, showing curves of degree of comfort in relation to several factors. An account of these would be beyond the scope of this chapter but Eissing compiled the most striking finding in 1995. We must confine ourselves to considering a few relationships between the climatic factors already mentioned which are important in evaluating the climate indoors.

What is the effect on *effective temperatures* of various combinations of air temperature and temperature of surrounding surfaces, air temperature and relative humidity, and air temperature and air movement?

Temperature of air and of adjacent surfaces

Physiological research has shown that the effective (perceived) temperature is essentially the average between the air and the adjoining surfaces. Expressed as a formula:

$$\text{Effective temperature} = \frac{T_A + T_S}{2}$$

where T_A = mean air temperature and T_s = mean temperature of adjacent surfaces.

It is important for comfort that the difference between T_A and T_s is small. Large areas of cold walls or windows are particularly uncomfortable, even if the air temperature is adequate. *It is a good rule of thumb that the mean temperature of adjacent areas should not differ from that of the air by more than 2 or 3 °C, either up or down.*

Temperature and humidity of the air

The effect of atmospheric humidity was prominent in early research by Houghten and Yaglou (1923); more recent work by Koch *et al.* (1960) and Nevins *et al.* (1966) has shown that after a prolonged stay in the same room, the impression of temperature is little affected by the humidity of the air. The following combinations of relative humidity (RH, in per cent) and air temperature produce equal effective temperatures:

70 per cent RH and 20 °C
50 per cent RH and 20.5 °C
30 per cent RH and 21 °C

Within the range 30 to 70 per cent, relative humidity has little influence on effective temperature. It can be assumed that at 18 to 24 °C the relative humidity can fluctuate between 30 and 70 per cent without creating thermal discomfort. The threshold at which the room begins to feel stuffy lies between the following pairs of values:

80 per cent RH and 18 °C
60 per cent RH and 24 °C

If the relative humidity falls below 30 per cent the air becomes too dry; we shall discuss this briefly later.

Table 20.1   Combinations of air movement and temperature readings that result in the effective temperature of 20°. *The two columns show dry-bulb and wet-bulb temperatures. The wet-bulb temperature is the reading of a thermometer with its bulb covered with a wick dipping into water. Its reading is lower than air temperature because of increased evaporation*

Air movement (m/s)	Dry bulb (°C)	Wet bulb (°C:100% RH)
0	20.0	20.3
0.5	21.0	21.3
1.0	22.0	22.2
1.5	22.8	23.0
2.0	23.5	23.8

Source: After Yaglou *et al.* (1936).

Temperature and movement of the air

The third factor to influence effective temperature is air movement. Yaglou *et al.* (1936) carried out experiments to determine pairings of air temperature and movement of the air which combine to give the same effective temperature of 20 °C. The results are set out in Table 20.1.

Fanger (1972) has shown that air movements in excess of 0.5 m/s are unpleasant even when the air is warm and that the discomfort depends on the direction in which the air is flowing and on the parts of the body exposed to it:

1 *Air currents from behind are more unpleasant than those meeting us frontally.*

2 *Neck and feet are particularly sensitive to draughts.*

3 *A cool draught is more unpleasant than a warm current.*

From general experience it can be said that seated people find air flows of more than 0.2 m/s unpleasant. Occasionally, when very precise work is being carried out, which keeps the operator motionless for long periods, even draughts of 0.1 m/s can be unpleasant. In contrast, standing work, especially if it involves strenuous physical effort, can be carried out without ill-effects in air flows of up to 0.5 m/s.

20.3 Dryness of the air

In temperate climates dryness of the air is mostly a winter problem, largely generated by heating, both at work and at home. For many years the tendency has been to prefer higher and higher temperature indoors during the period of the year when buildings are heated and this has led to lower values of relative humidity.

Grandjean's research team carried out random tests in a number of offices, both in winter and in summer, to measure the relative humidity of the air. The results, reported in 1988, are shown in Figure 20.4.

Medical effects

Most ear, nose and throat specialists think that the present-day tendency towards very dry air in heated rooms causes an increased incidence of catarrhal ailments and chronic irritation of the nasal and bronchial passages. They often note a desiccation of the mucous membranes of the air ducts, which they think obstructs the flow of mucus over the ciliary tracts, resulting in a diminished resistance to infection. These views of the effects of dry air are shown diagrammatically in Figure 20.5.

These observations may be summarised by saying that relative humidities of 40–50 per cent in heated rooms are desirable for comfort and below 30 per cent become unhygienic because they adversely affect the mucous membranes of the nose and throat.

It may be mentioned here that most of the humidifiers sold commercially are inadequate for their purpose. As a rough guide, a workroom of $100 \, m^3$ volume requires a minimum water output of 1 litre/h.

20.4 Field studies on indoor climate

We have already seen that there is considerable individual variation in what people consider to be a comfortable temperature

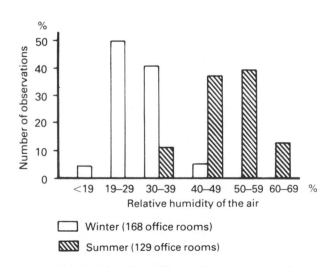

Figure 20.4 Relative humidity of the air in summer and in winter. *Winter samples taken in 168 offices without air conditioning. Summer samples taken in 60 offices with and 69 without air conditioning.* After Grandjean (1988).

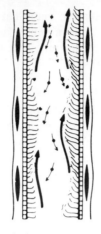
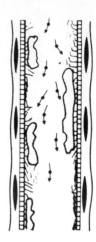

Normal air passage   Desiccated air passage

Dust particles adhere to the   Mucus has coagulated into
mucus and are carried away   clumps to which dust particles
with it by the ciliated epithelium   do not adhere

Figure 20.5 Diagram showing the effects of dry air on the
self-cleaning capacity of the mucous membranes in the throat.
*Left: normal mucous membrane with a ciliated epithelium which
rids the membrane of any contaminating particles. Right: the
ciliated epithelium is desiccated and parts of it are no longer
visible – the mucus forms into clumps and does not get rid of the
dust particles.*

and field studies emphasise this. Figure 20.6 shows the results of
McConnell and Spiegelman (1940) who recorded the views of 745
office workers on temperatures in New York.

The most striking result was the wide range of individual
responses. At 24 °C only 65 per cent of those questioned found
this temperature comfortable, the others finding it either too hot
or too cold. Grandjean (1973) carried out a study in the winter
of 1964/5 when he took measurements in 168 offices without
air-conditioning and at the same time questioned 410 employees
(140 men and 270 women) about their impressions of temperature.
The results are shown graphically in Figure 20.7.

Air temperatures were high, varying between 22 °C and 24 °C
most of the time. The reply 'comfortable' became less frequent
as the temperature rose, while 'too warm' took its place. Yet there
were many replies of 'comfortable' when the temperature was only
21 °C.

Similar studies by Grandjean et al. were carried out in the
summers of 1966 and 1967. This survey comprised 311 offices, 122
of them air-conditioned and 189 not air-conditioned. A total of
1191 employees were questioned and the results are summarised
in Figure 20.8.

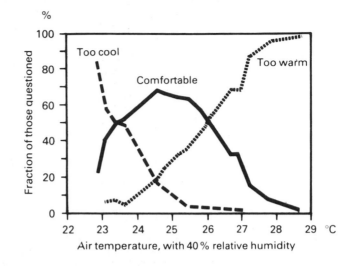

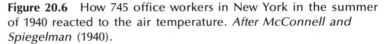

Figure 20.6 How 745 office workers in New York in the summer of 1940 reacted to the air temperature. *After McConnell and Spiegelman* (1940).

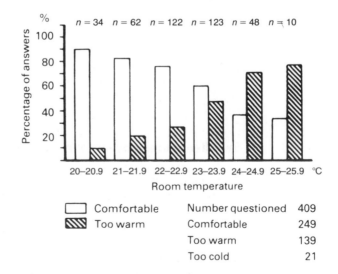

Figure 20.7 Reactions of 409 office workers to temperature in the winter 1964/5. *n = number of people questioned and also of temperature records. The question asked was: 'Do you find the indoor climate in the office today comfortable, too warm or too cold?' After Grandjean* (1973).

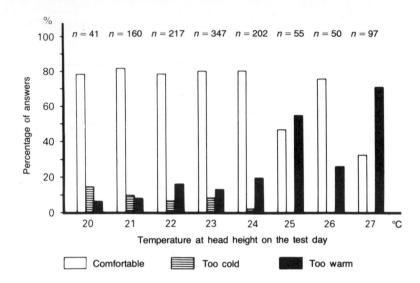

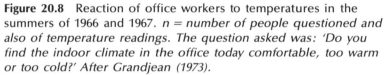

Figure 20.8 Reaction of office workers to temperatures in the summers of 1966 and 1967. *n = number of people questioned and also of temperature readings. The question asked was: 'Do you find the indoor climate in the office today comfortable, too warm or too cold?' After Grandjean (1973).*

In the offices without air conditioning, the air temperatures mostly ranged between 20 °C and 27 °C; in the air-conditioned offices the upper limit of temperature was usually 24 °C. As can be seen from the diagram, temperatures above 24 °C were assessed as 'too warm' by most people, an observation similar to the one made in the survey of McConnell and Spiegelman (1940) in air-conditioned offices in the USA. So we may conclude that *in summer room temperatures are comfortable as long as they do not exceed 24 °C.*

20.5 Recommendations for comfort indoors

Sedentary office work

The following guidelines may be applied to sedentary work which involves little or no manual effort:

1 *The air temperature* in winter should be between 20 and 21 °C and in summer between 20 and 24 °C.

2 *Surface temperatures of adjacent objects* should be at roughly the same temperature as the air – not more than 2 or 3 °C different. No single surface (e.g. the outside wall of the room) should be more than 4 °C colder than the air in the room.

3 *The relative humidity of the air in the room* should not fall below 30 per cent in winter, otherwise there will be a danger of desiccation problems in the respiratory tract. In summer the natural relative humidity usually fluctuates between 40 per cent and 60 per cent and is considered comfortable.

4 *Draughts* at the level between head and knees should not exceed 0.2 m/s.

It must be pointed out that the preferred air temperatures might deviate slightly from one country to another, mainly because of clothing differences and habits. For example, it is well known that in the winter room air temperatures are generally higher in the USA and often lower in the UK. Obviously, the conditions in tropical areas would be very different.

Air temperatures for physical work

With physical activities the internal production of heat rises steeply. The relevant considerations are set out in Figure 20.9.

The more active people are, the more heat they generate. If they are to remain comfortable, the air temperature within the room must be lowered so that it is easier to get rid of the surplus heat. At a relative humidity of 50 per cent various kinds of activity require the room temperatures listed in Table 20.2.

20.6 Heat in industry

Effects of heat

As long as a hot climate is not 'uncomfortable' (as explained in more detail below), neither light physical work nor mental task

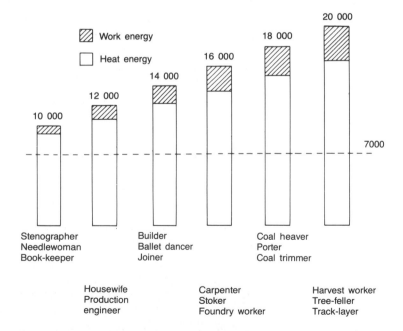

Figure 20.9 Heat production during various occupations. *The height of the columns, and the adjacent numbers, indicate the overall energy consumption in kJ per 24 h; the shaded portion of each column indicates the work energy and the white portion the heat production.*

Table 20.2 Recommended room temperatures for various activities

Type of work	Temperature (°C)
Sedentary mental	21
Sedentary light manual	19
Standing light manual	18
Standing heavy manual	17
Severe work	15–16

performance affected by a warm environment, as Ramsey (1995) concluded from a review of past studies. However, as the temperature rises above the comfort level, problems arise: first of a subjective nature and then physical problems which impair workers' efficiency. Extremely hot conditions can lead to serious health dangers. Some of these problems and their symptoms, in the range between a comfortable temperature and the highest tolerable limit, are listed in Table 20.3.

Importance of sweating

We have seen in Figure 20.3 that the range of temperatures between a comfortable level and the upper limit of tolerance, some 10–15 °C, is a zone in which heat regulation is achieved by the evaporation of sweat from the skin. As the external temperature rises, the body can lose less and less heat by convection and radiation (because of the lower temperature gradient) so that *sweating becomes the only way in which excess heat can be lost.* In fact, a point is soon reached at which convection and radiation convey heat *into* the body and this heat, together with that produced internally, must be lost by the evaporation of sweat.

When working in high temperatures, therefore, secretion and evaporation of sweat are of paramount importance for the preservation of physiological heat balance.

Mechanisms for physiological adaptation

If the ambient temperature rises, the following physiological effects may be produced:

1 Increased fatigue, with accompanying loss of efficiency for both physical and mental tasks.

2 Rise in heart rate.

3 Rise in blood pressure.

4 Reduced activity of the digestive organs.

5 Slight increase in core temperature and sharp rise in shell temperature (temperature of the skin may rise from 32 °C to 36–37 °C).

Table 20.3 Effects of deviations from a comfortable working temperature

20 °C	1. Comfortable temperature	Maximum efficiency
	2. Discomfort; increased irritability; loss of concentration; loss of efficiency in mental tasks	Mental problems
	3. Increase of errors; loss of efficiency in skilled tasks; more accidents	Psycho-physiological problems
	4. Loss of performance of heavy work; disturbed water- and salt-balances; heavy stresses on heart and circulation; intense fatigue and threat of exhaustion	Physiological problems
35–40 °C	5. Limit of tolerance	Exhaustion Physical danger

6 Massive increase in blood flow through the skin (from a few ml/cm^3 of skin tissue/min to 20–30 ml).

7 Increased production of sweat, which becomes copious if the skin temperature reaches 34 °C or more.

The effect of these adaptive changes is clearly *to transport more heat to the skin, by means of an increased blood flow*. This increased flow is at the expense of the blood supply to the musculature (hence the reduced performance and efficiency) and to the digestive organs (which also reduce their activity). Since thermal regulation is now the overriding problem, the other systems must take second place; the muscles work less effectively and the stomach refuses food (nausea).

Similarly, the heart and circulatory system adapt themselves. The rise in blood pressure, coupled with the dilation of the blood vessels of the skin (and the simultaneous constriction of the blood vessels to the internal organs) and the increased pumping action of the heart all contribute to the increased flow of blood for transport of heat to the skin. If the skin temperature should reach 34 °C, reflex action of the heat centre produces a copious flow of sweat, which is poured out from about 21 million sweat glands in the human skin.

Effects of overheating

If these control measures are insufficient, then the core temperature of the body begins to rise, leading to a heat accumulation which can soon be fatal. Clinical surveys have shown that during military exercises, core temperatures of around 39 °C have resulted in heat stroke, followed by death, although Robinson and Gerking (1947) reported that with rectal temperatures of 39–40 °C heat collapse was not necessarily fatal.

Heat stroke

As the body heat rises, the first alarming symptoms include a general feeling of listlessness, loss of performance in spite of every effort, bright red skin and an increased heart rate with a feeble pulse. These are followed by severe headache, giddiness, shortness of breath, perhaps vomiting and muscular cramps as a result of loss of salt. The final stage is unconsciousness, which may end in death within 24 h, in spite of medical attention. Death from sunstroke is really a special case of heat stroke in which direct heat from the sun on the head may be the decisive factor.

Liability to heat stroke is an individual matter and varies greatly from one person to another. The risk is much greater for a fat person than for a slim one and may be increased six times if a person is 25 kg overweight. Other factors involved include capacity for heat adaptation, age, intake of food and particularly the amount of physical exercise taken. The work of Wyndham *et al.* (1953) may be mentioned here. They recorded deaths from heat stroke among South African miners working in temperatures above 30 °C, with 100 per cent relative humidity; at 34.5 °C the mortality was 1 in 1000. Similar results were recorded for non-fatal heat collapse. Even with lower relative humidities, heat stroke sometimes occurs among the general population if the temperature outdoors rises above 35 °C.

Fevers and sport

It is interesting to note that the core temperature may rise very high (41 °C) during a fever, and up to 39.5 °C in strenuous sporting activities, without provoking collapse through heat stroke. Under these conditions the heat centre is able to cope with the rise in core temperature, preserving the vital functions during a fever and maintaining high performance during sport. Heat stroke is a consequence of excessive heat from outside the body, where it is passively accumulated. The brain seems to have some sort of protective mechanism against active heat accumulation generated within the body itself but this does not operate against heat from outside.

Heat tolerance

The important question in any problem of working in excessive heat is the tolerable heat load or heat tolerance.

From a physiological point of view, most authors agree that *rectal temperature should not exceed an upper limit of 38 °C* and several authors have compiled indices of climatic factors which would ensure that this physiological limit is not exceeded;

examples are Belding and Hatch (1955) and Dukes-Dobos (1976). Parsons reviewed the international heat stress standards in 1995. Since many indices of heat load are very complicated, taking into account the movement and temperature of air, humidity and radiant heat, we may adopt a simpler system after Wenzel (1964). These limits of heat tolerance are shown as shaded bands in Figure 20.10.

Practical limits

The limits in Figure 20.10 are valid for unclothed young men who are physically highly capable and well accustomed to heat; hence they cannot be applied directly to conditions in industry. There have been no detailed investigations under working conditions but it can be assumed that the curved bands in Figure 20.10 should be moved at least 5–10 °C to the left to make them applicable to working conditions in industry, and even further to the left for older workers or for jobs which require the use of protective clothing or breathing apparatus. Table 20.4 sets out the estimated figures for acceptable temperature ranges for work in the daytime, which are applicable to capable, healthy men wearing clothing appropriate to the job.

If the heat load is greater than listed in Table 20.4, and cannot be significantly reduced by technical means, then *the working time in the heat must be shortened*. Suggestions are given in Table 20.5.

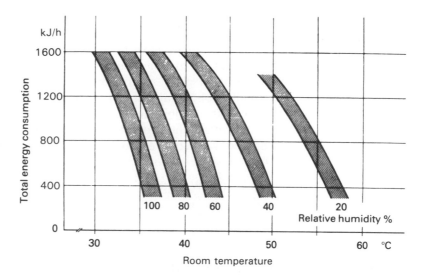

Figure 20.10 Limiting values for heat load in relation to physical effort (energy consumption), the relative humidity of the air and the air temperature. *Air temperature approximately the same as radiant temperature; air movements between 0.1 and 0.9 m/s. Working time 3–6 h. Modified from Wenzel (1964).*

Table 20.4 Proposed temperature limits for acceptable heat loads during daytime work

Overall consumption of energy (kJ/h)	Examples	Upper limit of temperature (°C)	
		Effective temperature	Temperature with 50% RH
1600	Heavy work; walking, with 30 kg load	26–28	30.5–33
1000	Moderately heavy work; walking at 4 km/h	29–31	34–37
400	Light sedentary work	33–35	40–44

Working under radiant heat

The limiting values quoted in Figure 20.10 and Tables 20.4 and 20.5 apply to conditions where the surrounding surfaces are at approximately the same temperature as the air. They cannot be used without modification for workplaces that are exposed to great radiant heat, such as, for example, in front of furnaces. Radiant heat is often measured by means of a *globe thermometer*. This usually consists of a hollow sphere of copper, about 150 mm in diameter, into which a normal mercury thermometer is inserted. The outer surface of the copper sphere is painted black, to absorb the radiation and to convert it into heat. *The globe thermometer usually shows an average value between the temperature of the air and that of surrounding surfaces.*

Various methods have been devised to integrate the various temperature factors into one. One of these is to combine the wet-bulb and globe thermometer readings into the WBGT (wet-bulb–globe temperature). Indoors, WBGT = 0.7 (wet-bulb reading) +0.3 (globe thermometer reading). Using this formula, the following values of WBGT for continuous work by acclimatised workers were proposed in the ISO Standard 7243 (1989):

Up to 117 W metabolic rate	WBGT of 33 °C
118 to 234 W	WBGT of 30 °C
235 to 360 W	WBGT of 28 °C
261 to 468 W	WBGT of 25 °C, no air movement
More than 268 W	WBGT of 23 °C, no air movement

However, Ramsey (1987) had expressed some concerns about the adequacy of the WBGT for combinations of very high humidity and little air movement.

Table 20.5 Permissible working times in hot, humid conditions for heavy work of 1900 kJ/h

Wet-bulb temperature (°C)	Permissible working time (min)
30	140
32	90
34	65
36	50
38	39
40	30
42	22

Note: The wet-bulb temperature is the reading of a thermometer with its bulb covered with a wick dipping into water. Its reading is lower than air temperature because of increased evaporation.
Source: After McConnell and Yaglou (1925)

Physiological limits

In practice it is usually very difficult to express the heat load and the degree of physical work exactly, so physiological means have to be sought to assess these. As we have already seen, heart rate, core temperature (rectal temperature) and perspiration can all be used for this purpose. Upper limits of these parameters for working in heat for an entire working day are:

Heart rate (daily average)	100–110 beats/min
Rectal temperature	38 °C
Evaporation of sweat	0.5 litres/h.

Acclimatisation to heat

Experience shows that it takes some time to become accustomed to working under very hot conditions and only after several weeks is performance equal to that of workers who are already 'heat-adapted'. This is a genuine acclimatisation of the body to heat, which proceeds by the following steps:

1 The body gradually increases its perspiration, losing more and more heat in the process. A worker who is heat-adapted can lose 2 litres of sweat per hour and up to 6 litres per working day.

2 As part of the acclimatisation process the sweat becomes more dilute, with a lower concentration of salts. The sweat glands 'learn' to conserve salts so that larger amounts of sweat can be produced without creating a salt deficit in the body. Such a deficit could lead to muscular cramps and eventually to exhaustion, and possibly death.

3 During acclimatisation there is a loss of weight, which helps heat loss by reducing the amount of insulating fat and reduces energy consumption.

4 As acclimatisation proceeds, the worker drinks more fluids to compensate for the greater amount of water lost by sweating.

5 The blood system and heart also adapt themselves to provide for the improved performance after acclimatisation.

After having become acclimatised to heat, a person feels thirsty whenever the body needs more liquid and so one tends to drink small amounts frequently. The question whether the person should take additional salt is debatable. Good results have been obtained from giving either salt tablets or salt-rich foods (meat broth and so on) both to workers and troops exposed to excessive heat while being physically very active. Generally, however, combined bodily exertions and heat stresses are not usually so great as to require the provision of additional salt. Most western diets are so rich in salt that they provide enough supply to replenish any loss through sweating.

Recommenda-tions

The following guidelines apply to work in very hot conditions:

1 *Acclimatisation to heat is necessary*. It is best accomplished in stages. One should try to begin by spending only 50 per cent of the working time in the heat and increase this by 10 per cent each day. The same procedure should be followed when a person returns to work from illness or a long absence.

2 *The higher the heat load and the greater the physical effort performed under heat stress, the longer and more frequent should be the pauses* (cooling periods). If the limit of heat tolerance is being exceeded, the working day must be shortened.

3 *A person should drink small amounts of fluid at frequent intervals*; not more than about 0.25 litres at a time, and a cupful every 10–15 min is recommended.

4 If large quantities of fluid are needed *it is best to drink plain water*, perhaps with an occasional drink of tea or coffee. Lukewarm or warm beverages are more quickly absorbed by the digestive system than cold drinks.

5 Iced beverages, fruit juices and alcoholic drinks are not recommended. Milk and mixed-milk drinks, too, are unsuitable for working in hot conditions because they put more stress on the digestive organs.

6 *Drinking water should be available close to the workplace so that a person can drink whenever needed.*

7 Where radiant heat is excessive (e.g. near blast furnaces) the worker must be safeguarded by eye protection devices, screens

and protective clothing against the risk of burning the eyes and skin.

In addition to these personal precautions, every attempt must be made to reduce the impact of heat on the worker. Some possible measures include improving ventilation, both natural and forced, perhaps artificial drying of the air and screens to protect against radiant heat.

20.7 Air pollution and ventilation

Deterioration of air quality

If a room has people in it, the air undergoes deterioration in various ways, changing its character due to:

- *release of odours*
- *formation of water vapour*
- *release of heat*
- *production of carbon dioxide*
- *air pollution*, either entering from outside or generated by activities within the room.

The first four effects arise mainly from the human body itself. The last, air pollution, depends on the situation of the building, what activities are carried on indoors and whether smoking takes place. Among the air changes that are of human origin, the odours given off by the skin are most important since even in very low concentrations they give rise to feelings of unpleasantness, disgust, distaste and revulsion. These odours are a mixture of organic gases and vapours, which are not toxic in the concentrations usually encountered but which are undesirable because of their subjective annoyance. When the air pollution in a room is primarily human in origin, it is personal odours that are most offensive, far more than the changes in carbon dioxide or water vapour. A general guideline for the quantity of fresh air needed in a room with people in it is *30 m³ of fresh air per person per hour*.

This is only a very rough estimate. Figure 20.11 summarises the results obtained by Yaglou and his colleagues (1949) in a long series of experiments. Based on these results we can make the recommendations listed in Table 20.6.

Environmental tobacco smoke

In recent years increasing importance has been attributed to environmental (or 'secondary') tobacco smoke. The concern is mainly due to the acknowledged long-term health risks, but on a day-to-day basis being in a smokey environment also often causes annoyance and even eye and throat irritations in workrooms as well as in vehicles and restaurants. As early as 1984 Weber reported on the effects of cigarette smoke on non-smoking subjects, so-called *passive smoking*. A field study on 472

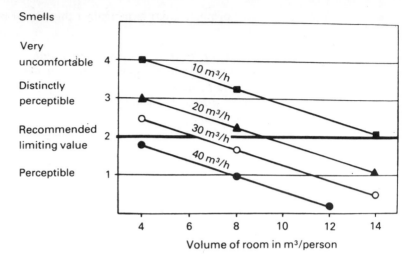

Figure 20.11 Guidelines for fresh-air requirements of sedentary workers in relation to the air space available to each person. *After Yaglou et al.* (1949).

Table 20.6 Recommended air and fresh-air volumes per person

Volume per person (m³)	Fresh air per person (m³/h)	
	Minimal	Desirable
5	35	50
10	20	40
15	10	30

employees in 44 workrooms revealed that 64 per cent of the non-smokers were 'sometimes or often' disturbed by the environmental tobacco smoke, whereas 36 per cent reported eye irritations at work. Approximately one-third of all employees qualified the air at work with regard to smoke as bad. The measured carbon monoxide concentrations due to tobacco smoke varied between 0 and 6.5 ppm with a mean value of 1.1 ppm.

In a second step, several experimental studies were carried out in a climatic chamber in which cigarettes were smoked by a smoking machine. The degree of air pollution due to tobacco smoke was evaluated by measuring the concentrations of carbon monoxide and several other pollutants such as nitric oxide, formaldehyde, acrolein, particles and nicotine. The degree of acute irritating and annoying effects was determined simultaneously on exposed subjects by means of questionnaires and measurements of the eye-blink rate.

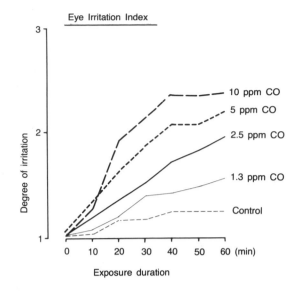

Figure 20.12 Average subjective eye irritation due to environmental tobacco smoke, related to smoke concentration and duration of exposure. *DCO = CO level during smoke production minus background level before smoke production. 32 to 43 subjects. 0 min = measurement before smoke production. Period 0–5 min = increasing smoke concentration. Period 6–60 min = constant smoke concentration. Degree of irritation: 1 = no irritation; 2 = little irritation; 3 = moderate irritation.*

The results revealed that the irritating effects were most pronounced on the eyes, followed by the effects on the nose and on the throat. Figures 20.12 and 20.13 illustrate the results obtained for subjective eye irritations and eye-blink rate of people being exposed to different smoke concentrations which were kept constant for nearly one hour.

It is evident that eye irritations as well as blink rates increase with higher smoke concentration and the duration of exposure. The mean incidences of subjects with eye irritations are reported in Figure 20.14. The results clearly reveal that there is a marked increase in strong eye irritations with smoke levels above 1.3 ppm carbon monoxide (CO). Weber (1984) concluded that a possible limit to protect healthy people in their everyday environment against adverse effects of environmental tobacco smoke should lie in this range, that is between 1.5 and 2.0 ppm CO. The concentration of 2.0 ppm CO, for instance, is already reached when two cigarettes are smoked per hour in a room of 80 m³ with a single air change.

Hence counter-measures to protect passive smokers are desirable when the CO level reaches 1.5 ppm and are necessary when it hits 2.0 ppm. The lower limit should be applied to

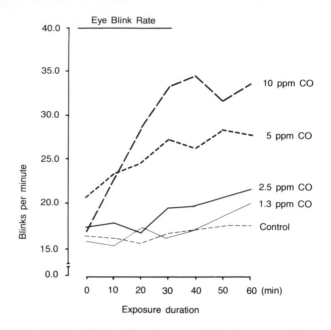

Figure 20.13   Mean effects of environmental tobacco smoke on eye blink rate. *CO = CO level during smoke production minus background level before smoke production. 32 to 43 subjects. The period 0 min = measurement before smoke production. The period 0–5 min = increasing smoke concentration. The period 6–60 = constant smoke concentration.*

workplaces where passive smokers cannot escape the exposure, and the upper limit to restaurants and other places where people usually go voluntarily and for a shorter lapse of time.

Fresh air supply in smoking rooms

Calculations show that a *fresh air supply of 33 m³/h per smoked cigarette is necessary to keep the CO concentration below the proposed upper limit of 2.0 ppm; for the lower limit, 50 m³/h of fresh air per smoked cigarette are required.* Depending on the number of people present in a room, a fresh air supply of 25–45 m³/h/person is necessary in order not to exceed the upper limit. In other words, the ventilation has to be two to four times higher than in a room where no one smokes (in which only 12–15 m³/h/person are required).

Organisational measures

To increase ventilation as a measure to protect passive smokers is of course not desirable from the energy and economic viewpoints. Therefore, whenever possible, organisational measures should be taken, such as separation into smoking and non-smoking rooms or, even better in all respects, elimination of smoking, rather than an increase of fresh air supply.

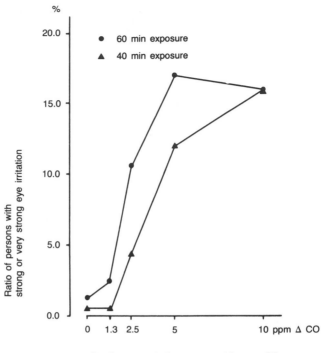

%

20.0 ● 60 min exposure

▲ 40 min exposure

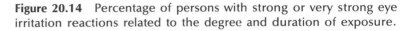

Smoke concentration expressed in ppm CO

Figure 20.14 Percentage of persons with strong or very strong eye irritation reactions related to the degree and duration of exposure.

Natural or forced ventilation

Especially if internal pollution is present, the location of the building and its windows are the most important factors in deciding whether it is necessary to have any form of forced ventilation or air-conditioning. If it is impossible to open the windows to generate a natural air flow, either because of traffic noise or pollution of the ambient air, then artificial ventilation of some kind is essential.

Windows and climate indoors

Modern buildings tend to have lower ceilings and more window area than buildings erected 50 or more years ago. These changes have important effects on the climate indoors. Lower ceilings make fresh air replacement even more necessary. Huge window areas act as cooling surfaces in winter and from spring to autumn they allow a great deal of heat to penetrate into the room. This becomes particularly obvious if objective measures of thermal conductivity (so-called K values) are used to assess the heat balance of a building. The windows of older buildings occupied 15–30 per cent of the external surface, whereas they are often more than 50 per cent of the surface of modern buildings. If we assume a K value of a window as being 3.5, and that of a wall as 0.8, then if the windows occupy 50 per cent of the outer walls, it can be calculated that 82 per cent of the total heat loss occurs from the windows and only 18 per cent from the walls and roof.

This means that in such a building four times as much heat is lost through the windows as through the walls and roof. Improving the thermal insulation of the window is four times as effective as insulating the roof and walls, and is strongly recommended.

Modern buildings admit more light and bring the occupants into closer contact with the outside world, with the sky and with nature; at the same time they raise serious problems of controlling the climate indoors and lead to a greater dissipation of energy.

Summary

The effects of the climate on human well-being and functioning are understood. All components (temperatures, humidity, air movement) of our personal 'microclimates' as well as clothing and workload are normally controllable, both indoors and out-doors.

CHAPTER TWENTY ONE

Daylight, colours and music for a pleasant work environment

21.1 Daylight

Daylight is naturally preferred to artificial light by most people but there are some special problems, which will be briefly discussed.

In addition to providing illumination, natural daylight penetrating into a room establishes contact with the world outside, giving a view of the surroundings and indicating the time of day and the state of the weather.

As a rule, it is psychologically and physiologically desirable to have as much light, as evenly distributed, as possible. The higher the daylight level, the less need there is for artificial lighting, especially in winter. On a cloudy day in December in Zürich, Switzerland, the daylight intensity indoors reaches 500 lx only for the four hours from 10.00 to 14.00.

Indoors, daylight illumination levels depend mainly on the position and type of windows; the height of windows is decisive for the penetration of light into the depth of the room.

One reservation must be made. Wide and high windows certainly help to distribute daylight within a room, but they have the drawback of admitting a great deal of sun heat, especially in the summer if they face south or south-west. Furthermore, in winter they act as cold surfaces, with an adverse effect on the climate of the room as discussed in Chapter 20. Bright windows can be a source of direct and indirect glare, as discussed in Chapter 18. The window sizes must not be decided solely in relation to daylight; there must be a 'balance sheet' of all the pros and cons.

Recommendations

The following rules of thumb relate to daylight indoors:

1 High windows are more effective than broad ones since the light penetrates further into the room. The lintel should not be deeper than about 300 mm.

2 Window sills should be at table height. If the window extends below the table top it may 'feel cold' in winter and cause glare if a person looks down as, for example, when reading.

3 The distance from window to workplace should not be more than about twice the height of the window.

4 For workrooms, the window area should be about one-fifth of the floor area. This is only a general rule which should be modified according to circumstances.

5 The glass should transmit plenty of light. Clear glass has a transparency of more than 90 per cent, whereas frosted glass, glass bricks or special heat-insulating glass may have transparencies from 70 per cent down to only 30 per cent.

6 Effective protections against glare, especially of direct sunlight, and against radiant heat transfer from and to the window are important in securing good visibility and comfort indoors. The most efficient method is an adjustable external sunshade, such as venetian blinds or shutters. Venetian blinds inside the window, or between the panes of double-glazing, are a mistake because they afford little protection against radiant heat.

   Insulating window panes block heat flow in either direction. In the summer the panes can trap heat inside the room (greenhouse effect) which requires air-conditioning to avoid. They often reduce the amount of light transmitted, which is undesirable in winter. Balconies, overhanging eaves and other building projections also create problems by cutting off some of the light in dull weather. In moderate latitudes (such as Switzerland) they should be restricted to south-facing walls.

7 Each window should receive direct light from the sky and it is desirable that a portion of sky should be visible from every workplace.

8 The nearest building should be at least twice as far away as its own height – a desire difficult to fulfil in modern cities.

9 Pale colours should be used, both in the room itself and in any courtyard outside, to reflect as much of the incident daylight as possible.

Glass houses and windowless factories

The trend in modern architecture has been to increase the window area of new buildings, sometimes ending up with a glass house. As we have already said, glass walls, or very large window areas, radiate heat away in winter and admit heat in summer. On the other hand, they admit natural light and allow occupants to see more of the outside world. The problems of regulating the room climate are both difficult and expensive to solve. However, it must be said that recent developments in window conductivity have reduced these problems considerably.

Equally problematical are factory buildings that are not

provided with windows at all. The arguments in favour of such buildings are that they can be given a uniform internal climate (through air-conditioning) and that the internal lighting, being entirely artificial, can be controlled by modern techniques. The more complete insulation of the walls reduces the cost of air-conditioning. The common opposing argument is that the absence of windows makes humans feel 'imprisoned' and 'cut off', and that people need some contact with the outside world when they are at work. Since the validity of these various arguments is not yet proved it would seem sensible to proceed slowly with building practices that are so unnatural and psychologically questionable. To provide at least some windows appears to be a sensible interim provision.

Skylights and fanlights

Skylights in the roof and fanlights in the walls or door are often useful accessories in single-storey buildings, in lofts and attics, and in other rooms with insufficient daylight. The following are the most important types of daylight panels in roofs:

1 *Pitched roofs*. If the axis of the skylight is no further away from the workplace than the height up to the ridge, the lighting is bright and even. It is particularly good if the workplaces are so arranged that each lies between two skylights at equal angles, and so is lit from both sides.

2 *Skylights in the roof ridge*. This provides glass above the longitudinal axis of the building. It is particularly suitable for high shops where the workers in the centre need the best illumination.

3 *Mansard roofs*. Here the skylights are placed in the sloping part of the roof, above the windows, giving a relatively high level of daylight all along the outer walls. However, they increase the risk of glare and do not provide much extra light to the people in the middle of the room.

4 *Zig-zag shed roofs*. The whole working area has clear glass above it, in panels which usually face north and slope steeply, often at an angle of 60°. The reverse face of each zig-zag is opaque, faces south and slopes at an angle of about 30°. This arrangement gives the highest and most uniform values of daylight and is especially recommended for large workshops.

21.2 Colour in the workplace

Physical principles

The visible colours of the spectrum cover certain bands of wavelengths, in nanometres ($1\,\text{nm} = 10^{-9}\,\text{m}$ = one millionth of a mm).

Violet	380–436
Blue	436–495
Green	495–566
Yellow	566–589
Orange	589–627
Red	627–780

Electromagnetic waves longer than 780 nm are infrared, felt as radiant heat; wavelengths shorter than 380 nm are ultra-violet rays, which are of critical importance for the synthesis of vitamin D in the body and for normal organic growth.

The colours that we see arise because the molecular structure of the surface of objects reflects only certain wavelengths from all that are incident. This reflection is what we perceive. For example, a green-painted machine absorbs all the incident light except the green. As discussed in Chapter 17, the colour receptors in the retina are the cones, which are capable of distinguishing more than 100 000 shades of colour.

Reflected colour

When deciding upon the colours in and around a workplace, it is necessary to consider reflectivity, as discussed earlier in Chapter 18. Table 21.1 gives a few examples.

Colours in and around workplaces have the following functions:

1 To achieve orderliness.

2 To indicate safety devices.

3 To generate contrasts that make work easier.

4 To psychologically affect people.

Orderliness

Sections of or whole machines, rooms, hallways and buildings can be coded by colour. This helps in maintaining a building by, for example, placing items into certain areas, identifying pipes that carry different fluids such as grades of fuel, guiding people to their destinations, such as showing patients their way in hospitals, and, in general, keeping the whole works on an orderly plan.

Safety colours

If the same colour is always used to indicate a particular danger, or place of help in an emergency, the correct association and reaction to them become automatic. This practice is now followed in most countries in a uniform manner according to international ISO standards.

Here are some of the common colour codes:

1 Red is the 'danger' colour: Halt, Stop, Prohibited. Red is also the warning colour for 'Fire' used on extinguishers and fire-fighting equipment.

2 Yellow means: Danger of collision, Attention, Look out, Risk of tripping. Yellow stripes on black are often used as warning colours in transport.

3 Green means: Rescue services, Safety exit and "things are in order". It is used to indicate all forms of rescue equipment and first-aid.

4 Blue is not actually a safety colour, but is used for giving directions, advice, and general indications.

Colour contrasts of large areas

When deciding on colour contrasts, large areas such as walls and furniture must be considered separately from small areas such as splashes of colour intended to attract attention on knobs, handles and levers.

The colours of large areas should be chosen so that they have similar reflectances (see Table 21.1 and Chapter 18), in order to have colour contrast without differences in luminance. The avoidance of large areas of contrasting brightness, close together, is an important factor in ensuring good visual acuity. Large areas and big objects should not be covered by pure colours, nor with fluorescent paint, since these cause local overloading of the retina, and lead to the production of after-images. Therefore walls, partitions, table tops and so on should be painted with unsaturated colours in a matt finish.

It is easier to lay out working materials and to select the one required if they are coloured differently from their immediate surroundings. This should be considered when the workplace is

Table 21.1 Reflectance as a percentage of the incident light

Colour or material	Reflectance (in %)
White	100
Aluminium; white paper	80–85
Ivory; deep lemon yellow	70–75
Deep yellow; light ochre; light green; pastel blue; pale pink; cream	60–65
Lime green; pale grey; pink; deep orange; blue-grey	50–55
Powdered chalk; pale wood; sky blue	40–45
Pale oakwood; dry cement	30–35
Deep red; grass green; wood; pale leaf green; olive green; brown	20–25
Dark blue; purple red; reddish brown; slate grey; dark brown	10–15
Black	0

being designed but unpleasant contrasts in luminance and intensity of colour should be avoided. For example, if the working materials are made of leather, wood or similar materials of an ochre yellow or brown colour, a suitable colour for the background would be matt green, olive green or matt bluish. Greyish-blue materials such as steel and other metals show up well against a background of dark ivory or light beige. The area surrounding the machine, or the work table, might be painted in cool, neutral colours, from yellow green to pastel blue.

Eye-catching colours

Eye-catching colours are the little spots of strongly contrasting colours that are used to attract attention, to 'catch the eye'. Colour is used in this way in nature: a red strawberry among green foliage, brilliant flowers which attract insects and other creatures by their colour contrast. On the other hand, nature also uses colours for concealment. Defenceless wild animals are often neutral in colour and merge into the background so that they are almost invisible.

It is a good idea to provide a few eye-catchers in a workplace, marking such things as the more important handles, levers, control-wheels, knobs and so on. If the eye-catchers are small, not more than a few square centimetres in area, they should contrast strongly, not only in colour but also in brightness. Colour codes make the controls easier to find, reducing the time taken to search for them and hence reducing the diversion of attention from the work itself.

The human eye sees the greatest colour contrast between yellow and black because the brain adds together the effect of colour and intensity. However, one of the greatest dangers in colour planning, and especially in the planning of eye-catchers, is excess. If there are too many visual stimuli in too many different colours, then the whole workplace becomes restless and distracting. Colour does not mean bunting. The most important physiological requirement in the use of colour is restraint, with three, or at most five, eye-catchers to each workplace. This applies, too, to colour in schoolrooms, restaurants, homes – everywhere, in fact, where people either work or relax. Less restraint is appropriate in shop-windows, display cases and exhibitions, where the customer is meant to be stimulated by eye-catchers.

Psychological effects

By the 'psychological effects' of colour we mean optical illusions and other phenomena that are triggered off by colour.

In part these are caused by subconscious associations with previous sights or experiences, and perhaps by hereditary factors. They influence mental effect and thereby a person's behaviour. ('Events' in art are phenomena of this kind. Abstract painting strives to produce such effects by colour and form alone, which,

Table 21.2 Psychological effects of colour

Colour	Distance effect	Temperature effect	Mental effect
Blue	Further away	Cold	Restful
Green	Further away	Cold to neutral	Very restful
Red	Closer	Warm	Very stimulating not restful
Orange	Much closer	Very warm	Exciting
Yellow	Closer	Very warm	Exciting
Brown	Much closer,	Neutral Cold	Restful Aggressive;
Violet	Much closer		unrestful, tiring

to the 'initiated', are at least as stimulating emotionally as realistic pictures.)

Psychological effects can be induced, too, by colours in a room, arousing feelings of like or dislike. Since, however, many rooms have to serve particular functions, their colours do not only have aesthetic consequences: their physiological and psychological effects must also be taken into account. However, there is always a good deal of latitude for aesthetic considerations.

Particular effects

Particular colours have special psychological effects, which are more or less similar in character for people with similar cultural background and upbringing, although with great individual variation. The most important particular effects concern distance, temperatures and the presumed effects on the mental state. Table 21.2 summarises these illusory effects of individual colours as felt by most Europeans and North Americans.

Broadly speaking, all dark colours are oppressive and tiring; they absorb the light and are difficult to keep clean. All light colours are bright, friendly and cheerful; they scatter more light, brighten up the room and encourage greater cleanliness.

Colours in a room

Before starting to plan the colour for a room, there must be a careful consideration of its functions and who is going to use it. After that it will be possible to plan its colours in relation to psychological and physiological factors.

Keeping the principal effects mentioned above in mind, careful consideration should be given to the work to be carried on. For example, is it likely to be monotonous or does it make heavy demands on concentration? If the work is monotonous, it is advisable to include a few areas of exciting colour, but merely a few items such as a pillar or column, a door or a partition wall, not large areas such as the main walls or the ceiling.

If the work going on in the room demands close concentration, the colours should be chosen carefully to avoid unnecessary distractions and unrestful items. In this case walls, ceilings and other structural elements should as far as possible be painted in light colours that do not attract attention.

If the workroom is very large, it can be divided visually by using different colours, thereby making it less anonymous.

Walls and ceilings painted yellow, red or blue may be very attractive at first glance but as time goes on they can become a strain on the eyes. Hence, such rooms often become unpleasant after a while.

More intense colours can safely be used in rooms that are mainly used briefly, e.g. entrance halls, corridors, lavatories, storerooms. Here, strong colours may help to brighten up the room and make it structurally more pleasing.

21.3   Music and work

Throughout the ages, music has been used to lighten human labour and many working-songs exist; some of the best-known, are soldiers' marching-songs or the *Song of the Volga Boatmen*. All music of this kind is melodious, with a well-marked rhythm, and the effect is to rouse the singers and urge them on to greater continued effort in a rhythmic fashion.

Physiological effects

An acoustic stimulus passes along the auditory nerve to the activating system and brings the entire conscious sphere of the cerebral cortex into a state of advanced readiness for action. Hence noise can have a stimulating effect, especially in boring situations. Music that is strongly rhythmical, with marked variations in loudness, affects the brain in similar ways, bringing the whole organism to 'action status'. Noise also has a distracting effect, however, so that activities that call for thought and alertness are disturbed. We might expect the same to be true of stirring, rhythmical music. In fact we do find that music is particularly welcome as a background to dull, repetitive work, but its effect on intellectual work is uncertain.

The extent of the distraction and disturbance provoked by music depends a good deal on its nature. Up to a point the distraction can be minimised by choosing suitable music.

Studies in industry

As part of the attempt to improve working conditions and worker output, for decades music has been used to relieve the boredom of certain jobs. In 1946, Kerr questioned 666 workers in a US factory about how they would like music distributed throughout the day; their jobs included spool-winding, operating presses and assembling radio tubes. The overwhelming majority wanted continuous music all day long. If it had to be restricted, then most of the workers preferred a series of between 10 and 16 music periods, equally distributed throughout the day. Mid-morning and mid-afternoon were the periods when music was particularly welcome. It appeared that the younger employees and female workers liked music more than the others.

Grandjean (1988) reported on similar research carried out in British, French and Swiss factories, which by and large confirmed these results. After a review of the existing newer literature on the topic, Kroemer *et al.* (1994) cautioned that the effects of 'music at work' depend decidedly, but rather unpredictably, on the kind of work, on the kind of music, on the time of the day or night, on the length of the playing time and on the characteristics of the listeners.

Background music

Originally, music at work was rhythmical, with a clear melody. The workers listened to it consciously and sometimes hummed the tune. More recently, starting in the USA, administrative offices, business houses, salerooms, railway stations, waiting-rooms, restaurants and even residential rooms have been provided with a different kind of music: persistent but very quiet, unobtrusive, hardly impinging on the consciousness. This is called 'background' or 'wallpaper music' or 'muzak', meant to surround one with a pleasant sound. It should have the advantage of not being distracting and therefore being suitable for work which demands concentration, such as designing or planning. Furthermore, it helps to mask some unwanted sound.

Recommendations

As far as present knowledge goes, the question of whether music at work is a good thing or not may be answered as follows.

Music at work can help to create a pleasant yet unobtrusive atmosphere, which stimulates the worker. This is particularly so if the work is boring, or repetitive, or makes few demands on thought or alertness. Such music is less helpful in big, noisy workshops or in jobs where mental alertness is essential.

When the music chosen is not of the quiet background type, but assertive or stirring, it attracts attention or close listening and should be played only for part of the working time. A short period of rousing music can begin the work day, which should end with friendly and festive tunes; the rest of the work shift may have four periods of 30 min each of light music. The tempo should be neither too slow and soporific, nor too fast causing irritation and haste. Well chosen, such attention-grabbing, stirring music can interrupt continuous work and stimulate emotional and physical activity. However, which, if any, music at work to use is hard to predict;

having the people involved participate in the decision, selection and presentation is important.

Summary

Among the conditions that make us content and active are daylight, colours and music. We are all influenced in similar ways by these conditions, although in our personal ways. Ergonomic recommendations are available that help to make our work environment pleasant.

References

AANONSEN, A. (1964). *Shiftwork and Health*. Norwegian Monographs on Medical Science. Oslo: Universitets Forgalet.

ABRAMSON, N. (1963). *Information Theory and Coding*. New York: McGraw-Hill.

AKERBLOM, B. (1948). *Standing and Sitting Posture*. Stockholm, Sweden: Nordiska Bokhandeln.

ANDERSSON, B. J. G. and ORTENGREN, R. (1974a). Lumbar disc pressure and myoelectric back muscle activity during sitting. 1. Studies on an office chair. *Scandinavian Journal of Rehabilitation Medicine*, **3**, 115–21. The same author with colleagues also in *Scandinavian Journal of Rehabilitation Medicine*, **3**, 104–14 (1974b), **3**, 122–7 (1974), **3**, 128–35 (1974).

AYOUB, M. M. and MITAL, A. (1989). *Manual Materials Handling*. London: Taylor & Francis.

BARNES, R. M. (1936). An Investigation of Some Hand Motions Used in Factory Work. University of Iowa, Iowa City, Studies in Engineering, Bulletin 6.

BARNES, R. M. (1949). *Motion and Time Study* (3rd Edn). New York: John Wiley.

BASCHERA, P. and GRANDJEAN, E. (1979). Effects of repetitive tasks with different degrees of complexity on CFF and subjective state. *Ergonomics*, **22**, 377–85.

BASMAJIAN, J. V. and DELUCA, C. J. (1985). *Muscles Alive* (5th Edn.) Baltimore, MD: Williams & Wilkins.

BELDING, H. S. and HATCH, T. F. (1955). Index for evaluating heat stress in terms of resulting physiological strains. *Heating, Piping and Air Conditioning*, **27**, 129–35.

BENDIX, T. and HAGBERG, M. (1984). Trunk posture and load on the trapezius muscle whilst sitting at sloping desks. *Ergonomics*, **27**, 873–82.

BENZ, C., GROB, R. and HAUBNER, P. (1983) Designing VDU workplaces. Deutsche Ausgabe: Gestaltung von Bildschirm-Arbeitsplätzen. Cologne: Verlag TÜV Rheinland Köln.

BILLS, A. G. (1931). Blocking: a new principle of mental fatigue. *American Journal of Psychology*, **43**, 230–9.

BJERNER, B., HOLM, A. and SWENSSON, A. (1955). Diurnal variation in mental performance. *British Journal of Industrial Medicine*, **12**, 103–10.

BLACKWELL, H. R. and BLACKWELL, O. M. (1968). The effect of illumination quantity upon the performance of different visual tasks. *Illuminating Engineering*, **63**, 143–52.

BLUM, M. L. and NAYLOR, J. C. (1968). *Industrial Psychology*. New York: Harper and Row.

BOFF, K. R., and LINCOLN, J. E. (Eds.) (1988). *Engineering Data Compendium: Human Perception and Performance*. Wright-Patterson AFB, OH: Armstrong Aeropsace Medical Research Laboratory.

BOFF, K. R., KAUFMAN, L. and THOMAS, J. P. (Eds.) (1986). *Handbook of Perception and Human Performance*. New York: John Wiley & Sons.

BONVALLET, M., DELL, P. and HIEBEL, G. (1954). Tonus sympathique et activité électrique corticale. *Journal of Electroencephalography and Clinical Neurophysiology*, **6**, 119–25.

BOUISSET, S. and MONOD, H. (1962). Etude d'un travail musculaire léger. I. Zone de moindre dépense energétique. *Archives Internationale du Physiologie et Biochimie*, **70**, 259–72.

BOUISSET, S., LAVILLE, A. and MONOD, H. (1964). Recherches physiologiques sur l'économie des mouvements. *Ergonomics*, **7**, 61–7.

BOUMA, H. (1980). Visual reading processes and the quality of text displays. In Grandjean, E. and Vigliani, E. (Eds.) *Ergonomic Aspects of Visual Display Terminals*. London: Taylor & Francis.

BRÄUNINGER, U., GRANDJEAN, E., VAN DER HEIDEN, G., NISHIYAMA, K. and GIERER, R. (1984). Lighting characteristics of VDTs from an ergonomic point of view. In Grandjean E. (Ed.) *Ergonomics and Health in Modern Offices*. London: Taylor & Francis.

BRITISH HEALTH AND SAFETY EXECUTIVE (1992). *Manual Handling*. Sheffield, UK.

BROADBENT, D. E. (1957). Effects of noise on behaviour. In Harris, C. M. (Ed.) *Handbook of Noise Control*. New York: McGraw-Hill.

BROADBENT, D. E. (1958a). *Perception and Communication*. London: Pergamon Press.

BROADBENT, D. E. (1958b). Effect of noise on an intellectual task. *Journal of the Acoustical Society of America*, **30**, 824–7.

BROUHA, L. (1967). *Physiology In Industry* (2nd Edn) Oxford: Pergamon Press.

BROWN, J. S. and SLATER-HAMMEL, A. T. (1949). Discrete movements in the horizontal plane. *Journal of Experimental Psychology*, **39**, 84–95.

BROWN, J. S., KNAUFT, E. B. and ROSENBAUM, G. (1948). The accuracy of positioning reactions as a function of their direction and extent. *American Journal of Psychology*, **61**, 167–82.

BRUNDKE, M. (1973). Langzeitmessungen der Pulsfrequenz und Möglichkeiten der Aussage über die Arbeitsbeanspruchung. In *Pulsfrequenz und Arbeitsuntersuchungen, Schriftenreihe Arbeitswissenschaft und Praxis*, Band 28. Berlin: Beuth-Vertrieb.

BUESEN, J. (1984). Product development of an ergonomic keyboard. Proceedings of Ergodesign 84, *Behaviour and Information Technol-*

ogy, **3**, 387–90.

CAKIR, A., REUTER, H. J., VON SCHMUDE, L. and ARMBRUSTER, A. (1978). *Anpassung von Bildschirmarbeitsplätzen an die physische und psychische Funktionsweise des Menschen, Bundesministerium für Arbeit und Sozialordnung*, Bonn: Referat Presse.

CALDWELL, L. S. (1959). *The effect of the special position of a control on the strength of six linear hand movements*, Report No. 411. Fort Knox, KY: US Army Medical Research Laboratory.

CAPLAN, R. D., COBB, S., FRENCH, J. R., HARRISON, R. V. and PINNEAU, S. R. (1980). *Job Demands and Worker Health: Main Effects and Occupational Differences*. Ann Arbor: Institute for Social Research, University of Michigan.

CARAYON, P. (1993). Job design and job stress in office workers. *Ergonomics*, **36**, 463–77.

CASALI, J. G. and BERGER, E. H. (1996). Technology advancements in hearing protection circa 1995: Active noise reduction, frequency/amplitude-sensitivity, and uniform attenuation. *American Industrial Hygiene Association Journal*, **57**, 175–85.

CHAFFIN, D. B. (1969). A computerized biomechanical model: development of and use in studying gross body actions. *Journal of Biomechanics*, **2**, 429–41.

CHAFFIN, D. B. (1973). Localized muscle fatigue — definition and measurement. *Journal of Occupational Medicine*, **15**, 346–54.

CHAFFIN, D. B. and ANDERSSON G. B. J. (1991). *Occupational Biomechanics*. (7th Edn). New York: John Wiley & Sons.

CHANEY, R. E. (1964). Subjective reaction to wholebody vibration. Boeing Company, *Human Factors Technical Report D3, 64–74*, Wichita, Kansas.

CHAPANIS, A. (Ed.) (1975). *Ethnic Variables in Human Factors Engineering*, Baltimore, MD: Johns Hopkins Univerity Press.

CHRISTENSEN, E. H. (1964). L'Homme au Travail. Sécurité, Hygiène et Médecine du Travail, Series No. 4. Geneva: Bureau International du Travail.

CLARKE, H. H., ELKINS, E. C., MARTIN, G. M. and WAKIM, K. G. (1950). Relationship between body position and the application of muscle power to movements of joints. *Archive of Physical Medicine*, **31**, 81–89.

COLLINS, J. B. (1956). The role of a sub-harmonic in the wave-form of light from a fluorescent lamp in causing complaints of flicker. *Ophthalmologica*, **131**, 377–87.

COSTA, G. (1996). The impact of shift and night work on health. *Applied Ergonomics*, **27**, 9–16.

COSTA, G. LIEVORE, F., CASALETTI, G., GAFFURI, E. and FOLKARD, S. (1989). Circadian characteristics influencing inter-individual differences in tolerance and adjustment to shiftwork. *Ergonomics*, **32**(4), 373–85.

COURTNEY, A. J. (1984). Hand anthropometry of Hong Kong Chinese females compared to other ethnic groups. *Ergonomics*, **27**, 1169–80.

COX, T. (1985). The nature and measurement of stress. *Ergonomics*, **28**, 1155–63.

CUSHMAN, W. H. and ROSENBERG, D. J. (1991). *Human Factors in Product Design*. Amsterdam: Elsevier.

DAMON, A., STOUDT, H. W. and McFARLAND, R. A. (1966). *The Human Body in Equipment Design*. Cambridge, MA: Harvard University Press.

DAVIES, D. R. and TUNE G. S. (1970). *Human Vigilance Performance*. London: Staples Press.

DAVIS, P. R. and STUBBS, D. A. (1977a). Safe levels of manual forces for young males. *Applied Ergonomics*, **8**, 141–50.

DAVIS, P. R. and STUBBS, D. A. (1977b). A method of establishing safe handling forces in working situations. *Report of the International Symposium on Safety in Manual Materials Handling*. Cincinnati, OH: National Institute of Occupational Safety and Health.

DICKINSON, C. E. (1995). Proposed manual handling international and European standards. *Applied Ergonomics*, **26**, 265–70.

DRURY, C. G. and FRANCHER, M. (1985). Evaluation of a forward-sloping chair. *Applied Ergonomics*, **16**, 41–7.

DUBOIS-POULSEN A. (1967). Notions de physiologie ergonomique de l'appareil visuel. In Scherrer, J. (Ed.) *Physiologie du Travail*, Tome 2. Paris: Masson, pp. 114–83.

DUKES-DOBOS, F. N. (1976). Rationale and provisions of the work practices standard for work in hot environments as recommended by NIOSH. In Horvath, S. M. and Jensen, R. C. (Eds.) *Standards for Occupational Exposures to Hot Environments*, US Dept. Health, Education and Welfare, NIOSH, No. 76–100, Cincinnati, OH.

DUPUIS, H. (1974). Mechanische Schwingungen, sowie: Messung und Bewertung von Schwingungen und Stössen. In Schmidtke, H. (Ed.) *Ergonomie*, Band 2. Munich: Hanser, pp. 211–36.

DURNIN, J. V. G. A. and PASSMORE, R. (1967). *Energy, Work and Leisure*. London: Heinemann Educational.

EASTMAN, M. C. and KAMON, E. (1976). Posture and subjective evaluation at flat and slanted desks. *Human Factors*, **18**, 15–26.

EGLI, R., GRANDJEAN, E. and TURRIAN, H. (1943). Arbeitsphysiologische Untersuchungen an Hackgeräten. *Arbeitsphysiologie*, **15**, 231–4.

EISSING, G. (1995). Climate assessment indices. *Ergonomics*, **38**, 47–57.

ELIAS, R. and CAIL, F. (1983). Exigences visuelles et fatigue dans deux types de tâches infomatisées. *Le Travail Humain*, **46**, 81–92.

ELLIS, D. S. (1951). Speed of manipulative performance as a function of worksurface height. *Journal of Applied Psychology*, **35**, 289–96.

ENGEL, F. L. (1980). Information selection from visual display units. In Grandjean, E. and Vigliani, E. (Eds.) *Ergonomic Aspects of Visual Display Terminals*. London: Taylor & Francis.

FANGER, P. O. (1972). *Thermal Comfort*. New York: McGraw-Hill.

FLETCHER and MUNSON, W. A. (1933). Loudness, its definition, measurement and calculation. *Journal of the Acoustical Society of America*, **5**, 82–108.

FLUEGEL, B., GREIL, H. and SOMMER, K. (1986). *Anthropologischer Atlas*. Berlin: Tribüne.

FOLKARD, S. and MONK, T. H. (Eds) (1985). *Hours of Work*. Chichester: John Wiley & Sons.

FRANKEL, V. H. and BURNSTEIN, A. H. (1970). *Orthopaedic Biomechanics*, Philadelphia, PA: Lea & Febiger.

FRANKENHÄUSER, M. (1974). *Man in Technological Society: Stress,*

Adaptation and Tolerance Limits. Report from the Psychological Laboratories, University of Stockholm, Suppl. 26.

FRANKENHAÜSER, M., NORDHEDEN, B., MYRSTEN, A. L. and POST, B. (1971). Psychophysical reactions to understimulation and over-stimulation. *Acta Psychologica*, **35**, 298–308.

FRASER, T. M. (1980). *Ergonomic principles in the design of hand tools*. Occupational Safety and Health Series No. 44. Geneva, Switzerland: International Labour Office.

FREIVALDS, A. (1987). The ergonomics of tools. *International Reviews of Ergonomics*, **1**, 43–75.

FRIEDMANN, G. (1959). *Grenzen der Arbeitsteilung*. Frankfurt: Europäische Verlagsanstalt.

FRY, G. A. and KING, V. M. (1975). The pupillary response and discomfort glare. *Journal of the IES*, 307–24.

GARRETT, J. W. and KENNEDY, K. W. (1971). *A Collation of Anthropometry* (AMRL TR 68-1). Wright-Patterson AFB, OH: Aerospace Medical Research Laboratories.

GIERER, R., MARTIN, E., BASCHERA, P. and GRANDJEAN, E. (1981). Ein neues Gerät zur Bestimmung der Flimmerverschmelzungs frequenz. *Zeitschrift für Arbeitswissenschaft*, **35**, 45–7.

GORDON, C. C., CHURCHILL, T., CLAUSER, C. E., BRADTMILLER, B., McCONVILLE, J. T., TEBBETTS, I. and WALKER, R. A. (1989). *1988 Anthropometric Survey of US Army Personnel: Summary Statistics Interim Report* (Technical Report NATICK/TR 89-027). Natick, MA: US Army Natick Research, Development and Engineering Center.

GRAF, O. (1954). Studien über Fliessarbeitsprobleme an einer praxisnahen Experimentieranlage. *Forschungsbericht d. Wirtschaftsund Verkehrsministerium Nordrhein-Westfalen*, Nos 114 and 115. Cologne: Westdeutscher Verlag.

GRANDJEAN, E. (1959). Physiologische Untersuchungen über die ner-vöse Ermüdung bei Telephonistinnen und Büroangestellten. *Internationale Zeitschrift Angewandte Physiologie*, **17**, 400–18.

GRANDJEAN, E. (1973). *Ergonomics of the Home*. London: Taylor & Francis.

GRANDJEAN, E. (1987). *Ergonomics in Computerized Offices*. London: Taylor & Francis.

GRANDJEAN, E. (1988). Fitting the task to the man. (4th Edn). London: Taylor & Francis.

GRANDJEAN, E. and BURANDT, H. U. (1962). Das Sitzverhalten von Büroangestellten. *Industrielle Organisation*, **31**, 243–50.

GRANDJEAN, E., EGLI, R., RHINER, A. and STEINLIN. H. (1952). Der menschliche Energieverbrauch der gebräuchlichsten Waldsägen. *Helvetica Physiologie et Pharmacologie Acta*, **10**, 342–8.

GRANDJEAN, E., BÖNI, A. and KRETSCHMAR, H. (1967). Entwicklung eines Ruhesesselprofils für gesunde und rückenkranke Menschen. *Wohnungsmedizin*, **5**, 51–6.

GRANDJEAN, E., HÜNTING, W., WOTZKA, G. and SCHÄRER, R. (1973). An ergonomic investigation of multipurpose chairs. *Human Factors*, **15**, 247–55.

GRANDJEAN, E., NAKASEKO, M., HÜNTING, W. and LÄUBLI, T. (1981). Ergonomische Untersuchungen zur Entwicklung einer neuen Tastatur für Büromaschinen. *Zeitschrift der Arbeitswissenschaft*, **35**, 221–6.

GRANDJEAN, E., HÜNTING, W. and PIDERMANN, M. (1983). VDT workstation design: preferred settings and their effects. *Human Factors*, **25**, 161–75.

GREINER, T. M. (1991). *Hand Anthropometry of US Army Personnel* (Technical Report TR 92/011). Natick, MA: US Army Natick Research, Development and Engineering Center.

GRIFFIN, M. J. (1990). *Handbook of Human Vibration*. San Diego, CA: Academic Press.

GUIGNARD, J. C. (1985). Vibration. In Crally, L. V. and Cralley, L. J. (Eds). *Patty's Industrial Hygiene and Toxicology*. New York: John Wiley & Sons, pp. 635–724.

GUTH, S. K. (1958). Light and comfort. *Industrial Medicine and Surgery*, **27**, 570–4.

HAERMAE, M. (1996). Ageing, physical fitness and shiftwork tolerance. *Applied Ergonomics*, **27**, 25–9.

HAGBERG, M. (1982). Arbetsrelaterade besvär i halsrygg och skuldra. *Swedish Work Environment Fund*, Report, Stockholm.

HAGGARD, H. W. and GREENBERG, L. A. (1935). *Diet and Physical Efficiency*. New Haven, CT: Yale University Press.

HARRIS, W., MACKIE, R. R. et al. (1972). *A Study of the Relationship among Fatigue, Hours of Service and Safety of Operations of Truck and Bus Drivers*. Report No. 1727–2, Santa Barbara, Goleta, CA 93017: Human Factor Research Inc.

HARRISON, R. V. (1978). Person-environment fit and job stress. In Cooper, C. and Payne, R. (Eds.) *Stress at Work*. New York: John Wiley.

HASHIMOTO, K. (1969). Physiological features of monotony manifested under high speed driving situations. In *Proceedings of the 16th International Congress of Occupational Health, Tokyo*, 85–8 Railway Labour Science Institute, Japan National Railways.

HAUBNER, P. and KOKOSCHKA, S. (1983). *Visual Display Units — Characteristics of Performance. International Commission on Illumination (CIE)*, 20th Session in Amsterdam, 52 Bd. Paris, France: Malesherbes.

VAN DER HEIDEN, G. and KRÜGER, H. (1984). *Evaluation of Ergonomic Features of the Computer Vision Instaview Graphics Terminal*. Report of the Department of Ergonomics. Zürich: Swiss Federal Institute of Technology.

VAN DER HEIDEN, G., BRÄUNINGER, U. and GRANDJEAN, E. (1984). Ergonomic studies on computer aided-design. In Grandjean, E. (Ed.) *Ergonomics and Health in Modern Offices*. London: Taylor & Francis.

DEN HERTOG, F. J. and KERKHOFF, W. H. C. (1974). Vom Fliessband zur selbständigen Gruppe. *Industrielle Organisation*, **43**, 21–4.

HESS, W. R. (1948). *Die funktionelle Organisation des vegetativen Nervensystems*. Basel: Schwabe.

HETTINGER, T. (1960). Muskelkraft bei Männern und Frauen. *Zentralblatt Arbeit und Wissenschaft*, **14**, 79–84.

HETTINGER, T. (1970). *Angewandte Ergonomie*. Frechen, Germany: Bartmann.

HETTINGER, T. and MÜLLER, E. A. (1953). Der Einfluss des Schuhgewichtes auf den Energieumsatz beim Gehen und Lastentragen. *Arbeitsphysiologie*, **15**, 33–40.

HEUER, H. and OWENS, D. A. (1989). Vertical gaze direction and the resting posture of the eyes. *Perception*, **18**, 353–77.

HILGENDORF, L. (1966). Information input and response time. *Ergonomics*, **9**, 31–7.

HILL, J. H. and CHERNIKOFF, R. (1965). Altimeter Display Evaluation: Final Report USN. *NEL Report 6242*, Jan. 26.

HILL, S. G. and KROEMER, K. (1986). Preferred declination of the line of sight angle. *Ergonomics*, **29**, 1129–34.

HORNE, J. A. (1988). *Why We Sleep — The Functions of Sleep in Humans and Other Mammals*. Oxford, UK: Oxford University Press.

HORT, E. (1984). A new concept in chair design. *Proceedings of Ergodesign 84, Behaviour and Information Technology*, **3**, 359–62.

HOUGHTEN, F. C. and YAGLOU, C. P. (1923). Determining lines of equal comfort. *ASHVE Transactions*, **29**, 163–71.

HOYOS, C. (1974). *Kompatibilität in Ergonomie Band 2*, Schmidtke, H. (Ed.). Munich: Hanser.

HÜNTING, W. and GRANDJEAN, E. (1976). Sitzverhalten und subjektives Wohlbefinden auf schwenkbaren und fixierten Formsitzen. *Arbeitswissenschaft*, **30**, 161–4.

HÜNTING, W., NEMECEK, J. and GRANDJEAN, E. (1974). Die physische Belastung von Arbeitern an der Gesenkschmiede. *Sozial- und Präventivmedizin*, **19**, 275–8.

HÜNTING, W., NAKASEKO, M., GIERER, R. and GRANDJEAN, E. (1982). Ergonomische Gestaltung von alphanumerischen Tastaturen. *Sozial- und Präventivmedizin*, **27**, 251–2.

ILLUMINATING ENGINEERING SOCIETY (1972). *IES Lighting Handbook* (5th Edn). New York.

IMRHAN, S. N., NGUYEN, M. T. and NUYEN, N. N. (1993). Hand Anthropometry of Americans of Vietnamese Origin. *International Journal of Industrial. Ergonomics* **12**, 281–7.

INTERNATIONAL ORGANIZATION FOR STANDARDIZATION TC 43 (1971). *Assessment of Noise-Exposure during Work for Hearing Conversation Purposes*. Geneva.

INTERNATIONAL ORGANIZATION FOR STANDARDIZATION 2631 (1974). *Guide for the Evaluation of Human Exposure to Whole-Body Vibration*. Geneva.

INTERNATIONAL ORGANZATION FOR STANDARDIZATION (1989). Hot Environments. Standard 7243. Geneva, Switzerland.

JASCHINSKI-KRUZA, W. (1991). Eyestrain in VDU users: viewing distance and the resting position of ocular muscles. *Human Factors*, **33**, 69–83.

JASPER, H. (1974). Quoted from W. F. Ganong: Lehrbuch der Medizinischen Physiologie, Deutsche Ausgabe. Berlin: Springer Verlag.

JENKINS, W. O. (1947). The tactual discrimination of shapes for coding aircraft type controls. In Fitts, P. M. (Ed.) *Psychological Research on Equipment Design*. Army Air Force, Aviation Psychology Program, Report 19.

JERISON, H. J. (1959). Effects of noise on human performance. *Journal of Applied Psychology*, **43**, 96–101.

JERISON, H. J. and PICKETT, R. M. (1964). Vigilance: the importance of the elicited observing rate. *Science*, **143**, 970–1.

JOHANSSON, G. (1984). In Cohen, B. G. F. (Ed.) *Human Aspects in Office Automation*, Elsevier Series in Office Automation, Vol. 1 New

York: Elsevier.

JOHANSSON, G. and ARONSSON, G. (1980). *Stress Reactions in Computerized Administrative Work*. Report from the Department of Psychology, University of Stockholm, Suppl. 50, November.

JOHANSSON, G., ARONSSON, G. and LINDSTRÖM, B. O. (1976). *Social, Psychological and Neuroendocrine Stress Reactions in Highly Mechanized Work*. Report from the Psychological Laboratories, University of Stockholm, No. 488.

JORNA, J. C., MOHAGEG, M. F. and SNYDER, H. L. (1989). Performance, perceived safety, and comfort of the alternating tread stair. *Applied Ergonomics*, **20**, 26–32.

JÜRGENS, H. W., AUNE, I. A. and PIEPER, G. (1990). *International Data on Anthropometry* (Occupational Safety and Health Series No. 65). Geneva: International Labour Office.

KALSBECK, J. W. H. (1971). Sinus arrhythmia and the dual task method in measuring mental load. In Singleton, W. T., Fox, J. G. and Whitfield, D. (Eds.) *Measurement of Man at Work*. London: Taylor & Francis.

KARRASCH, K. and MÜLLER, E. A. (1951). Das Verhalten der Pulsfrequenz in der Erholungsperiode nach körperlicher Arbeit. *Arbeitsphysiologie*, **14**, 369–82.

KEEGAN, J. J. (1953). Alterations of the lumbar curve related to posture and seating. *Journal of Bone and Joint Surgery*, **35**, 567–89.

KELLY, D. L. (1971). *Kinesiology: Fundamentals of Motion Description*. Englewood Cliffs, NJ: Prentice Hall.

KERR, W. A. (1946). Worker attitudes toward scheduling of industrial music. *Journal of Applied Psychology*, **30**, 575–8.

KLOCKENBERG, E. A. (1926). *Rationalisierung der Schreibmaschine und ihrer Bedienung*. Berlin: Springer.

KNAUTH, P. (1996) Designing better shift systems. *Applied Ergonomics*, **27**, 39–44.

KOCH, K. W., JENNINGS, B. H. and HUMPHREYS, C. H. (1960). Is humidity important in the temperature comfort range? *ASHRAE Transactions*, **66**, 63–8.

KOGI, K. (1991). Job content and working time: the scope of joint change. *Ergonomics*, **34**, 757–773.

KOGI, K. (1996). Improving shift workers' health and tolerance to shiftwork: recent advances. *Applied Ergonomics*, **27**, 5–8.

KRÄMER, J. (1973). *Biomechanische Veränderungen im lumbalen Bewegungssegment*. Stuttgart: Hippokrates.

KROEMER, K. H. E. (1964). Uber den Einfluss der räumlichen Lage von Tastenfeldern auf die Leistung an Schreibmaschinen. *Internationale Zeitschrift für Angewandte Physiologie einschl. Arbeitsphysiologie*, **20**, 240–51.

KROEMER, K. H. E. (1965a). Uber die ergonomische Bedeutung der räumlichen Lage kreisbogenförmiger Bewegungsbahnen von Betätigungsteilen. Dissertation, Fakultat für Maschinenwesen. Hannover: Technische Hochschule.

KROEMER, K. H. E. (1965b). Vergleich einer normalen Schreibmaschinentastatur mit einer 'K-Tastatur'. *Internationale Zeitschrift für Angewandte Physiologie einschl. Arbeitsphysiologie*, **20**, 453–64.

KROEMER, K. H. E. (1971). Foot operation of controls. *Ergonomics*, **14**, 333–9.

KROEMER, K. H. E. (1972). Human engineering the keyboard. *Human Factors*, **14**, 51–63.

KROEMER, K. H. E. (1989). Cumulative trauma disorders: their recognition and ergonomic measures to avoid them. *Applied Ergonomics*, **20**, 274–80.

KROEMER, K. H. E. (1995). Alternative keyboards and alternatives to keyboards. In Greco, A., Molteni, G., Occhipinti, E. and Piccolo, B. (Eds) *Work with Display Units '94*. Amsterdam: North Holland, pp. 277–82.

KROEMER, K. H. E. (1997). *Ergonomic design of material handling systems*. Boca Raton, FL: CRC Press.

KROEMER, K. H. E., SNOOK, S. H., MEADOWS, S. K. and DEUTSCH, S. (Eds.) (1988). *Ergonomic Models of Anthropometry, Human Biomechanics, and Operator-Equipment Interfaces*. Washington, DC: National Academy Press.

KROEMER, K. H. E., KROEMER, H. B., and KROEMER-ELBERT, K. E. (1994). *Ergonomics: How to Design for Ease and Efficiency*. Englewood Cliffs, NJ: Prentice Hall.

KROEMER, K. H. E., KROEMER, H. J. and KROEMER-ELBERT, K. E. (1997). *Engineering Physiology: Bases of Human Factors/Ergonomics*. (3rd Edn). New York: Van Nostrand Reinhold.

KRUEGER, H. (1984). Zur Ergonomie von Balans-Sitzelementen im Hinblick auf ihre Verwendbarkeit als reguläre Arbeitsstühle. Report of the Department of Ergonomics, 8092 Zürich: Swiss Federal Institute of Technology.

KRUEGER, H. and HESSEN, J. (1982). Objective kontinuierliche Messung der Refraktion des Auges. *Biomedizinische Technik*, **27**, 142–7.

KRUEGER, H. and MÜLLER-LIMMROTH, W. (1979). Arbeiten mit dem Bildschirm–aber richtig! 8000 Munich 40: Bayerisches Staatsministerium für Arbeit und Sozialordnung.

KUORINKA, I. and FORCIER, L. (Eds) (1995). *Work-related Musculoskeletal Disorders (WMSDs): A Reference Book for Prevention*. London: Taylor & Francis.

LÄUBLI, T. (1981). Das arbeitsbedingte cervicobrachiale Überlastungssyndrom. Thesis, Medical Faculty, University of Zurich.

LÄUBLI, T. and GRANDJEAN, E. (1984). The magic of control groups in VDT field studies. In Grandjean, E. (Ed.). *Ergonomics and Health in Modern Offices*. London: Taylor & Francis.

LÄUBLI, T., HÜNTING, W. and GRANDJEAN, E. (1981). Postural and visual loads at VDT workplaces, Part 2: Lighting conditions and visual impairments. *Ergonomics*, **24**, 933–44.

LÄUBLI, T., SENN, E., FASSER, W., MION H., CARLO, T. and ZEIER, H. (1986). Klinische Befunde und subjektive Klagen über Beschwerden im Bewegungsapparat. *Sozial- und Präventivmedizin*, **31**(2).

LAZARUS, R. S. (1977). Cognitive and coping processes in emotion. In Monat, A. and Lazarus, R. S. (Eds) *Stress and Coping., New York: Columbia University Press*.

LECRET, F. (1976). *La Fatigue du Conducteur. Cahier d'etude de l'Organisme National de Sécurité Routière* (ONSER) Bulletin No. 38, Paris.

LEHMANN, G. (1962). *Praktische Arbeitsphysiolgie* (2nd Edn). Stuttgart: Thieme Verlag.

LEHMANN, G. and STIER, F. (1961). Mensch und Gerät. *Handbuch der gesamten Arbeitsmedizin*. Vol. 1. Berlin: Urban und Schwarzenberg, pp. 718–88.

LEPLAT, J. (1968). Attention et incertitude dans les travaux de surveillance et d'inspection. *Sciences du Comportement*, No. 6, Paris: Dunod.

LEVI, L. (1975). *Emotions–Their Parameters and Measurement*. New York: Raven Press.

LILLE, F. (1967). Le Sommeil de jour d'un groupe de travailleurs de nuit. *Le Travail Humain*, **30**, 85–97.

LIND, A. R. and McNICOL, G. W. (1968). Cardiovascular responses to holding and carrying weight by hand and by shoulder harness. *Journal of Applied Physiology*, **25**, 261–7.

LUCKIESH, H. and MOSS, F. K. (1937). *The Science of Seeing*. New York: Van Nostrand.

LUEDER R. and NORO, K. (Eds). (1995). *Hard Facts about Soft Machines: The Ergonomics of Seating*. London: Taylor & Francis.

LUNDERVOLD, A. (1951). Electromyographic investigations of position and manner of working in typewriting. *Acta Physiologica Scandinavia*, **84**, 171–83.

LUNDERVOLD, A. (1958). Electromyographic investigations during typewriting. *Ergonomics*, **1**, 226–33.

MACKAY, C., COX, T., BURROWS, G. and LAZZERINI, T. (1978). An inventory for the measurement of self reported stress and arousal. *British Journal of Social and Clinical Psychology*, **17**, 283–4.

MACKWORTH, J. F. (1969). *Vigilance and Habituation*. Harmondsworth: Penguin Books.

MACKWORTH, N. H. (1950). *Research on the Measurement of Human Performance*. London: HMSO.

MAEDA, K., HORIGUCHI, S. and HOSOKAWA, M. (1982). History of the studies on occupational cervicobrachial disorder in Japan and remaining problems. *Journal of Human Ergology*, **11**, 17–29.

MAIRIAUX, P. and MALCHAIRE, J. (1995). Comparison and validation of heat stress indices in experimental studies. *Ergonomics*, **32**(1), 58–72.

MALCHAIRE, J. (1995). Methodology of investigation of hot working conditions in the field. *Ergonomics*, **38**(1), 73–85.

MALHOTRA, M. S. and SENGUPTA, J. (1965). Carrying of school bags by children. *Ergonomics*, **8**, 55–60.

MANDAL, A. C. (1984). What is the correct height of furniture? In Grandjean, E. (Ed.) *Ergonomics and Health in Modern Offices*. London: Taylor & Francis.

MARIC, D. (1977). *L'Aménagement du Temps de Travail*. Geneva: Bureau International du Travail.

MARRAS, W. S., KING, A. I. and JOYNT, R. L. (1984). Measurements of loads on the lumbar spine under isometric and isokinetic conditions. *Spine*, **9**, 176–88.

MARRAS, W. S. and MIRKA, G. A. (1989). Trunk strength during asymmetric trunk motion. *Human Factors*, **31**, 667–77.

MARRAS, W. S., LAVENDER, S. A., LEURGANS, S. E., FATHALLAH, F. A., FERGUSON, S. A., ALLREAD, G. A. and RAJULU, S. L. (1995). Biomechanical risk factors for occupationally related low back disorders. *Ergonomics*, **38**, 377–410.

MARTIN, E. and WEBER, A. (1976). Wirkungen eintönig-repetitiver Tätigkeiten auf das subjektive Befinden und die Flimmerverschmelzungsfrequenz. *Arbeitswissenschaft*, **30**, 183–7.

McCONNELL, W. J. and SPIEGELMAN, M. (1940). Reactions of 745 clerks to summer air conditioning. *Heating, Piping, Air Conditioning*, **12**, 317–22.

McCONNELL, W. J. and YAGLOU, C. P. (1925). Work tests conducted in atmospheres of high temperatures and various humidities in still and moving air. *Journal of the American Society of Heating and Ventilation Engineers*, **31**, 217–21.

McCORMICK, E. J. and SANDERS, M. (1987). *Human Factors in Engineering* (5th Edn). New York: McGraw-Hill.

McFARLAND, R. A. (1946). *Human Factors in Air Transport Design*. New York: McGraw-Hill.

McGRATH, J. E. (1976). Stress and behaviour in organisations. In *Handbook of Industrial and Organisational Psychology*. Chicago, IL: Rand McNally.

McMILLAN, G. R., BEEVIS, D., SALAS, E., STRUB, M. H., SUTTON, R. and VAN BREDA, L. (Eds.) (1989). *Application of Human Performance Models to System Design*. New York: Plenum.

MITAL, A., NICHOLSON, A. S. and AYOUB, M. M. (1993). *A Guide to Manual Materials Handling*. London: Taylor & Francis.

MOLBECH, S. (1963). Average percentage force at repeated maximal isometric muscle contractions at different frequencies. *Communications of the Danish National Association for Infantile Paralysis*, **16**.

MONOD, H. (1967). La dépense energétique chez l'homme. In Scherrer, J. (Ed.) *Physiologie du Travail*. Paris: Masson.

MOTT, P. E., MANN, C., McLOUGHLIN, C. and WARWICK, P. (1965). *Shiftwork: The Social, Psychological and Physical Consequences*. Ann Arbor, MI: University of Michigan Press.

MÜLLER, E. A. (1961). Die physische Ermüdung. In *Handbuch der gesamten Arbeitsmedizin*, Band 1. Berlin: Urban & Schwarzenberg.

MÜLLER-LIMMROTH, W. (1973). Sinnesorgane. In Schmidtke, H. (Ed.) *Ergonomie* Vol. 1, Munich: Hanser.

NACHEMSON, A. (1974). Lumbal intradiscal pressure. Results from *in vitro* and *in vivo* experiments with some clinical implications. 7. *Wissenschaftliche Konferenz. Deutscher Naturforscher and Aerzte*. Berlin: Springer.

NACHEMSON, A. and ELFSTRÖM, G. (1970). Intravital dynamic pressure measurements in lumbar discs. *Scandinavian Journal of Rehabilitation Medicine*, Suppl. 1.

NAKASEKO, M., GRANDJEAN E., HÜNTING, W. and GIERER, R. (1985). Studies on ergonomically designed alphanumeric keyboards. *Human Factors*, **27**, 175–87.

NASA (1989). *Man-System Integration Standard* (NASA Standard 3000, Revision A). Houston, TX: LBJ Space Center SP34-89-230.

NATIONAL INSTITUTE FOR OCCUPATIONAL SAFETY AND HEALTH (1981). *Work Practices Guide for Manual Lifting*. Cincinnati, OH.

NEMECEK, J. and GRANDJEAN, E. (1971). Das Grossraumbüro in arbeits-physiologischer Sicht. *Industrielle Organisation*, **40**, 233–43.

NEMECEK, J. and GRANDJEAN, E. (1975). Etude ergonomique d'un travail pénible dans l'industrie textile. *Le Travail Humain*, **38**, 167–74.

NEUMANN, J. and TIMPE, K. P. (1970). Arbeitsgestaltung. *Psycho-physiologische Probleme bei Ueberwachungs- und Steuerungstätig-keiten*. Berlin: VEB Deutscher Verlag der Wissenschaften.

NEVINS, R. G., ROHLES, F. H., SPRINGER, W. and FEYERHERM, A. M. (1966). A temperature-humidity chart of thermal comfort of seated persons. *ASHRAE Journal*, **8**, 55–61.

NISHIYAMA, K., NAKASEKO, M. and UEHATA, T. (1984). Health aspects of VDT operators in the newspaper industry. In Grandjean, E. (Ed.) *Ergonomics and Health in Modern Offices*. London: Taylor & Francis.

NORTHRUP, H. R. (1965). *Hours of Work*. New York: Harper and Row.

NORWEGIAN MONOGRAPHS ON MEDICAL SCIENCE (1964). Oslo: Universitets forlaget.

OEZKAYA, N. and NORDIN, M. (1991). *Fundamentals of Biomechanics*. New York: Van Nostrand Reinhold.

O'HANLON, J. F. (1971). *Heart Rate Variability: A New Index of Drivers' Alertness/Fatigue*. Report No., 1812–1. Santa Barbara, CA: Human Factors Research Inc.

O'HANLON, J. F., ROYAL, J. W. and BEATTY, J. (1975). *EEG Theta Regulation and Radar Monitoring Performance in a Controlled Field Experiment*. Report No. 1738. Santa Barbara, CA: Human Factors Research Inc.

OLESEN, B. W. and MADSEN, T. L. (1995). Measurement of the physical parameters of the thermal environment. *Ergonomics*, **38**, 138–53.

PARSONS, K. C. (1995). International heat stress standards: A review. *Ergonomics*, **38**, 6–22.

PEACOCK, B. and KARWOWSKI, W. (Eds) (1993). *Automotive Ergonomics*. London: Taylor & Francis.

PEARSON, R. G. and BYARS, G. E. (1956). *The Development and Validation of a Checklist for Measuring Subjective Fatigue*. Report 56–115. Randolph AFB, TX: School of Aviation Medicine, USAF.

PHEASANT, S. (1986). *Bodyspace: Anthropometry, Ergonomics and Design*. London: Taylor & Francis.

PHEASANT, S. (1996). *Bodyspace: Anthropometry, Ergonomics and the Design of Work* (2nd Edn). London: Taylor & Francis.

PROCEEDINGS OF THE INTERNATIONAL CONFERENCE ON ENHANCING THE QUALITY OF WORKING LIFE (1972). Harriman, New York: Arden House.

PROKOP, O. and PROKOP, L. (1955). Ermüdung und Einschlafen am Steuer. *Deutsche Zeitschrift für gerichtliche Medizin*, **44**, 343–50.

PUTZ-ANDERSON, V. (Ed.) (1988). *Cumulative Trauma Disorders: A Manual for Musculoskeletal Diseases of the Upper Limbs*. London: Taylor & Francis.

RAMSEY, J. D. (1987). Practical evaluation of hot working areas. *Professional Safety*, 42–8.

RAMSEY, J. D. (1995). Task performance in heat: A review. *Ergonomics*, **38**, 154–65.

REY, P. and REY, J. P. (1965). Effect of an intermittent light stimulation on the critical fusion frequency. *Ergonomics*, **8**, 173–80.

ROBINSON, G. S. and CASALI, J. G. (1995). Audibility of reverse alarms under hearing protectors for normal and hearing-impaired users. *Ergonomics*, **38**, 2281–99.

ROBINSON, D. W. and DADSON, R. S. (1957). Threshold of hearing and equal-loudness relations for pure tones and the loudness function. *Journal of the Acoustical Society of America*, **29**, 1284–8.

ROBINSON, G. and GERKING, S. D. (1947). The thermal balance of men working in severe heat. *American Journal of Physiology*, **149**, 102–8.

ROEBUCK, J. A. (1995). *Anthropometric Methods. Designing to Fit the Human Body*. Santa Monica, CA: Human Factors and Ergonomics Society.

ROHMERT, W. (1960). Statische Haltearbeit des Menschen. Special issue of REFA-Nachrichten.

ROHMERT, W. (1966). Maximalkräfte von Männern im Bewegungsraum der Arme und Beine. *Forschungsberichte des Landes Nordrhein-Westfalen No. 1616*. Cologne: Westdeutscher Verlag.

ROHMERT, W. and HETTINGER, T. (1965). Ergebnisse achtstündiger Untersuchungen am Kurbel- und Fahrradergometer. In Hill, J. H. and Chernikoff, R. (Eds.) *Altimeter Display Evaluation: Final Report*. USN, NEL Report 6242, Jan 26.

ROHMERT, W. and JENIK, P. (1972). *Maximalkräfte von frauen im Bewegungsraum der arme und beine*. (Series Arbeitswissenschaft und Praxis). Berlin: Beuth.

ROHMERT, W., RUTENFRANZ, J. and ULICH, E. (1971). *Das Anlernen sensumotorischer Fertigkeiten*. Frankfurt. Europäische Verlagsanstalt.

ROSEMEYER, B. (1971). Elektromyographische untersuchungen der Rücken-und Schultermuskulatur im stehen und sitzen unter berücksichtigung der haltung des autofahrers. *Archiv orthopaedische Unfall-Chirurgie*, **69**, 59–70.

RUPP, B. A. (1981). Visual display standards: a review of issues. *Proceedings of the Society for Information Display*, **22**, 63–72.

RUTENFRANZ, J. and KNAUTH, P. (1976). Rhythmusphysiologie und Schichtarbeit. Vienna: Sensenverlag,.

RYAN, A. H. and WARNER, M. (1936). The effect of automobile driving on the reactions of the driver. *American Journal of Physiology*, **48**, 403–9.

SAITO, H., KISHIDA, K., ENDO, Y. and SAITO, M. (1972). Studies on bottle inspection task. *Journal of Science of Labour*, **48**, 475–525.

SALVENDY, G. (1984). Research issues in the ergonomics, behavioural, organizational and management aspects of office automation. In Cohen, B. G. F. (Ed.) *Human Aspects in Office Automation*. New York: Elsevier.

SANDERS, M. S. and MCCORMICK, E. J. (1993). Human factors in engineering and design. (7th Edn). New York: McGraw-Hill.

SAUTER, S. L. (1984). Predictors of strain in VDT users and traditional office workers. In Grandjean, E. (Ed.) *Ergonomics and Health in Modern Offices*. London: Taylor & Francis.

SAUTER, S. L., GOTTLIEB, M. S., JONES, K. C., DODSON, V. N. and ROHRER, K. M. (1983). Job and health implications of VDT use: Initial results of the Wisconsin-NIOSH study. *Communications of the ACM*, **26**, 284–94.

SCHERRER, J. (1967). Physiologie Musculaire. In Scherrer, J. (Ed.) *Physiologie du Travail*. Paris: Masson.

SCHOLZ, H. (1963). Die physische arbeitsbelastung der giessereiarbeiter. Forschungsbericht des Landes Nordrhein-Westfalen No. 1185. Cologne: Verlag.

SCHMIDTKE, H. (1973). *Ergonomie*, Band 1. Munich: Hanser.

SCHMIDTKE, H. (1974). *Ergonomie*, Band 2. Munich: Hanser.

SCHMIDTKE, H. (Ed.) (1981). *Lehrbuch der Ergonomie*. Munich: Hanser.

SCHMIDTKE, H. and STIER, F. (1960). Der Aufbau komplexer Bewegungsabläufe aus Elementarbewegungen. *Forschungsbericht des Landes Nordrhein-Westfalen*, No. 822. Cologne: Westdeutscher Verlag.

SCHOBERTH, H. (1962). *Sitzhaltung — Sitzschaden — Sitzmöbel*. Berlin: Springer.

SELYE, H. (1978). *The Stress of Life*. New York: McGraw-Hill.

SHACKEL, B. (1974). *Applied Ergonomics Handbook*. Reprints from Applied Ergonomics Vols 1 and 2. Guildford: IPC Science and Technology Press.

SHAHNAVAZ, H. (1982). Lighting conditions and workplace dimensions of VDU operators. *Ergonomics*, **25**, 1165–73.

SHANNON, C. E. and WEAVER, W. (1949). *The Mathematical Theory of Communication*. Urbana, Ill: University of Illinois Press.

SHUTE, S. J. and STARR, S. J. (1984). Effects of adjustable furniture on VDT users. *Human Factors*, **26**, 157–70.

SLEIGHT, R. B. (1948). The effect of instrument dial shape on legibility. *Journal of Applied Psychology*, **32**, 170–88.

SMITH, M. J., STAMMERJOHN, L. W., COHEN, B. G. F. and LALICH, N. R. (1980). Job stress in video display operations. In Grandjean, E. and Vigliani, E. (Eds.) *Ergonomic Aspects of Visual Display Terminals*. London: Taylor & Francis.

SMITH, M. J., COHEN, B. C. F., STAMMERJOHN, L. W. and HAPP, A. (1981). An investigation of health complaints and job stress in video display operations. *Human Factors*, **23**, 387–99.

SNOOK, S. H. and CIRIELLO, V. M. (1991). The design of manual handling tasks: revised tables of maximum acceptable weights and forces. *Ergonomics*, **34**, 1197–213.

SODERBERG, G. L. (Ed.) (1992). *Selected Topics in Surface Electromyography for Use in the Occupational Setting: Expert Perspectives*. (DDHS-NIOSH Publication 91–100). Washington, DC: US Department of Health and Human Services.

SWINK, J. R. (1966). Intersensory comparisons of reaction time using an electropulse tactile stimulus. *Human Factors*, **8**, 143–5.

TAYLOR, C. L. (1954). The biomechanics of the normal end of the amputated upper extremity. In Kloptleg and Wilson, (Eds.) *Human Limbs and Their Substitutes*. New York: McGraw Hill, pp. 169–221.

TEPAS, D. I. and MAHAN, R. P. (1989). The many meanings of sleep. *Work and Stress*, **3**, 93–102.

THIBERG, S. (1965–70). *Anatomy for Planners, Parts I–IV*. Statens Institut für Byggnadsforskning. In Grandjean, E. (1988). *Fitting the task to the man* (4th Edn). London: Taylor & Francis.

THIIS-EVENSON, E. (1958). Shiftwork and health. *Industrial Medicine*, **27**, 493–7.

TICHAUER, E. R. (1968). Potential of biomechanics for solving specific hazard problems. *Conference of the American Society of Safety Engineers*, pp. 149–87, Park Ridge, Il: ASSE.

TICHAUER, E. R. (1973). *The Biomechanical Basis of Ergonomics*. New York: Wiley.

TICHAUER, E. R. (1975). *Occupational Biomechanics* (Rehabilitation Monograph No. 51). Center for Safety: New York University.

TICHAUER, E. R. (1976). Biomechanics sustains occupational safety and health. *Industrial Engineering*, **27**, 46–56.

TIMMERS, H. (1978). An effect of contrast on legibility of printed text. *IPO Annual Progress Report*, No. 13, 64–7.

VAN COTT, H. P. and KINKADE, R. G. (Eds.) (1972). *Human Engineering Guide to Equipment Design* (Rev. Edn). Washington, DC: US Government Printing Office.

VERNON, M. H. (1921). *Industrial Fatigue and Efficiency*. New York: Dutton.

WAKIM, K. G., GERTEN, J. W., ELKINS, E. C. and MARTIN, G. M. (1950). Objective recording of muscle strength. *Archives of Physical Medicine*, **31**, 90–9.

WARGO, M. J. (1967). Human operator response speed, frequency and flexibility: a review and analysis. *Human Factors*, **9**, 221–38.

WARRICK, M. J., KIBLER, A. W. and TOPMILLER, D. A. (1965). Response time to unexpected stimuli. *Human Factors*, **9**, 81–6.

WASSERMAN, D. E. (1987). *Human Aspects of Occupational Vibrations*. Amsterdam: Elsevier.

WEBER, A. (1984). Irritating and annoying effects of passive smoking. In Grandjean, E. (Ed.) *Ergonomics and Health in Modern Offices*. London: Taylor & Francis.

WEBER, A., JERMINI, C. and GRANDJEAN, E. (1973). Beziehung zwischen objektiven und subjektiven Messmethoden bei experimentell erzeugter Ermüdung. *Zeitschrift für Präventivmedizin*, **18**, 279–83.

VAN WELY, P. (1970). Design and disease. *Applied Ergonomics*, **1**, 262–9.

WENZEL, H. G. (1964). Möglichkeiten und Probleme der Beurteilung von Hitzebelastungen des Menschen. *Arbeitswissenschaft*, **3**, 73–83.

WILSON, J. R. and CORLETT, E. N. (Eds). (1995). *Evaluation of Human Work* (2nd Edn). London: Taylor & Francis.

WIRTHS, W. (1976). Ist eine Zwischenverpflegung während der Arbeitszeit ernährungsphysiologisch notwendig? In *Ernährungspädagogisches Colloquium*, Bericht 10. Bonn: Mühlenstelle.

WISNER, A. (1967a). Audition et Bruits. In Scherrer, J. (Ed.), *Physiologie du Travail*, Vol. 2. Paris: Masson.

WISNER, A. (1967b). Effets des vibrations sur l'homme. In Scherrer, J. (Ed.) *Physiologie du Travail*, Vol. 2. Paris: Masson.

WOODSON, W. E., TILLMAN, B. and TILLMAN, P. (1991). *Human Factors Design Handbook* (2nd Edn). New York: McGraw-Hill.

WYATT, S. and MARRIOTT, R. (1956). A study of attitudes to factory work. *MRC Special Report Series*, 292, London.

WYNDHAM, C. H. *et al.* (1953). Examination of heat stress indices. *Archives of Industrial Hygiene and Occupational Medicine*, **7**, 221–33.

YAGLOU, C. P., RILEY, E. C. and COGGINS, D. I. (1936). Ventilation requirements, *ASHVE Transactions*, **42**, 133–58.

YAGLOU, C. P., RILEY, E. C. and COGGINS, D. I. (1949). *Ventilation Requirements and the Science of Clothing*. Philadelphia, PA: Saunders.

YAMAGUCHI, Y., UMEZAWA, F. and JSHINADA, Y. (1972). Sitting posture: an electromyographic study on healthy and notalgic people. *Journal of the Japanese Orthopedics Association*, **46**, 51–6.

YLLÖ, A. (1962). The biotechnology of card-punching. *Ergonomics*, **5**, 75–9.

ZEIER, H. and BÄTTIG, K. (1977). Psychovegetative Belastung und Aufmerksamkeitsspannung von Fahrzeuglenkern auf Autobahnabschnitten mit und ohne Geschwindigkeitsbegrenzung. *Zeitschrift für Verkehrssicherheit*, **23**, 1.

ZIPP, P., HAIDER, E., HALPERN, N. and ROHMERT, W. (1983). Keyboard design through physiological strain measurements. *Applied Ergonomics*, **14**, 117–22.

Index

abduction angle 149
absenteeism 203, 224
  monotony 233, 235
  working hours 243, 244, 245
accidents
  fatigue 192, 205, 208–9
  monotony 233, 236
  night work and shift work 263
  vibrations 348, 349
  visual strain 291–2
  working hours 243
accommodation 278–81, 291, 336
action potential 6, 18–19
adaptation 190, 221, 222–3
  skilled work 147, 148–9
  vision 282–3, 284
adenosine triphosphate (ATP) 3, 19
adrenalin 195, 200–1, 224, 229
  stress 200, 211, 216
afferent sensory system 197–8, 199, 200
age 302, 372
  body size 34–45
  efficiency 25–6, 30
  hearing loss 329, 331
  heavy work 105
  lighting 301, 302, 308
  night work 268, 272
  vision 280–1, 286, 291
agriculture 121, 123, 165, 171
  back problems 129
  heavy work 101, 120, 121, 123, 124
  vibrations in tractors 348, 351, 352, 353
  working hours 244
air movement 355, 358–9, 362–4, 369, 373, 382
air pollution 377–82
alarm 325, 338
alertness (including vigilance) 103, 178, 183–90
  boredom 219, 221, 223, 225, 226
  fatigue 195, 196–8, 201, 202, 208–9
  monotonous tasks 236
  music 391
  night work and shift work 261
  noise 324, 325, 332, 335, 336
  rest breaks 249
  stress 211
  temperature 360
  vision 278
alpha rhythms 195, 205, 209, 226
altimeters 160–1
ankles 50
annoyance 332, 339–40, 350
armrests 76
arms
  body size 33, 36–45, 46, 48, 49–50
  controls 164, 168
  efficiency 28, 29, 30–1
  fatigue 193
  grasp and reach 66
  handling loads 130, 131
  heavy work 120, 125
  keyboard design 95–9
  muscular work 9–12, 14–15
  sitting 76, 78
  skilled work 149–50, 151, 152, 153, 154–5
  VDT design 84, 86–8, 90–2, 95–9
  vibrations 346, 348, 349, 351
  working height 58–9, 61
arousal or activation theory 189, 190
arthritis 13, 351
atrophy 351
audiometry 330, 345
auditory nerve 319, 322–3, 390
automation 22–3, 148, 173, 176
autonomic systems 198, 199
autonomy 234–5

back and back problems 71–81
  biomechanical models 134–6
  efficiency 26, 29
  grasp and reach 66
  handling loads 129–36, 142, 143, 146
  heavy work 107, 120, 125
  muscular work 9–10, 11, 13, 14, 15
  VDT design 84, 85, 87, 92, 93, 94, 98
  vibrations 347, 350, 351
  working height 54, 54–5, 57–8, 60–3
backrests 75, 76, 77, 78–83
  head and neck posture 65–6
  pedal controls 171
  VDT design 87, 93, 94
  working height 56, 57–8
blinking 287, 288

blocking theory 185
blood and circulation
  body temperature 355–7, 360–1, 371, 376
  heavy work 101–2, 116–17, 120
  muscles 4–5, 7, 8, 10, 14, 16
  nervous system 17
  see also heart; heart rate
blood pressure
  fatigue 199
  heat in industry 370–1
  heavy work 116, 117
  muscular work 5
  night work 260
  noise 338
  stress 211, 213, 215, 216
body size 33–51
  grasp and reach 66–7
  handling loads 138–41
  VDT design 89
  working height 53–7, 60, 61
body temperature 355–9
  comfort 359–64
  fluid intake 255
  heat in industry 370–7
  heavy work 116–17, 118
  night work 260
boredom 219–29
  control panels 175
  mental activity 187, 227–8
  monotonous tasks 219–20, 223–6, 229,
      231–6, 239
  music 390–1
  skilled work 154
  stress 217, 224, 228–9
brain 2, 180–3, 196–200
  body temperature 355, 356, 372
  boredom 221–2, 223, 226, 228
  fatigue 192–3, 195–200
  mental activity 177, 179, 180–3, 189, 190
  music 390
  nervous system 17, 19–20
  noise 319, 323, 324, 325, 332, 338
  skilled work 147
  stress 211
  vision 275, 276, 278
breathing see respiratory system
building industry 101, 171

CAD workstations 93–4
carbon monoxide 378–81
carpal tunnel syndrome 96
carrying loads 108, 110–14, 123, 129
  technique 145–6
catecholamines 211, 215, 224
cervical spine 78, 130, 347
cervical syndrome 78
cervicobrachial disorders 78
chairs and seats 57, 69–83
  distance to desk 58, 60
  seat angle 75–7, 79–83
  VDT design 83, 87, 93–4
  working height 57, 58, 60, 62
channel capacity theory 179–80, 189
characters in display 162–4
  light 296, 302, 312

vision 289–91, 293–4
chronic fatigue 194, 202–3, 266–7, 340
circadian fatigue 194, 202
circadian rhythm 259–60, 263, 264, 266–7
climate indoors 213, 355–82
  comfort 359–64
climbing 111
coding human–machine systems 165
colour 384, 385–90, 392
  light 296, 297–8, 301, 305–7, 309, 385–90
  monotony 236, 389
  vision 275, 277–8, 290–1
  walls and ceilings 297, 304, 305
complexity and efficiency 232
compression force 135, 136
computers 83–95
  noise 327–9, 341
  repetitive strain injury 352
  stress 215–17
  vibration 352, 354
  see also VDT operators
concentration 103
  colour 389–90
  fatigue 208
  mental activity 178, 186, 189–90
  music 391
  noise 324, 335, 342
  rest breaks 249, 250
  skilled work 147, 148, 153–4
  temperature 360
  vision 291
conduction of heat 357, 358
cones (retina) 275, 277
contrast 161, 300, 301–14, 387–8
contrast ratio 303–4, 306, 310, 312, 313
contrast sensitivity 285, 286–7, 288, 291
  lighting 302, 303, 309
controls 157, 158, 164–76
  panels 174–6
convection of heat 357, 358, 359, 370
coordination 147, 192
cornea 276, 277, 278, 279
cranks 169–70

damping vibrations 347, 348, 350
daydreaming 231–2
day–night rhythm 194, 198
decibels 320
decision making 178, 190, 233
delta rhythms 195, 261
design
  body size 33, 46, 48, 51
  control panels 174–5
  lighting 305
  monotonous tasks 231–9
  skilled work 152–5
  workstations 53–100
desk top tilt 63
desynchronisation 195, 200, 266
digestive problems 17, 71
  heat in industry 370, 371
  night work 265–7, 272
  noise 338
  stress 212, 213, 216, 217
digital display 158, 159

display 157–64, 172–6
document holders 87, 88
dragging 129
draughts 364, 369
driving
    boredom 226–7
    fatigue 205, 207, 208–9
    monotony 236
    posture 92, 95
    sleep 263–4
    vibration 348–9, 353
drop forging 120, 122
drugs 206–7, 266
dynamic effort 7, 8, 9, 11–13
    handling loads 129, 136
    heavy work 106, 115–17, 119, 120, 123

ear–eye line 64–6, 99
earplugs 344–5
ears 322–3, 344–5, 365
    damage to hearing 329–32
    noise 319, 320, 322–3
effective temperature 362–4, 374
efficiency 25–31, 105–15, 232
elbows 149, 153
    body size 36–45
    efficiency 27, 29–31
    grasp and reach 68
    VDT design 88, 90, 91, 95
    vibration 351
    working height 53–5, 61
electrical phenomena in muscles 5–6, 16,
    192–3
electroencephalograms (EEG)
    boredom 226
    fatigue 195–6, 204, 205, 209
    sleep 261–3
electromyograms (EMG) 5, 6–7, 74–5, 151
    keyboard design 97
    muscle fatigue 193, 196
    working height 58–9, 63
electrooculograms (EOG) 261
energy consumption 102, 103–6, 204
    heat 369, 374, 376
    heavy work 101, 102, 103–6, 107–10,
        115–17, 121
    nutrition at work 251–3
    rest breaks 246
    skilled work 148
    vibration 348
energy expenditure 103–4, 121, 151
engrams 180
equivalent noise level 325–8, 332, 341
ergotropic setting 199, 211, 260
errors
    display reading 160–1, 175
    fatigue 192
    monotonous tasks 233, 236
    night work 263
    nutrition at work 254, 255
    temperature 360
    vibration 349
    visual strain 291
evaporation 357, 358, 359, 361
    heat in industry 370, 375

evoked potentials 195–6
eyes 275–94
    air pollution 377, 379, 380, 381
    body size 36, 38, 40, 42, 44
    colour 388, 390
    efficiency 29, 30
    fatigue 194
    flicker–fusion 204, 205–7, 209, 225, 226–8,
        298
    head and neck posture 64–6
    light 295, 298, 300, 302, 303, 305, 308
    movements 284–5, 291
    protection from heat 377
    skilled work 151, 154
    sleep 261
    stress 216
    VDT design 89
    vibrations 347, 348
    working height 55

factories 9, 384–5
    boredom 221, 225
    monotonous tasks 234
    music 391
    night work 265
    noise 326, 328, 341, 343, 345, 331, 334,
        337, 341
    working hours 241, 244
fanlights 385
far point 280
fat 3–4, 101, 102, 250, 252, 253
fatigue 191–209
    boredom 220, 223, 225, 226
    control panels 175
    efficiency 28–9
    general 191, 194–201
    heat 370, 371
    heavy work 112, 118, 123, 125, 127
    lighting 314
    measurement 203–9
    mental activity 185, 194–201, 202, 204, 208
    monotony 206, 233, 236, 239
    muscles 2–4, 6, 8, 10, 13–14, 16, 28, 191–4
    neck and head posture 64–6
    night work and shift work 209, 266–7, 271
    noise 336, 340
    rest breaks 201–2, 209, 247–50
    sitting 78–9
    skilled work 150, 154, 194
    vibrations 352
    visual 65, 276, 280–1, 282, 285, 291–4, 314
    working hours 241, 243, 245
feet see legs and feet
filament lamps 297, 310
fine and delicate work 306–8
fingers 12
    controls 164, 166–9
    skilled work 147, 148, 152–3
    vibrations 347, 351–2
filter theory 189
flexible working hours 245–6
flicker 291, 293, 294, 298–9, 305
flicker–fusion 298
    boredom 225, 226–8
    fatigue 204, 205–7, 209

fluid intake 255, 256, 376
fluorescent lighting 298–9, 306, 310, 311
 phase–shifting 298, 310
food *see* nutrition
footrests 31, 56–8, 60, 82
forestry 101, 105, 120, 352
fovea 277, 278, 283, 288, 289
fragmentation of work 223, 231–3, 237–8, 239
frequency
 sound 319–26, 329–31, 335, 336, 339, 345
 vibrations 346–8, 349, 350, 351, 353

gender
 body size 34–45, 47, 49, 50
 boredom 220
 efficiency 25–6, 28, 30
 energy consumption 103–6
 grasp and reach 66–8
 handling loads 114, 137, 138–9, 140–4
 hearing loss 331
 heart rate 119
 heavy work 105, 123, 125
 monotonous tasks 231
 night work and shift work 268
 nutrition at work 251–2
 working height 54, 56, 60
 working hours 244

glare 283, 284, 302, 306–7, 383–5
 lighting 296–7, 299–300, 302, 305–9, 312,
  314–16
glass houses 384–5
glucose 3–4, 101, 116
grasp and reach 66–9, 150

habituation 190, 221, 222–3
hand levers 167–8, 173–4
hands 48–50
 body size 37, 39, 41, 43, 45, 48–50
 controls 164, 166–71
 efficiency 26–7, 29
 grasp and reach 66, 67, 68
 grip design 152–4
 handling loads 130, 131, 135
 heavy work 125
 keyboard design 95–8
 muscular work 12, 13, 15
 protection from heat 377
 skilled work 147–9, 152–5
 VDT design 84
 vibrations 346–9, 351–3
handwheels 170–1
Hawthorne Effect 337
head 9, 63–6, 78
 body size 37, 39, 41, 43, 45
 ear–eye line 64–6, 99
 handling loads 1331
 skilled work 153
 VDT design 86, 87, 91, 92
 vibrations 347
 working height 57, 63
hearing loss 329–32
heart
 body temperature 355, 371, 376
 heavy work 101, 115–21

muscular work 5, 8, 10–12, 14
heart rate 117–20
 fatigue 199, 209
 heat in industry 370, 372, 375
 heavy work 115–27
 mental activity 186
 night work and shift work 260
 noise 338
 stress 211, 215, 216
 vibration 348
heat 5, 369–77, 384–5
 acclimatisation 375–6
 from lighting 310
 heavy work 101–2, 106, 117, 120–1
 working conditions 103, 115–16, 120, 122,
  126, 250
 see also body temperature; temperature
heat stroke 361, 372
heavy work 101–27
 heat 370, 371, 374
 nutrition 251–2, 256
 rests 103, 248–9, 250
hippocampus 181, 182
hoeing 108, 109
human–machine systems 157–76
humidity 355, 364–7
 heat in industry 372, 373, 375
 temperature 355, 358, 359, 362–9, 382
humoral control 200
hypothalamus 181

illumination 295, 297, 299–302, 305, 308–12,
  316–17
 daylight 383, 385
incandescent lighting 297–8
industry 121, 369–77
 alertness 183
 boredom 219, 220, 221, 223
 heavy work 120, 121, 124
 human–machine systems 165, 171
 monotony 234
 music 390–1
 noise 325, 326, 336
 nutrition at work 254, 255
 working hours 244, 246, 249
information processing 177, 178–80
information theory 178–9, 182
information uptake 178–80
inhibiting systems 189, 198, 199, 201
inner ear 319, 320, 322–3
intervertebral discs 71–6, 94, 98, 130–4
 handling loads 129, 130–6, 142, 143
 muscular work 13, 14
 pressure 131–4
 sitting 71–6, 78
 vibration 351
 working height 62
intraabdominal pressure 131, 136–7, 143
iris 277, 281
ischaemia 112

job control 213, 214, 237–8, 239
job satisfaction
 boredom 220–1
 monotonous tasks 231, 232, 237–9

stress 213–14, 215, 217
joints 13, 14, 85
  angles of rotation 49–50

keyboards 11–12, 16, 95–100
  computers 83–95
  noise 326, 328, 341
  repetitive strain injury 239
  skilled work 152
  working height 60–1
  see also VDT operators
knobs for controls 165, 168, 172–3
kyphosis 72–4, 130

lactic acid 3–4, 10, 117, 192
legs and feet 31, 50, 60–1, 364
  body size 33, 36–45, 47, 50
  controls 164, 166, 171–2
  muscular work 9, 14, 15
  pedals 69, 154, 166, 171–2
  reach 67, 69
  sitting 71, 73–4, 82
  VDT design 86, 87, 92
  vibrations 346, 347, 350
  working height 56, 57, 58, 60–1, 62
lens 276, 277, 278, 279, 280, 282
lifting 129, 131–4, 136–8, 142–6
ligaments 13
light and lighting 295–317
  colours 296, 297–8, 301, 305–7, 309, 385–90
  controls 165
  display 161, 162, 163
  measurement 295–9
  monotonous tasks 236
  natural daylight 383–5, 392
  stress 213
  VDT operators 83–4, 296–7, 302, 305,
    310–17
  vision 275–8, 281–4, 286–8, 291–2
  visual strain 291–2
limbic system 181–2, 190, 198–9, 200
  boredom 222, 223
line of sight 64–6, 99, 308
load distance 135–6, 142, 143, 144
load handling 101, 129–46
lordosis 72–4, 76, 78, 130
loudness 319, 320–1, 329, 333, 336–7, 339
lowering 129, 139, 143, 145
lumbar pads 76, 77, 79, 81, 82
luminance 295–6
  colour 387–8
  lighting 295–305, 310, 311–13
  vision 286, 287, 293, 294

machinery
  noise 326–9, 336, 337, 341–3
  vibration 348, 349, 351, 352, 353
manual handling 101, 129–46
memory 178, 180–3, 190, 208
mental activity 177–90
  boredom 187, 227–8
  colour 389
  fatigue 185, 194–201, 202, 204, 208
  music 391
  noise 335–6, 342

rests 250
  temperature 360, 369, 370, 371
mental load 84, 178, 185, 190
mental stress 103
metabolism
  basal 103
  fatigue 192, 199
  heavy work 101–2, 103, 116–17, 118
  night work 260
  noise 338
  rest breaks 250
  skilled work 149
  stress 211
  temperature 360
  vibrations 348
mining 101
monotony and monotonous tasks 231–9
  boredom 219–20, 223–6, 229, 231–6, 239
  colour 236, 389
  fatigue 206, 233, 236, 239
  mental activity 178
  skilled work 150, 154
  snacks 257
  stress 213, 214, 215, 217
motivation 190, 232, 242
  boredom 220, 227
  fatigue 198–9, 208
muscles 1–16
  contraction 1–3, 5–7, 9–10, 19–21, 191–2
  controls 164, 169
  efficiency 2, 4, 25–8
  eyes 277, 279, 280, 284–5, 287, 291
  fatigue 191–4, 199, 209
  handling loads 131, 134
  head and neck 64
  heavy work 103, 115–17, 119, 120, 125
  keyboard design 97
  nerves 2, 19–23
  shoulders 58–9
  sitting 74–7, 78–9
  skilled work 147, 148–52, 154–5
  static effort 2, 7–16, 28–30, 58, 71
  strength 1–2, 25–7, 30
  temperature 355–6, 360, 371, 372
  underload 223–4
  VDT design 85, 92, 94
  vibration 348, 350
musculoskeletal problems 13–16, 64, 84, 152
  handling loads 134
  working height 56, 60
music 226, 236, 345, 390–2

National Institute for Occupational Safety and
  Health (NIOSH) 142–3, 145
near point 278, 279, 280, 336
neck 14, 63–6, 78, 153, 364
  handling loads 131
  heavy work 125
  keyboard design 98
  posture 63–6, 305
  VDT design 84, 86, 87, 92, 93
  working height 53, 56, 57, 60
nervous disorders 265–7
nervous system 17–23
  boredom 222, 229

nervous system (*continued*)
  fatigue 192–3, 194, 201, 209
  mental activity 179, 182
  skilled work 147, 152
  vibration 348
neurones 17–22, 180, 194, 195, 275
night work 259–74
  boredom 220
  fatigue 209, 266–7, 271
  health 264–9, 271, 274
  monotonous tasks 236
  sleep 259, 260–74
noise 325–9, 354
  damage to hearing 329–32
  music 390–2
  perception of sound 319–25
  physiological and psychological effects
      332–40
  protection 340–5
  side effects 324
  stress 213, 324, 338, 340
noise-induced hearing loss (NIHL) 329–32
no noise 341
noradrenalin 200, 211
nuisance of vibration 349, 354
nutrition 3, 16, 250–7
  night work and shift work 264–7, 269, 273

odours 377
optic nerve 275, 276, 277
optimal working field 150–1, 153
oscillations 346–9, 352, 353
overexertion 129–30
overload 224
overtime 241, 243
oxygen 3–4, 8, 11–12, 102, 118, 151

pain
  back problems 129, 131
  grasp and reach 66
  head and neck posture 64
  heavy work 123, 125, 127
  keyboard design 96
  muscles 8, 10, 13–14, 28
  nervous system 20–1
  sitting 73, 78
  VDT design 85, 86, 87
  vibration 350
  working height 56–7, 58, 62, 63
pedals 69, 154, 166, 171–2
perception
  fatigue 196, 202, 204, 207
  mental activity 178
  sound 319, 322
  speed 285, 287–8, 291
  vibration 348
  vision 275, 285, 287–8, 291
person–environment fit 212–13
phon curves 320–1
pitch 320
pointed bar knobs 169, 170
posture
  body size 33
  efficiency 25, 29–30
  energy consumption 103, 104–5

grasp and reach 68
handling loads 131, 132, 133, 136, 144–5
head and neck 63–6, 305
heavy work 113, 125
keyboard design 95–8
lighting 305
muscular work 7, 12, 14–16
sitting 57, 71–6
skilled work 149–50, 152
VDT design 83, 84, 86, 88–95, 100
working height 55, 56–7, 61, 63
precision
  human–machine systems 157, 164, 166, 168,
      169, 172
  skilled work 147, 150, 155
presbyopia 280–1
printers 327–9, 341
productivity
  fatigue 203–5
  night work and shift work 263–4
  nutrition 254
  rest breaks 247–8
  visual strain 291, 292
  working hours 241–3
Profile of Mood States (POMS) 212
protective caps (earmuffs) 344–5
proteins 1, 3–4, 102, 250, 252, 253
pulling 9, 27–8, 129, 141, 145
pupil 276, 277, 281–2, 284, 286, 305
push-button controls 166–7
pushing 9, 27–8, 129, 140, 145
psychomotor tests 204, 207–8
psychosomatic disorders 203, 212, 217, 224,
      266, 272
pyruvic acid 3–4

radiation of heat 357, 358–9, 370
  industry 373, 374, 376–7
Raynaud's disease 351–2
reaction times 183–5, 207, 224
reading 288–91
  lighting 297, 310, 311, 317
  vision 278, 280, 285, 288–91, 293–4
recovery pulse 118, 119–20, 123
reflectance 296, 300–1, 304, 306–9, 313–15
  colour 386–7
reflex arc 20, 147
reflexes 20–3, 173, 185
repetitive strain injury 12, 15, 239, 352
repetitive work
  boredom 219–20, 222–9
  fatigue 194, 206
  monotony 231–9
  music 390, 391
  rest breaks 249
  stress 213, 214, 215, 217
  VDT design 84
resistance of controls 166
resonance 347, 349
respiratory system
  air pollution 377, 379
  fatigue 199
  heavy work 101–2, 116–17, 118
  indoor climate 365–6, 368
  sitting 71

vibration 348, 351
response time 183, 185
rest and rest breaks 10, 246–50
 fatigue 201–2, 209, 247–50
 heat in industry 376
 heavy work 103, 248–9, 250
 monotonous tasks 236, 237, 238
 night work 268–73
 nutrition 254
 working hours 243, 246–50, 257
resting potential 5–6, 18
reticular activating system 196–200, 222, 223
retina 282–4, 296, 298, 305
 colour 386, 387
 vision 275–81, 282–4, 285, 289, 291
ribonucleic acis (RNA) 182
rods (retina) 275, 277–8
rotating knobs 168, 172–3
rotating switches 168–9

saccades 288–9, 290
safety and colour 386–7
satiation 220, 233
sawing 108, 109
scales 158, 159, 161, 162
screen height and inclination 87, 89, 93, 95
screen reflection 313–14
 light 297, 302, 312–15, 317
 visual strain 293–4
shift work 195, 221, 241, 245, 259–74
shivering 356, 357
shoulders 58–9
 body size 36–45
 efficiency 30
 fatigue 193
 grasp and reach 66–8
 handling loads 131
 keyboard design 97, 98
 muscular work 9–11, 14, 15
 sitting 78
 skilled work 149–52
 VDT design 84–7, 90, 92, 93
 vibration 347, 351
 working height 53, 56, 57–9, 61, 63
shovelling 107–8
sickness
 night work 264–9, 271, 274
 noise 340
 vibration 350–1, 354
 working hours 243, 245
sinus arrhythmia 186
sitting 69–83, 368–9
 air pollution 378
 body size 35, 36, 38, 40, 42, 44, 46–7
 climate 364, 368–9, 370, 374
 efficiency 26–7, 29
 grasp and reach 67
 nutrition 251–2, 256
 vibration 349–50, 351
 working height 53, 55–7, 60–2
skilled work 20–3, 30, 103, 147–55
 fatigue 150, 154, 194
 vibration 348
skylights 385
sleep 260–4, 338–9

drowsiness 189
 fatigue 194–5, 197, 199–201, 205–6
 night work 259, 260–74
 noise 324, 332, 338–9, 340
 stress 216
 temperature 360
smoking 377–81
social contacts 235–6, 238, 268–9
 mental activity 178
 monotonous tasks 235–9
 night work and shift work 259–60, 266–71
 rest breaks 249
 stress 213
sodium–potassium pump 19
sound see noise
sound absorption 343
sound dampening 343
speech 327, 332–6, 337–8
spinal cord 20–1
spine
 cervical 78, 130, 347
 handling loads 130–1, 133, 134–6, 142, 143
 sitting 71–7
 vibrations 347, 351
 working height 63
staircases 111, 115
standing 9, 14–15, 16, 71, 72–4
 body size 34, 36, 38, 40, 42, 44, 47–8
 climate 364, 370
 efficiency 27–8
 grasp and reach 66–8
 pedals 171, 172
 vibration 350, 351
 working height 53–5, 61–2
static effort 2, 7–16, 28–30, 58, 71, 129, 172
 heavy work 103, 106–7, 115–17, 120–1, 123, 125
 skilled work 151, 153, 155
stereotype control and display 172–6
stress 103, 211–17
 adrenalin 200, 211, 216
 boredom 217, 224, 228–9
 fatigue 194, 195, 200, 201–4, 206, 208
 heat 373, 376
 heavy work 116, 121
 mental activity 186, 188
 night work and shift work 268
 noise 213, 324, 338, 340
 vision 291
sugars 3, 8, 250, 252
 fatigue 192, 199
 heavy work 101, 116
 stress 211
summated frequency level 326
supervisory work design 236
supports 29–31, 54, 56–7
 keyboard design 98
 VDT design 86, 87–8, 91
sweating 117, 357, 358
 body temperature 356–9, 361
 fluid intake 255
 heat in industry 370–1, 375, 376
switch levers 167–8

tachistoscopy 287

Taylorism 223, 231, 232
teeth 256–7
temperature 355, 358, 359–60, 362–9, 381–2
  colour effect 389
  heat in industry 369–77
  see also body temperature; heat
temperature of surfaces 355, 358, 359, 362–3, 368
temporary threshold shift 330–1, 336
tendons 1, 12, 13–14, 20–1, 96, 239
  keyboard design 95
  skilled work 152
  VDT design 85, 86
  vibration 351
textile industry 123, 125, 126, 127, 307
thalamus 181
theta rhythms 195, 205, 261
toggle switches 167
training 147–9, 249, 250
transport
  alertness 183
  boredom 219, 223
  fatigue 208–9
  heavy work 101
  human–machine systems 165
  noise 325, 326–7, 342
  vibration 348, 351, 352, 353
  vision 287
  see also driving
tremor 284
trophotropic setting 199–200, 260
trunk
  body size 33, 36–45
  efficiency 29
  grasp and reach 66, 68
  handling loads 131, 134, 136, 142, 144
  head and neck posture 66
  heavy work 123
  muscular work 11
  sitting 74–5, 78, 80
  skilled work 149, 151
  VDT design 90, 91, 92, 94, 95
  working height 56, 57, 58, 61, 63

ulcers 212, 213, 265
underload 223–4

Valium 206–7
varicose veins 14, 15
ventilation 377–82
vergence 285, 291
vibrations 320, 342, 346–54
viewing distance 55, 61, 89, 161, 163
vigilance see alertness
vision 275–94, 336
visual acuity 282, 285–6, 287, 288, 291
  colour 387
  lighting 309
  noise 336
  vibration 348
visual control 147

visual display terminal (VDT) operators 83–95, 310–17
  design of workstations 83–95
  display characters 163
  head and neck posture 66
  keyboard design 95–100
  lighting 83–4, 296–7, 302, 305, 310–17
  monotony 237–8
  noise 327, 328, 329
  sitting 74, 78
  stress 215–17
  vision 290, 291, 292–4
  working hours 249
visual distance 74, 89, 94, 153
visual fatigue 65, 276, 280–1, 282, 285, 291–4, 314
visual field 278–9, 284, 287, 289–90
  lighting 302–6, 308, 311
visual strain 84, 291–4

walking 108, 110
waste products 3–4, 8, 117, 192
weariness
  boredom 219, 221, 228
  fatigue 194–6, 202, 205, 206
  temperature 360
weekly hours 243–5, 257
weighted sound level 321–2
windows
  coverings 315
  daylight 383–4
  lighting 304, 305, 314, 315, 316
  ventilation 381–2
working conditions
  boredom 225
  colours 297, 304, 305, 385–90
  heat 103, 115–16, 120, 122, 126, 250
  indoor climate 355–82
  monotony 231, 233, 237
  music 390–2
  see also light and lighting
working height 53–63, 83, 87–9, 93–5, 149–50
working hours 241–6, 257
  heat in industry 376
  night work and shift work 269–70
  nutrition 254, 256
  rest breaks 243, 246–50, 257
workstations 53–99
  computers 83–95
  head and neck posture 63–6
  keyboards 95–100
  room to grasp and move 66–9
  sitting 69–83
  working height 53–63
wrists
  controls 168
  keyboard design 95–9
  mobility 49–50
  skilled work 152
  VDT design 84, 86, 87–8, 91, 92
  vibrations 348, 351, 352
writing 22–3, 148

Sheila Norton lives near Chelmsford in Essex with her husband, and worked for most of her life as a medical secretary, before retiring early to concentrate on her writing. Sheila is the award-winning writer of numerous women's fiction novels and over 100 short stories, published in women's magazines.

She has three married daughters, six little grandchildren, and over the years has enjoyed the companionship of three cats and two dogs. She derived lots of inspiration for her animal books from remembering the pleasure and fun of sharing life with her own pets.

When not working on her writing Sheila enjoys spending time with her family and friends, as well as reading, walking, swimming, photography and travel. For more information please see www.sheilanorton.com

Also by Sheila Norton:

The Vets at Hope Green
Oliver the Cat Who Saved Christmas
Charlie the Kitten That Saved a Life

SHEILA NORTON

THE PETS AT PRIMROSE COTTAGE

EBURY
PRESS

3 5 7 9 10 8 6 4 2

Ebury Press, an imprint of Ebury Publishing
20 Vauxhall Bridge Road,
London SW1V 2SA

Penguin
Random House
UK

Ebury Press is part of the Penguin Random House group of companies
whose addresses can be found at global.penguinrandomhouse.com

First published in the UK in 2018 by Ebury Press

www.penguin.co.uk

A CIP catalogue record for this book is available from the British Library

ISBN 9781785034213

Typeset in India by Integra Software Services Pvt. Ltd, Pondicherry

Printed and bound in Great Britain by Clays Ltd, St Ives PLC

Penguin Random House is committed to a sustainable future for
our business, our readers and our planet. This book is made
from Forest Stewardship Council® certified paper.

For all my friends and readers in my adopted county of Devon. Crickleford isn't a real place, of course – but I think it should be!

ACKNOWLEDGMENTS

With grateful thanks once again to Sharon Whelan, this time for her advice about rescuing a pony. And to Sue Viney for her first-hand knowledge about keeping house rabbits! And as always, to everyone at Ebury for all their hard work in bringing my stories to the readers.

PART 1

A PLACE TO HIDE

CHAPTER ONE

I hopped off the bus, pulling my suitcase after me, and stared around, taking in all the sights I remembered so well, despite the many years that had passed. The market cross, the town hall with its ornate black and gold clock that chimed loudly every half hour, the humpback bridge over the river, and the view, between the stone-built shops and cottages in the Town Square where I stood, of the big, square castle on the hill. So here I was, after all these years, back in Crickleford. Apart from the fact that I was now seeing everything through a January snow shower instead of in summer sunshine, nothing seemed to have changed. And that was exactly what I'd been hoping.

I'd stayed in this little Devon town, tucked away on the edge of a fairly remote part of Dartmoor, several times for family holidays from when my sister and I were about ten or eleven. Our parents had fallen in love with its charm and peacefulness, whereas we children, after the initial novelty of being in the country had worn off, found it too quiet and dull. No cinema! No swimming pool! No bowling alley! What were we supposed to do all week? By the time Kate and I were teenagers, Mum and Dad had given into pressure from us and started taking us somewhere livelier for our holidays.

Now, though, peace and quiet were exactly what I needed, and it couldn't have been much more tranquil than it was now, on this cold, snowy afternoon. I pulled my woolly hat down over my ears, checked the address I'd tapped into my phone's memos, then grabbed the handle of my suitcase, hoisted my rucksack onto my shoulders and set off up the lane. I'd found the advert on an internet search, but I had no idea what to expect. I'd never been a lodger before, never expected to be one, either. When I thought about the life I'd been living, just a few short weeks earlier, it seemed incredible that it had all come to this. But I knew I mustn't think about that. I just had to get on with it, now, whether I liked it or not.

It was only a ten-minute walk to Primrose Gardens, which was a small turning off Lavender Lane. As I trudged along through the falling snow, I felt I could almost smell the perfume, in the chill winter air, of those spring flowers in the road names. They seemed to hold a promise of better days ahead – and I wasn't disappointed when I arrived at my destination. Primrose Cottage was right at the end of Primrose Gardens, and it was a one-off, a little jewel of a pastel pink cottage, in a road of fairly ordinary semi-detached houses. Presumably the cottage was there long before its neighbours were built, giving its name to the road. There was a neat little front garden and a fairly old Peugeot parked outside. I walked up the path and rang the doorbell, suddenly feeling nervous, and it was answered by a woman of perhaps about thirty-five with short, curly fair hair and bright blue eyes.

'Hi.' I gave her a smile. 'I'm Emma Nightingale. We spoke on the phone—'

'Emma! Yes, of course, we're expecting you. I'm Lauren Atkinson. Come in, quickly, out of the snow. Just drop your bags there. How was your journey? You didn't have to get a taxi all the way from Newton Abbot, did you?' she asked, ushering me through the hallway to the kitchen.

'No, I got the bus,' I said, following her. The house smelt of polish. Had she been cleaning up for me? I was only the lodger!

She turned to look at me in surprise. 'The bus! You were lucky, then. There only are two a day.'

'Yes. I researched that on the internet, and planned my train time to coincide with it.'

She looked impressed, as if this wasn't a perfectly normal thing to do for a long journey.

'You might be disappointed with the internet connection around here,' she said sadly. 'Well, with the mobile phone signal too, to be honest. They both tend to come and go. You can normally get a decent phone signal up at the Town Square, though.'

I had a mental picture of the entire population of Crickleford congregating on the Town Square to send their text messages.

'Thanks for the warning,' I said. I'd have to call Mum and Dad later, to let them know I'd arrived safely. I sighed, remembering the looks on their faces when I told them I was leaving. I'd only been home from America for a few weeks, but my homecoming had caused them nothing but aggravation. They said they were sorry I wasn't staying, but their faces told me otherwise. They were relieved. I wasn't the kind of daughter a family would want to have living with them. I was a liability. When I said I was coming all the way to Devon, they didn't offer to drive me. I guess even having me in the car with them would have been more trouble than it was worth.

My sister had been more sympathetic, but I could tell that even she thought it would be better for me not to stay at home in Loughton.

'You can come back when things have calmed down,' she said, at least having the decency to look distressed on my behalf. 'It's just ... right now ... well, all this fuss and attention is just as bad for you as it is for us, isn't it.'

Actually, it was surely *worse* for me, since I was the cause of all the fuss and attention. But I could understand Kate's concerns. Married to the lovely Tim, with their nice home, good jobs and two perfect little children, Kate was my twin, but I often thought she must have inherited the entire stock of our parents' combined genes for sensible behaviour, leaving me with just the stupid, irresponsible ones. I couldn't stay at home. Everyone around there would be talking about me. It wasn't fair. And hopefully, here in this rural backwater away from most vestiges of civilisation, and having reverted back to my real Christian name, I could be anonymous. The very idea of anonymity, right now, was bliss.

While I'd been thinking all this, Lauren had pulled out a chair for me at the kitchen table, boiled the kettle and got mugs down from a shelf. I felt strangely like an honoured guest instead of a paying boarder.

'Tea or coffee?' she asked brightly, putting a biscuit tin on the table in front of me.

'Tea would be great, thanks – but I can do it!'

'Oh, don't worry, I'll show you where everything is and you can help yourself in future. But I thought you'd probably appreciate having a cuppa made for you, after your long journey. From London, you said?'

'On the outskirts, yes.'

She shook her head in wonder, as if I'd said I'd flown there from the moon.

'So, what made you want to leave there and come all the way down here?' she went on, sitting down opposite me. 'Have you got a job lined up here?'

I couldn't blame her for asking. After all, she needed to be sure I was going to pay my way. It would be hard to explain that I'd had money in American bank accounts, which, by now, would certainly have been made unavailable to me. Other than that, I only had enough for the first month or so here, and that was thanks to the generosity of my parents. Or their eagerness to see me gone.

'I've got a couple of irons in the fire,' I lied vaguely. 'Interviews lined up.'

'That's good. What is it you do, then?'

Do? I resisted the urge to laugh. I hadn't actually had to do anything much apart from float around looking glamorous, since I'd left England at the age of nineteen with Shane, the love of my life (at the time). Before that, well, there'd been a brief spell of—

'Caring!' I said. 'Working in a care home.'

'Really? With elderly people? That must have been very rewarding.'

I'd only done it for about a year. Between leaving school with no qualifications, moving in with Shane, and him getting his big break. But I remembered there'd been talk of me needing to take NVQs.

'Yes,' I said now, my fingers crossed under the table. 'I've got my NVQs and everything.'

'Well in that case, you shouldn't have any trouble. There's a massive shortage of carers everywhere, isn't there.'

Was there? How would I know? But I nodded sagely as I sipped my tea. And then, fortunately, before I could add any further lies, there was a shout from the next room:

'Mummeeee! What are you doing? Can I watch TV?'

Lauren raised her eyebrows at me.

'That's my little one, Holly. I did warn you, didn't I? She's not normally too noisy.'

'Oh, that's fine, I like children. How old is she?'

'Three. Here she is. Holly, this is Emma. Remember I told you? She's going to be living here.'

A little girl with blonde curls and blue eyes like her mother was watching me suspiciously from the doorway.

'I'm *not* three,' she told her mother crossly. 'I'm nearly four.'

'Hello,' I said, smiling at her. 'I hope we're going to be friends.'

The look of suspicion intensified.

'Let's take Emma upstairs and show her her room, shall we?' said Lauren brightly. 'Can you manage your bags up the stairs, Emma – let me take one.'

'No, that's fine, I've got it.' I grabbed the case and the rucksack again and followed her up the slightly rickety stairs, with Holly stomping up behind me.

'Here you are,' Lauren said, throwing open a door, revealing a room that I could only describe as very blue. Blue walls, blue curtains, blue duvet, even a dark blue carpet. Fortunately I like blue. The little lattice window looked out over the garden, where the snow was beginning to settle on a couple of small trees and a child's swing. I suddenly felt sure I was going to feel happy here, in this little blue room.

'It's really nice,' I said.

'It's my grandad's room,' Holly said in a mutinous tone.

'Oh!' I looked at Lauren, confused. Another door had a sign on it in the shape of a teddy bear with the name HOLLY painted in pink, and I presumed the other two doors belonged to the bathroom and my host's own room. The cottage wasn't exactly big enough to be hiding another wing.

'Yes, darling,' Lauren was saying patiently to her daughter. 'But Grandad's not here any more, is he?'

'Oh!' I said again. 'I'm so sorry to hear—'

'No, no, I didn't mean that!' She laughed. 'Not quite. We've had to move my dad into Green Pastures. The nursing home. He'd started wandering. And, well, he thought we were still at war with the Germans. It was getting difficult. Still, I suppose you're used to that kind of thing!' she added. 'Maybe they'll have a vacancy for you there.'

'Yes, maybe,' I said faintly.

'So, of course, what with the fees for the nursing home ...' She sighed. 'That's why we decided to start letting the room. You're our first lodger, actually, so I'm not at all experienced with all this. I hope everything will be all right for you.'

She looked at me anxiously, and her uncertainty made me warm to her. I was glad I'd found this place, this cottage, this little family. This blue room of her poor demented father. It could have been a lot worse.

'I'm sure it will be fine,' I said. 'It's my first time of being a lodger, too, so we can work it out together, can't we?'

After I'd unpacked – my every move followed by the slightly unnerving stare of Holly, who remained just outside my door, arms folded as if my being in her grandad's room was an affront to her sensibilities – I slipped my coat and hat back on, to take my phone as far as I needed in order to get a signal and call home.

'Where are you going?' Holly demanded as I headed for the front door.

'Just to make a phone call,' I said, giving her a smile. 'I won't be long.'

'Holly!' Lauren remonstrated, rushing out of the kitchen. 'You mustn't ask questions. Emma doesn't have to tell us where she's going.'

'Why not?' the child demanded sulkily.

'Because she's a grown-up, and ... oh, look, sorry, Emma.' She turned to me, shaking her head. 'I'll have to explain all this a bit more to her. She doesn't quite understand.'

'Of course she doesn't, don't worry!' I laughed. 'It's not a problem.'

'Well, you're entitled to your privacy. Speaking of which, here are your keys. This one's for the front door and the other one's for your room.' She winked. 'I'm sure you don't want a little visitor popping in and out whenever she fancies it.'

'Are Romeo and Juliet allowed in her room?' Holly asked.

I blinked in surprise. I'd thought Holly was the only child. And – Romeo and Juliet? Really?

Lauren was laughing at my expression. 'They're our cats, you haven't met them yet – they're probably outside some-where. My husband's an English teacher so everything has to have some kind of Shakespearean connection. And no, Holly, I'm sure Emma won't want the cats in her room, lying on her bed—'

'Oh, I won't mind at all, if you don't,' I reassured her. 'I love cats, we had one in ...' I stopped, swallowing. I'd nearly said *in New York*. I missed Albert, my beautiful

Ragdoll house cat. Would Shane be looking after him? I doubted it. 'We had one at home,' I finished quickly.

'Ah, well, that's OK then. But just shoo them out if they annoy you.'

I walked back towards the town centre, snowflakes now blowing in my face, holding out my phone every now and then to see if I had a signal yet. Having Holly around would be a pleasant distraction from my worries, once she'd got used to me anyway. And I didn't care about the phone or the internet. It suited me not being easily contactable. I wasn't even going to give anyone apart from my family my new address. But talking to Lauren about my supposed job interviews had made me realise that, as well as looking for work, I'd have to concoct some kind of background for myself. People were always curious about newcomers, especially in a small town like this, so I needed to be prepared.

It was late in the afternoon, and already dark, by the time I'd stopped outside the library to call my parents.

'Glad you arrived safely,' Mum said, sounding unhappy. 'You know we wish you didn't have to—'

'I know.' I swallowed, determined to sound brave. 'But honestly, Mum, I think I'm going to like it here. I just want to hibernate for a while. Is it ... any better at home now?'

'Not yet, no. I'm not even sure they realise you're not here.'

'Well, sorry, but if I hadn't slipped out the back way before it got light this morning, I'd have been mobbed again and, even worse, they'd have tried to follow me. They'll soon give up once they realise I've disappeared.'

'I hope so.'

'Well, look, if you ever need me urgently you'll have to call that landline number I gave you for my landlady. Lauren Atkinson. She's really nice. I've got no mobile signal at the house,' I warned her before we said goodbye.

I walked back to Primrose Gardens through the snow, feeling guilty all over again. My parents didn't need this hassle, and it was all my fault. I'd turned up on their doorstep just before Christmas and the whole festive season had been ruined because of me. Of course, I'd been a disappointment to them for most of my life, but I'd really excelled myself this time. It would probably do them a favour if I stayed away for good. All I'd done was bring them trouble and shame. Why couldn't I have been a better daughter, a more sensible, dutiful girl like Kate?

Well, I decided, this was my chance. If I couldn't turn things around now, when would I get another opportunity? I made up my mind there and then to make a go of things, here in Crickleford. I'd become a model citizen here, and perhaps even save up, in due course, for a nice little place of my own. I imagined myself with a husband – one who was grown-up and normal, who wore a suit and worked in an office – and a nice, well-behaved little child (I pictured one a bit like Holly but with a less hostile stare), and living in a pretty little cottage with its own garden. I'd work hard – at something – and earn a proper salary, for the first time in my life. And when my parents came to visit me, they'd finally have that look on their faces, the one they normally reserved for Kate. *Well done*, they'd say. *We're proud of you, Emma.*

As I let myself back into Primrose Cottage, I had tears in my eyes, and it was nothing to do with the onions Lauren was frying in the kitchen.

CHAPTER TWO

The trouble with making good resolutions, of course, is that they're hard to keep. Instead of getting up early the next morning, ready and determined to start my new life as a model citizen by doing a full-on job search, I forgot to set the alarm on my phone and slept in till nearly ten o'clock. The house was in silence, and outside a weak winter sun was now making the fallen snow from the previous day sparkle so brightly I had to squint as I looked out at the garden. But the heating was on and I showered, dressed, went down to the little warm kitchen and fixed myself some breakfast, feeling strangely lost and lonely. Where was Lauren? With a guilty start, I realised I didn't even know if she went out to work, or what she did.

I took my cereal and coffee into the lounge and put on the TV to catch up with the news, sat on the sofa and within a few minutes I heard the rattle of the cat-flap in the kitchen door, and a chorus of meows, and two fluffy black and white cats came rushing into the lounge, shaking their cold paws and then stopping in surprise when they saw me there.

'Hello,' I said, bending down to give them both a stroke. 'You must be Romeo and Juliet. I'm sorry I don't know which of you is which, but I'm pleased to meet you.'

The plumper of the two had white socks and ears, while the other one was almost all black, with just a white splodge by his nose. They both responded to the stroking with deep purrs of pleasure, and almost simultaneously made a leap for my lap, where they turned around a few times, nudging each other out of the way, and then settled down, curled together with their heads tucked under each other's paws. It was so comforting to have them with me, feeling their warm breath and the vibration of their purrs, that I couldn't bear to move, and ended up sitting there, watching a repeat of some ancient inane sitcom, until nearly lunch-time. When there was a ring at the doorbell, I jumped up as if I'd been shot, scattering the cats in two different directions.

'Hello?'

There was a stout, elderly woman on the doorstep, looking at me in confusion.

'Is Lauren not home yet?' she said.

'No, I'm afraid she's not.'

I looked back at her, not sure what to say. *Can I help?* seemed a bit pointless, given my circumstances.

She frowned. 'Sorry, I don't mean to be rude, but you are ...?'

I grappled with my uncertainty. I still hadn't worked out my story. Why I didn't just say 'the lodger', I have no idea, but just then one of the cats shot past my legs into the front garden, and, flustered, I gabbled: 'I'm just looking after the cats.'

'Oh! I see.' The woman stared after the retreating cat. 'I didn't realise Lauren employed someone ...'

'It's a new thing. I mean, I just started yesterday. She doesn't like them being left on their own, while she's ... out.'

'I see. Well, perhaps you'd just tell her Mary called round with this.'

She handed me a plastic bag that felt like it had books in it. As she turned to go, she looked back at me over her shoulder and added:

'Do you look after dogs too?'

'Um ... yes. Cats, dogs, rabbits, whatever.' I was already regretting the lies but they seemed to just be jumping off my tongue of their own accord.

'I see,' she said. 'And your name is?'

'Emma. Emma Nightingale,' I said. Maybe she'd forget, I thought as she walked off down the path. I hoped she wouldn't mention it to Lauren.

It was, in fact, only another ten minutes before I heard the front door opening, and the sound of Holly's chatter. I turned off the TV and rushed into the kitchen with my dirty breakfast crockery, trying unsuccessfully to look busy.

'What are you doing?' Holly said, fixing me with her blue-eyed stare.

'Just washing these things up.' I turned to smile at her.
'And where have you been?'

'Preschool of *course*,' she said scornfully. 'I go every morning.'

'Oh, right. And did you have a nice time?'

'Yes.'

She obviously wasn't ready yet to trust me with the intimate details of preschool life. Lauren, meanwhile, was chattering about making sandwiches and asking if I wanted one.

'Thanks, but no, I'll go out and get myself something.'

Our arrangement included breakfasts and evening meals but I'd expected to be out at lunchtimes – where, I hadn't quite worked out yet.

'OK, if you're sure.'

'Oh, and someone called Mary came, and asked me to give you this.' I nodded at the bag of books.

'Aha. That's my latest lot of reading.' Lauren laughed. 'Mary's a retired colleague of my husband's, from the school. Ever since I told her I'm not very well-read, she's been trying to educate me. It's nice of her, but I don't get a lot of free time for reading, unfortunately. It's just as well she only brings these bagsful about once a month.'

So at least I wouldn't have to be interrogated by her too often.

Lauren smiled at me. 'I didn't want to wake you this morning. I should have explained, Holly and I always go out at half past eight and get back about this time. I work mornings while she's at preschool. I'm a teaching assistant at the infants' school.'

'Oh, I see.' I smiled sadly to myself. It must be lovely to work with children, but I'd never be able to do anything like that. Lauren had said she wasn't well educated, but I was pretty sure she wasn't a complete dimwit like me.

'Yes, and with Jon being a teacher, it works well,' she was saying. 'We both get the school holidays. Speaking of which,' she added, 'I forgot to mention: we're going away in a few weeks' time, for the February half-term holiday. I'm sorry it's a bit soon after you arriving, but I hope you'll have settled in by then. You won't mind being on your own here, will you?'

'Not at all,' I said, trying not to sound too eager.

'Good. We've found a bargain sunshine break. Tenerife will be lovely in February. The thing is, I'd normally book Romeo and Juliet into the cattery, but the nearest one has just closed down. I wondered ... as you said you like cats ...'

'Oh, of course I'll look after them!' I said.

What a stroke of luck! I closed my eyes and offered up a silent prayer to the god of stupid lies. I was actually going to *be* a cat sitter, even if for only a week!

'That'd be marvellous,' Lauren was saying. 'We'll pay you, of course.'

'Oh, I wouldn't expect that. I'm living here anyway, so—'

'We'll be saving the cattery fees, which are really expensive, and the cats will be much happier at home. So of course we'll pay you, or else knock something off your rent, or whatever. We'll come to an arrangement, definitely.'

'Well, OK, thank you. I've been having a cuddle with them this morning, actually. They're lovely cats.' I nodded at the corner of the kitchen, where they were both happily wolfing down some food Lauren had dished up for them. 'Which one is which?'

Holly gave me a pitying look.

'Romeo is the boy, and Juliet is the girl, of course.'

'Yes, but—'

'Holly, don't be rude,' Lauren chided her gently. 'Romeo's the one with the white paws, Emma. He's greedier, so he's getting a bit fat. But watch out for Juliet. She's the one who's more likely to bring in mice and things, I'm afraid.'

'I'll bear that in mind.' I looked at my watch. 'Well, I'd better be off,' I said, making a great play of having somewhere to go.

'Have you got one of your interviews today, then?'

'Um, yes, that's right.'

'At Green Pastures?' she asked, looking excited.

'No,' I said quickly. I'd have to kill that idea stone dead, I realised. She'd soon find out whether I'd been there or not, when she visited her father. 'No, somewhere else.'

'OK. I understand, you don't want to jinx it by telling us too much. Well, good luck, Emma.'

I felt guilty all over again as I strolled down the road into town. I didn't like lying to Lauren, when she was being so kind to me. I'd have to try to get some real interviews lined up quickly, and with no reliable internet at home, the library would be the best option. It would be warm in there, and I could use one of their computers.

Despite the winter sunshine there was a sharp wind blowing as I walked up Fore Street towards the Town Square. People were scurrying from shop to shop, eager to keep out of the cold, but everyone who passed me gave me a nod and said 'Good afternoon', some of them pausing for a moment to give me a look of surprise, which unnerved me a little. Did they recognise me? Surely the stories hadn't reached this far? I pulled my warm hat down further over my ears. Having red hair could be a curse. It was far too distinctive. Why on earth hadn't I thought of colouring it before I came here? I stopped outside a hairdresser's, staring through the steamy window and wondering whether I could afford to have my colour changed professionally. Once again I was overcome with gloom, remembering how little I'd needed to worry about such things in the past. I'd had my own stylist when I lived in New York, and my own manicurist. No expense spared. I sighed, reminding myself that those things hadn't ended up making me happy. Was I going to be happy living in penury here in Crickleford, making up stories about myself and thinking about disguising my appearance? Well, it was worth a try. I slipped into Superdrug and bought myself a DIY hair colour kit. *Cheeky Chestnut*. How cheeky was a chestnut? That remained to be seen.

I needed lunch before I did anything else, and next door to the library was a small cosy-looking café with lead light windows of spun glass. It looked as old as the castle on the hill. The heavy oak door had two large signs on it, one declaring the establishment to be *Ye Olde Crickle Tea Shoppe*, the other, in huge bold type, warning: MINDE YE STEPPES! Ridiculously, I was then too busy smiling at the sign-writer's sense of humour to actually mind the steps, which were immediately inside the door, and tripped down them, landing on my knees beside the table of a couple of giggling women with a baby. So much for trying not to be noticed.

'You all right, my lovely?' said a lady I presumed to be the proprietor, in a heavy Devonian accent. 'Don't thee be frecking, my lovely, everyone does that, first time they comes in.'

So would it not have been a good idea, I seethed to myself as I got up, rubbing my knees, to amend the sign to something like STEEPE STEPPES RIGHT INSIDE DOORE? Or do away with the steppes altogether and build a rampe?

'New here, are you, my luvver?' the woman asked as I seated myself at a table away from the door. She seemed to carry out all her business from behind the counter – customers happily yelling their orders across the café. I wondered whether she was actually unable to move. I was beginning to wish I'd gone somewhere else – the pub, or the inevitable pizza bar over the road. I'd chosen this place because I thought it would give me more privacy, and I'd ended up being stared at by the entire clientele.

'Yes,' I muttered, scanning the menu as quickly as I could, to try to avoid further interrogation.

'Thought as much,' she said, looking satisfied. 'What be dwain round 'ere, then?'

'Pardon?'

'Annie asked what you're doing here,' someone at another table translated for me. 'It's not holiday season, see. We don't get too many down from Up Country this time of year.'

I was tempted to tell them to mind their own business, but that was obviously only going to provoke even more interest, not to mention making me unpopular.

'I'm staying with a family in Primrose Gardens.'

Back home in Loughton, even if anyone had been (to my mind) rude enough to ask a stranger what they were doing there, this would have been more than enough to satisfy their curiosity. But the proprietor, Annie, merely waited, arms folded, watching me, and I could feel the hum of expectation all around me until she finally went on, as they all seemed to know she would:

'Why be that, then?'

'I'm cat sitting for them,' I said, exasperated. What the hell? At least it wasn't exactly a lie any more.

This information seemed to take a while for everyone to digest, judging by the muttering going on around the place – long enough for me to order a coffee and toasted sandwich, anyway, but not long enough to allow me to consume them in peace when they were finally plonked on the counter by Annie, whose manner of service was to bellow 'White coffee an' a cheese toastie!', so that I had to go and collect it from her.

'Staying here permanent-like, be ye?' she demanded loudly after I'd seated myself again.

'Um ... perhaps,' I said, ducking my head, trying to avoid the stares.

Annie nodded thoughtfully to herself and a fresh murmur of interest broke out around me. The toastie was hot, but I ate it so fast, swilling it down with scalding coffee, that I had a sore mouth for the rest of the day. Ye Olde Crickle Tea Shoppe was one place to avoid like the plague, I decided as I stumbled back up ye olde steppes into the street, if I wanted any chance whatsoever of being anonymous around here!

And just to round off the day, the library was closed. I stared at the sign on the door, which managed, without a single *Ye* or extraneous *e* on the end of anything, to explain that it was open from 9 till 12 on Mondays and Wednesdays, closed Tuesdays and Thursdays, open again on Fridays from 12 till 3, and then on Saturdays, a comparatively full day of service from 9 till 3.

'Why?' I muttered to myself, trying without much success to understand and memorise these peculiarly diverse opening hours.

'Because it's a part-time library,' said a voice from behind me.

It was one of the women who'd been in the olde tea shoppe. I felt like thanking her for stating the bleeding obvious.

'The council made cuts, see,' she went on. 'So the library can't stay open all week now. That's if you can still call it a library. What with sessions for toddlers banging tambourines, and meetings for grannies doing their knitting. So anyway, I was wondering if you look after dogs as well as cats.'

'Um – yes,' I said, nodding and turning away, hoping to end the conversation as fast as possible. But she put a hand on my arm to detain me.

'So will you look after my dog while I'm on my holiday? I know you're new here, but I'm desperate, you see, what with the kennels closing down – bit of a shock, just before my holiday. Last week of February, first week of March, it is. How much do you charge?'

I gulped. The lie was getting serious now. How was I going to get out of this? But then again, I found myself thinking, what was the harm? It wouldn't stop me looking for a proper job, of course, that's if I ever actually got started looking for one, but in the meantime, why not? It'd give me something to do (mostly in the warm, I hoped), and earn me a few quid at the same time. And maybe, if I'd understood right that the local boarding kennels and cattery had closed down, a few other people might ask me to help them out till they found somewhere else.

'Um ... five pounds per hour,' I said, grabbing what I hoped was a realistic figure out of thin air. I had absolutely no idea, of course. Should I charge more for a dog than a cat? More for walking it twice a day than once a day? Who knew? 'What kind of dog is it, anyway?' I asked. I imagined, hopefully, a fluffy little mongrel with short legs who'd prefer to doze by the fire than go on long romps over the fields.

'An Alsatian. He's energetic, but a bit kind-of nervy.'

Oh.

'But very intelligent and loving,' she added quickly. 'If you give him a couple of good long walks a day he'll be no trouble at all.'

'Right.'

'Best to stay with him all day, though. He cries and barks a lot if he's left on his own, and all the neighbours complain. Miserable buggers.'

It sounded like I'd be earning my five pounds an hour. But nevertheless I agreed, and I was rewarded, to my surprise, by a hug of gratitude.

'You're a life-saver,' she said. 'I was beginning to think I'd have to cancel the holiday. I'm Pat, by the way. Sorry, I didn't even catch your name.'

'Emma,' I said. We exchanged contact details and I walked home feeling slightly bemused. I had a *booking* – a client! How odd. Well, it was nice to be able to help that poor lady out of a predicament. But I'd definitely need to get myself back online when the part-time library opened up, in case she changed her mind and cancelled me. I had considered bringing my laptop to one of the cafés instead, if they had Wi-Fi or had even heard of it. But after today's experience, I wasn't so sure about that.

Back in Primrose Cottage, I celebrated by locking myself in the bathroom and smothering my hair in Cheeky Chestnut.

'Oh! You look *very funny*,' Holly told me without a hint of amusement, after I'd finally washed the stuff off and blow-dried my hair into what I thought was an attractive style. 'Put the other hair back on, it was nicer.'

I'd always thought I liked children. But I don't think I'd ever realised how much they can dent your confidence.

CHAPTER THREE

The next morning I set off for the library straight after breakfast, determined to get in there while it was open and get started on my job hunt.

'I'd like to use one of the computers,' I told the library assistant cheerfully. He was a young lad who didn't look old enough to have left school.

'Oh. Sorry,' he said, idly fiddling with his earring while he shuffled some papers.

'Sorry for what?' My cheerfulness was dissipating.

'Well, not that it's our fault, but the computer's down.'

'You only have one?'

'Yeah. The other one's gone away.'

He hadn't even looked up at me yet. I wondered, for a moment, how much his tone and manner would change if I suddenly announced my real identity. He'd probably pull his bloody earring out with shock. Then again, he'd never believe me, not with this hair, to say nothing of my woolly hat and lack of make-up.

'What do you mean, *gone away*?' I said. 'Did it walk out? Go on its holiday?'

Now he looked up, clearly wounded.

'No need to be like that, lady. It's gone away to be fixed, is where it's gone.'

'So they've *both* gone wrong, then.'

He thought about this. 'Well, yeah. But this one—' he indicated the culprit with a point of his thumb, 'hasn't gone away yet.'

'I see.' I sighed. 'Well, let's hope it goes away soon, then. And the other one comes back even sooner.'

He just stared at me. I gave up.

'Can I join the library?' I asked instead.

I took out two books on dog care, from the children's section. They looked easier to understand than the ones on the adult shelves, and had nice pictures. Then I went across the square to the pizza bar to get a cup of coffee. At least it was warm in there and nobody demanded to know who I was or what I was doing. I perched on a stool in the corner and started swotting up on looking after large, energetic dogs. There was rather too much mention of exercise for my liking. It wasn't so much that I was unfit; I'd had my own personal trainer in New York. But I wasn't used to exercising in the countryside, with no gym or health spa. I couldn't remember the last time I'd ever walked anywhere that wasn't paved. I'd never have dreamed of jogging *off-path* in Central Park, for instance – my trainers were far too expensive to get mud on them. I'd come home from the States in just the clothes I stood up in, and because of the situation back in Loughton, I'd had to resort to online shopping to equip myself with a wardrobe suitable for my new life. And I could see the Alsatian was going to expect long, healthy romps through the woods and across the moor. I'd have to order some walking boots before my stint with him.

To start getting myself in the right frame of mind, and because I couldn't think what else to do, I set off for a brisk walk around the town. I didn't remember too much from my childhood holidays, apart from the town centre, so it wouldn't hurt to find my way around. I started off heading towards Castle Hill. It was snowing lightly again, but the walk was warming me up nicely so I went on to climb the footpath up the hill, and walked round the outside of the castle walls. A vague memory came back to me, of going inside the castle with Mum and Dad and Kate, and of people inside being dressed up as Saxons or Normans or whatever. I really ought to swot up on local history too when I was next in the part-time library. But at this time of year, the castle appeared to be closed to the public. Never mind, this would be somewhere to bring the Alsatian, surely – I passed a couple of dog walkers – and perhaps climbing the hill would tire him out.

Back on Fore Street, I wandered down to Crickle Bridge and strolled along the path by the river for a while. Another good dog walk, I decided, as long as I kept him on a lead. I didn't want him jumping in for a swim and frightening the ducks. Or were they geese? I'd better swot up on wildlife, too, while I was at it, if I wanted to fit in around here.

The river walk became a bit monotonous after a while. There were some houses along the riverbanks, some of them quite smart, with nice gardens extending right down to the water. I noticed they all had sandbags piled up next to their doors, and wondered if they often had to worry about the risk of flooding. When the houses petered out, I turned and headed back into town, but instead of going straight home, turned into a narrow little road called Moor View Lane, that wound gently away from the town centre. The houses and cottages on either side of the road looked so pretty with their light dusting of snow. I recognised the name of the road from somewhere, and suddenly realised it was the address the Alsatian's owner had given me. I still had the slip of paper she'd written it on, in my coat pocket, and I fished it out now. She was at number thirty-two. I slowed down, staring at the houses as I passed them, beginning to entertain a little fantasy that involved me owning a cottage, something like Primrose Cottage but in a nice country lane like this, quiet and rural but only five minutes' walk from the Town Square. I pictured myself (for some bizarre reason) cleaning the windows of the cottage and polishing its door knocker, and doing things in the garden – I had no idea what, as I'd never done anything in a garden in my life, but this was a fantasy after all.

Then I rounded a bend in the road, and it was there, in front of me: the very cottage I'd been imagining. It had the blue door I'd pictured, the cream-coloured walls, the red slate roof and the apple trees in the front garden. It even had the same kind of door knocker I'd dreamed up, although nobody was currently polishing it. How was this possible? Had I walked down this road before, on one of those long-ago family holidays, and somehow retained a memory of this very cottage? I stopped outside the rickety garden gate, and stared at the wooden sign next to the front door. BILBERRY COTTAGE. I had no idea what bilberries were, unless they were the same as blueberries, but it was a pretty name for such a pretty cottage.

I stood staring at Bilberry Cottage for so long, I suddenly realised that if anyone inside glanced out of the window they'd probably think I was planning a break-in. But in fact, there was no sign of life. No car outside, no muddy boots left on the doorstep, no children's bikes or wheelbarrows in the front garden. And from what I could see, no curtains at any of the windows. Looking even more closely through the nearest downstairs window, I could see a bare lightbulb hanging from the ceiling, and what looked like dust sheets over the furniture. Perhaps the place was being redecorated, but if so, the decorators must be on their lunch break! Or perhaps the cottage was unoccupied.

I walked on again, looking for the Alsatian owner's house. Right: there it was, a neat little bungalow on the other side of the road. At least I knew where I'd be going when I had to pick up the key from her. I turned to walk back into town. Not that it was any of my business, of course, but I decided that if I ever managed to get on the internet again I'd look at Rightmove or Zoopla and try to see whether Bilberry Cottage was for sale. *And then what?* I thought to myself. Try to buy it for peanuts – literally? Put in an offer, with the money I hadn't even earned yet for looking after an Alsatian? *Get real, Emma.* The cottage would have to stay a fantasy, obviously.

Meanwhile I was hungry, and I'd seen a pub – The Riverboat Inn – back on the other side of the bridge, that looked a lot nicer than The Star, on the Square, so I was going to treat myself to a ploughman's lunch. I could imagine The Riverboat Inn being busy during the spring and summer months, perhaps with hikers who'd lost their way on Dartmoor and stumbled across Crickleford by accident. The little town wasn't much more than a dot on the map, with no major roads nearby to encourage motorists to drop in for a look around. But this pub was exactly what a country pub in a small town should be like: big, old and sprawling, with little rooms off bigger rooms, low ceilings you had to bend double to walk beneath, oak beams, high-backed bench seats and a massive inglenook fireplace. I ordered my lunch and carried a small glass of red wine to a seat near the fire, pleased to be told that the food would be brought to me when it was ready, rather than being announced across the bar.

I'd just opened my library book to read a little more about dog care when a loud voice hailed me:

'Well, hello, I was hoping I'd bump into you again.'

I looked up, startled, and for a moment didn't recognise the elderly woman smiling down at me.

'You're the pet sitter. Emma, isn't it?' she went on, sitting herself down opposite me without waiting for an invitation. And because I must have still been looking at her blankly, she reminded me: 'Mary. Friend of Lauren.'

'Oh yes, of course, sorry,' I said. 'I gave Lauren the bag of books. She said she was pleased.'

'Did she?' Mary laughed. 'Good, although I don't suppose she'll get through them all. She means well, bless her, but the reality is, she's too busy to do much reading.'

'I'm sure she is,' I said. 'Working at the school, and looking after Holly.'

I didn't do too much reading myself, but I didn't have the excuse of being busy.

'Well, anyway, that's not what I wanted to see you about,' she said. 'Will you take on my Scrap?'

'Your ... *scrap*?' I frowned. Was this another piece of Devon dialect I hadn't learned?

'He's no trouble, well, not as long as you don't wear slippers. For some reason, he can't abide slippers. He seems to want to tear them to shreds – he'll rip them off your feet to get at them. Growls something terrible. But other than that, he's as quiet as a baby.'

'Oh. He's a dog?'

'Of course he's a dog.' Mary stared at me. 'What did you think he was? He's a Cairn terrier, two years old, still just an adorable little puppy really.' She sighed. 'I expect you know about Dribstone.'

I frowned again. I was beginning to feel like I'd just arrived in a foreign country without a phrase book. 'Dribstone?' I echoed, shaking my head.

'Oh – you're new here I suppose. You haven't heard. Dribstone Boarding Kennels and Cattery – just outside Dribstone. The village further down the river?' she added, looking a bit exasperated at my lack of local knowledge. 'Well, it's closed down.'

'Oh, yes. Someone else mentioned that.'

'It's not surprising, really,' Mary said. 'The couple who ran it didn't even like animals. No idea why they took it over in the first place. Nobody really liked leaving their pets with them.'

'I can understand that!'

'There was a woman here in Crickleford who sometimes looked after cats and rabbits and so on,' she went on, 'but she was a bit funny about taking dogs. She claimed they made her sneeze.' Mary snorted. 'Load of nonsense. Anyway, she's gone.'

'Gone?' I said, alarmed.

'Moved. Got a job in Bath, apparently. You know what it's like in small towns. No jobs.'

'Oh.' I swallowed. That wasn't exactly what I wanted to hear.

'So all the younger people move out to the cities,' she went on, 'and we end up with a population of pensioners and pets. And all the pensioners have nothing better to do than swan off on lots of holidays.' She nodded at me. 'That's why I'm glad you're here. Are you all right for the next two weeks?'

I thought quickly. The half-term holiday, when Lauren needed me for the cats, was three or four weeks off yet. This would fill some of the intervening time nicely, and once Lauren and Jon were back from their holiday I'd be having the Alsatian. After that, hopefully, I'd have a proper job. I nodded.

'Yes, I can do that.'

'Good. I'm off to my sister's in Torquay and she doesn't like dogs. Last time I went, she shut the poor baby in the shed.'

'The *baby*?' I gasped, imagining a madwoman in Torquay keeping children locked up in an outhouse. My ploughman's lunch had just been delivered to our table but every time I thought about picking up my knife and fork to get stuck in, Mary was giving me a fresh shock.

'Yes, and he barked his little head off, poor little chap. So I'd rather take a chance on you, to be honest, even though I know nothing about your credentials. Are you experienced?'

'Of course,' I lied, trying to slide *Looking after Your Dog* out of sight under the table.

'Got references?'

'Um, I'm afraid I left them behind. I had to move here suddenly, you see, because, um ... my house burned down.'

'Oh my goodness.' Mary's eyes were wide with shock, and I cursed myself for not thinking of something a bit less extreme. 'How absolutely dreadful for you.' She paused, placing a hand over mine and lowering her voice. 'I do hope you didn't lose anyone.'

'No, nobody got lost,' I said, puzzled, and then: 'Oh! I see what you mean. No, thankfully I was alone at the house at the time of the fire.'

'Thank God. Still, a terrible ordeal for you, having to move away and start again. I suppose you're waiting for the insurance to pay out?'

'Yes.' I nodded sadly. I was getting into this story. I quite liked imagining myself as the victim of a shocking disaster. 'The house was razed to the ground – a blackened heap of rubble. All my most treasured belongings gone, up in flames.' I could feel tears threatening. Any minute now, I'd be blubbing out loud.

'You poor dear girl.' She shook her head. 'What a thing to happen. And your parents couldn't take you in?'

'No. They haven't got the room. They look after my grandparents. And my bedridden auntie. And four orphaned refugee children from Syria.'

Where the hell had *that* come from? I gave myself a shake. I needed to rein in these lies quickly before I lost track of what I'd told her. I hated to admit it, but I'd actually been enjoying myself there, inventing this ridiculous sob story, but it had to stop, right now!

Mary was patting my hand again. 'Your parents must be saints,' she said softly. 'True saints.'

'Oh, I don't know about that ...'

'Well, my dear, I'm sure the people in Crickleford will open their hearts and their arms to you when they hear about your dreadful situation.'

'I don't think anyone needs to know!' I said, sitting up straight in panic. 'Really, I'd rather keep it to myself – I shouldn't have told you—'

'But surely people need to be aware, so that they can look out for you, help you out a bit. I presume Lauren knows all this?'

'No. I didn't want to worry her, or anyone. Honestly, I don't want sympathy, so please don't spread it around that I need help or anything like that.'

She smiled and patted my hand yet again. 'So brave,' she murmured. 'I admire your courage, Emma, and your independence. Don't worry, I'm not one to gossip, anyone will tell you that.'

'Oh, good. Well, then.' I just wanted to get rid of her now. 'I'll see you next week, then, shall I, if you still want me to look after your dog?'

'Of course I do. Scrappy will love you. Just don't forget about the slippers.'

She wrote down her address for me and we arranged that I'd go round there the next day, to meet the slipper-hating Cairn terrier and be given instructions in his care. At least I'd be earning some money. If I'd known I was going to make her feel so sorry for me, I'd have doubled the amount I was charging.

While I ate my lunch I tried to memorise the details of the stupid story I'd told Mary. What on earth had come over me? It was one thing making up a background story for myself, but surely I could have come up with something a bit less far-fetched? I just hoped she *didn't* gossip about it. Nobody would believe that load of nonsense for very long, and then they'd be wondering why I was lying, what I had to hide. At this rate, I'd be rumbled before I'd even been in Crickleford a week!

CHAPTER FOUR

That night I dreamt I was back in America. Not in the swanky apartment in New York, but in our beach house in California. We lived there permanently when we first moved to the States, when things were good between Shane and me, when life was on the up, and everything was new and exciting. When we still shared everything together. In more recent years it had become our holiday home, our escape from the city, but it never felt quite the same as it did during those early years.

I woke up feeling unhappy, and unsure why. I didn't miss my life in New York, and I certainly didn't miss Shane. I suppose the dream had unsettled me, though, reminding me as it did of how sure I'd been, at the beginning, that I was doing the right thing, leaving behind my family, going against their wishes and advice, to follow the man I believed would always love me. How silly and naïve I'd been!

I got out of bed, went to the bathroom, and on returning to my room the first thing I saw was my impulse buy from the previous afternoon – a bag of doughnuts, the grease now leaking unattractively through the brown paper. I'd been hungry, but I'd only managed to eat one doughnut before Lauren had dinner ready. I peeked inside the bag now. They might already be a bit stale, but waste not, want not. I was midway through the first one when I looked up and saw Holly watching me from my open doorway, her arms folded, giving me a disapproving frown. I wiped my mouth quickly and held the bag out to her.

'Would you like one?'

'No, thank you,' she said primly. 'Mummy says they're very bad for your teeth.'

'That's true,' I admitted. 'But I need the sugar, actually.'

'Why?'

'Because ... um, I've been ill,' I fabricated quickly.

Holly considered this for a moment, looking at me doubtfully.

'Is that why you're not going to work? My mummy couldn't go to work after she had the flu. Did you have the flu as well?'

'Er ... no. Something else,' I said vaguely. 'But I'll be going to work very soon.'

She ran off downstairs. I breathed a sigh of relief, closed the bedroom door and got stuck into another doughnut. By the time I'd showered and dressed and gone downstairs myself, Lauren and Holly were just about to leave the house. But Lauren turned to me as she opened the front door and said, sympathetically:

'You should have told me you'd been ill. I hope it wasn't anything serious?'

'Oh, no, not really,' I said, trying to sound brave about it. 'I'm on the mend now, anyway.'

'Good. Well, take care, and don't overdo it.'

Overdo what? I thought to myself as I made myself a cup of coffee and stared out at the snow. I wasn't doing a damned thing. I felt guilty all over again now, especially for making up yet another lie. What was wrong with me? I couldn't seem to stop myself – making things up was getting to be a habit with me.

At least I had somewhere to go that day. It was my appointment with Mary, to meet Scrap the Cairn terrier. Mary lived in Church Hill, yet another of the steep little lanes off Fore Street, between the church and the hardware shop. Her house was right at the top of the lane, which finished abruptly with a stile over a five-bar gate, and an ancient signpost indicating a footpath to Windy Tor. I had no idea what Windy Tor was, but the view across the snowy expanse of the moor from this high point was so fabulous that I felt sure Scrap and I would be following that footpath during the forthcoming couple of weeks.

'The tors are outcrops of rock,' Mary told me over coffee, while Scrap sniffed around my legs to make sure I wasn't wearing slippers. 'They're all over Dartmoor. Windy Tor's not a huge one but you'll discover lots more if you go for walks around here. Take a map with you if you don't want to get lost.' She got up and handed me a map and a Dartmoor guidebook from her shelf. 'Here, borrow these.'

'Oh, thank you.' I looked at the small print in the guidebook without much enthusiasm, but added out of politeness that it certainly would be a good idea for me to find my way around. 'I'll probably take him for walks along the river, too.' I remembered the sandbags, and added, 'Does it flood along there sometimes?'

'The water meadows just out of town will flood occasionally, if we've had a lot of wet weather,' she said. 'But don't worry, the Crickle hardly ever bursts its banks this far down. I can only remember it happening a couple of times. Now, 1963 – that was a terrible year. Houses almost completely underwater, that part of town completely cut off for weeks – nobody's ever seen anything quite like it since, thank God.'

'Yes, thank God,' I agreed, thinking of the lovely houses along the riverbank. They looked like they'd been built a long while after 1963. I wondered how they'd ever got planning permission for building so close to the river. But at least it sounded like I'd be safe taking Scrap for walks along there, even if it rained!

I hadn't had any personal experience of looking after dogs – we'd only had cats at home – but I liked all animals, and Scrap seemed a nice little thing. Mary seemed to think he liked me. He kept running out to his bed in the kitchen and bringing me things – a squeaky toy, a half-eaten chew stick, a blanket, a rubber bone. I stroked him and tickled him behind his ears and told him we were going to be

friends, and he licked my hand with a surprising degree
of enthusiasm. I promised Mary I'd be back the next
morning to start my dog sitting, and left feeling quite warm
and fuzzy and pleased with myself. And to complete my
morning, I took another stroll down Moor View Lane. This
time I found a spot on the opposite side of the road from
Bilberry Cottage, where I could pretend to be looking at
Mary's Dartmoor guidebook while surreptitiously staring
into the windows again. Nothing seemed to have changed;
it still looked unoccupied, the dust sheets still in place.
What if it was for rent, rather than for sale? Perhaps the
landlord was in the process of doing it up. Would the
money from dog walking and cat sitting cover the rent for
a pretty little cottage like this? I didn't need anyone to tell
me the answer to that one!

It being a Friday, the part-time library would be open
from midday so I walked back to the Town Square slowly
and got there just as it opened. I didn't hold out a lot of
hope for a computer, but to my surprise the boy with the
earring told me the one that had gone away was now back.
I logged on, with the secret password on the slip of paper
the boy passed me furtively across the desk, making me
feel like a Russian spy, and went straight to the Rightmove
website, followed by Zoopla and then the listings of several
Devon letting agencies. But none of them had Bilberry
Cottage on their books. I wasn't sure why I felt so disappointed. It wasn't as if I was likely to be in a position to
buy, rent or even pay admission to anything other than

my one room in Primrose Cottage any time this decade. So perhaps that was the end of my little fantasy. Instead I looked at a couple of job vacancy sites, before giving up and going to get some lunch.

Lauren made a nourishing chicken casserole for dinner that night.

'Eat up,' she said, smiling at me. 'This'll soon have you feeling well again.'

It was delicious, but my guilty conscience completely ruined it for me.

Next morning, I started out early for my first day with Scrap. Mary had given me full instructions on his care, and after giving him his breakfast I took him for our inaugural walk along the footpath to Windy Tor. I'd bought myself a good pair of walking boots from a little shop called 'The Moor Outdoor Store' – which, being on Fore Street, I thought was a masterpiece in rhyme and was only disappointed it wasn't number four – and I couldn't wait to try them out. My first hurdle, however, turned out to be the stile, namely getting over it while holding onto Scrap's lead. I was too nervous about losing him, to let go of it or let him off the lead. But I soon discovered he was used to the stile, squeezing neatly underneath it and waiting for me to climb over and join him. Half an hour later, I was beginning to realise the tor was further than it looked on the map. It was also starting to snow again, and the lovely rolling expanse of Dartmoor countryside I'd been so entranced by, the previous day in the sunshine, now looked bleak and uninviting under a dark, threatening sky.

'Come on, let's go back,' I said to Scrap. 'We'll try again tomorrow.'

By the time we got back to Mary's house, it was snowing heavily, and we were both cold and wet. The legs of my jeans were soaked and my feet ached from the unaccustomed exercise. I took off my new boots, which were caked in thick red Devon mud, rubbed Scrap down and turned on the gas fire in the lounge for him to dry off, and before long he was asleep. Then I stripped off my wet jeans and anorak and hung them around the fire, put on a dressing gown of Mary's I found behind her bedroom door – making a mental note to bring a change of clothes the next day – and sat down to eat the sandwich I'd brought with me. Before long, the warmth of the fire had me dozing in the armchair, and when there was a ring at the doorbell I nearly jumped out of my skin.

'Hello,' said the lady standing on the doorstep, staring at me suspiciously. 'Who are you?'

'Um, I'm Emma, the dog sitter,' I said.

'Oh.' She looked me up and down, and I remembered about the dressing gown. 'I got wet,' I explained, 'walking Scrap in the snow.'

'OK. Well I'm sorry, but I live next door and I knew Mary was away, so when I noticed you walking up the path and you didn't come back again, I thought I'd better check. Jackie Johnson,' she introduced herself, holding out her hand. 'I didn't realise Mary used a dog sitter.'

'It's the first time I've worked for her.'

'I see. You're new around here, aren't you?'

'Yes. I'm Emma. I'm lodging over in Primrose Gardens.'

There, I'd managed to meet someone new without telling any outright lies. It could be done.

Jackie gave me a friendly smile and said she hoped I was settling down well in the town.

'We've been in desperate need of a new pet sitter around here,' she commented as she said goodbye.

I was beginning to realise it was hopeless to expect to live like a complete hermit here in Crickleford, where everyone seemed to be so involved in each other's business. It was nice that they all looked out for each other, but I'd just have to keep being careful not to reveal myself to anyone.

'I presume you've got a job now, Emma,' Lauren said the next day as I set off again for my stint with Scrap.

'Yes,' I said brightly. 'Thank goodness!'

'A care home, is it?'

I hesitated. I didn't want to admit I was just pet sitting. After all, the money was dire, and as for job security, I'd have been better off on the game. I didn't want Lauren to chuck me out because she couldn't imagine how I'd pay my rent.

'Kind of,' I said, slipping out of the door quickly.

It was Sunday. Carers worked at weekends. She'd assume that was what I was doing. It wasn't exactly a lie, either – I was caring for a dog, wasn't I? In a home? So it was, kind of, a care home, wasn't it?

Scrap and I were becoming good friends already. That second day, he ran to meet me with his lead in his mouth, and by halfway through the week we'd finally made it to Windy Tor and back. I doubted whether it was the most exciting sight on Dartmoor, but I was pleased with myself, not to say knackered, at the end of the walk. Scrap and I both slept for the whole of the afternoon. Looking after dogs was a doddle, I decided. The secret was obviously just to tire them out completely, then give them a big meal and leave them to sleep it off.

In fact I'd already fallen in love with the little terrier and was enjoying myself immensely. Every now and then I remembered that I ought to be looking for a proper job. But what was the rush? It was only my second week in Crickleford, after all. I needed time to adapt, didn't I? Time to recover from the trauma of having to leave the family home, to say nothing of the trauma that had brought me back there from America. A little spell of rest and relaxation and healthy walks with my new little doggy friend was surely just what the doctor ordered.

I was beginning to recognise people, too, on my twice-daily walks with Scrap. We didn't always go to Windy Tor, of course. Sometimes we walked along the riverbank, sometimes along Moor View Lane to dawdle for a while outside Bilberry Cottage, and a couple of times we went up Castle Hill. And there were usually quite a lot of other dog walkers around. I'd noticed how friendly they seemed to be with each other, stopping to talk together about their dogs and the cold weather and the state of the footpaths, and to be honest I tried to avoid these cheery little gatherings. It was nice that everyone was so friendly, but after all, I'd come to Crickleford to hide away from people, not to chat to them.

But on our second time up on Castle Hill, I noticed someone I'd seen the previous day and had instantly had to look away from, for no other reason than there was something about him that reminded me of Shane. I couldn't quite put my finger on it: yes, he was tall, slim and dark-skinned – Shane was mixed-race, with an African-American father and white British mother. But this guy's colouring looked kind of Mediterranean. No, it was more about the way he loped along, with long strides, head held high and an air of confidence about him. For a moment that first time, I'd felt physically sick from the memories, and actually turned round and took Scrap in a different direction, away from the lookalike. But on this second occasion I almost walked straight into him and, to my discomfort, he gave me a wide grin and said 'Good morning!' I nodded and muttered something back, but he just stood there, deliberately, right in front of me, and said again, 'Good morning!'

This was awkward.

'Hello,' I muttered, trying to manoeuvre Scrap around him. The stranger smiled, bent down and patted him. He must have been the only person walking on Castle Hill who didn't have a dog with him.

'Nice morning, isn't it?' he persisted.

'Yes, it is.' I glanced at him quickly. He was smiling, just a friendly smile. He didn't look like Shane after all, close up, and to my relief he didn't appear to be staring at me in a way that indicated he might know who I was. I risked a smile back. 'Excuse me, then,' I said. 'But we're in a bit of a hurry.'

'To get to the top of the hill?' he said, looking surprised.

'Yes. I'm ... timing the walk. For ... fitness purposes.'

'Oh, right. Very commendable.' He smiled again and stepped aside to let me pass. 'Well, good luck, then, with the timing. And the fitness.'

'Thanks. Bye.' I tugged at Scrap's lead – he seemed reluctant to part from his new friend. 'Come on, Scrap.'

We made our escape without me having to make up any more stories. But something about that guy's smile had unsettled me. Why had he insisted on talking? Surely he didn't recognise me?

That night I gazed at myself in the mirror for a long time, wondering about myself and what I'd become. Who was I? I used to be someone, I thought mournfully, someone who was recognised everywhere I went, someone who turned heads, who was photographed for the newspapers and asked for autographs. And now I was just a nobody, a stranger in a strange town, someone with DIY Cheeky Chestnut hair who made up stories about house fires, care homes, illnesses and fitness regimes.

But on the other hand, shouldn't I be asking myself who, really, was that 'somebody' I used to be? If I'd only been 'somebody' because of Shane, then in actual fact, hadn't I really been a nobody all along? It wasn't the nicest thought. But perhaps, sadly, it was the most honest one I'd had for a long while.

CHAPTER FIVE

By the time my two weeks with Scrap were up, we were well into February and I felt as if I'd been living at Primrose Cottage forever. Holly seemed to have got used to me being around now, and by the time her parents were packing their bags for their winter sunshine holiday, I was able to help out by keeping the overexcited little girl occupied with a game.

'Thank you, Emma,' Lauren said gratefully when they'd finished the packing. She hesitated for a moment. 'We're going to be leaving early in the morning. Are you absolutely sure you're OK to look after the cats for us?'

'Of course! I'm looking forward to it.'

It was the truth. It was good to have an excuse to stay in the house as much as possible, safe from the curiosity and stares of people in the town.

'I mean, I hope it'll fit in OK with your new job,' she said anxiously. 'It's not as if you have to be with Romeo and Juliet all the time. Just as long as you're here to give them their meals, and a bit of a cuddle now and then.'

'Don't worry about that, Lauren, I'll give them lots of cuddles, and my ... um ... hours are quite flexible so it'll be fine.'

'Oh good. Well, I've stocked up the freezer with some ready meals for you. I'm sure you won't want to be rushing home from work to cook for yourself from scratch, will you, not after being on your feet all day looking after those elderly folk. I know how tiring that can be.'

'Thanks, Lauren, I appreciate that,' I said, my heart feeling heavy with shame. Since my two weeks of looking after Scrap, I'd done nothing with my time apart going to the library, when it deigned to open, to trawl the job vacancy websites. And my vision for the week Jon and Lauren were away actually involved far more lounging on sofas with warm cats on my lap watching rubbish on TV, than any kind of exhausting work. As for cooking from scratch – all that stuff was a complete mystery to me.

I waved them off early the next morning and sat straight back down with a cup of coffee, savouring the thought of a whole week indoors to myself. But within a couple of days, even with the pleasant companionship of two warm purring moggies, I was beginning to suffer from cabin fever. The cottage was probably half the size of my parents' house in Loughton, which was itself only a fraction of the size of my apartment in New York. *Shane's* apartment in New York, I corrected myself quickly. I knew I'd been spoilt.

But lying around all day every day with nothing to do didn't feel quite the same now that I knew it was a luxury I couldn't afford.

On the third day, I gave in and went out to the library again. I'd finished with my books on dog care and anyway, looking after Scrap had been such a doddle that I didn't feel like I needed any further study. So I took them back and found some books on wildlife instead.

'You're not from round these parts, are you?' said the boy with the earring, who I'd noticed staring at me whenever I was in there. 'You're down from up-country.'

'Maybe I am,' I said defensively, starting to edge away from the desk with my new books.

'Are you married?' he asked casually.

I stared at him. Was that a requirement for borrowing books from the library these days?

'No.'

'Right. Got kids, though?' he said.

'What makes you say that?'

'You're taking books out of the kids' section,' he said, grinning.

'Oh. Yes, they're for my ... um ... little girl,' I said, feeling flustered. 'Holly.'

'No husband, though?' he persisted. 'So do you fancy coming out for a drink with me Friday night, then?'

'No!' I said, horrified. 'I mean, no, I can't, I'm not single, I've got, um, a boyfriend, back in, um, London—'

'You never said,' he complained. His ears had gone red. 'So, you're kind of on your own anyway, sort of thing, while you're down here.' He eyed me speculatively. 'Is he the jealous type, your boyfriend?'

I stared at him. 'I'm not available to have a drink with you,' I said firmly. 'Thank you.'

I was glad to get out of there. It was one thing being propositioned by someone younger than myself, and God only knows I could have used a bit of a confidence-boost at this point – but this particular gawky teenager looked like he shouldn't even be let out on his own. If I took him up on his offer, I'd probably have his mother turning up, berating me because he was underage.

And not only that but, damn it, now I'd acquired a mythical boyfriend as well as a mythical job, illness, house fire and keep-fit regime – to say nothing of pretending Holly was my own child. I needed to start writing down all the lies, and who I'd told them to, or I'd never keep track. If my situation wasn't so depressing it would actually be quite funny, I thought to myself grimly. To cheer myself up, I strolled down Moor View Lane and stared at the windows of Bilberry Cottage again. The dust sheets were still in place, and now there was a stepladder up in the middle of the room – but still no sign of anyone working there. Sighing to myself, I carried on down the lane a bit further this time, past the bungalow where I'd be going to look after the Alsatian, round a couple of bends, and suddenly there were no more cottages, the high hedges gave way to open land, and there before me once again, almost taking my breath away, stretched the wild expanse of Dartmoor.

'Wow,' I breathed, stopping where I stood, and staring into the distance. As far as the eye could see, were the gentle rises of hills, some dotted with sheep, some sporting rocky crags where streaks of snow still remained, others scattered with patches of bright yellow gorse. The little lane narrowed ahead of me, twisting and turning until it disappeared into a valley where I could see the river glinting in the winter sunshine.

As I strolled back past Bilberry Cottage I glanced up at the windows again and wondered whether the upstairs windows afforded that same view. How wonderful it would be to get up in the morning, open the curtains and see those rolling hills, and watch the sunshine chasing the shadows across the moor. It surely wouldn't be possible to wake up in a bad mood with that kind of view to sustain you, would it? Or perhaps it was just the thought of something so completely different from the view from my bedroom in New York that appealed to me. I realised I was already falling in love with this little town, its situation here on the edge of the moor, its charm and especially its remoteness. I'd been right to choose it as my hiding place. It suited me perfectly. Spending time alone, going for country walks in the fresh cold air – it couldn't have been more different from the life I'd been living before, but it was giving me what I needed: time to calm down, lick my wounds and decide what I was going to do with the rest of my life.

That evening, I'd just given the cats their evening meal, when Mary turned up with a bunch of flowers and a box of chocolates.

'Just to thank you for looking after Scrap at such short notice,' she said as I held the door open for her.

'Oh! I didn't expect ... I mean, you've paid me for looking after him, there was no need for anything else.'

'Your charges are ridiculously low and, besides, I heard from my neighbour how long you spent with Scrappy every day and all the long walks you took him on. I'm really grateful that he was looked after so well.'

'Oh, it was no trouble. I enjoyed it, honestly.'

I wondered if her neighbour also mentioned me wearing Mary's dressing gown.

'Well, anyway, I'll certainly be recommending you.'

'Oh. Well, thank you.' I smiled at her. I didn't for a minute imagine anyone else would be asking me to look after their dogs – it had been surprising enough to get the two bookings I had – but it was nice of her. 'Would you like to stay for a cup of tea? Lauren and Jon are away for the week.'

'Yes, I know. Tenerife, isn't it. Lucky them, getting away from this cold weather. I hope they're enjoying it. I won't stop; I can smell your dinner cooking.'

I didn't enlighten her. It wasn't cooking, of course; it was being thawed and reheated in the microwave.

'I just wanted to say I heard from young Josh in the library that you had to leave your boyfriend behind when you moved down here.' She gave me such a sympathetic look that I felt myself shrivelling up inside with embarrassment.

'Um, yes,' I muttered.

'I suppose you were too traumatised to stay in the area, after the fire?'

I nodded, trying to look suitably traumatised. Thank God Josh didn't seem to have mentioned that Holly was supposed to be my daughter.

'Well, I do hope you'll soon be reunited with him. It must be a terrible strain – living apart.'

'Yes. It is,' I muttered. 'Well, thanks again for these.' I waved the flowers and chocs at her.

'My pleasure. Now, if there's anything I can do to help you, Emma, any time, you just give me a call, you understand?'

'I will. Thank you.'

I closed the door and leant on it. Damn that stupid young Josh – it seemed he was a worse gossip than anyone else around here. I'd have to be wary of him; I might have hurt his pride by turning down his offer of going for a drink, and if he got too curious about me he might start digging. If necessary, I thought grimly, it might even be worth going out with him just to get him on my side and shut him up.

With these thoughts crowding my brain, I dished up my sad, over-microwaved lasagne-for-one and scoffed it in front of the TV, following it up with nearly the whole box of Mary's chocolates before having an early night in bed with a stomach ache.

At least Romeo and Juliet were a pleasure to look after, although Juliet did spoil things slightly by bringing home two dead birds on the same day. They were lying side by side on the front doorstep, waiting for me like an Amazon delivery when I got back from another walk. I made such a fuss, squealing pathetically as I picked them up on a shovel from the garden shed and tried to bury them in the frozen garden without looking at them too closely, that both cats appeared behind me to watch, meowing crossly at me. When Juliet then tried to recover her prize while I was still doing my best to bury it, I turned and hissed at them both to clear off, and they loped away across the frosty grass with an offended air.

'I'm sorry, babies,' I told them later, dishing up their food. 'I know it's just your instinct, but the problem is, I'm not used to things like that. My Albert lived indoors in the apartment, you see. He didn't go outside, he was ...' I swallowed. 'I mean he *is*, a Ragdoll. That's a special breed, they're happy to live inside, so he never hunted birds or anything ...' I tailed off again, tears coming to my eyes. I'd loved Albert so much. He was a beautiful, docile, affectionate cat. How could I have left him, walked out on him without a backward glance? I'd been so distraught I hadn't even given him a thought – I'd just had to get away. And once I'd gone, I couldn't go back. I'd had to make myself scarce. Poor Albert. If only I knew whether or not he was being looked after.

I sat down at the kitchen table, watching Romeo and Juliet demolishing their dinner. Juliet looked up at me when she'd finished eating, and, as if she'd realised I needed consoling, trotted over and jumped up onto my lap, where she made herself comfortable and began to wash her face, purring happily. Within a few minutes Romeo was curling himself around my legs, joining in the chorus of purrs, and I had to laugh despite my tears.

'Well, at least I've got you two to keep me company,' I said, wiping my eyes.

But I knew that, somehow, when all the fuss had died down, I was going to have to get in touch with Shane, whether I wanted to or not. I hated having to admit it, but there was nobody else who could tell me if Albert was all right. Despite all the attention I'd received during my time in New York, and despite the fact that the people I used to hang around with would be agog to hear my side of the story, to get their hands on any salacious details and lap up all the scandal, I knew none of them could really care less about me. The real me, the girl behind the tabloid photos and the celebrity lifestyle. The girl from Loughton who got in too deep, fell in love and made a mess of her life. The girl who was desperately trying to reinvent herself in a little Devon town by telling everyone a fistful of lies. The truth was, that girl didn't actually have a friend in the world apart from the cat she'd left behind.

CHAPTER SIX

On the last day of the Atkinsons' holiday, I was gazing idly out of the window at a shower of sleety rain, when I noticed a man hanging around outside. He seemed to be staring straight back at me. To say this was unnerving would be an understatement. I actually dived away from the window and ducked down behind an armchair. After a couple of moments waiting for my heartbeat to return to normal, I poked my head cautiously out from my hiding place. He was still there, strolling backwards and forwards outside. It was late afternoon and a dull, murky day so it was hard to see exactly what he looked like, especially from behind the armchair. But there was definitely something furtive about his movements as he looked around him, up and down the road, back at the cottage and then slowly walked away. I felt my pulse beginning to return to normal.

'Calm down,' I told myself crossly, out loud. 'For God's sake, he was probably just waiting for someone. He's not looking for you. Nothing to do with you whatsoever. Get out from behind this armchair and stop behaving like an idiot.'

There was a clatter and a thump from the kitchen, and I nearly yelled out in fright, until Romeo came trotting into the lounge, meowing, and I realised it had just been him jumping through the cat-flap.

'Is anyone out there, boy?' I asked him. 'Anyone prowling around outside?'

I crept cautiously to the window again and peered out into the gloom. I knew I was being ridiculous. Nobody had followed me to Crickleford. Nobody knew I was here. Nobody was watching the house. I went around drawing all the curtains and put the television on to drown out my stupid fears. I was safe, I reminded myself. But all the same, I was relieved when the Atkinsons arrived home the next day. It felt better having everyone else around.

They were eager to tell me all about their holiday, and little Holly was hopping up and down with excitement, desperate to relate how she'd learned to jump into the swimming pool wearing her armbands. She sat on my lap and chattered about how the *next* exciting thing, now the holiday was over, was her fourth birthday, which was coming up soon. It seemed perhaps absence had made the heart grow fonder and Holly had really started to like me. It was nice to know somebody did.

The next morning I was due to start looking after the Alsatian in Moor View Lane, and I felt slightly nervous about it. When I'd been introduced to him he'd barked his head off at me, but Pat, his owner, had just laughed and said he did that to everyone at first, and it was only because he was a bit scared of strangers.

When I arrived at her bungalow for my first day, I could hear him crying and barking as I walked up to the front door. I let myself in and called as loudly as I could above the noise:

'All right, Bingo, calm down, it's me, Emma.'

Not that he'd remember who I was, but he'd have to get used to me if I was going to be in charge. Pat had left him shut in the kitchen, which was quite a big room, but it was evident from the way he was charging around in there, whining and barking himself silly, that he was frustrated at being shut in. When I opened the door to the room, he bounded out, with his tongue out and his tail swishing – before he evidently realised I wasn't his beloved Pat but this strange new person that he wasn't too sure about. He stopped in his tracks, lowering his head and looking at me suspiciously, giving a little whimper.

I laughed. 'Don't be nervous, Bingo!' I said. 'I'm not going to hurt you, you silly boy.' It seemed really funny that such a big, tough-looking dog would be such a wimp. 'Come on,' I said. 'You'll soon get used to me. I'll let you out in the garden first and then we'll go out for a walk.'

He flew out of the back door as soon as I'd opened it, but once he'd relieved himself he promptly began barking loudly again. What on earth was wrong with him now? I opened the door to let him back in but he was way up the end of the garden.

'Bingo!' I yelled, but he completely ignored me. He seemed to be staring at something halfway back down the garden. First he'd bark at it madly, then run forward a few steps, before stopping, whining loudly as if in fear. My first thought was: who's out there? But I forced myself to dismiss this as paranoia again, and instead began to worry about all the neighbours getting angry about the noise.

'Bingo!' I shouted again, still with no effect whatsoever. I sighed and strode out down the garden path. 'Bingo, what's the matter?'

There was a cat, a large, cross-looking tabby, halfway down the path. It had its back arched and all its fur standing on end, and was hissing at Bingo contemptuously. Surely *that* wasn't what Bingo was scared of? Was he a dog or a mouse?

'Shoo!' I said to the cat. 'Go on, off with you.' The cat turned to glare at me, tossed his head and, with a final hiss at Bingo, ambled off across the garden, jumping up onto the fence and down the other side in one leisurely movement, as if to prove how unbothered he was.

Bingo looked at me with something like respect and wagged his tail a couple of times.

'OK, I've seen him off for you. See? I'm your friend,' I said. 'Now then. Walkies?'

This magic word seemed to get the desired effect, and he bounded ahead of me back into the house, where I fastened on his lead. 'OK, are you up for a brisk walk to Windy Tor?'

That should tire him out, I thought. It had worked with Scrap. But, of course, I was underestimating the amount of exercise an Alsatian needed, compared with a little terrier. Pat had assured me that he would come back obediently when he was called, but after the episode in the garden just now, I wasn't too confident about this. And she'd also added that it would be best to keep him on his lead if we were in an area of the moor where there were any sheep or livestock.

'That's really important,' she'd said, 'especially during the lambing season. Farmers are entitled to shoot dogs who chase or worry livestock.'

After the episode in the garden, I was wondering frankly whether Pat was kidding herself. If he was that nervous of a spitting tabby cat, I couldn't quite see him chasing sheep! But I wasn't going to take any chances, even though I didn't have a clear idea of when lambing season was. The idea of being confronted by an angry farmer with a gun was enough to decide me that Bingo was going to stay on his lead for the duration of the walk. Having apparently now decided to trust me sufficiently to come out with me, though, he reacted to being on the lead by setting off like a bullet, pulling me along behind him helplessly as if I were a child's toy on the end of a string. I wondered what would happen if he decided to chase a rabbit and pulled me into the undergrowth. That hadn't been covered by my dog care book.

'Bingo!' I yelled, puffing and gasping as I stumbled along the footpath after him. 'Slow down. Bingo, for God's sake!'

So much for him responding obediently to being called back. One word from me, and he did exactly as he liked. We reached Windy Tor in about half the time it had taken me to get there with Scrap, even on our best day. Fortunately for me, I suddenly got the answer to my debate about what would happen if he saw a rabbit. A little one happened to dart across the path in front of us just as we arrived at the Tor, and he skidded to a halt, whining in fear and turning back to run behind my legs.

'What are you like, Bingo?' I laughed out loud. 'It's just a tiny little thing. Come on, it's gone now. Whatever made you such a scaredy-dog?'

At least it gave me a chance to tighten up his lead, forcing him to stop for a rest. I had a pain in my side and I could hardly breathe. How had I become so unfit, so soon after stopping my workouts with my personal trainer in New York?

'It's all very well for you,' I told Bingo. 'But I'm not used to these conditions.' I was referring to the red mud that had once again covered my boots and splashed up the legs of my jeans. Of course it was exhausting, running over this kind of ground! There was no comparison with the gentle paths of Central Park, or the treadmill in the private gym in our apartment block. And glancing again at Bingo, I could see that my work wouldn't be over once we got back home. His paws, legs and flanks were red with the Dartmoor soil too.

'Come on, then. Home,' I said, trying to sound authoritative. I kept him on a tighter lead this time, and he seemed happy enough to walk at a slightly more reasonable pace. Probably scared of meeting another rabbit, I thought with a giggle – and I felt a surge of affection for this big softie of a dog.

We were nearly home, and I was beginning to hallucinate about a mug of hot chocolate and a lie down, when I was stopped in the lane by a cross-looking woman wearing a purple bobble-hat.

'Are you looking after him while Pat's away?' she demanded, nodding at Bingo.

'Yes,' I said, wondering what I'd done wrong already.

'Well, do us all a favour and stop him bloody barking all the bloody time, will you?' she said.

'Oh. Um ... well, I'll do my best.'

'He's always outside barking at that bloody tabby cat,' she said. 'I don't know whose cat it is, but it's obviously scared of him.'

'Actually, I get the impression it's him that's scared of the cat,' I offered.

'Don't be ridiculous,' she scoffed. 'Great big bloody dog like him. Pat ought to keep him under control.'

So this was the next-door neighbour. I remembered now that Pat had told me about this – the 'miserable' neighbours complaining about his barking.

'Just keep Pongo indoors, will you, if you can't bloody well shut him up?' she went on. 'I tell Pat all the time, but she doesn't bloody do anything.'

'Right,' I said faintly. 'Um ... who's Pongo?'

But she was already walking away, leaving me standing there staring after her.

When we got indoors, I shooed Bingo into the shower in the bathroom, soaped him all over and rinsed the mud off him. He seemed quite used to this and appeared to enjoy it, but when he came out of the shower, of course, he shook himself all over me and all over the floor. I sighed to myself as I rubbed him dry. I'd need a shower and a change of clothes myself, and then I'd have to wash the bathroom floor before I could finally have a rest. Looking after Scrap might have been a doddle, but I could see Bingo was going to be more of a challenge.

When I picked up his collar and lead from the floor, I caught sight of his identity disc.

P – O – N ... I read, frowning at the name engraved on it. Oh dear. He wasn't even called Bingo. No wonder he'd ignored me when I called him.

'Pongo?' I tried, quietly – and he came trotting over to me, wagging his tail like a good doggy.

Note to self: next time, write down the name of the pet as soon as you've been told it.

I gave Pongo another quick walk before I left him to go home for my dinner. Pat had suggested that this would make sure he'd had enough exercise and should then rest quietly on his own for the evening and through the night. We stayed on the road, though, for the second walk – a brisk march up to along Fore Street up to Town Square and back again. I didn't see the point of going through the whole mud-shower-clean-up process twice in one day. I slowed down slightly past Bilberry Cottage, of course – I just couldn't stop staring at it – but as nothing ever seemed to look any different there, I just sighed and went on my way.

Lauren was looking worried when I arrived home.

'Is everything all right?' I asked her, my heart missing a beat at the thought that someone had told her about my lies. The rest of them, that is – the ones I hadn't already told her myself.

'Yes,' she said, giving me a half-smile. Then she shook her head. 'Well, not really. I'm a bit concerned about something I've just heard.'

Oh no, I thought. Here we go. She's going to ask me why I've been lying about Holly being my daughter. Or about my house burning down, and my parents' commitment to aged relatives and refugee children.

'What?' I asked shakily. She wouldn't throw me out of her house, would she, just for making up stupid stories? Not unless I couldn't pay my rent?

'Well, it seems there's been another break-in in town.'

'Another one?' I said, frowning. This was the first I'd heard about it.

'Yes. Sorry, you wouldn't know, but there's been a lot about it in the local paper. It's been going on since Christmas. They think it might be the same person – or people – responsible for all the break-ins. They seem to follow the same pattern. An empty house, no lights on, a back window forced—'

'And does much get stolen?'

'So far, mainly jewellery and cash, so the police think the burglar is on foot and gets in and out of the house as fast as he can. That's how he gets away with it. But it's a horrible shock for the people to come home to find their house ransacked – drawers pulled out, things thrown around, and sometimes expensive jewellery stolen, or things that have sentimental value.'

'Yes, I can imagine.' I had a sudden flashback to seeing that man outside in the dusk the other day. 'Where was this latest break-in?' I asked her, holding my breath.

'Over the other side of the river.'

I let the breath go. Nothing whatsoever to do with the stranger in the street. I really mustn't let my overactive imagination run away with me.

'The poor old dear who lived in the house was in hospital at the time,' Lauren continued. 'Can you imagine coming home to that, after having an operation? Poor old thing.'

'That's awful,' I agreed. 'Well, I hope the police catch him soon.'

'Yes, and we all need to be careful. Make sure you lock all the doors properly when you go out, won't you, Emma.'

'Of course.'

Dinner was ready then, and everyone was talking about other things – their days at work, what their pupils were doing, Holly's birthday and how many children she'd invite to her party – and the subject of the break-ins was dropped. And it says a lot about my selfish concerns with my own situation that I didn't even give it another thought.

CHAPTER SEVEN

The neighbour in the purple hat pounced on me again at Pat's bungalow the next day, before I'd even got the key in the door.

'He's been at it again,' she said. 'Bloody barking and howling all night. Bloody nuisance, that's what he is. Doesn't like being on his own, if you ask me. Didn't Pat ask you to stay with him at night?'

'No, she didn't,' I said. 'She said he'd be OK as long as I gave him two good walks during the day. And I did,' I added defensively. 'I don't know why he'd be barking at night.'

'Well, I'm warning you, I'm going to call the police about Noise Nuisance if you can't put a bloody stop to it,' she said with a satisfied nod, and with that she turned and went back inside her own house, banging the door after her.

Great. *Not* the start to the day I'd been hoping for. But as I let myself in, I could already hear Pongo howling loudly, and I suspected the angry neighbour was probably right – if he wasn't used to being left in the house on his own at night, it was obviously upsetting him. Poor Pongo, I thought, feeling a pang of sympathy for his sensitive doggy soul. I wondered whether I could leave some lights on when I left that night, and maybe a radio playing. It was gratifying, at least, that he stopped his noise as soon as he saw me, and bounded straight up to me, lying at my feet as if he was asking for a hug.

'All right, boy,' I said softly, crouching down to stroke him. 'I'm here now. You're all right.'

I went through the same procedure as the day before, letting him out in the garden and calling him back in when once again he was cornered by the big scary tabby cat that had to be shooed away. It seemed like the cat made a point of waiting for him in the middle of the path, just to upset him. But at least Pongo did come back in when I called him, now I'd got his name right.

I played with him for a while with his toys, thinking about the long night he'd endured on his own, crying for company. I could understand the neighbours being annoyed, to be honest, if he'd been howling and barking non-stop, but I decided it really wasn't fair for poor, timid Pongo to be left alone. I grabbed his lead, and he ran around me, barking with excitement now, and despite the noise and my anxiety about the police, I had to laugh. He was getting used to me. Perhaps he even liked me. Perhaps I was becoming an animal-whisperer, one of those people animals instinctively trust. Well, I seemed to make a mess of my relationships with people so perhaps I'd concentrate on dogs and cats from now on.

'Come on, then. Let's go up Castle Hill today,' I said, fastening Pongo's lead.

The walk went much better than the previous day, now he was recognising my commands. He actually *was* a very obedient dog, if he was called by his right name, and perhaps if I worked on imitating a bit of a Devon accent he'd feel even happier. I felt more confident, and more in control, and yes, definitely like a proper dog person. To my relief there was no sign of the guy who'd reminded me of Shane this time. At least, I told myself I was relieved, but a little voice in my head asked why, if I so much wanted to avoid seeing him again, I hadn't chosen a different walk.

If I *had* seen him, he'd certainly have been impressed with my fitness regime, anyway. Pongo and I not only walked up Castle Hill, we walked twice round the castle too before coming back down. I felt fit and strong and pleased with myself. I was so thirsty after my exertion I even contemplated going into Ye Olde Crickle Tea Shoppe on the way home, but I wasn't sure whether Annie allowed dogs in, and the thought of falling down ye olde steppes and having Pongo tumbling down on top of me decided me against it.

When we arrived back in Moor View Lane, I lingered, as usual, outside my favourite cottage for a while, but Pongo was pulling me on, so I'd actually crossed the road and was almost back at Pat's house before I heard a car pull up behind me. I turned round, and saw in surprise that it had parked right outside Bilberry Cottage. Slowing down again, pulling Pongo back in towards me, I watched as a young man got out of the car and started taking photos. Not just snaps on a phone, but proper, careful photographs with what looked like a real camera. Intrigued, I stopped a little way round the bend, behind a hedge, and peeked back down the lane. Yes, he was definitely taking pictures of Bilberry Cottage. He crossed the road, took another picture from further away, crossed back, and then suddenly turned and aimed his camera down the lane – lowering it just as quickly when he caught sight of me peering around the hedge, watching him. I jumped back out of sight, my pulse racing again. It was *him*! The guy from Castle Hill who'd reminded me of Shane. I felt my face flaring hot with embarrassment. What the hell was I thinking of, watching someone taking photos? Wasn't that exactly what I'd come here to hide from? What was wrong with me?

Pongo was pulling on his lead again, and I hurried the rest of the way along the lane back to his home. *Calm down, Emma*, I told myself, as I let myself back into the bungalow. From what I'd seen, it was fairly obvious that this guy must be an estate agent. I was pretty sure I'd been right – Bilberry Cottage was being put on the market. Not that it made any difference to me, of course. Unless I got a proper job instead of hanging around with people's pets every day, I'd be lodging with the Atkinsons for the rest of my life. But he'd seen me watching him. Would he have recognised me from our meeting on Castle Hill? Perhaps not, as I had a different dog with me now. I sighed. Short of going out in public wearing a mask, as well as dyed hair and a woolly hat, I didn't know how else I could escape people's attention.

I went home for dinner again that evening, as I'd done the previous day, but told Lauren I had to go back to 'work' to do a night shift. It was no good, I couldn't risk leaving Pongo on his own all night again, keeping the neighbours awake. The last thing I wanted was that woman calling the police, and anyway, I felt sorry for Pongo. I'd have to sleep on Pat's sofa.

'A night shift, after you've worked all day?' Lauren said, looking worried.

'Yes. It's just ... kind of an emergency. Lots of staff are off sick. I'll be doing it for the rest of this week and next week.'

'Oh, dear. That doesn't sound good, Emma. You'll be exhausted.'

'I've got the mornings off, though,' I lied. What was one more little lie added to the rest? Lauren and Holly were always out all morning, so they'd never know I wasn't coming home. On the other hand, if I did want to come back briefly to shower and change and have breakfast, she'd never know that either. I could even bring Pongo with me. That way he'd spend even less time on his own.

'Well, that's something, I suppose,' she said. 'Make sure you go to bed and sleep for the whole of the morning, then.'

She still looked worried, and I felt guiltier than ever now. Why didn't I just *tell* her I was actually pet sitting, not working in a care home? I could pretend it had all just been a misunderstanding, or that the care home job hadn't worked out. I took a deep breath.

'Actually, Lauren—' I began, but at that moment Holly started crying upstairs about some toy she'd broken, Lauren went off to sort her out, and the moment was lost.

Fortunately, spending time with Pongo was helping to take my mind off my worries. I was beginning to appreciate now what an intelligent and affectionate dog he was. He watched me, tail wagging, as I moved around his house, his eyes bright with curiosity, his expression alert and attentive. Already he seemed to have accepted me as his substitute human, and every time he came trotting back to me obediently when I called him, I felt a rush of pleasure and pride. As well as our two walks every day, I made a point of playing with him in the garden as often as possible now that the remainder of the snow had melted. He'd fetch a ball over and over again, never tiring of the game. To my surprise, I found that now he was responding to my commands (and I was using his correct name), if he got overexcited and started barking during our games I only had to say 'Quiet, Pongo!' and he'd stop. Another victory to me. And another good reason to stay with him overnight.

The night times were no problem at all. I found a spare duvet and pillow in Pat's airing cupboard and snuggled down on her comfy sofa in the lounge. The first night, Pongo whined for a little while out in his bed in the kitchen, and eventually gave a couple of frightened yelps. I opened the door to the room, and softly called out, 'Quiet, Pongo!', and he looked up at me in surprise, obviously having assumed I'd gone home and left him alone again. He got up and trotted out to join me, tail wagging, and after a moment's hesitation I gave in and brought his bed into the lounge, next to my sofa. We both slept peacefully for the rest of the night and I wondered if I should ask Pat, when she came back from holiday, whether she'd considered having him sleep next to her bed, to save the neighbours' complaints.

When we woke up in the mornings, I'd feed Pongo and then take him for his first walk of the day – back to Primrose Cottage, making sure I arrived after the Atkinsons had all left. Fortunately Romeo and Juliet made a dive for it out of the cat-flap as soon as they heard the snuffly sounds of a dog entering their home, so I didn't have a fight on my hands. I couldn't believe how well behaved he was now, while I showered, changed and had a quick breakfast – then we were on our way again, completing our walk with a march up Castle Hill or along the river.

By the second week, I felt completely at home in the bungalow in Moor View Lane with Pongo. I'd even made up a little fantasy for myself that this was actually my own home, and Pongo was my own dog. I got so carried away with it in my head, I sometimes almost believed it, and managed to forget Pat would be back soon to reclaim her life. In my fantasy, my imaginary husband (who unfortunately always looked a bit like Shane – why couldn't I get his image out of my mind yet?) was at work but would be thinking about me, of course, and would be desperate to get home to me at the end of the day. It must be so nice to have someone desperate to see you, I thought, wistfully. It had been years since Shane had made any pretence of still feeling like that about me.

Even my parents and my sister seemed to have shaken me off without much regret. I'd sent emails to them every now and then since I'd been in Crickleford, but had only had a few brief responses. Occasionally when I was in a part of town where there was a decent phone signal, my mobile would suddenly come to life announcing a text message from my sister. Usually just a dutiful 'Hi, how are you?' But I got the distinct feeling that they were all as relieved now that I was hidden away in Devon, as they were when I was out of sight, out of mind, in America.

'That's what comes of being the black sheep of the family', I told Pongo, stroking his ears as he gazed at me adoringly.

Who needed humans anyway, I thought. It seemed to me that animals were far more loyal.

That night, I was in the middle of another dream about the California beach house, when I woke up to the sound of Pongo, beside me, growling deep in his throat. I sat up on the sofa, disoriented for a moment.

'Quiet, Pongo!' I muttered, reaching out to stroke his head. I thought he must have been dreaming too – perhaps having a nightmare about that big tabby cat – but to my surprise, I quickly realised he was not only awake but sitting upright, his head down, his ears flat against his head.

'What's up, boy?' I said, reaching out to turn on the lamp – but suddenly there was a sound from the kitchen, a thud followed by ... surely that wasn't *footsteps*? And Pongo was up in a flash, throwing himself against the closed door of the lounge, head back, barking fit to bust. Still half befuddled by sleep, all I could think was that for some reason, Pat had come home early, and that I had to stop Pongo barking before Purple Hat next door called the police. I jumped up, ran to open the lounge door, and Pongo flew out ahead of me towards the kitchen, still barking furiously. I ran after him, calling out as I ran. 'Quiet, Pongo! For God's sake, what's the matter?'

And then I saw. The kitchen window hanging open. The young man, white-faced in the light from his own torch, which was lying where he'd apparently dropped it on the floor. He was shaking with fear as Pongo, up on his back legs, held him against the wall, growling fiercely and threateningly.

'Get 'im off me!' he yelled. 'He's gonna kill me! Get 'im off me, I don't like dogs!'

And then we blinked at each other in surprise. It was young Josh from the library.

CHAPTER EIGHT

The police arrived quickly, which was just as well for Josh, as I had no intention of calling Pongo away from him until they were there. I took pity slightly – but only slightly – when I saw that Josh had wet himself in fright.

'I'm scared of dogs,' he said, actually starting to cry.

'You shouldn't break into people's houses, then,' I said, then added quietly to Pongo: 'OK, leave him alone. Sit. Stay,' – and, giving me a look of disappointment, Pongo obediently sat – on the boy's feet. I was very impressed by this, but I didn't think Josh would have tried to run away at this stage anyway.

In fact I was very impressed by everything about Pongo's reaction. As we waited for the police, I patted his head to reassure him, but there was no trace now of the cowering scaredy-dog I'd come to love and pity. How amazing, I thought to myself wonderingly, that his cowardice had completely vanished in the face of an actual threat.

'Good boy, clever boy,' I repeated to him gently as he continued to rumble with angry growls at the terrified Josh. To Josh himself, I said nothing. I just wanted him to be scared of Pongo for as long as it took for the police to arrive. It served him right.

'Are you all right, miss?' one of the police officers asked me in concern as soon as they'd got hold of Josh and handcuffed him. 'I expect you're a bit shaken up.'

'No, I'm just cross,' I said, shaking my head at Josh. 'You're just a kid! What's wrong with you? Breaking into people's homes, causing all this trouble and upset to people?'

'It's my first time!' he protested, despite the police officers reminding him about his right to remain silent. 'It wasn't me all the other times, honest! I just … kind of helped. It was Andy!'

'We'll talk about all that down at the station, lad,' the police officer said. 'Come on. Well done, miss,' he added to me. 'You and your dog have been a big help.'

'Oh, he's not my dog, exactly—' I started, but they were already halfway out of the door.

It was nearly four o'clock in the morning and I was too wired now to get back to sleep. I gave Pongo a chew stick as a reward and made a big fuss of him, and had a look at the damage to the kitchen window, where Josh had forced it open. I'd have to call someone in to fix that. Pat would be home in a couple of days' time – what a shock she'd have. But it would have been so much worse if it hadn't been for Pongo.

Once it was a reasonable time in the morning, I went to see Purple Hat next door.

'Ah, come to apologise, have you?' she said as soon as she opened the door. 'That bloody dog was at it again in the early hours this morning. Well, I've had enough – I did warn you – I'm calling the police—'

'Actually the police already know about it,' I said, smiling at her sweetly, and gave her a quick résumé of Pongo's actions. Her face was a picture.

'So they've caught the burglar?' she said. 'You're telling me it was him – that stupid young Josh – all this time?'

'He *said* it was his first time. He was trying to pin the blame for the other times on some friend of his called Andy.'

'Not bloody Andy from the paper shop?' she squawked.

'I've no idea, sorry. I expect it will all come out in court.'

'Oh, it won't take that long, not around here,' she said with a nod of satisfaction. 'Everyone knows everyone in this town. We'll soon get to the truth.' She paused then, and continued more quietly, 'Well, I suppose I should take back what I've said about that dog of Pat's. It seems like he's done us all a bloody favour.'

'He certainly saved Pat's house from being burgled, anyway,' I agreed.

'Bit of a bloody hero, isn't he,' she said, nodding again. 'And you, too, girl – what's your name, anyway?'

'Emma,' I said, smiling.

'Well, Emma, if it does turn out young Josh – or that bloody idiot Andy – have been behind all these break-ins, I think you'll find everyone here in Crickleford will want to shake you by the hand and thank you for putting a stop to them. So let me be the first.'

She grasped my hand and pumped it up and down, finally giving me a grin that transformed her face.

'I'm Hattie, by the way,' she added as I turned to go.

I had to smother a grin. Purple Hat Hattie. Who'd have thought we'd end up friends?

The conversation had unnerved me somewhat, though, and I couldn't stop thinking about it as I walked back with Pongo to Primrose Cottage for my shower and breakfast. I was pleased to think everyone in the town would be grateful to Pongo, but I didn't like Hattie's assertion that they'd also want to thank *me*. For a start, I hadn't really done anything, apart from calling the police, but more to the point, I'd come here to avoid being the centre of attention, not to be thrust back into it.

I also needed to find a way of explaining myself to Lauren, before the story began to circulate. During the rest of the day, as I played with Pongo and then sat in the house with him while a handyman, recommended by Hattie, came to fix the forced window, I gradually worked out how I could do it with the least possible number of lies to add to my repertoire.

'I've got a confession to make,' I began, when I was seated at the dinner table with the family that evening. 'What I told you about working night shifts – it was slightly stretching the truth.'

'What do you mean?' Lauren asked, looking at me in surprise.

I crossed my fingers and ploughed on, telling her that I had been working nights – kind of – but not at the care home. That I'd been doing some pet sitting, on the side, which had started accidentally, with Mary hearing about me looking after Romeo and Juliet, and that I now had another dog to look after, who had needed me to stay with him overnight. I said I'd kept quiet about it because she and Jon had been so kind to me. I hadn't wanted them to think I was taking on too much.

'But you're doing this pet sitting on top of the care home job?' she said.

'That's the thing.' I had to look down at my plate now. They were both looking so worried – what an idiot I'd been for inventing that job in the first place. They were such nice people, and I felt so stupid and guilty about it now. 'I've enjoyed the pet sitting so much, I think I might concentrate on that, and hand in my notice at the care home. I realise I might not earn so much, but I don't want you thinking I'm going to default on my rent or anything like that. I'll make sure I can pay—'

Somehow.

Lauren put down her knife and fork and reached out to put a hand over mine, saying that I mustn't worry, that she and Jon trusted me and would help me all they could by recommending me to their friends.

'I'm glad you're not working nights and days at the care home,' she went on. 'I've been really worried about it. I feel kind of responsible for you, Emma. You're here all alone, a long way from your family and I think it's very brave of you to start up something like this on your own. It's a great idea, though – you're starting in time for when people book their summer holidays, and realise there's no cattery or kennels around here any more. I really hope you get lots more bookings and make a real success of it.'

My eyes filled with tears. I couldn't bear her being so nice to me. I didn't deserve it. But I was relieved finally to be telling the truth, or a very small part of the truth, anyway, to at least two people. The *actual* truth, of course, was that I hadn't even thought too seriously about looking for more pet-sitting work. I'd planned to go back to the proper job hunt and just look after pets in the meantime, if anyone asked me. But Lauren had made a really good point. There was apparently a pet-sitting gap in the market here in Crickleford at the moment, and people were getting to know about me. Why not go for it and make it my real job? The thought was exciting as well as a little bit scary.

I went on to tell Lauren and Jon what had happened the previous night with the break-in, and everyone, even Holly, gave up on dinner altogether. Lauren and Jon sat open-mouthed, staring at me as I described Pongo's barking, finding Josh in the kitchen, and the police taking him away.

'Good lord!' Jon said when I'd finished. 'You'll be the local heroine when this gets out.'

'But I don't want anything like that,' I told him hurriedly. 'Please, don't let anyone make a big thing about it.'

'It could help your new business, Emma,' Lauren pointed out.

'Maybe. But ... well, I'd find it embarrassing.' I looked down at my plate again. 'I don't like fuss.'

They shook their heads but said they understood.

'If it *was* that Andy from the paper shop behind it all,' Lauren said thoughtfully, 'it does make sense. He'd have access to the newspaper delivery accounts. He'd know when people cancel because they're going to be away on holiday or in hospital, so he could assume their house was going to be empty.'

'Oh!' I said, suddenly remembering. 'I saw someone hanging around outside here when *you* were on holiday. I should have told you, but I thought I was just being silly.'

'That was probably one of the Neighbourhood Watch people,' Jon said with a shrug. 'We're members, and we always let the others know when we're going away, so they can keep an eye on the house.' He stopped, raising his eyes at Lauren. 'Josh's parents are members too.'

'Yes.' She nodded slowly. 'So he'd know which houses are being watched, and tell Andy to avoid them. It's suddenly all very obvious, isn't it? Well, don't worry, Emma, we won't spread it around, about you being at Pat's the other night, if you don't want us to.'

However, as Hattie had predicted, I wasn't going to get away with it that easily in Crickleford. The next morning as I walked Pongo, three different people stopped me because they recognised him.

'It's Pongo, the dog who caught the burglar!' they all said, and before I could nod and smile and make my escape, they were telling other people around them, shaking me by the hand, asking my name and congratulating me, however much I tried to explain that I really hadn't done anything. Lauren had been right, too: by the time I headed back to Pat's house, two more people had asked me about my pet sitting – how much I charged and whether I took on cats as well as dogs – and said they'd let me know their holiday dates.

'Come on in, my luvver, have a cup and a cake on the house,' Annie from Ye Olde Crickle Tea Shoppe said when I passed, just as she was changing her menu in the window. 'Ah, don't thee freck 'bout 'im!' she added when I pointed out that I had a muddy dog with me. ''Ee be a hero round here now. Mind the ruddy step this time, don't be falling arsy-varsy like thee did backalong.'

I must have been in Devon longer than I thought – I could almost understand her now. Of course, it was a mistake to let my eagerness for a hot chocolate and a toasted teacake get the better of me, because once we were inside and safely down ye olde bloody steppe, the entire clientele let out a cheer, and people were knocking over their chairs in their haste to pet Pongo, pat me on the back, and congratulate us. Poor Pongo had reverted to type, pulling on his lead to slink away from the noisy crowd and cower under a table – which, of course, caused hilarity all around.

'I thought he was supposed to be brave!' one woman laughed.

'He's more scared of us lot than he was of the burglar!' someone else said.

'Have you heard the latest?' another woman called out, after I'd sat myself down and done my best to calm my nervous canine hero. The woman raised her voice so that everyone could hear her. 'They've arrested that Andy from the paper shop too. They're saying he was the one behind it all, and he roped in poor young Josh to help him.'

Straight away, other people were joining in, saying it was no surprise, that this Andy had always been in trouble, even when he was still at school, and 'poor young Josh' was, unfortunately, easily led. While I demolished my drink and my teacake, I was conscious of people nudging each other and muttering together about their dogs who'd been booked into the ill-fated Dribstone Boarding Kennels, their cats and their rabbits and guinea pigs who were, in the past, looked after by the woman who'd moved to Bath. People were raising their eyebrows and nodding to each other. I had a feeling that I might be getting some more bookings before too long. And that if I actually *had* a job in a care home to give up, I might even be looking to do so sooner rather than later.

Back at the house, I'd only just finished showering Pongo and myself and getting dressed again when there was a knock at the door. I opened it cautiously, and nearly fainted with surprise to find the young man I'd encountered before on Castle Hill, and again outside Bilberry Cottage taking photos, standing on the doorstep giving me a devastatingly charming smile.

'Emma?' he said brightly.

'Er ... yes.' I frowned.

'Sorry to bother you like this. I got your name from the lady next door – I knocked there when you weren't in—'

'What's this about?' I interrupted, wondering if he, too, wanted a pet looking after. It was one thing being approached by people for work when I was out and about, but I didn't want them coming to look for me at other people's houses.

'Well, I expect you realise you've become something of a celebrity around here, since you—'

'No!' I almost shouted it. 'I'm absolutely *not* a celebrity! I've *never* been a celebrity!' My voice was shaking. I must have sounded like a madwoman. 'And if you don't mind, I've got a dog to look after, so—'

'Pongo. Yes. I've heard all about how he caught the burglar breaking into the house.'

I suddenly realised he was holding a notebook. He'd glanced at it to check Pongo's name. What the hell ...?

'What do you want?' I asked more quietly. 'What's this got to do with you?'

'I'm Matt Sorrentino,' he said, holding out his hand to me and treating me to another lovely smile, 'and I'm a local—'

I was so distracted by the undeniable appeal of that smile, that I accepted his handshake without thinking. Close up, he wasn't like Shane at all – his build was similar but his colouring was lighter, and he was actually even better-looking.

'Estate agent. I know,' I interrupted him.

'What? No!' He laughed. 'Not guilty! What made you think that?'

'You were taking pictures. Outside Bilberry Cottage.' I nodded down the lane. 'And I just assumed ...'

'Oh.' His smile faltered a little. 'No, I was just ... well, I just kind of like that cottage, that's all.' I thought he looked a bit uncomfortable for a moment, but he shrugged and went on: 'Look, I wondered if we could have a little chat? Would that be OK? I've obviously heard all about your amazing episode last night with the break-in, and ... um ... Pongo ...' He tailed off. 'Can I come in?'

'What for?' I felt totally confused now. He wanted to chat? Was he *interested* in me, or something? I felt a brief flicker of pleasure. It had been a while since anyone of the opposite sex had paid me any attention – unless you counted Josh. And this Matt Sorrentino *did* seem nice, not to say very fit and attractive. I remembered how he'd lingered, seeming to want to talk, on our first meeting up on Castle Hill. Had I got a new admirer? It would be nice to be asked out on a date ... although I'd have to be careful about all the lies I'd told people ...

His smile had wavered a little. 'Well, as I said, I'd just like to chat about what happened last night. Don't worry, I realise we can't give the names of any suspects at this stage, but people around here have been so concerned about all the burglaries, they're really chuffed somebody's been caught. So I'm sure they'd love to hear your side of the story, Emma.'

'Oh.' I tried to hide my disappointment. Not a new admirer, then. 'You're with the police, are you?'

'No.' He gave me a puzzled look. 'Didn't I say? Sorry. I'm a local journalist, working for the *Crickleford Chronicle*, and—'

He didn't get any further. I not only slammed the door in his face, I locked it, bolted it, ran and hid in the back bedroom, and stayed there for the rest of the day, cuddling Pongo and fighting back tears. I just couldn't believe it. After everything I'd done – leaving home, cutting myself off from my family, hiding away in the most remote little town I could think of, making up stories about myself and even dyeing my bloody hair – all it had taken was one little episode with a dog and a stupid kid trying to be a burglar, and once again I had journalists sniffing around me, wanting to ask questions. How long would it take before a *little chat* about last night became a probe into my background? How long before I tripped up and said something that aroused his suspicion, and the next thing I knew, it wouldn't just be the local papers, it'd be the regional ones, then the nationals, then it'd be the TV cameras, and my poor family back home would be bombarded all over again – just because I'd been taken in by a handsome face and a cute smile?

It was no good. The damage was done. Matt Sorrentino wasn't going to take no for an answer. I knew only too well what these people were like – I shuddered at the memory of previous occasions in the States when the paparazzi had hung around our apartment for days, intruding into every aspect of our lives, following me down the street, hiding in bushes deliberately to take photos of me when I wasn't looking my best. And now they had something really juicy to get their teeth into – I knew from the way they'd surrounded my parents' house in Loughton over Christmas how desperate they were for details of my break-up with Shane. I couldn't bear it happening again here in Crickleford. I was going to have to move on, and the sooner the better. Just as I was settling down here, just as things were looking good for me, just as I'd finally started doing something with my life, something worthwhile, something I enjoyed, something that was *my own* at last – I was going to have to give it all up and start again.

I gave Pongo an extra cuddle and stroke as I said goodbye that evening and crept out into the dusk, looking nervously around me in the shadows. I wasn't going to come back tonight – I hoped Pongo would be all right on his own just this once. Purple Hattie seemed to be his friend now, so I didn't think she'd call the police if he howled – and Pat would be home the next morning. I'd phone her and make some excuse for not calling in; hopefully she could pay my money directly into my account. Then I'd have to pay Lauren for the next month – it was the least I could do, letting her down like this, when she'd been so kind to me.

As I trudged back past Bilberry Cottage for the last time I sighed, thinking how ridiculous it had been to imagine I might end up living somewhere like that. I needed to wake up and realise I was never going to shake off my past unless I completely changed my identity and, perhaps, went to live even further afield. The highlands of Scotland, perhaps, or the west coast of Ireland. Or somewhere in Africa. That'd do it.

'Goodbye, Bilberry Cottage,' I whispered.

And goodbye to my dreams.

PART 2

NEW BEGINNINGS

CHAPTER NINE

As soon as I got back to my little blue bedroom in Primrose Cottage, I started packing my bags. I was so upset it was hard to keep from crying, but it seemed I had absolutely no alternative but to give up my new life here in Crickleford and start again somewhere else. I hadn't yet decided where. I thought perhaps I'd just get the morning bus back to Newton Abbott the next day, jump on the first train that came along and see where I ended up. I grabbed the last few of my tops off their hangers, chucked them into the suitcase and pulled open the first drawer of the chest. And then I paused, staring at a parcel wrapped in bright red and yellow paper with an elephant design.

It was my birthday present for Holly: the new book by her favourite author, Julia Donaldson. I'd picked it out from the Crickle Bookshop just a few days ago, and I'd been looking forward to seeing the smile on her face when she opened it, together with the elephant-themed birthday card I'd chosen, with its bright red 'I AM 4' badge pinned on the front. I took the parcel out and held it against my heart, blinking back tears. I'd already become so fond of little Holly, and her parents. How could I even consider leaving before her birthday? But how could I stay, when there was now a local journalist pursuing me? It was only going to get worse.

To give myself a few minutes to think, I put the parcel back in the drawer and slammed it shut – only just in time, as Holly suddenly appeared in my doorway, frowning at the suitcase and general mess in my room.

'What are you doing?' she said. 'Are you going on holiday?'

'Um ... possibly,' I said, with a sigh.

'Where to?'

'I haven't quite decided yet.'

She put her head on one side, considering this.

'But you can't go until after my birthday. You said you were going to come to my party.'

I managed a smile. 'Of course I'm going to come,' I said, and held out my arms to her for a hug. 'I'm looking forward to it.'

Holly trotted off happily to her bedroom to play with her toys, and I closed my eyes and took a deep breath. Her birthday was this Monday, and the party was going to be in the afternoon, straight after preschool: her little friends were coming for lunch. Three more days. I'd finished looking after Pongo, and I had no more definite bookings yet – just a list of people who were going to give me their holiday dates. If I stayed at home as much as possible, perhaps the fuss would start to die down. I could use the time to make my excuses to Lauren and Jon, and cancel all my other pet-sitting bookings. Matt Sorrentino wouldn't give up easily – journalists never did, I knew that from past experience – so I'd still have to move away if I didn't want to be found out. But at least I wouldn't be leaving in such a hurry.

Decision made, I unpacked my bags again and went down for dinner. Afterwards, I tried to catch Lauren on her own in the kitchen so I could tell her I was going to have to leave after Holly's birthday, but every time I started trying to bring up the subject, I got cold feet and couldn't go through with it. I told myself it could wait until the morning, but when the morning came, I felt even more reluctant to start the conversation.

'No pets to look after today, Emma?' she asked me cheerfully as I helped her clear away the breakfast things. It was Saturday, and Jon had already gone out somewhere with Holly.

'No, nothing booked for this week.'

I felt a twinge of guilt for not going to Pat's house for the last time. Poor Pongo had been on his own all night – was he missing me? Was he barking his head off again? Pat had told me she'd be getting home early this morning, and I'd been too nervous to go back there, in case Matt came hanging around outside again. How long before he found out where I was living? It would be awful if he turned up here.

'Are you all right?' Lauren asked me a little later, giving me a concerned look. 'You're very quiet.'

'Sorry, I'm fine, thanks.' I forced a smile. Now was my opportunity: I should tell her now about leaving. I cleared my throat, took a breath – but the words stuck in my throat. 'Perhaps I could give you a hand with some housework or something?' I said, instead. Anything to keep my mind off my worries.

'There's no need for you to do that,' she laughed.

'I'd like to, honestly. You're always so busy, and I'm free today, so give me a job I can take over from you.'

'Well ...' She hesitated. 'OK, I'll tell you what would be very helpful. Could you possibly pop into town for me, post a letter and pick up some icing sugar and a pack of butter? Are you sure you don't mind?'

'Oh.' I felt the smile freeze on my face. 'Wouldn't you prefer me to do the Hoovering? Or clean the bathroom, or something like that?'

'I did all that yesterday, thanks, Emma. I want to make Holly's birthday cake, while I've got the opportunity – Jon's taken her to the park this morning and promised to keep her out of the way for a few hours. That's why I need the icing sugar,' she added with a smile. 'If I get the cake baked this morning, I can ice it while she's in bed tonight.'

Well, I obviously wouldn't be able to do *that* for her. I had about as much idea of how to bake or ice anything, as I knew how to fly. And I sensed Lauren was looking forward to having the house to herself while she concentrated on the birthday cake.

'OK,' I said, trying not to sound reluctant. 'I'll go.'

In fact it was good to be outside in the weak March sunshine, despite the fact that I was looking around me and behind me with almost every step along the road. The little town was busy with Saturday shoppers, and although I tried to keep my eyes down as I hurried along, every now and then someone recognised me and gave me a wave or a nod. I rushed on, anxious not to give anyone a chance to come and talk to me. I realised now that I'd made a big mistake by letting so many people here get to know me. Wherever I chose to go to next, I'd hide myself away in a cave if necessary, and avoid all human contact.

I'd posted Lauren's letter, bought the icing sugar and the butter, and I was just starting to head back home when, just my luck, I saw him. *Matt Sorrentino*, the one person I'd wanted to avoid above all others. He was walking towards me with the jaunty stride that had caught my eye the first time I saw him, up on Castle Hill, his dark, floppy hair falling forward over his forehead, his hands in his pockets, his eyes bright with the confidence of someone who knows what they want and how to get it. I immediately tried to turn around, to cross the road, to break into a run – but it was too late. He'd already seen me.

'Emma!' he shouted.

I kept going, away from him, but within seconds he'd caught me up.

'Please leave me alone,' I muttered, trying to shake off the hand he'd put on my arm to stop me.

'But I wanted to apologise,' he said, keeping pace with me as I kept on walking, fast, in the wrong direction, wanting only to get rid of him. 'I obviously said something to upset you the other day. I don't know what, but whatever it was, I'm sorry!'

I stopped abruptly, turning to face him.

'You're a journalist. I don't want to talk to you, OK? Now, please, leave me alone.'

I strode off again, faster still, but again he walked with me, silently at first, and then, just as I thought I'd actually have to give him a shove to make him clear off, he said, panting slightly:

'All right, you've made your point.'

'About what?' I retorted, swinging round to face him before I could stop myself.

'The fitness regime. I give in. You're much fitter than me.' And he pretended to bend double, wheezing and coughing, and despite everything I struggled to keep a straight face.

'Like I said, I'm sorry if I've upset you, Emma,' he went on now. 'I know I'm a journalist, but I'm not a complete pig.' And he finished with loud piggy snort, making other people in the street turn and smile at us. To my annoyance, when he made the snorting noise again, I actually burst out laughing.

'OK, so you're not a pig, even if you snort like one,' I said. 'Now we've got that clear, will you go away?'

'Only after you've let me buy you a coffee. To prove it.'

'Prove what?' Why was I even talking to him? I should be running away, as fast as I could. Did he think I was stupid enough to let the offer of a cup of coffee loosen my tongue?

'That I'm not a pig, of course. Come on.' He nodded at The Star pub over the road. 'It's eleven o'clock, they've just opened, and honestly, their coffee's better than their décor – or their beer, come to that. It's either that or Annie's Olde Gossipe Shoppe,' he added with a grin.

I smiled, despite myself. It had to be said, I did fancy the idea of stopping for a coffee. And I might have been a lot shallower than I wanted to admit, but I fancied the idea even more of stopping for a coffee in the company of a very charming and good-looking man. After all, it would definitely be the last time I would see him; I'd made it very clear I wasn't talking to him and he surely wasn't going to find out who I was in the ten minutes it would take me to drink my coffee.

'OK,' I said. 'Thank you. But I can't be more than ten minutes. My landlady's waiting for her icing sugar.'

'Never let it be said that I kept a lady waiting for her icing sugar,' he said seriously, taking my elbow to steer me across the road and into the pub.

It was dark inside, and to my relief, almost completely empty. He was right about the décor – it was grubby and dated, the carpet looked as though it had the beer of centuries soaked into it, and even the wooden tables were slightly sticky. I sat in an alcove at the back, away from the door, and waited while Matt brought me a cappuccino.

'How can this dump possibly compete with The Riverboat Inn?' I asked.

He shrugged. 'The locals prefer it. It's authentic.'

'That's one word for it,' I said, taking a sip of my coffee. He was right about that too, though – it was delicious.

'The Riverboat's a grockles' pub.'

'A what?'

'Grockles: tourists, holidaymakers.' He grinned. 'That's you, I suppose?'

I put down my cup and shook my head. I wasn't getting caught out that easily.

'I wasn't probing, Emma,' he said, surprisingly gently. 'Look: no notebook today. No voice recorder.'

'No questions, then,' I said.

'OK, fair enough.' We sat in silence for a moment, and then he added: 'I didn't intend to intimidate you, you know. I only wanted a couple of quotes from you about the break-in.'

I stared down at the table.

'There wasn't a question in that sentence, by the way,' he said, making me laugh again, against my will. 'Look, whatever it is you don't want to talk about,' he continued, more seriously, 'it's fine by me. I'm not interested in you, OK? Well, not from a professional point of view, anyway,' he added with a little grin that – again, against my will – made my insides go slightly weak.

'That's good,' I said firmly, trying to get a grip on myself, 'because I won't be here much longer anyway. I'm moving away in a few days' time.'

'Sorry to hear that. Holiday over?'

'It wasn't a—' I stopped, annoyed with myself. Despite everything I'd said, here I was, starting to answer his bloody questions! 'This is precisely why I'm not staying here!' I snapped. 'People are so nosy. I came here for peace and quiet, but nobody will leave me alone.'

'Nosy? Or just friendly?' He put his head on one side and looked at me sadly. 'I really hope you're not going because I tried to talk to you for the newspaper. I'd never forgive myself if I thought I'd driven you away.'

'Don't kid yourself,' I said, no less snappily. And then, to spoil it, I went on, in the exact same tone I used to adopt back in New York when the paparazzi were bothering me too much: 'I simply don't want to speak to the press.'

He looked away, a smile flickering at the corner of his mouth.

'I understand,' he said, and gave a little shrug. 'OK, looks like I'll just have to wait a bit longer for my big break.'

'What big break?' I said, frowning.

'Ah, who's being nosy now?' He laughed. 'Well, it's no secret – I've been trying for years to get a big enough story to impress my boss at the paper, and persuade him to give me a promotion. The trouble with little towns like this is that nothing much happens. If a tree blows down in a gale it makes the front page. No, I'm not kidding!' he said when I started to laugh. 'So when I heard a girl and a dog had foiled the guys behind this spate of break-ins, you can't blame me for thinking I'd finally got something worth writing about. Never mind.' He shrugged again and nodded at my empty coffee cup. 'Nice talking to you, Emma. You'd better get back to your landlady with the sugar.'

Ridiculously, I now felt sorry for him. And even more ridiculously, I didn't want to leave it like that.

'Look, I wish I could help you, with your story,' I said. 'But I can't. I don't want people talking about me.' I hesitated. 'Not that there's anything to talk about. I just came here to get away from an ex-boyfriend who was harassing me, all right?'

Well, it was only a little white lie. Shane had never, in fact, done anything to harass me. Towards the end he hardly even seemed to notice me. But perhaps that would stop Matt probing.

'And now you're moving away again?' He frowned. 'Look, *I'm* not going to harass you, I promise. How about I write up the story without mentioning your name? Although, to be honest, I think most people in the town already know it was you.'

'Exactly.'

'But, Emma, it's the *detail* of the story they'd be interested in: did the dog hear the intruder first, or did you? Were you scared? Did the dog bite him? Did he cause any damage? How did you stop him from escaping? That kind of thing.' He paused. 'What if I promised not to reveal anything whatsoever about you, *apart* from your name?'

'Just my first name, then,' I insisted. I shook my head. I couldn't believe I was agreeing to this. 'No photos,' I said, sternly. 'And no questions about my background – where I'm from, why I'm here – nothing. Or I walk away.'

'I promise,' he said, his soft brown eyes gazing into mine. 'Cross my heart and hope to die.'

Afterwards, I wondered what the hell had come over me. Even before I'd had to run away from America, I'd *never* speak to journalists if I could help it. They always used to make me nervous, and I'd end up saying something stupid, something that they could twist and turn and make into an embarrassing headline in the following day's papers. At first, in the early days in California, when Shane was just a rising star and I was still his soulmate, he'd laugh off my silly faux-pas with the press.

'Don't worry about it,' he'd say, kissing me on the nose as if I were a sweet, helpless little kitten. 'Any publicity is good publicity.'

I didn't see how it could be good publicity that, in my agitation at being cornered by journalists as I came back from the beach, I'd forgotten the name of Shane's new record, but I loved the fact that he was so patient with my mistakes. But it all changed when he became really famous. Then I had to learn to keep my mouth shut. And yet, here I was with a hack from a little local paper – with more to keep quiet about than ever – and after ten minutes and a cappuccino, I was eating out of his hand. Unbelievable.

CHAPTER TEN

Matt and I had agreed to meet again the following morning, and that I'd only talk to him about the details of the burglary if I still felt happy to do so then. The fact that he'd given me twenty-four hours to reconsider, did make me trust him a little more. Lauren, Jon and Holly went out to visit some relatives in Totnes for the day, so I was alone in the house, struggling with my indecision, when there was a knock at the door. Still feeling somewhat nervous about the possibility of being found by hordes of paparazzi, I peeked through the front room window first – but it was only Pat, with Pongo on his lead. It was so nice to see him, I rushed to open the door and make a fuss of him.

'I just wanted to thank you,' Pat said, beaming at me. 'I've heard all about what happened, obviously.'

'It was Pongo who saved the day, not me.'

'Well, a lot of people would have been useless in a situation like that, but from what I hear, you dealt with it superbly, keeping that young idiot there until the police came, *and* getting my window fixed afterwards. Thank you, Emma. I'll recommend you to everyone, of course.'

'Oh.' I looked away. 'Well, I'm not sure. I mean, I might not be staying here for much longer, so ...'

'Oh. I'm sorry to hear that. What a shame. Pongo obviously adores you, and I've heard quite a few people saying they were going to ask you to look after their pets.' She glanced behind her. 'She'll be sorely missed around here if she goes, won't she, Mary?'

I'd been crouching down, fussing over Pongo, and hadn't even noticed Mary coming up the garden path. She was staring at me as I stood up.

'You're talking about leaving Crickleford?' she said. 'Already?'

'Um ... yes.' They were making me feel bad about it now. 'Sorry.'

'That's a real shame,' Mary said, shaking her head. 'You've made yourself very popular here.'

She smiled at me, and I felt a sudden warmth, unlike anything I'd ever experienced before. I was *popular*? Really? People actually *liked* me? For myself?

'Oh!' I stared at her. I couldn't quite take it in. Surely, she only meant that people liked me because I'd helped to catch the burglars. It they knew what I was really like, behind all the lies and stories, I wouldn't be popular then, any more than I'd ever been back home in Loughton. But ... how nice it would be if I *could* be that person they thought I was. If I could fit in here, and make friends with people – real friends, people who liked me, not hangers-on who took selfies with me and followed me on Twitter but laughed at me behind my back. 'Oh, do you really think so?' I said.

I must have sounded even more needy than I felt. The desire to hold onto that moment of warmth, that sweet, cosy moment of being told people would miss me, was almost overpowering. Mary met my eyes and nodded slowly, and I had the distinct feeling she somehow understood.

'Of course. You've fitted in really well here.'

'Oh,' I said again. 'Well, thank you.'

'And if you did decide you could stay, after all – perhaps if you didn't need to move back to your house yet? – I hope we could be friends.'

I was now close to tears. Friends? Mary must have been in her late sixties. She could have been my grandma. Why would she want to be friends with someone like me? And – maybe I'd imagined it – but did she *know* I'd been lying about the house fire? There was something about the way she'd mentioned moving back to my house, but without actually mentioning the fire, in front of Pat ... something about the look she'd given me ... and I realised what I would really have liked to do was completely unburden myself to her, apologise for all my lies and tell her the truth. I was sure, somehow, that she'd understand, and forgive me.

But what was I thinking? If I *was* going to stay in Crickleford, I couldn't possibly start blabbing the truth to people. I had to stay on my guard! Keep my secrets! Keep my distance! I wiped my eyes, gave Mary a smile, and told her I'd think about it.

'I suppose I might be able to stay for a bit longer,' I said, as casually as I could.

'I do hope so. Scrap would miss you so much if you went.'

'So would Pongo,' Pat agreed. 'Please come and see us, if you decide to stay.'

'And us,' Mary said with another smile. 'We could take Scrappy for walks together.'

'I'd like that,' I said wistfully. I took the bag of books Mary handed me for Lauren, and watched them both walking away before closing the door and leaning on it. Maybe I *could* give it a bit longer here, and see how things panned out. I'd just have to be careful around Matt. Meet him, perhaps, but hardly talk to him at all. I could do that, surely.

That was what I told myself, anyway, as I walked into town later to meet him in The Star for coffee, as arranged. But as soon as I saw him I had mixed feelings again. In one way I was wishing I'd stuck to my guns and just refused completely to talk to him for the newspaper. On the other hand, by the time it was published I could be out of town, on my way to the Outer Hebrides and/or Timbuktu. But I wasn't sure now about leaving! And I felt sorry not to help Matt out with his big story, now he'd turned out to be ... well, quite nice to me.

In the end, we stuck to our agreement. No photo. No surname. No mention of my age or where I might have come from. I guessed he was being sympathetic to my supposed harassment from an ex-boyfriend. Fortunately I did manage not to mention being made homeless in a raging inferno, having a little girl called Holly or any of my other ridiculous lies. He was simply going to refer to me as: *Emma, who was staying overnight at the house to look after Pongo the Alsatian.*

'I'm not sure that this is really going to be my big story, to be honest,' Matt said with a sad smile, when he'd finished jotting down the few sparse details I'd given him.

'Maybe you'll get a better story after the trial. More stuff might come out then,' I tried to console him.

'Mm, maybe. Well, thanks anyway, Emma,' he said. 'And, well, good luck – wherever you end up. Pity you have to go, though.'

'Thank you.' I felt myself blushing slightly. Did he mean that, or was he just being polite? He'd looked at me in a certain kind of way, as he said it, making me wonder if we might perhaps have seen more of each other if I didn't have to move away. But maybe I was just imagining it. 'I'm not a hundred per cent sure now, actually, about leaving,' I admitted. 'I don't really want to go.'

'Don't, then.'

I definitely wasn't imagining it. But then again, he was probably just hoping I'd stay around so he could get more material out of me.

'I'm not just saying that because I want you to tell me more about the break-in,' he said. What was he – a mind-reader? 'I just don't understand what you're running away from – apart from your ex, I mean. And I presume he hasn't found you here yet.'

I didn't reply. He continued to look into my eyes. It was disconcerting, to put it mildly.

'I understand that you don't want to tell me anything,' he went on after a moment. 'But I haven't I badgered you for more information, have I?'

'No,' I admitted.

'Haven't I, in fact, stuck to our deal, kept my word and promised faithfully to only write what you're allowing me to write?'

'Yes, but—'

'I'm not going to hassle you, Emma. I can see how worried you are about talking to the press, and I understand that you don't want your ex to find out where you are. I think everyone's entitled to their privacy anyway.'

'You don't sound much like a journalist.'

'Perhaps that's why I still haven't got my promotion,' he said with a rueful smile. He paused, and then went on: 'I meant what I said yesterday. I'd feel bad if I thought you were going because of me. I'm grateful that you've agreed to talk to me at all, but I won't be chasing you for more.'

I thought about it, of course, all the way home. Was I overreacting? Was it ridiculous to threaten to move away just because one journalist had wanted to talk to me about the break-in? He did seem a completely different type of journalist from any I'd had the misfortune to be bothered by before, but could I really believe his assurances? I'd been fooled often enough before by journalists who pretended to be my friends, who'd buy me a couple of drinks and encourage me to confide in them – tell them stuff about my relationship with Shane that I absolutely shouldn't have talked about.

The bottom line was, why would I trust Matt any more than those others? I couldn't let my guard down just because he was nice to me, or because … well, let's be honest, I fancied him! But I did want to stay in Crickleford, especially after Mary saying all those things about how well I'd settled in. Could I risk it?

I still hadn't completely made up my mind the next day, but as it was Holly's birthday I tried to put everything else out of my mind until after she'd gone off to preschool.

'Happy birthday, sweetheart.' I handed her the wrapped present and she tore the paper off excitedly. 'Mummy, it's another one of my favourite books!' she squealed.

'Ah, thank you, Emma,' Lauren said, smiling at me. 'That was a perfect choice.'

I smiled, pleased to see the little girl was so happy with it. I couldn't remember ever being thrilled with a book, of any sort, when I was a child. But, then, I was a different kind of child altogether.

'Will you read it to me, Emma?' Holly asked me, looking through it at the pictures.

I hesitated. 'I thought you were going to preschool in a minute?'

'Yes, come on, Holly, get your shoes on or we'll be late,' Lauren said. 'Perhaps Emma will read it to you at bedtime, after your party.'

I waved them off, promising Lauren I'd get everything ready for the lunchtime party while she was out, as I wasn't working. It was hard to concentrate on sandwiches and decorations while my mind kept wanting to return to my problem. Was I going to move away? And if I did, would I ever work as a pet sitter again? I'd been enjoying it, and I was looking forward to taking on the new work that I'd been promised. This had been a new start for me, the possibility of an actual new career. I sighed. It would probably be best to go. I should know better than to trust Matt Sorrentino, just because he was charming and good-looking with an exotic-sounding surname.

By the time Lauren came home with a very overexcited Holly, I'd set the table for the lunch party, blown up all the 'Happy Birthday' and '4 Today!' balloons, arranged sandwiches, sausages and crisps on plates and put on one of Holly's CDs of kids' songs.

'Ah, thanks, Emma,' Lauren said. 'What would I do without you?'

'It was no trouble,' I said. 'But where are all the guests?'

'Just following us down the road with their mummies. OK, here's the first one at the door! Let the fun begin!'

And yes, it was fun, but it was also exhausting! I developed a whole new respect for mothers everywhere during the course of that afternoon. Lauren and the other mums who stayed to help dealt, apparently calmly and effortlessly, with drink spillages, food mess, squabbles, tears (nobody liked being the first one out in Musical Chairs), a toilet accident and a panic over a lost hairband, which the child in question viewed as a major catastrophe ('*It was my bestest, favouritest pink spotty one*'). It finally transpired one of the other little girls was wearing it around her wrist as a bracelet, which needless to say caused some more fuss but was handled, I thought, with admirable diplomacy. Jon managed to get home early that evening, in time to calm down his overexcited daughter and put her to bed.

'Have a glass of wine with us, Emma,' Lauren said, as I was helping her dish up our dinner. 'You've been such a help today.'

'But I really enjoyed it,' I said truthfully. The party had left me feeling wistful, in a way I didn't quite recognise. I'd found myself thinking about my niece and nephew back in Loughton. Little Jeremy's third birthday would be coming up soon. He and his baby sister Rose were both born while I was in the States, and I'd sent them lavish, expensive gifts to make up for the fact that I wasn't seeing them. I couldn't afford the expensive gifts now but I'd make sure I didn't forget their birthdays, even though I still couldn't be there with them.

Then I started thinking about Kate, my twin, and our own birthday parties when we were children. We'd started off having a happy childhood – hadn't we? But from a very early age I'd been aware of the differences between us. We weren't even very similar in appearance. Kate was dark, like our mum, and shorter than me, whereas I'd inherited Dad's red hair and freckles. But of course, the other differences were much more of an issue, and by the time we were at senior school, we were so totally unlike each other that people who didn't know us often refused to believe we were twins.

Lauren was smiling at me now. 'Are you OK? You look like you're miles away.' She gave me a thoughtful look. 'I expect you miss your family, don't you?'

I felt as if she'd been reading my mind.

'Yes, I do, of course,' I admitted. And then I flushed, wondering yet again whether she might have heard my dreadful stories about the house fire and my parents looking after refugees.

'Well, if it's any consolation, you're already like one of the family to us,' she said. She put down the serving spoon and gave me a quick hug. 'You know, I always used to wish I had a younger sister. I'd have wanted her to be just like you.'

I sniffed. She was making me feel tearful now.

'You might not think that if you really *were* my sister,' I said, trying to make it sound like a joke.

'Rubbish. Honestly, Jon and I couldn't have asked for a nicer person to have living with us.'

And that decided it once and for all. How could I leave, now? Not only had Mary said we could be friends, and that she thought I'd fitted in well, but I'd actually acquired a new surrogate family, one that seemed to like me more than my own did. And as long as they never found out who I really was, then perhaps my place, now, really was here in Crickleford with them.

CHAPTER ELEVEN

Within days, I'd had my next pet-sitting booking confirmed, and it was for the following week. It was for one of the ladies who'd been in Ye Olde Tea Shoppe when Annie had made such a fuss of me after the break-in. A strikingly beautiful and smartly dressed woman of Asian appearance, aged I guessed about forty, she'd been sitting on her own with a cup of coffee and a newspaper, keeping herself rather more to herself than the other customers, but as she was leaving she'd stopped at my table, introduced herself as Vanya Montgomery, shook my hand rather formally and asked me how experienced I was with cats. When I told her about Albert, and the cats my family used to have, she seemed highly delighted.

'You've had pedigree cats yourself. Ah, you'll understand my concerns, then.'

Her concerns seemed to be considerable. Her champagne Burmese, Sugar ('*That's obviously a pet abbreviation of her pedigree name, you understand'*), was a show champion who apparently required meticulous care and attention. I soon gathered that Vanya and her husband had no children, and Sugar was not only a champion but also Vanya's baby. It seemed there'd been a local woman who could just about be trusted to look after Sugar when necessary, but she'd now divorced her husband and moved to Bath for work, which was extremely inconsiderate of her and had left Vanya in a difficult position with regard to poor Sugar. I remembered hearing about this woman from other people – she was the one with an apparent allergy to dogs. She must have let down a lot of cat owners by her thoughtless divorce and move to Bath!

'If I find I can trust you to look after my baby, I'll pay you well, and there'll be a lot of work for you,' Vanya had continued.

'Do you have lots of holidays, then?' I asked.

'Holidays?' She looked at me in surprise. 'My goodness no, I wouldn't leave poor Sugar just to go flitting off on *holidays*. Unfortunately, though, I have to make frequent business trips. That's the price you pay for being the CEO of a multinational business.' She smoothed her hair and gave a weary sigh. She may have made her job sound like a burden, but she still managed to sound pretty impressed with herself.

 'I'd love to look after Sugar for you,' I'd said. 'Just let me know the dates as soon as you have them.' And we'd arranged that I'd call around and meet the cat this week.

I set off, feeling pleased and positive about my decision to stay in Crickleford. Vanya's house was set on the outskirts of town, down a long drive, and was by far the largest I'd seen in Crickleford. I'd felt slightly intimidated by Vanya and her haughty manner, but Sugar soon put me at my ease. She was not only as beautiful and loveable as Vanya had said, but she also seemed to like me straight away, which was quite a relief. One of my parents' cats had been a Burmese, so I was used to their loud vocalisation, and was well practised in engaging them in conversation. By the time Vanya had gone through the four A4 pages of large-print instructions on Sugar's care, giving extra emphasis and finger-underlining to the parts typed in bold capitals, Sugar had curled herself comfortably on my lap, purring happily, and Vanya was beginning to relax.

'It seems you're a natural,' she said. 'I'd like to book you for my trip next week, if you're still available. It'll only be a two-day conference in Brussels, so look upon it as a trial.'

I had a feeling Sugar and I were going to get along well. But the downside of that meeting was that it made me think with longing, again, about my Albert. Although he was a pedigree cat too, I'd only ever treated him as a pet – a companion – and we'd had a particularly close bond during the last few years of my time in New York, when I'd often felt lonely and in need of a friend.

I'd tried several times to contact Shane's agent, Leo, to find out whether Albert was being looked after. I really, *really* didn't want to talk to Shane. But my emails to Leo were going unanswered, and on the one occasion I'd managed to get through on the phone, sitting as usual on the bench outside the library on the Town Square, the line had gone dead as soon as I'd said hello.

After the meeting with Vanya and Sugar, I tried again to call Leo, but this time the call went straight to voicemail. I had an uncomfortable feeling he'd probably been told by Shane not to talk to me. But, desperate to find out about Albert, I left a message anyway, asking him to please call me back about a matter of some urgency. I was hoping that might make him more likely to respond, but after I hung up, I regretted it in case he thought I was after money – in which case he'd not only never call me back, he'd probably also delete my message and block all future calls from me. Well, if I didn't hear anything soon, I'd have to email Shane. Anything would be better than hearing his voice. How depressing it was to realise that after spending so long in the USA, being feted as a kind of celebrity and chased around by the media, there was nobody there I now felt I could even contact to ask about the welfare of my cat.

The following week, as arranged, I turned up early at Vanya's house to look after Sugar. I was surprised, as I fumbled in my bag for her key, to have the door suddenly opened for me by a large blond man in a bright lime-green T-shirt and shiny black tracksuit bottoms.

'Yes?' he barked.

'Oh. Um … Mr Montgomery?' I asked hesitantly. Vanya hadn't said her husband was going to be there, so I hoped I wasn't making the wrong assumption.

'That's me. And you are …?'

'Sorry, I'm Emma. The pet sitter. I've come to look after Sugar.'

'Oh, right.' He smiled and held the door open wider for me. 'Yes, Vanya did mention something about that. You'd better come in, then, Emma. I'm Rob. Make yourself at home.' He waved an arm vaguely in the direction of the kitchen. 'I'm working from home today but I'm just off to the gym for an hour before I start work.'

'Oh.' I frowned. 'Did you not really need me today, then? If you're going to be at home?'

He laughed. 'You must be joking. I don't get involved with looking after the cat. Vanya doesn't trust me with her precious baby.'

There was something about his tone I wasn't sure about. Although he was laughing and making it sound like a joke, I got the impression Rob didn't share his wife's affection for the *precious baby*. Maybe he just wasn't a cat lover – fair enough. But there was also something about the way he was now looking me up and down, a slight spark of interest in his sharp blue eyes, that made me feel a little uncomfortable and self-conscious. I was probably imagining it, I told myself. He had a beautiful, dynamic and successful wife so why on earth would he be looking like that at someone like me?

Just then, we were interrupted by the appearance of the precious baby herself, who meowed loudly as she trotted straight over to me and then walked round and round my legs, purring happily. While I bent down to stroke her,

Rob said he'd see me later, picked up a gym bag and disappeared out of the door. I soon realised he wasn't joking about not getting involved with Sugar. He quite obviously hadn't even fed her that morning. I'd fully expected to do so myself, of course, but I found it rather surprising that, being there in the house, he wouldn't have responded to Sugar's hungry demands as she walked around her empty food bowl. She seemed to know exactly what she wanted and how to get it – but that was only to be expected, having obviously been treated as a little princess all her life!

'Doesn't your daddy love you as much as your mummy does?' I crooned to her as she got stuck into the Gourmet cat food I'd dished up and forked over lovingly, exactly as Vanya had instructed me. 'Never mind, Sugar. I'll make it up to you.'

She looked up from her bowl and gave me a haughty meow as if to tell me off for interrupting her while she was eating.

'So sorry, your highness,' I giggled, walking away to give her the privacy she seemed to require for her breakfast.

Vanya's detailed timetable had set aside times for grooming, and even times for playing, although personally I just wanted to play with Sugar all the time – she was so intelligent and responsive, and obviously loved human company – even if she did seem to think of me as her servant rather than her friend. Considering how much Vanya adored her, to say nothing of how valuable she must be, I'd been surprised that she was allowed outside of the house at all, but on my introductory visit Vanya had shown me her outdoor run: a huge, tailor-made framework that took up a large portion of the garden. The grass inside the run was dotted with safe plants to sniff, shrubs to hide behind, boxes and tubes to climb into and even half a tree for her to climb up. The run was not only large but high enough for us to walk inside with her, and I was surprised to see a comfy garden chair set up in one corner.

'I don't ever leave her alone out here,' Vanya had explained. I'd got the impression Sugar had no idea what it was like to be alone!

So, outdoor exercise (accompanied) was also scheduled every day, unless it was raining of course, or too cold for the little darling. Part of the run's steel mesh roof had a heavy-duty green cover over it, but I was firmly told that this was not to keep rain off her – she hated water and absolutely mustn't be allowed to get wet under any circumstances – but to give her shade in the event of the sun becoming too hot for her. There wasn't too much danger of that happening in the middle of March, but I said I'd bear it in mind.

I enjoyed my first time of being out in the garden run with Sugar so much, I didn't realise how much time had passed until I noticed Rob watching me out of the kitchen window. I had no idea how long he'd been back from the gym, but when I took Sugar back into the house, he smiled at me and said he was just making a coffee if I'd like one.

'You looked like you were enjoying yourself out there,' he commented.

'I was. We were having a lovely game with your toys, weren't we, Sugar?' I said to my little charge, and she meowed back at me, rubbing her head on my leg.

'The other woman just used to sit on the chair reading her magazines, looking fed up,' he said. 'Not that I blamed her. Can't see the fun in it, myself – playing with a cat.'

'You don't share your wife's passion, then, I take it?' I said.

'Not in the slightest. Still, it keeps her happy, I suppose.'

He sounded so dismissive, his tone so scornful and sarcastic, that I thought it best to change the subject. I was getting the definite impression that there wasn't a great deal of marital harmony in the Montgomery household, and I had no wish to get involved. Fortunately, Rob spent the rest of the day in his study, only coming out when I knocked on the door at six o'clock to let him know I was leaving.

'Is that the time already?' he exclaimed, getting up and having a stretch. Although he was heavily built, I could see that it was mostly muscle, and judging by the tight-fitting tops he seemed to favour, he was apparently proud of his physique. 'Well, thanks for today, then, Emma. See you in the morning.'

'Are you working from home again?' I asked, somewhat surprised.

'Yep. Easier to concentrate, when the wife's not here getting on at me with her constant demands and complaints – you know how it is,' he said, again laughing as if he was joking, but I had the definite feeling he wasn't. And again

I thought how odd it was that he couldn't keep an eye on Sugar himself for Vanya, if he was going to be around all the time she was on her business trip. It seemed she *really* didn't trust him!

'So will you be OK with Sugar overnight?' I said, beginning to wonder if I should actually be leaving her with Rob at all. 'I've fed her and changed her litter again.'

'Yeah, yeah, the cat will be fine,' he said, sounding completely disinterested. 'Bye.'

It was nice, now that I'd come clean to Lauren and Jon about my pet sitting, to be able to talk to them about the animals I was looking after.

'Sugar reminds me so much of a Burmese cat my parents used to have,' I told Lauren wistfully as we were clearing up the dinner things together that evening. 'She's so friendly and intelligent, although it's also really funny – because she's been so spoilt, she's such a diva! I'm hoping Vanya will use me a lot more if she's satisfied with me this time. She said she goes away on loads of these business trips.'

'That's the lady who lives in the big house on the way out of town, isn't it?' Lauren said. She shut the dishwasher door with a clunk. 'Yes, I believe she has a very high-powered job. I've met her husband – he's a parish councillor.'

'Really? I did wonder what his job was. He seems to like working from home.'

'Oh no, Emma, parish councillors are just volunteers. I don't know what he does for a living, but I suspect it isn't as important as his wife's job.' She grinned. 'At the parish council meeting I went to, he was doing nearly all the talking, shouting everyone else down. I thought he seemed very pompous and up-himself, to be honest, and he gave the impression he was definitely hoping to be elected as the next chairman. Jon says he often sees this on school parents' committees too.' She chuckled.

'Sees what?' I asked, intrigued.

'Well, he thinks it happens with men who have less influential jobs than their partners. They get involved in something in a voluntary capacity, and suddenly they're acting like Very Important Persons.'

I smiled. 'Yes, I can imagine that, too.'

Jon had joined us in the kitchen now, and was putting on the kettle for coffee.

'I've been meaning to say to you, Emma,' he said, suddenly sounding serious, 'Now you're getting more book-ings for the pet sitting, you need to get your business affairs set up properly – unless you've already done that?'

'How do you mean?' I asked, looking at him in panic. 'I don't have any business affairs.'

'Exactly.' He smiled. 'Don't worry, I just mean you ought to register yourself as self-employed, pay your National Insurance and start keeping proper records, for the taxman.'

'Oh.' I went hot and cold all over. 'But I don't know how to go about anything like that.'

'Well, I'd help you, of course, but it's not something I've got any experience with myself,' he admitted, frowning. Then he smiled again. 'Ah, but I know someone who *would* be able to advise you. She was a colleague of mine; she retired early from teaching and is self-employed herself now. Would you like me to have a word with her for you?'

'Emma's already met her, Jon,' Lauren intervened. 'I presume you're talking about Mary?'

'Oh. Yes, of course, I know Mary, she's lovely. I looked after her little dog. I thought she was completely retired; I didn't realise she still worked.'

'She just does bits and pieces,' Jon said with a shrug. 'But I'm pretty sure that however little you might earn, you need to register as self-employed. Mary will be able to tell you for certain, though. I'll give her a call for you now.'

I liked Mary, and since she'd talked to me about being friends, I'd been hoping to see her again soon. So I'd certainly be grateful if she'd be able to talk me through the minefield of these business matters that I'd had absolutely no idea I had to deal with. And hopefully she wouldn't mention my house fire any more. Or my parents looking after refugees!

CHAPTER TWELVE

That same evening, I found I had a good internet connection at home – it didn't happen too often – so, as I still hadn't heard anything back from Shane's agent, I took a deep breath and emailed Shane himself. Trust me, there was no 'Dear Shane' or 'Love from Emma' on my message, just a couple of short lines asking him whether Albert was OK, and who was looking after him. Even so, I felt sick as I pressed 'Send'. I hadn't wanted, ever, to be in touch with him again and if it wasn't for my longing to know about Albert, I wouldn't be hoping for a reply. It seemed almost impossible, now, that he and I used to be so much in love that I couldn't bear to be apart from him. Was it all just a figment of my imagination? Did he never really feel the same as I did?

I first met Shane when I was fifteen and still at school. I was a difficult teenager, a worry to my parents and a pain in the neck to my teachers. I did all the things my twin sister Kate never did: staying out late, drinking, smoking, mixing with friends my parents didn't approve of, and going out with boys they definitely wouldn't have liked, if they'd found out. I used fake ID to get into bars and clubs with the older girls I hung around with, and it was at one of these that Shane first chatted me up. He was gorgeous – several years older than me, of course, and with his dark skin, sexy brown eyes and good looks, I fancied him at first sight. He bought me a drink, we danced and kissed for the rest of the night, and I couldn't believe my luck when he asked to see me again. From then on, we became almost inseparable.

He lived in a grotty little bedsit on the worst estate in town. I knew my parents would have had a fit if they'd known I was going over there at all, which simply added to the thrill, as did the fact that I soon found out he had a reputation locally as a bit of a *bad boy*. He hung around with a crowd who tended to get thrown out of places for being rowdy. He'd had a police caution for being involved in a fight in the street, and he'd already been banned from driving, for clocking up so many speeding convictions, roaring around town in his noisy old banger, annoying pedestrians. But with me, he was gentle and funny. We spent hours just sitting around in his room, drinking cheap beers while he played his guitar and sung to me, and then we'd lie together on his single bed until I had to make a dash for the last bus home. I told my parents I was at a friend's house revising for my GCSEs and pretended her father dropped me home.

But I did confide in Kate. Despite our differences, we were always close, and had shared a bedroom all our lives. She might have chosen to stay in every night studying French and chemistry and all the other things I had absolutely no interest in, but she enjoyed hearing about my exploits. One evening we were giggling together as I described, in quite intimate detail, my blossoming love life with Shane when, unknown to us, our mum was listening outside the bedroom door.

Of course, I can understand completely now why she screamed the house down at me, and why my dad tried to ground me. But it didn't work. Short of locking me up, there was no way they were going to stop me from seeing Shane. I left school at sixteen, as soon as I'd finished my exams (all failed – no surprise there), and found a job as a care assistant in a local care home. When I said I was going to move in with Shane, my parents sat me down and gave me a long lecture about ruining my life. I think, when they saw it wasn't going to make any difference, they probably decided to let me go so that I'd find out how hard it was and come running back. Mum cried when I packed my bag. Even Kate (now working hard for her A-levels) tried to talk me out of it. I ignored them all. Shane was all I cared about.

Of course, I always thought he was talented. He used to perform in some of the bars we went to, and I'd always be there at the front of the crowd, watching him with pride. But it wasn't until he performed one of his own compositions on YouTube – a belter of a song called 'Baby No

Chance' – that everyone else started realising how talented he was too. Within days the YouTube recording had gone viral and, within months, he had a recording deal. And the rest, as they say, was history.

After sending my email that night, I lay on my bed in my little blue room, unable to stop thinking about all this, and wondering how the hell it had gone so terribly wrong between us. Since I'd been here in Crickleford, I'd tried my hardest not to think about Shane too much at all, but it was difficult – he'd been my whole life for nine years. I'd had no adult life without him, which was probably why I didn't seem to be very good at doing anything for myself now I was on my own.

But I was so glad I'd started the pet sitting, even though it had happened more or less by accident. The pets I'd looked after had really helped to cheer me up. The second day with Sugar was as pleasant as the first. I'd already fallen in love with her, and it was nice to know she seemed to feel the same, even if I did think she looked down her nose at me somewhat. She came trotting across to meet me when I arrived in the morning, meowing a friendly greeting that seemed to be saying *Oh, you're here at last, are you? Come along, I'm waiting for my breakfast!* Then she purred and rubbed herself around my legs as I dished up her food. I was really going to miss her after we'd said goodbye that evening, but I hoped Vanya would book me again.

Rob was working from home again that day, and fortunately he spent most of his time shut away in his study, but I was somewhat perturbed by the fact that, whenever he came out to make himself a coffee or a snack, he'd get uncomfortably close to me while he chatted, looking me up and down with his slightly flirtatious smile. He must have been in his mid-forties, just about old enough to be my father really, and he had an air of wealth and respectability about him that was peculiarly at odds with the way he was eyeing me up. I couldn't quite make up my mind whether I liked him or not, but I instinctively tried to keep my distance and I was quite glad when Vanya came home that evening.

'How has she been?' she cried, almost elbowing her husband aside in her haste to get to Sugar.

'Absolutely fine,' I said. 'I've really loved looking after her.'

I'd been playing with her at the time, and I passed her straight over to Vanya, who cradled her in her arms, crooning at her and fussing over her.

'Well, I can tell she's taken to you, and she seems very happy,' Vanya said, giving me a grateful smile. 'Thank you, Emma. Now, your fees are obviously completely ridiculous—'

'Oh!' I said, startled, but she laughed and went on:

'I mean ridiculously low, of course. I'm doubling them.'

'Oh,' I said again. 'But it's what I'm charging everyone else. I can't let you—'

'Yes you can. Sugar's a very special cat, a champion, and I expect to pay well for her care, to make sure she's properly looked after. What you charge everyone else is up to you. Do you have an invoice ready for me?'

I blinked. Invoice? Of course I didn't. My other clients hadn't asked me for one – I'd just told them the hours I worked and they gave me the money.

'Er ... no, sorry,' I said, feeling somewhat nervous of her businesslike manner.

'Not to worry. Let's just do it in cash this time, then.' She winked at me. 'Keep it off the books, eh?'

I had no idea what she was talking about. Books? What books? I definitely needed to have that talk with Mary.

Vanya counted out a wad of notes and passed them to me.

'Now then,' she went on. 'Have you got your diary? I'd like to book you for a week at the end of the month. There's a conference in Berlin I can't manage to get out of.'

How strange, I thought, to have such a high-powered position but be unable to get out of something you didn't want to do. Couldn't she have sent someone else? But I wasn't about to argue. A whole week of looking after Sugar would be lovely. Especially with double fees. I was humming happily to myself as I walked home. It didn't strike me until later that Vanya and Rob hadn't even said hello to each other by the time I left.

I still felt cheerful and optimistic the next morning. I had no more pet-sitting work that week so I was free to go for a walk, mooch around the shops and take a stroll down Moor View Lane for another longing look at my favourite cottage. The weather had suddenly turned milder, and the Town Square looked bright and clean, with purple and yellow crocuses in bloom in the planters, and people smiling and waving as they recognised me. Colourful bunting, hanging from the lampposts, was waving gently in the breeze and I saw from posters in the shop windows that there was to be a Spring Fayre in the Castle Fields on Easter Monday. The posters were decorated with pictures of bright yellow chicks and startled-looking rabbits, which looked like reproductions of children's school art work.

I decided I'd definitely go to the Spring Fayre. Perhaps Lauren and Jon would let me take Holly. Although I'd come to Crickleford to hide away from people, I couldn't help feeling pleased that, since the break-in at Pat's house, I seemed to have been so quickly accepted by the local population. And as long as they never found out who I really was, that was fine, wasn't it? I was beginning to feel that maybe I had nothing to fear from people here. Most importantly, Matt Sorrentino had been true to his word, only referring to me in very vague terms in his article in the *Chronicle*.

Because Matt had come to my mind now, I started puzzling again about why he'd been outside Bilberry Cottage taking photos, that day when I was staying at Pat's house. He'd made some excuse about just liking the look of the cottage, but it still seemed very odd to me. If I got into more of a conversation with him at some point, I'd definitely ask him about it again. He might even know who owned the cottage and why, every time I walked past it – which I was doing now – it still looked as if it was being done up, although nobody ever seemed to be inside. Through the front window I could see the stepladder was up, but there was no sign of life. Yes, I'd ask Matt about it. He must know something.

But what was I thinking? If I didn't want Matt asking *me* awkward questions, trying to get information out of me, perhaps I shouldn't be considering giving *him* the third degree about something he'd seemed reluctant to talk about. I was probably just overthinking the whole thing because I loved the cottage so much. What was so strange about him liking it too?

Although … taking *photos* because he just liked the place? No. No matter how much I tried to rationalise it – that *was* odd. I decided, reluctantly, that he must have been lying to me. The photos must be for his newspaper, and he didn't want to tell me why. But I was the last person to complain about someone telling me lies, wasn't I? I seemed to have become an expert in lying since I'd come to Crickleford. I just wish I knew how to stop, now I'd started.

CHAPTER THIRTEEN

I was in Ye Olde Tea Shoppe later, having a cheese and pickle sandwich shoved across the counter at me by Annie, when my phone pinged with an email. I waited till I'd sat back down at my corner table before looking at it; I'd learnt by now that there was no such thing as privacy, as far as Annie's customers were concerned. As it was, I could feel a dozen pairs of eyes on me as I took my phone out of my bag and opened the email. My heart nearly missed a beat. It was from Leo, Shane's agent. I frowned and made the text bigger, preparing myself for a lengthy read. But, in fact, it was not only short, it was also very much to the point.

I am instructed to tell you that the animal you are enquiring about has been given to a good home. Please do not contact my client again. If you persist in doing so, we may be forced to consider legal action.

I gasped out loud, causing several heads to turn in my direction and one old lady to ask if I was all right. I wasn't.

I wanted to cry. Given to a good home? Whose home? And how did he know it was a good one? How dare Shane give my lovely cat away without asking me? And as for *Please do not contact my client again* – anyone would think I'd been making a habit of pestering him with emails and texts. As if I'd even want to! If Leo had bloody well answered my calls or emails himself, I wouldn't have needed to at all. I was steaming by now, too cross and upset to eat my sandwich. I emailed Leo straight back, tapping the keys of my phone so hard I broke a nail, and telling him in equally terse terms what I thought of his message, adding that I'd rather boil my head than contact his bloody client ever again, but that if I found out my cat had gone to anything other than the most loving home in America, I'd take sodding legal action myself. I felt better after I'd sent it, but only for a moment. I sipped my coffee, trying to calm down, trying not to picture Albert living with some stranger somewhere who didn't love him the way I did.

'All right, my lovely?' Annie called across the café to me, looking concerned, and I realised I'd screwed up my face and both my fists.

'Yes, thank you,' I said, getting to my feet. If I stayed here, I'd only get asked questions, and I'd probably end up in tears, as well as saying more than I should. 'Sorry, I've just realised I'm late for ... um ... the library.'

I couldn't even think whether it was one of the library's days for opening, but I didn't care. I left, leaving my sandwich untouched on the table and almost tripping up ye olde steppe on the way out. I walked home and watched children's cartoons on the TV with Holly until I started to feel marginally better. But I was never going to forgive Shane for this – on top of everything else. That much was definite.

The next day I had an appointment with Mary to discuss putting my little business on a proper footing. I was on my way there when I bumped into Matt Sorrentino loitering outside the library.

'I was hoping to get some background on your burglar,' he said when I stopped to say hello. He sounded despondent. 'He and his mate have been bailed until their case comes to court. But it seems he's been given the push from his job here, and none of the staff are willing to talk about him.'

'More than they dare do, I suppose,' I said, sympathetically.

He shrugged. 'Oh well, never mind, I guess I'll get another stab at a big story eventually.'

'I hope so.' I gave him a smile. Although I was still wary of talking to Matt, I felt I had to add, 'I'm sorry I wasn't more helpful.'

'Not to worry.' Then he grinned. 'You didn't end up moving away, then?'

'No. You seem to have talked me out of it,' I said with a little laugh.

What was I doing? *Flirting* with him? Too late, I put a hand over my mouth as if to stop myself.

'Well, I'm glad,' he said. 'You'll love it here now spring's finally on its way. Have you been out on the moor much, yet?'

'Only to get myself thoroughly muddy with dogs,' I said.

'Oh yes, of course, the dog walking! I love animals too. I'm hoping to get a dog myself one day. When I'm ... in a position to have one.'

He suddenly looked so sad that I felt the need to keep talking, to try to make him smile again.

'Me too. It's perfect for dog walking around here. I'm a bit nervous of getting lost out on the moor though, to be honest.'

'That's completely understandable. The weather can turn nasty out there at the drop of a hat. You can start out on a hike in warm sunshine and suddenly the clouds come rolling in and the rain comes down in sheets, scuppering the visibility. We've had to send out search parties for missing grockles more often than I can count. They get disorientated in fog or rainstorms and lose their way.'

'Must be frightening. Remind me not to go too far on my own, then.'

'Do you drive?' he asked.

I shook my head. 'No car.'

'Would you like me to take you out for a run across the moor one day? Just ... to help you find your way around a bit more, you know. And as an animal lover, I'm sure you'd like to see the Dartmoor ponies.'

I hesitated. 'That's good of you, and I would like that, but ... um ... I'm a bit busy right now, you know, with people's pets.'

I was making excuses, of course. The truth was that I was far more tempted than I should have been, to accept Matt's offer. Touring Dartmoor, seeing the ponies – in the company of a friendly and good-looking man – was not only an appealing proposition, but would be a welcome distraction from my upset about Albert. But I couldn't let go of my fears; how could I allow myself to trust him? I'd made the mistake of trusting a journalist too much, once before – a female one who'd appeared to be particularly kind and understanding when she interviewed me soon after we moved onto the New York scene. Like so many others afterwards, she'd acted like she was my friend, encouraging me to talk too intimately about my life with Shane, promising most of it was 'off the record' – and then published a horrendous article misrepresenting almost everything I'd said. Needless to say, Shane wasn't pleased. In fact it was probably the first time he'd joined the rest of the world in thinking me a complete idiot.

Matt was giving me a puzzled smile now, and I realised I was scowling, as I remembered how I'd felt back then.

'No pressure, Emma,' he said gently. 'Just a guided tour of the moor. I promise I won't take you hostage in some remote shepherd's hut and leave you there till you agree to tell me all your secrets.'

He'd meant it as a joke of course, but at the mention of my *secrets*, the idea of going out for a ride in his car with him completely lost its appeal.

'I'll let you know if I get a free afternoon,' I said in a hurry. 'Now, I really ought to dash.'

'Yes, sure. Keep up the good work.'

'Sorry?'

'Your fitness regime,' he said, laughing. 'And don't forget, you can contact me anytime at the *Chronicle* office.' He pointed back down Fore Street at a squat stone building next to the doctors' surgery. 'I hope you'll get that free afternoon soon.'

Mary welcomed me with coffee and biscuits, and Scrap ran around me with his tongue out, giving little yaps of excitement. It was good to see them both.

'Thank you for offering to explain all this stuff to me,' I said, once we'd sat at her table together. 'I'm so stupid, I really didn't have any idea that I had to do anything like this.'

'You're not stupid at all,' she said firmly. 'How would you know, if you've never done it before? You said you were worried about invoices? Well, you don't need to do anything special for them, just a sheet of paper with your details on, and the amount you're charging, will be fine for now. The most important thing is to keep a copy for yourself. Have Lauren and Jon got a printer you can use?'

'Yes, if they don't mind.'

'I'm sure they won't. So that'll be the easiest way to do it. You could print off some nice leaflets, too – I take it you're going to advertise?'

'Oh, I'm not sure about that. So far people have just approached me themselves.'

'Well, that's good, as long as you get enough work that way, but if not I'll help you design a leaflet, or you could even do something online. A website, or a Facebook page.'

'No, I don't really want that,' I said a bit too quickly.

'OK.' She smiled gently. 'But you do still need to decide on a name for your business – unless you just want to use your own name.' She gave me a thoughtful look. 'Your surname is quite memorable, so—'

'No. I don't think so.' She must have thought I was being a bit negative. No advertising, no Facebook, no spreading around of my *memorable* surname. I didn't want people pondering my name too much. Once again, I wished I'd changed it, but everything had happened so quickly, I hadn't really had time to think about things like that. At least using my real Christian name, rather than my assumed one, seemed to be throwing people off the scent.

'Perhaps a pet-related name, then,' Mary suggested. 'Something catchy but not too corny.'

'Yes, I think I'd prefer that,' I said. 'I'll give it some thought.'

'And don't forget you need to keep a record of every job and what you get paid. That's it, really.' She smiled.

'Apart from the taxman,' I said, grimacing.

'Well, don't worry, you'll probably find you don't even earn enough yet to pay tax. But Jon's right, you do still have to register as self-employed, anyway. You can do it online.' She looked at me with her head on one side. 'Would you like me to help you with that now, Emma?'

'Oh. Are you sure you wouldn't mind?' I had to admit, I'd been dreading it. 'But can you get a decent Wi-Fi connection here, though? It's awful at Lauren and Jon's, goes off more often than it works.'

She laughed. 'I know. Luckily, ours seems to be a lot better in this part of town.' She opened her laptop and tapped something into the browser. 'Here we go.' She glanced at me. 'How about you call out your details and I'll fill in the online form for you?'

'Thanks so much,' I said. 'Sorry for being so rubbish with all this stuff.'

'Don't keep putting yourself down,' she remonstrated. 'I'm just helping you get started, that's all.'

And somehow, with her calm, patient manner, she did make me feel more capable.

'Right, first you need to choose a password,' Mary said. She looked up at me and I smiled.

'Primrose,' I said.

'Right. But it needs to contain a number.'

'OK. Primrose4. For Holly's age.'

She nodded and typed it in, and I suddenly sat up, smiling again.

'I know what I'll call my business! *Primrose Pets*.'

'Brilliant!' she said. 'Yes, that's perfect, Emma – simple but catchy. I really like it.'

Primrose Pets. I repeated it to myself over and over in my head as we went through the form for HMRC. I was going to run my own business: Primrose Pets. Me, a businesswoman, an entrepreneur! How about that? Wouldn't that make my parents proud? Would they even be able to believe it? I decided I'd send them an email later to tell them about it.

While we had the laptop out, Mary showed me how to make a spreadsheet so that I could type in my earnings for each job, and how to set out a simple invoice. I was beginning to feel more confident about managing it all.

'I didn't realise you had a business yourself,' I said, as she finally closed the computer. 'I thought you were retired.'

'Well, yes, I retired from teaching a couple of years ago. But I didn't feel ready to stop altogether. I'm on my own, you see, and it can be lonely if you don't keep yourself occupied. So I just do a bit of private tutoring.'

'Oh, I see.' I nodded to myself. Mary seemed very clever. I could imagine her teaching something like art or music. Or maybe Spanish or Italian. 'What do you tutor, then? Foreign languages?'

'No – English literacy. My speciality at the school was teaching the less able children, and I've always found that very rewarding. So I do one-to-one tutoring for kids who're struggling at school. And adults who don't read or write too well. You'd be surprised how many people get right through school without ever managing to learn the most basic things most of us take for granted.'

I swallowed and looked away. I could feel my ears burning. I glanced down at the notes I'd made while Mary had been talking to me. They weren't too bad, were they? Was she dropping hints about my spelling or my scruffy handwriting?

'Reading is the most important skill anyone ever learns,' she was going on. 'It's so sad when people miss out on that.'

'Yes, absolutely,' I said, still not looking at her. 'I *love* reading. I read so many books, the library can hardly keep up with me.'

'Really? That's wonderful, Emma, it's nice to hear that. So many young people nowadays spend all their time online and never think to pick up a book. Who is your favourite author?'

'Um ...' Dammit. I'd done it again, hadn't I – dropped myself into another big fat lie. Favourite authors? I couldn't think of a single one. Could I count Holly's favourite, Julia Donaldson? I actually really did like those books – I loved hearing Holly giggling and joining in with the words, as Lauren read the stories to her. But perhaps Julia Donaldson only wrote children's books? Hang on: who was the one Jon was always going on about? The one I was supposed to study at school, but never did? Ah yes: 'Shakespeare,' I said.

'A girl after my own heart.' She beamed. 'So which is your favourite of Shakespeare's works?'

I was wishing like mad now, of course, that I'd never opened my stupid mouth. And then I suddenly remembered: the Atkinsons' cats!

'*Romeo and Juliet*,' I said, crossing my fingers.

'Ah! You like the tragedies, then.'

'Yes.' I smiled back at her now, relieved. 'And his thrillers are pretty good too.'

Fortunately, at that point Mary's phone started ringing in the kitchen, and by the time she'd taken the call and come back, no more was said about Shakespeare or reading until I was getting up to leave.

'Don't rush off, Emma,' Mary said. 'Come and choose some books to borrow.'

'Oh, no, I couldn't possibly—' I said, but she took me by the hand and led me to her bookshelves, which I already knew were so high we needed a chair to reach the top ones, and were stuffed full of books of every size and description.

'I know our little library isn't always open, and it can be maddening if you've run out of reading material,' she said. 'So please, help yourself to a couple for now, and you can come back for more any time.'

'That's very kind of you.' I stared at the rows of spines. The lettering on some was tiny. I felt panic-struck just looking at them. 'But honestly, I don't think ...'

'Would you like me to choose for you?' she suggested. 'How about another Shakespeare? *Othello*? Or did you do that at school? No? Take that, then, and ... have you read *King Lear*? Ah, well, you'll enjoy this.' She handed me the two slim hardbacks, and I passed them awkwardly from one hand to the other.

'Thanks so much,' I said.

'You're very welcome. I've enjoyed our chat today. Come and see me again soon, won't you. And I'll be needing you again in the summer to look after Scrappy, if that's OK. I'll check the dates with my sister and let you know.'

'That'll be great. Thanks again, Mary.'

I walked home, cursing myself for my stupidity. Just as I'd been feeling relieved that Mary hadn't once today, thank God, mentioned my house fire or my sainted parents who looked after refugee children, I'd had to go and tell her another whopper of a lie and get myself into this situation of being her one-woman bloody book club. Reading Shakespeare? Me, of all people? How the hell was I going to get out of this one?

CHAPTER FOURTEEN

The next day I spent a long time with my laptop, starting my spreadsheet for my pet-sitting earnings and setting up a template for my invoices. I wanted to do it before I forgot everything Mary had taught me. I was beaming with pride as I typed the words *PRIMROSE PETS* at the top of the page, and felt so pleased with myself when I'd finished, I wrote an email to my parents, copying in my sister, telling them that I was now an entrepreneur, having single-handedly set up my own successful business.

'It's doing very well. Loads of customers. I have to turn people away every day. I'm earning lots of money and I'm thinking about buying a cottage here.'

In my desperation to impress my family, I'd clicked 'SEND' before I could change my mind about the exaggerations. I didn't want to think of them as more lies. Normally, emails from Primrose Cottage didn't go straight away because of the terrible internet connection, but typically, on this occasion it whizzed off immediately, leaving me wishing I'd taken more time and been more truthful. It wasn't until a reply from my sister pinged on my phone a little later –*That's great, Emma, but what kind of business?* – that I realised I'd also left out some vital details. I tried to respond, but of course, the internet had now gone down completely so, realising I'd just have to reply another time, I set off to get some shopping before Lauren and Holly came home.

Holly was in an excitable mood and, as it had started to rain, she couldn't run around in the garden to let off steam so I spent the afternoon helping to keep her occupied, much to Lauren's relief.

'I'll be working again the next couple of weeks,' I said, 'so it's nice to spend some time with Holly while I can.'

'I'm glad you've got so many bookings lined up,' she said. 'Who is it next week?'

'A couple of house rabbits, for Mary's next-door neighbour.'

'Oh yes – Jackie. I know her.' Lauren smiled. 'And her rabbits. Flopsy and Mopsy, though – not the most original names, are they?' she giggled.

I laughed. 'Mary was quite scathing about the rabbits, actually, when she heard I was going to look after them. She thinks Scrap can smell them, as he's always sniffing around under Jackie's hedge. She said he probably thinks it's a tin of rabbit flavour Pedigree Chum.'

'That's awful!' Lauren protested. 'I'm sure they're perfectly lovely bunnies – although I never quite understood why she doesn't just have cats.'

'I suppose rabbits don't catch birds,' I pointed out. We'd had a mangled woodpigeon to clear up earlier, after Juliet's latest hunting expedition, which neither of us had been very happy about. We both agreed that rabbits might have their advantages!

As it turned out, they *were* nice rabbits, too. Jackie introduced me to them on my first day, before she left for her holiday.

'They're both girls,' she explained, picking one of them up and stroking her silky ears. 'You can probably imagine, one of each would be asking for trouble unless you want to start your own rabbit club. And two males together is a recipe for aggressive testosterone displays.'

'Typical,' I commented, and we both laughed.

'Flopsy is the black one with white splodges,' she said, pointing to the rabbit hopping around her feet. 'Mopsy's white with black ears.'

Black and white cats at home, black and white bunnies here – I should be getting good at telling negative images from each other.

'OK,' I said, and squatted down to give Mopsy an experimental stroke. She bit me. 'Ouch!' I protested, trying not to sound too aggrieved. Jackie laughed, which I found a bit irritating. She could have warned me they might do that.

'It's only because she's not used to you,' she said. 'Give them an hour or so and they'll be leaping onto your lap.'

Three hours after I'd waved Jackie off, Mopsy was still sinking her teeth into my hand at every opportunity. Flopsy didn't seem to mind me, but there certainly wasn't any leaping onto my lap, by either of them. I talked to them sweetly, fed them their chopped cauliflower and carrots and their BunnyBits rabbit feed, cleaned out their litter boxes, and made a game out of chasing them into their bedtime cages before I left in the evening. I liked them. They just didn't seem to like me very much.

I was disappointed that this didn't seem to improve much over the next couple of days. Perhaps I was only a dog-whisperer and cat-whisperer after all, not a bunny-whisperer. But everything changed in the middle of the week. Jackie had given me a few pointers about keeping rabbits safe in the house. The electric wiring, for instance, was all encased in heavy-duty conduit so they couldn't chew it.

'All rabbits love to chew,' she said. 'So I leave them bits of old carpet they can work their teeth on to stop them attacking my rugs. And lots of newspaper in their cage – they like to shred it and make bedding with it. Top the newspaper up every day. There's a stack of it in the kitchen. Keep an eye on them, though. If you catch them chewing anything they shouldn't, just tell them *No*. They're very well-trained.'

She had no house plants, because so many leaves are toxic for rabbits, and apparently the bunnies had been trained not to chew the legs of Jackie's furniture. I found it quite amazing that they were so well trained, better than a lot of dogs were, by the sound of it. But there was one thing Jackie had forgotten to mention: the bathroom.

I was normally in the habit of keeping bathroom doors closed, so this particular danger didn't become evident until, to my surprise, I was using the toilet on the Wednesday afternoon when I heard my name being called. For a minute I thought Jackie must have come back early. Then I realised the call was coming from the back garden. Flustered, I hurriedly flushed the toilet, washed my hands, and ran through the house, leaving the bathroom door open. I was careful, of course, to make sure neither of the rabbits was following me as I went out of the back door into the garden, and closed the door carefully behind me. Although Flopsy and Mopsy had an outdoor run where they could run and dig to their fluffy little hearts' content, the weather had been so wet, I hadn't yet been able to let them out.

'Emma!' the voice was still calling me and, with a sigh of relief, I realised it was Mary, looking over the top of the fence between her bungalow and Jackie's. 'Ah, there you are. I remembered you saying you'd be here this week with the dreaded rabbits.'

'Oh, they're very sweet, actually. I don't think they like me, though.'

'Well, I'm glad they're not outside, making Scrap's mouth water—'

'Stop it!' I protested. I shivered. It was chilly and starting to drizzle again. 'It's nice to see you, Mary, but we're going to get wet standing out here. Do you want to come in for a cuppa?'

'No – thanks, but I can't stop. I just wanted to ask how you're getting on with *Othello*?'

'Othello?' For a moment I thought it must be the name of somebody's pet. Quite a good one, as it happened. Then I remembered, and felt my face flush with embarrassment and my heart start to race. 'Oh, *Othello*,' I said. 'Yes, it's fantastic, isn't it. I've nearly finished it.'

'I'm glad you're enjoying it. How about *King Lear*?'

'Yes, that's good too. But *Othello*'s one of his best, I think,' I invented quickly. 'Very ... um ... tragic.'

'Yes, it's one of my personal favourites. Which was your favourite character?'

I started to panic. Quite apart from the fact that we were both now getting wet, I really could do without this unexpected Shakespeare test. I obviously hadn't opened either of the damned books and had no idea who any of the characters were.

'Um ... Othello?' I tried.

'Really? I always thought he was so *weak*. I could shake him for not trusting Desdemona. All Iago's fault, of course, but still ...'

'Yes, he really should have trusted him, I agree.'

'Her. Desdemona.'

'Yes. Her, too.' It was time to put a desperate stop to this. 'Look, sorry, Mary, but I left the rabbits running around in there—'

'Oh, of course – I'm so sorry. It's pouring with rain now, too. I hardly noticed. I could talk about Shakespeare all day, couldn't you, Emma?'

'Absolutely. All day, and all night.'

'Well, please do borrow a couple more when you bring back those two, won't you.'

'Sure. Thanks, Mary.' I bolted indoors, grabbing a towel to dry my head and shoulders and cursing myself for not having the guts to admit I hadn't read either of the bloody Shakespeares. 'Flopsy!' I called, looking around for my charges. 'Mopsy! Where are you, girls?'

There was a strange, scrabbling noise coming from somewhere on my left. I followed it down the hallway – and into the bathroom, where the first thing I saw was Flopsy on her back legs in the bath, desperately trying to climb out. And the second thing was Mopsy perched on the toilet seat – in my hurry I'd even forgotten the basic common sense of closing the lid – looking down into the water for all the world as if she was about to dive in.

My heart in my mouth, I crept up behind Mopsy in complete silence. I was terrified of startling her into falling in. With one swift movement I grabbed her firmly around her tummy, lifted her clear of danger and shut the lid. To my surprise, for once she didn't bite me. I put her down and bent to retrieve her sister from the bathtub. Thank God there was no water in *that*. Flopsy seemed to be so relieved to be out of the bath, she lay gently in my arms like a big soppy pussy cat. I carried her out of the bathroom, with Mopsy following, and closed the door behind us.

'Phew,' I said, collapsing onto the sofa with Flopsy still stretched out in my arms. 'Oh my God, girls, that was a close shave. And all my stupid fault. Don't tell your mummy, will you? If anything had happened to either of you, I'd never have forgiven myself.' I actually felt a tear come to my eye. They were lovely bunnies, and Jackie was right, they were well trained and affectionate. I should have been more on the ball. 'Sorry, babies,' I said, and instinctively reached down to stroke Mopsy, who was sitting placidly by my feet. Again, she didn't bite me, so I decided to risk picking her up. There were only so many fingers I could lose, after all. But she calmly and willingly allowed herself to be lifted and placed on the sofa next to where Flopsy sprawled across my lap.

'Well, look at you two!' I said, thrilled and amazed. 'So are we friends now, then?'

It seemed all it had taken was for them to be rescued from a dangerous situation of my own stupid making, and I'd been accepted as their new favourite human. By the time Jackie came home, they were leaping onto my lap just as she'd predicted.

'Any trouble?' she asked brightly as she put her suitcase down in the hallway.

'Not at all,' I said. 'They were great.'

I could have sworn I saw Flopsy winking at me. But that, of course, was ridiculous.

When I arrived home, that final evening of looking after the rabbits, Lauren greeted me, looking a little anxious.

'I had a call for you while you were out,' she said. 'Your mum.'

'Really?' I said, staring at her in surprise. My mum had only called me once before, the whole time I'd been in

Crickleford, and had sounded so put out at having to leave a voicemail message for me to find when I was somewhere with a signal, she never bothered again. She'd certainly never before used Lauren's home number – the number I'd asked her to use if ever there was an emergency – so needless to say, I was pretty worried. 'Is it OK if I call her straight back?' I asked Lauren.

'Of course.'

The phone rang and rang, and eventually went to Mum and Dad's voicemail. I left a message, saying I hoped everything was all right, and asking Mum to call me back and let me know what she wanted. I tried twice more that evening, but again the calls went to voicemail. When I'd still heard nothing by the next morning, I walked into town and tried to call from my mobile, then tried my sister's number instead. Nothing. Everybody seemed to be out, or refusing to speak to me. I decided that if there really was anything wrong, my sister would have let me know by now, and that Mum would eventually try me again. I didn't exactly forget about it, but I did put it to the back of my mind. I suppose it's fair to say that I'd never been a particularly dutiful daughter, and after five years of living on the other side of the Atlantic, I'd got used to not knowing what was going on at home.

The following week, I was excited to be going back to looking after Sugar the Burmese cat. I was used to cats – more so than rabbits, anyway – and despite her diva-like behaviour it wasn't Sugar who was the problem. This time around, Rob seemed to be making no pretence whatsoever about his intentions towards me. I was beginning to wonder whether he really had a job at all. He made vague, airy references to the importance of his position but he seemed to spend his entire life 'working from home'. And when he wasn't ensconced in his study, he'd be prowling the house, following me around, making little flirtatious comments that were only just short of being actually lewd. I'd be dishing up Sugar's food and an arm would suddenly come round me from behind, squeezing my waist.

'Just making a coffee,' he'd excuse himself, as he deliberately brushed too close to me at the sink.

'There you go,' he'd say as he put a steaming mug down in front of me, stroking my hair or my neck or my shoulder as he did so.

I always protested and moved away, but I suppose I wasn't doing it forcefully enough. I knew I should have done, obviously, but ... it had been a long time since an attractive man had made advances to me. Looked at me with *that* look, undressing me with his eyes. Said I looked particularly stunning, on a day when actually I felt like crap. Touched my arm with that lightest but most meaningful of touches, that made my skin tingle with a forgotten kind of anticipation. I knew, too, that part of the attraction was the fact that he was older, and that there was absolutely no semblance of similarity to Shane to put me off. All the things that had drawn me to Shane had been wild and rebellious, free and dangerous and exciting. In comparison, Rob seemed settled, mature, sensible and safe. Which was ridiculous really, when he was behaving so badly.

I tried to concentrate on Sugar. It was easy enough to block out Rob when I was playing with her, cuddling her, grooming her – she was such a delight, so affectionate and beautiful, and by now we had a lovely rapport. But when she lay dozing in an armchair, her head on her paws, making those soft little grunty noises cats sometimes make in their sleep, and I went out to the kitchen to make my lunch, he always seemed to appear from his study at the same time. Then the arm strokes and waist squeezes seemed to be getting firmer and lasting longer, the flirtatious comments more serious. I knew I had to put a stop to this. He was married and, not only that, he was also my employer's husband. Nothing could be allowed to happen. No way.

Despite the fact that nothing did happen, I could hardly look at Vanya when she returned at the end of the week. I was still slightly in awe of her anyway, and I felt guilty even for harbouring any inappropriate thoughts about her husband. But I *had* harboured them. And when I accepted the next booking for Sugar, I knew I was getting into dangerous territory and I'd have to make sure a line was drawn, very firmly, with Rob Montgomery in future, if I didn't want to lose my best client.

CHAPTER FIFTEEN

On Easter Saturday, there was no rain for the first time
for what felt like weeks. The river had become higher
than I'd ever seen it, and I began to understand exactly
why local people lived in dread of it bursting its banks
– even if it was a rare occurrence. But suddenly now,
the weather had turned surprisingly mild, the sun was
shining as if it meant it, and I left off all my winter
wrappings of thick anorak, warm scarf, gloves and woolly
hat to walk up to the Town Square. When I caught sight
of my reflection in a shop window in Fore Street, I was
momentarily shocked. Was that really me? When had I
last paid any attention to the way I looked? I'd got out
of the habit of putting on more than the bare minimum
of make-up, and my face looked pink from all the fresh
air and recent cold winds. I'd put on weight from eating
Lauren's lovely home-cooked meals instead of living on
a strict diet of fashionable grains, salad, pumpkin seeds

and probiotic yoghurt, like everyone I knew in New York. And worst of all, my hair was beginning to grow back red again.

I was just about to dash into Superdrug for another pack of DIY Cheeky Chestnut, when I glanced at my reflection again and realised I also needed to have a proper cut. After years of having my hair coloured, cut and styled regularly by one of the top New York hairdressers, I'd now abandoned it to its own cheeky chestnut fate. It was sticking up in places it ought to lie flat, and hanging limply where it used to curl softly around my neck. In short, I looked such a mess that I wondered what was wrong with Rob Montgomery's eyesight if he really did fancy me.

'I can do you next Thursday at six o'clock. It's our late evening,' said the bored-looking stylist in Heads Up – the only hair salon in town.

'Nothing today?'

She stared at me. 'It's Saturday. We were booked up weeks ago.'

'Monday?'

'Bank Holiday.'

'Oh yes. Tuesday, then?'

'We're closed Tuesdays.'

I sighed. 'So what about Wednesday?'

'Pensioners' day.' She looked me up and down and thankfully appeared to decide I didn't qualify. 'Thursday's the best I can do. Well, not me, actually – our apprentice, Jade, will do you. I'm fully booked for the next three months.'

My hairdresser in New York would have fitted me in at barely a moment's notice, because even the *tip* I gave him, every week, was more than twice what I'd be paying Jade the apprentice to 'do' me next Thursday. I sighed to myself as I turned to walk home, thinking again about how spoilt I'd been back then. I was beginning to dislike the person I'd been in New York almost as much as I disliked Shane. But it wasn't all his fault, was it? I'd been eager enough to go off to America with him when he got his first number one hit over there.

'You can't,' my mum had said, as soon as I told her. Shane and I had been living together for two years by then. We'd moved out of the grotty bedsit into a proper house – he'd already followed up 'Baby No Chance' with another big hit here in the UK and now it looked like worldwide fame was on the cards. 'They won't let you in. You're only nineteen and you haven't even got a job.' I'd given up the care-home work, of course. I didn't need to work now. I played at keeping house for Shane.

'Shane's sorted it all out,' I told her. 'He's got dual citizenship because his dad's American.'

'I think you'll find that doesn't apply to you,' my dad said. 'You're not related to him.'

I argued the point, of course, but I was losing my certainty. When I told Shane, he just laughed and said there was nothing he couldn't fix. And the way he chose to fix it was typical of him – as flamboyant and outrageous as ever. Two days later he took me out to dinner and slid a diamond engagement ring on my finger.

'I've checked out the rules,' he said with a grin. 'You'll be allowed to enter the States with me now, as my fiancée.'

I was too thrilled, too carried away by my idealised dream of a happy-ever-after, to stop and think that getting around US immigration rules wasn't, perhaps, the most romantic reason ever for an engagement. That ring stayed on my finger for the whole six years we were in America. It was last seen flying across our bedroom on the day I walked out on him.

Thinking about this had spoilt the good mood the sunshine had induced in me earlier. It was Easter weekend, everyone in town looked happy and spring-like but I suddenly felt down in the dumps. I was cross with myself for being tempted to respond to Rob's flirtations. What was wrong with me? Why didn't I have more self-respect? And I was miserable about my lovely Albert. I missed him, and I still didn't know who Shane had sold him to. Suddenly, on an impulse, I decided who I needed to see to cheer me up.

Before I could change my mind or decide it was the last thing I should be doing, I marched across the road to the *Chronicle* office next to the doctors' surgery. I half expected to be disappointed. I didn't even know whether Matt worked on Saturdays, but there he was, apparently the only person in the office, sitting at his desk looking as much in need of cheering up as me. It was gratifying to see his face light up with a smile when he looked up and saw me.

'Emma! This is an unexpected pleasure.' He got up and came over to greet me. 'Is there something I can help you with?'

'Well, um, yes, kind of.' I looked back at him, feeling nervous now and beginning to regret my impulsiveness. Had I already *forgotten* that he was a journalist? 'You mentioned about going for a drive on Dartmoor,' I ploughed on regardless, despite my doubts. 'I wondered ... if you're still up for it ...?'

He brightened even more. 'Yes, I'd love to do that. It can't be today, though, sadly – I'm working all day. Tell you what, though: why don't we leave that till next week? I was thinking of going to the Spring Fayre. We could go together.' He hesitated. 'Er ... sorry, only if you'd like to, of course. It's on Monday.'

But by now my initial enthusiasm for the idea of going out with him had started to wear off, being beaten back by my worries about his occupation. Being here in the newspaper office had heightened my anxiety and made me wonder what the hell I was doing, voluntarily exposing myself to the kind of situation I'd done my best to avoid for so long. Why had I come here? OK, so I was feeling sad and upset about Albert, I was cross with myself for not rejecting Vanya's husband more firmly, perhaps I needed cheering up and was a bit lonely for male company, but surely there were other guys in Crickleford I could get to know if I made a bit of an effort, guys who weren't likely to start harassing me for stories and digging into my past?

He was looking at me, waiting for an answer. I saw the eager expression on his face slowly change to one of disappointment.

'You don't want to come to the Fayre?' he said. 'OK, no worries, I'll give it a miss and take you for the Dartmoor drive instead—'

'No!' I said, a little too quickly. 'I mean, I don't want you to change your plans for me. But I'm already going to the Fayre, you see.'

'With someone else?' he said, looking even more disappointed.

'Yes. I've promised to take my landlady's little girl. Sorry.'

He laughed. 'Oh, *that's* no problem. I like kids. Bring her along!'

'Really?' Now I'd used up my excuses and didn't know where else to go. 'Um ... she's only four.'

'That's fine, Emma. I know how to talk to four-year-olds. I was one myself once.'

I laughed, despite myself, and he seemed to take that as agreement, suggesting where he'd meet us at the start of the Fayre. It would have been difficult to duck out of it by then and, besides, deep down I was looking forward to spending time with him. Surely he wouldn't be able to worm any personal information out of me during one afternoon at the Fayre, with a child in tow?

Easter Sunday was, of course, another cause of great excitement for young Holly. She was awake early, shouting about the Easter Bunny, and nothing would satisfy her until we'd been out in the garden with her to search for eggs to put in her little basket.

'I had to get up at the crack of dawn to hide them,' Jon complained to me in a whisper. 'I can't think why we don't reinvent the Easter Bunny's role and get him to leave the eggs in children's bedrooms. Father Christmas and the Tooth Fairy seem to manage it.'

Holly had soon found all the eggs, though, and when we went inside I gave her the chocolate bunny I'd bought her.

'I didn't get one for you, though,' Holly said mournfully to me.

'That's all right. I've had enough of bunnies recently,' I joked.

When I told Lauren and Jon that I was going to the Fayre with Matt the next day, they looked worryingly excited about it.

'It's nothing like *that*,' I said hastily, feeling myself going a bit red. 'We're just friends.'

'OK.' Lauren smiled at me. 'But you won't want Holly hanging around you, will you, while you're with your *friend*. We can take her.'

'No, honestly, it's fine. I've told Matt I'm bringing her along. I've been looking forward to taking her. And it'll give you both a break.'

'Well, if you're sure. Thank you. She's going to love being with you.'

And it seemed she did. Once again her excitement was infectious as we set off the next day to meet Matt.

'Is he your boyfriend?' she asked with a little giggle.

'No. Just a friend. But he's very nice.'

'*I've* got a boyfriend,' she confided. 'His name's Luke.'

She went on to give me a detailed description of Luke – where he lived, what his mother's name was, what colour his hair and eyes were, and (bizarrely) what he liked to eat for breakfast. It crossed my mind that she knew more about Luke than I knew about some of the people who were supposed to be my closest friends (hangers-on) in New York.

The Fayre was held in the Castle Fields. I'd expected it to be all about dodgems, helter-skelters and scary rides for teenagers, where they got turned upside down and flung around, but it wasn't that kind of fair at all. There was one merry-go-round, where Matt and I spent a pleasant fifteen minutes watching Holly have four rides before we insisted on moving on, but apart from that it consisted of stalls selling home-made jam and chutney, knitted jumpers and baby clothes, and all manner of craft items.

'It's a *country* fayre,' Matt said, laughing, when I described the kind of fair I'd been expecting. 'People come from all over Devon to sell their wares. In the past there'd be live-stock too. But look: there's a coconut shy, if you want something more entertaining.'

He led me and Holly over to the stall.

'Who likes coconut?' he asked as he paid for a go.

'Mm. I love it,' I said, laughing at Holly, as she pulled a face and said 'Yuck', before admitting she didn't know what it was.

'OK. One coconut coming up for the lady in the blue jumper,' Matt said, taking aim. His first ball shot straight between two coconuts. He turned to Holly with a pretend frown. 'Did you nudge me?' he teased.

'No!' she protested. 'Have another go! Get Emma a coky-nut, go on!'

With greatly exaggerated concentration he took aim again and – much to his own surprise – smashed straight into a coconut on the front row.

'Wow!' he said, as the stallholder placed the coconut in my arms, Holly doing a little dance of excitement around me. 'That's the first time I've ever hit anything! Unless you count that tractor, of course, on the road to Newton Abbot that time,' he added, half under his breath.

'And you want me to come with you for a drive!' I laughed.

'It was the tractor driver's fault. Wouldn't let me pass him on that blind bend,' he joked, winking at me. 'I'm a good driver, me. Never knowingly overtaken.'

We were laughing together as Holly tugged us towards the hoopla stall.

'Can you win me that rabbit, Matt? *Pleeeese*?' she begged him, pointing at a huge, lurid purple stuffed toy at the back of the table. 'I *love* him!'

'Well, in that case, Madam, I'll have to see what I can do,' he said, giving her a little bow. 'Although I'm not sure whether my rabbit-hooping is quite as amazing as my coconut-bowling.'

All three of us were holding our breath as Matt skimmed the first of his five hoops through the air. Plonk. It landed with just one edge over a box of chocolates.

'Can I have half the box, then? The half the hoop went over?' he asked the stallholder after she'd explained he had to cover it completely, and pretended to be disappointed when she shook her head.

'Try again, Matt, try again!' encouraged Holly, jumping up and down beside me.

'Hang on a minute,' I said, as his second hoop landed upright against a bottle of wine. 'How are you supposed to get a little hoop like that to completely cover a huge stuffed rabbit?' I turned to the stallholder. 'It's impossible!'

'The rabbit's just for display, not to aim for,' she said, in a tone that suggested it should have been obvious. 'To get those big prizes, you have to get three of your five hoops over the smaller ones.'

'She might've told us that,' Matt muttered, looking doubt-fully at the three hoops left in his hands. 'I bet nobody ever does it. Oh well, Holly, sorry to disappoint you, sweetheart.'

He chucked the hoop carelessly at the display without aiming it.

'Oh my God, Matt – you've won a packet of fruit drops!' I laughed. 'Do the next one without trying too!'

He did – and the hoop landed neatly over a five-pound note. We looked at each other in surprise.

'Should I try a bit harder with this last one, do you think?' he whispered to me.

'Nah. Just chuck it again. You seem to be better when you don't try!'

And when the hoop settled over a jar of marmalade, we both screamed with disbelief and hugged each other. It was quite a fierce hug, and I think we both realised what we were doing at the same moment. We pulled apart without looking at each other, a bit embarrassed.

'Here are your prizes, then,' said the woman, looking put out about it, as she handed Matt the sweets, the money and the marmalade. 'Want to choose your big one?'

'The purple rabbit, please,' Matt said without hesitation.

Holly's face was a picture. She took the rabbit from him, managing a squeaky little *thank you*, almost speechless with pleasure. It was nearly as big as her.

'That was amazing!' I said, turning back to him. 'You've made her day. Thank you so much!'

'I don't even know how I did it!' he said. 'It must've been that little prayer I said to the God of Purple Rabbits.'

I giggled and put my arm through his. He was such a nice guy, so sweet and funny. Surely he wasn't going to turn out to be like those other journalists.

'Well, the prayer must have worked,' I said, smiling up at him. 'Look how happy you've made Holly.'

We both turned to look at her. And she'd gone.

'Holly? Oh my God!' I gasped. 'She was right here! Where is she?'

'OK, don't panic. She can't have gone far.' Matt spun around, peering through the crowds. 'I think I can see her. Holly!' he called, starting to run towards the next stall.

Thank goodness, almost immediately he was back, leading Holly by the hand. She'd only been a few yards away, only been out of my sight for two minutes, but my heart was pounding like a sledgehammer and as I ran to grab hold of her I actually thought I was going to faint with relief.

'I couldn't see you!' I told her, my voice trembling. 'We thought we'd lost you.'

'I saw Luke over there,' she said, looking a bit frightened.

'She was showing him her new rabbit,' Matt explained.

'Are you cross? Are you going to tell Mummy?' Holly asked, anxiously.

I sighed. 'No. But please stay with us now, and hold my hand, OK?' I turned to Matt. 'Thank you.'

He gave me another of his warm hugs. 'She was fine – it's just that you panicked when you couldn't see her. But I'm glad I was here to help.'

'Me too,' I said. And I realised I meant it – and that I liked the hug. I hugged him back, and we exchanged a smile, and I knew with a sudden certainty that I wanted to see him again. Journalist or not.

'Hello my lovelies!' a familiar voice interrupted me as I was still smiling at Matt and holding his arm. Needless to say it was Annie from Ye Olde Tea Shoppe, with another woman who looked like she might be her sister. "Er's a priddy liddle maid. She your tacker, then?'

'Sorry?'

'She's asking if Holly's your daughter,' Matt interpreted for me.

'Oh, no!' I laughed. 'Holly's my landlady's little girl.'

'Tha' be so?' She looked surprised. 'Chitter round 'ere is, you got a youngun yersel, name of Holly.'

I laughed again, but this time the laughter was tinged with embarrassment. I'd murder that stupid young Josh if I ever saw him again. He must have spread around the story about Holly being mine. I couldn't even remember why I'd said it, but I needed to kill it off right now.

'No, that must be a misunderstanding,' I said firmly. 'She's not mine. I don't have a child.'

'I see,' Annie said, not looking convinced. 'Well, I'd bedder be off now, beddern I.' And off she went, muttering to her sister about me, no doubt.

Matt was giving me a strange look. 'Why on earth did Annie think Holly was yours?'

'I've no idea. Someone must have got the wrong end of the stick. I wish people around here weren't such gossips,' I said, a little too abruptly.

'Well, it's a country town. That's what people do,' he replied in a similar tone.

We let go of each other's arms. And after the earlier feeling I'd had when he hugged me, it suddenly felt as if the sun had gone behind a cloud.

CHAPTER SIXTEEN

My next booking with Sugar came around quickly, and I turned up at Vanya's house bright and early.

'Now, I must tell you, Emma,' she said, as she pulled on her smart jacket ready to jet off to Edinburgh for another conference, 'it's especially important that Sugar is kept in tip-top condition this week. She's entered into the South West Counties Cat Show next week, and a hell of a lot rests on her winning Best Cat in Show. I mean, her reputation, not the cash prize, although that *is* significant, of course. If she does well in this one, I'll enter her into the national show again this year. So please keep a very close eye on her, won't you. I'll try my best to get home before the end of the week but it might be difficult. I'm relying on you, Emma. Don't let me down.'

She air-kissed me so as not to spoil her lipstick and, completely ignoring Rob as usual, flew out of the door and trotted off in her high heels to her waiting taxi. I sighed, remembering how it had felt to dress glamorously like that, before glancing down at my serviceable jeans and sweat-shirt, and smoothing my newly coloured and shorn hair – a short cut, easy to maintain without regular and expensive trips to the hairdresser.

'What's up?' Rob asked, leaning too close to me as always, his hot breath tickling my ear.

I inched away. 'Nothing. I'm fine. Right, where's our little princess? Time for cuddles, Sugar!' I crooned.

Rob sighed with boredom and sloped off to his study. I couldn't imagine him caring less whether Sugar won Best in Show or not. But Vanya had made it very clear it was important to her; she was my best client, and I was still a bit nervous of her. I was going to make damned sure I did everything I possibly could to ensure her furry baby's success. And even more sure that I ignored Rob's heavy breathing, his muscular arms, strong hands and expensive-smelling cologne. I was *absolutely* not interested – right?

For most of that week, it worked well. He'd actually started to ramp up his flirtation to the next level, which was freaking me out a bit, so I had no problem rebuffing him. I avoided being in the same room as him as much as possible. But when I arrived on the Friday morning, he was in his study, talking loudly on the phone, and when I called out to Sugar as usual, I was surprised and a little concerned that she didn't respond with her normal loud vocalisation or come running to meet me.

'Where are you, baby?' I called out. 'Playing hide and seek, are you?'

I checked all around the house. Not asleep on any of the beds upstairs. Not hiding behind the sofa or playing games with me from behind the curtains. Had she somehow got herself locked in a cupboard? I heard Rob hang up the phone, and I charged into the study, breathless with anxiety.

'Where's Sugar?'

He looked up at me lazily. 'The cat?'

'Well of course the cat. Who else here goes by the name of Sugar?' I was feeling more and more worried. Had he done something to her? I honestly wouldn't have put it past him. 'I can't find her, Rob. I've looked under all the beds and—'

'Oh,' he said, giving me a suggestive smile. 'I'd have come to help if I'd known you were in my *bedroom*—'

'Oh, shut up, for God's sake!' I snapped. 'I'm worried about your wife's cat, all right? Do you know where she is?'

'In the garden, unless someone's done me a favour and taken her,' he said as if it were a joke. 'Bloody thing wouldn't stop yowling while I was trying to talk on the phone.'

'You let her *out*?' I squealed, already running to the back door.

'All right, calm down, I put her in her *run*, obviously,' he retorted, following me. 'She'll be fine, for God's sake, what's the matter with you?'

'You *know* she's not allowed out there unless somebody's with her.' I flung open the back door and rushed into the garden – and then stopped, gasping, my hand over my mouth. The door of Sugar's run was open. And she wasn't inside.

The next half hour was one of the longest of my life. Rob finally had the decency to look pretty worried too, probably thinking about his wife's reaction, just as I was.

'You couldn't have shot the bolt on the gate,' I said, my voice shaking. 'For God's sake! Sugar can open doors with her paw as easily as anything. She could be anywhere by now. Sugar! Sugar!' I yelled, starting to run around the garden. It was huge, and there were flowerbeds, plants and shrubs everywhere. I had no idea where to start.

'You search this side, I'll start on the other side,' he suggested. But almost as soon as we'd started, I heard her. She was yowling, and it wasn't a happy sound.

'She's in pain,' I muttered, my heart thumping with fear. And then there was more yowling – a different voice, a different cat, and then Sugar's yowl of pain again, accompanied by some loud hissing. I knew what that meant.

'She's being attacked!' I yelled, racing towards the sound of the fight.

Behind a pergola further down the garden there was a big, angry-looking tabby standing menacingly over the prostate form of poor Sugar, who was in the submissive position and crying pitifully. I recognised that tabby straight away. He was the same cat who'd bullied poor, nervous Pongo.

'Leave her alone!' I shouted, holding the tabby back with one hand as I made a dive to pull Sugar away from him with the other. Before I could stop him, the big cat had dug his teeth into my wrist. 'Ow! Get off!' I yelled, wrestling myself free, getting a scratch on my arm in the process. 'Go on – scram!' He glared at me balefully and turned reluctantly to slink away. I picked Sugar up in my arms. 'All right, baby,' I soothed her. 'You're OK, I've got you now.'

She was limp with shock and trembling. She was also bleeding from her mouth, and from a wound on her ear. She meowed pitifully as I tried to inspect the wounds. There was quite a lot of blood, making it difficult to see whether the injuries were serious or not, but she was certainly in a state.

'You won't tell Vanya about this, will you?' Rob said, sounding shaken.

'Are you joking?' I retorted. 'Of course I have to tell her! But right now, you need to take us to the vet's. Now!' I repeated more emphatically. 'It's an emergency!'

'OK, OK.' He led the way back to the house. 'Grab a towel or something,' he said. 'I don't want her bleeding all over the car seat. Or you,' he added, glancing at the scratch on my arm and the bite on my wrist, which were both dripping blood.

'Get me a couple of plasters, then. And Sugar needs to be in her travelling basket,' I said, trying to restrain my anger. I got the basket out of the cupboard where Vanya kept it, spread an old towel inside and gently placed Sugar onto it. She didn't protest – she was still too dazed and scared. The poor thing had probably never even had to face another cat without being safely on the other side of her protective fencing. 'I'll sit on the back seat with her,' I said tersely to Rob as we went out to the car. 'Hurry up, can't you? I'm worried that she might be going into shock.'

Luckily, the vet's was only a few minutes' drive away.

'I'll wait in the car,' Rob said, turning on the radio.

'No, I need you to come in,' I said. 'You're paying the bill.'

I got out and slammed the door of his precious car. Sugar was still ominously quiet in her basket, her breathing shallow. I was worried about her suffering shock, obviously, but, although her wounds weren't bleeding too heavily, I was also anxious about scarring. Normally that would have been the least of my worries, but Vanya had drummed into me how important the damned cat show was to her, and I was aware that if Sugar had any visible 'defects' on the day, her chances would be scuppered. I was terrified at the thought of her blaming me – she had a scary way with her and I didn't like to imagine her really angry. And how would something like this affect my new little business? She could make things bad for me if she was cross enough.

As I carried Sugar into the vet's surgery, she let out a couple of feeble cries, and I felt ashamed for worrying about myself when she was obviously in pain.

'OK,' said the vet after examining her carefully, 'she's obviously very shaken and scared, but she's not actually in shock. That could have been very serious. These wounds look superficial and the bleeding seems to have stopped now; if she'd lost a lot of blood I'd be more worried, but to be honest the one on her ear is just a nasty scratch, and the gum should heal well. I'd say the other cat caught her with his claw when her mouth was open.' He glanced at me. 'She's very lucky. It could have been a lot worse. I know Sugar very well. Mrs Montgomery brings her here for all her treatment. She's a very well-cared-for cat, of course, but she wouldn't have had a clue how to defend herself.'

'Poor Sugar,' I said, tears in my eyes. 'Well, I'm relieved it's nothing serious. But do you think these wounds will heal in time for next week? Mrs Montgomery's entered her into a cat show and I know how important it is for her that Sugar is at her best.'

'Quite.' He gave me an appraising look. 'And you're supposed to be looking after her for Mrs Montgomery?'

'Yes,' I said, flushing red. 'But it wasn't me who let her get out. Mr Montgomery put her in the run before I arrived this morning, without fastening—'

'Ah.' The vet nodded, his mouth a tight line. 'Say no more.'

'I would never … I mean, I love Sugar! I look after her properly. I would never let any harm come to her.' I went on anyway, my anger with Rob now mixing with my upset about Sugar and my determination to protect my reputation.

'I understand,' he said quietly. 'Look, the gum will certainly heal without any problem. And the scratch on her ear is superficial – it *shouldn't* scar, but I can't promise. I'd like to say she'll be fine in time for the show, but it's too soon to say for certain. Give her plenty of TLC and keep bathing the wound. I'd suggest Mrs Montgomery brings her back to me before the show, if she's worried. But only time will tell.'

I thanked the vet and went back out to the waiting room, where I left Rob to take care of the bill.

'The cat's fine, isn't she?' he said as he joined me outside. 'Seriously, Vanya doesn't need to know, does she?'

'Yes, Rob, she does,' I said, crossly. 'The vet can't be sure whether or not Sugar will have a scar from the wound on her ear. Vanya will be furious if she can't be entered for the show, and I'm *not* taking the blame for it. I've got the reputation of my business to think about.'

'Bloody cat,' he muttered as he started the car. 'More bloody trouble than it's worth. It's all Vanya cares about – she's obsessed with the damned thing.'

'Perhaps Sugar's the one who gives her the affection she needs,' I retorted. I was past caring about being rude to him. Vanya was my client, not Rob.

'Affection? She certainly never shows *me* any,' he retorted. 'I only stay with her because of my reputation – I'm a parish councillor, you know. I don't want the scandal of a divorce, but we live completely separate lives, if you follow my drift,' he added in his usual suggestive tone. 'She has her cat, and I have ... my own ideas of fun.' He caught my eye in the mirror and winked at me.

I ignored him. The state of their marriage didn't surprise me, but I was disgusted that even now, having come so close to letting Sugar get seriously injured, he was smarming up to me like this.

I spent the rest of the day making a fuss of Sugar, and by the time I left that evening, she was trotting around the house in her usual, confident little-princess manner. I could only hope the wound on her ear was going to heal without a scar – otherwise I couldn't even imagine how upset and cross Vanya was going to be. Sugar couldn't win any more cat shows if her perfect looks were ruined by a scar, and although I didn't really approve of animals being judged on their appearance, my scary client was the one who made those decisions.

It had turned surprisingly warm, and on the way home, I pulled up the sleeves of my jumper and enjoyed the sunshine. Turning into the Town Square, I bumped into Matt.

'Ah, good, I'm glad I've seen you,' he said, smiling. 'Are you still OK for a drive over Dartmoor this Sunday? This weather is supposed to last for the weekend.'

I was relieved that the awkwardness I'd felt between us, after the conversation with Annie at the Fayre, seemed to be forgotten. I really wanted us to be friends, now that I was getting over my worries about him.

'Yes, I'd like that, thank you,' I said, and then noticed he was staring at my arm. I looked down. The scratch from my earlier encounter with the tabby cat had bled through the plaster, and there was also an angry weal on my wrist where he'd bitten me.

'What happened?' he asked, looking concerned. 'Did the cat you're looking after do that?'

'No. Another cat, who was attacking her. Long story. She's OK, though, thank God.'

'But *you* might not be. Seriously, you know cat scratches and bites can be dangerous, don't you?'

'Yes, I do, but this is very minor. It's fine—'

'Has it bled a lot? Did you wash it thoroughly?'

'No, it's only bled a bit and no, I didn't exactly have time ... what is this? Are you a doctor as well as a journalist now?' I teased him.

'I've done a first aid course,' he said, smiling. 'And I don't want you getting ill.'

'Well, that's very nice of you.' I must admit, it felt good to have someone care about me. 'But I'm absolutely fine. I just hope Sugar's going to be all right for the cat show her owner is counting on winning. Her utter dick of a husband couldn't even manage to bolt the door of her run.'

'You saved the day, by the sound of it – getting yourself wounded in the process, too. Quite a story!' He smiled again. 'Can I write it up for the paper? Everyone loves an animal story, and it has that lovely feel-good factor—'

I stared at him. 'No! No, Matt, you *can't* write it up for the bloody paper. I can't believe you! I had a horrible experience there this morning. I was really frightened we'd lost Sugar, and then I was scared she was badly hurt. She still might be scarred for life – and the first thing you think about is getting your damned story!'

'Actually, the first thing I thought about was the possibility of you getting a nasty infection,' he retorted.

'Really? It seems to me that you only care about getting *stories* out of me. Well, I wish I'd never told you about it. Look somewhere else for your stories, Matt, because you won't get another bloody word out of me.'

'I see.' He turned away from me crossly. 'Well, thank you so much for your support, Emma. I thought we were friends, but it seems to me it's all a bit one-sided. I *will* look elsewhere for my stories, then. And you can bloody well look elsewhere for your taxi ride over the moor.'

'Suit yourself!' I shot back, and walked away, seething. As if I'd *asked* him to take me for a drive! What was the point of a friendship with someone who turned every conversation into an opportunity for a news story? One-sided? He was right about that, but it seemed to me it was all on *his* side and nothing on mine!

I felt miserable all evening. After what had happened with Shane, it had been nice to think that someone like Matt was taking an interest in me, being kind to me, or so I thought. Now it seemed he was just using me for his own agenda after all. Was I doomed to be treated like crap by every man I met?

The following day was the last of my current stint of looking after Sugar, and Vanya was due home at lunchtime. Again I kept my distance from Rob, spending the whole morning playing gently with Sugar. Perhaps I'd give up with men altogether, as I should surely have done already, and devote myself to animals. I felt sad, tearful, tired and fed up. When Vanya came home and asked how Sugar was, I suggested she talk to her husband. I had my jacket on, ready to leave, by the time she came back out of his study, her face like thunder.

'I don't want him anywhere near Sugar in future, Emma,' she said. 'Next time I go away, I'd like you to stay here overnight, if you don't mind. I just can't trust him here on his own with her. I'll make it worth your while, of course,' she added. 'And to show my appreciation for what you've done this time—' She got her wallet out of her bag and began to count out banknotes. 'I'm giving you a bonus, in cash, so it counts as a gift and you don't have to declare it on your books.'

'Oh, no – honestly, you don't have to do that. I'm glad we saved Sugar from being badly hurt, of course, but we still don't know if she'll have a scar or not.'

'I've had a look, and it's healing nicely, Emma. I'll take her back to the vet for another check before the show, but I feel fairly confident. If it hadn't been for your quick thinking – to say nothing of getting scratched and bitten yourself, I hear – I shudder to think what might have happened. So please accept my gift. Treat yourself. You deserve it.'

I thanked her profusely and asked her to let me know whether Sugar was going to be OK for the cat show, and how she got on. I was relieved she wasn't cross with me, and touched that she was so grateful, that she seemed to think so highly of me. But it hadn't lifted my mood. I still felt hurt and upset about the argument with Matt. I didn't want to go straight home. I had a headache that wouldn't shift, and I thought perhaps a walk in the fresh air would help, and maybe even cheer me up a little. And it had been quite some time since I'd walked past Bilberry Cottage. I wanted to see whether there was any sign of anyone living there yet.

Instead of easing it, the bright sunshine seemed to make my headache even worse. I trudged slowly down Moor View Lane, wondering if it would have been better after all to go straight home and take some paracetamol. I was nearly at Bilberry Cottage now, though. I rounded the bend in the road and slowed down, staring in at the windows of my favourite little house. At first glance, nothing seemed to have changed. The stepladders were still in the middle of the downstairs room, the windows bare, the place seemingly empty and deserted. Then I noticed the car parked at the side of the house. And, squinting through the sunshine, I thought I saw movement at one of the upstairs windows. Trying to ignore the throbbing of my temples as I lifted my head to look more closely, I watched as a shadowy figure came into view at the window. My head swam. I held onto the gate, everything beginning to go hazy. My mind must be playing tricks on me. Surely it couldn't be ... I must be imagining it! I tried to focus, but the figure at the window blurred and swam in front of my eyes. Panicking, I grasped the gate and leant over it, my head down now, waiting for the dizzy spell to pass, but instead it intensified, the world beginning to spin in circles around me. And the last thought I had before everything went black, was: *Shane! Shane, here! Here in Crickleford, here in Bilberry Cottage. Oh my God. He's found me.*

PART 3

TRUST YOUR HEART

CHAPTER SEVENTEEN

Where was I? What had happened? I held my hand up to my head. I was burning up. I tried to sit up, but everything started to spin around again.

'Lie still,' someone said. The voice seemed to come from a million miles away. 'I'm taking you to hospital.'

Hospital? Why? I shook my head, trying to make sense of it, but it just made my head hurt even more. I blinked and my surroundings suddenly became clearer. I was lying on the back seat of a car. And we were moving. Who ...? And then I remembered.

'Shane!' I gasped. I sat up, trying to ignore the dizziness and nausea. 'Stop the car! Let me out!' I made a grab for the door handle, and the car screeched to a halt. Through a haze I saw the driver jump out of the car.

'Hey, what are you doing?' he yelled, wrenching the back door out of my grasp, but I'd already fallen back onto the seat, unable to move. 'For God's sake, Emma, just lie still. You fainted. You're ill. We're going to A&E, OK?' He paused. 'Who's Shane?'

I stared back at him. Blinked again.

'Oh. Matt.'

'Yes. I found you collapsed in the street. It was lucky I just happened to be passing. Now, could you possibly just stay in the back there and not throw yourself out of the car?'

'OK,' I said, weakly. 'Sorry.'

I couldn't work it out. I felt too ill. I closed my eyes again, and when I next came to, I was being lifted into a wheelchair and taken into the hospital. And it wasn't until I was, eventually, waking up again in a bed on a ward, with a drip pumping intravenous antibiotics into me, that I noticed the state of my arm. It was red and swollen from elbow to wrist, the wounds where the cat had scratched and bitten me looking twice as big and twice as nasty as before.

'Cat scratch fever is bad enough,' a nurse told me a little later. 'But their bites are often really dangerous. Cats have all kinds of horrible bacteria in their mouths from killing and eating their prey. You'll be all right now, thankfully, but I've nursed patients before who needed emergency surgery because their infections went right down to the bone.'

I flinched. That bloody vicious tabby cat! 'Sorry to have caused so much trouble,' I muttered.

'Not at all. It was lucky that young man found you. Apparently you'd passed out at the side of the road.'

'Matt,' I remembered. 'Yes. Lucky he was there.'

But as the nurse walked away, I frowned, trying to clear my head. I'd been at Bilberry Cottage. What a coincidence that Matt had turned up there, just as I collapsed. And – was that a *dream* I'd had about Shane appearing in the window? It must have been! A hallucination, perhaps, because I'd been running a fever. I shuddered. It had scared the life out of me to think he might somehow have found me here in Devon. Not that he'd ever want to see me again, of course, but he might be looking for revenge, after what I'd done. No, I had to think logically – it couldn't have been him. *Someone* had been in the cottage, though, so perhaps it was occupied, after all. So that certainly put an end to my dream of buying it myself one day!

I had no idea where the hospital was – the journey had been a blur – but at least I had a mobile signal there, so I was able to call Primrose Cottage and tell Lauren and Jon what had happened. Once the doctor had pronounced me fit to be discharged on oral antibiotics, Lauren arrived to take me home.

'You poor thing!' she exclaimed, looking at me anxiously. 'What an awful thing to happen. Are you sure you're going to be OK?'

'Yes, I feel much better now. Sorry to drag you all this way.'

'Don't be silly. I'm just glad Matt was kind enough to rush you here when you collapsed.'

'Yes.' I frowned, remembering the argument I'd had with him the day before I was taken ill. Although he'd still been good enough to bring me to the hospital, I guessed he'd probably left as soon as he'd deposited me in A&E, and I supposed he didn't want to see me again in a hurry. I sighed. I ought to get in touch with him, if only to thank him. I wished we hadn't fallen out, but I didn't see how we could stay friends if he was going to keep on wanting to write stories about me, probably digging into my past despite all his earlier promises.

But of course, this being Crickleford, even though I'd refused to talk to Matt for the paper about the incident with Sugar, it was soon being gossiped about all over town.

'I hear you've been the Good Samaritan again!' Mary said when I met her while I was walking a Scottie dog for a man who'd gone to London for the week. She raised her eyebrows at me. 'I'm surprised Vanya hasn't thrown her old man out, putting her precious cat in danger like that.'

I shifted from foot to foot, feeling uncomfortable. 'I don't think I ought to be discussing my clients with each other,' I said.

Mary gave me a smile. 'And that's a very commendable attitude, too. But well done, anyway, for the rescue. People are saying it was that nasty tabby cat, the one we all complain about, who attacked poor Sugar. Is that right?'

'Well, it was a big angry tabby, yes – and I'm pretty sure it was the same one who was tormenting Pongo, Pat's Alsatian, when I was looking after him. Do you know who he belongs to, then?'

'He lives on Collier's Farm up on the moor. Roams for miles, he does, hanging around people's doors looking for food and scaring off anything that gets in his way. Used to try to get through other cats' cat-flaps, until he got too fat and got stuck halfway through one. Wish they'd left him there. Nasty-tempered thing.'

'I know,' I said ruefully. 'I ended up in hospital because of him.'

'So I heard, love. Glad to see you're OK now. Is Vanya aware of it?'

'I don't know,' I said. But it seemed everybody else was! I'd only got a bit further into the Square when Annie came flying out of Ye Olde Tea Shoppe, having spotted me from the window.

'I 'eard tell you got took bad, maid. Anguish in your arm, from a cat bite, was't? That Montgomery woman, 'er be like a hen with one chick, all she care about be that buggering cat. Bettering now, bist thee?'

'Annie's asking if you're feeling OK now,' a woman who'd followed her out of the café translated for me.

'Oh, yes, I'm fine now, thank you,' I said. 'Really, it's not a big deal. I was just relieved Sugar wasn't injured any worse.'

'That buggerin' tabby be a nasty cradded thing, oughta be kep' up there on the farm instead of aggravating folk around town. Even sneaks in the café 'ere sometimes when I baint lookin', on the sniff for crumbs drop' by folks. An' by all accounts 'ee do fritten the livin' daylights outa that gurt cow-baby dog of Pat's.'

'I know,' I said. 'Poor Pongo. Well, I'll be watching out for that tabby in future, Annie.'

'Good fer you.'

The other woman had been listening to our conversation. 'You're a good girl,' she said, grabbing my hand and pumping it up and down. 'Local heroine, that's what they're all calling you. You should get your picture in the paper—'

'No!' I said, a little too loudly, and she dropped my hand, recoiling slightly. 'Sorry,' I went on. 'But I'm ... a bit shy. I don't like a lot of fuss. Honestly, it was nothing out of the ordinary, I was just looking after my client's cat, that's all.'

'Well, I know one thing: it's you I'll be asking to look after *my* cats when I go on my holidays now,' she said. 'And I'll tell all my friends too.'

'Thank you.' I smiled at her. My diary for the summer was already filling up. The publicity certainly wasn't what I'd wanted, but there was no denying it was good for business. And that evening I had a surprise visit at Primrose Cottage. It was Vanya Montgomery, bearing flowers and chocolates.

'But you've already given me extra money!' I protested.

'That was before I knew you'd been ill and ended up in hospital because of what happened,' she said. 'I'm *so* sorry, Emma. I can't tell you how angry I am with Rob. But you'll be pleased to hear Sugar has made a full recovery, thanks to you. There's no sign of her injuries whatsoever now. Just in time for the show tomorrow.'

'So she's on course to win Best in Show?'

'Fingers crossed! I'll let you know.'

I watched Vanya walk back to her car, wondering which version to believe: was she, despite her rather supercilious and overbearing manner, really a pleasant lady who was frankly a saint for putting up with her husband? Or was he right – did she love Sugar to the exclusion of everything and everyone, including him? Although I was still angry with him for the incident that could have ended up with Sugar being so much more seriously hurt, I had to admit to a certain degree of sympathy with him. She seemed to view him with complete contempt. I just wasn't quite sure whether he deserved it. Relationships were far more complicated, I realised now, than I'd naively assumed when I fell in love with Shane, when I was little more than a child.

Over the next few days, the weather stayed fine and I went for long walks with Jock, the stout little Scottie dog I was looking after now. The fresh spring air helped to clear my head. The trees along the lanes were sprouting bright new green leaves and some were already wearing their pretty new outfits of pink and white blossom. Cottage gardens were coming alive with flowers, their walls suddenly garlanded with baskets of colourful blooms. During my years in America, I'd forgotten how beautiful England could be in late spring and early summer – and I was already aware that it didn't get much more beautiful than Crickleford on a sunny day. Now that I'd recovered from my brief spell in hospital, I was feeling fit and healthy again. The regular exercise was doing me good; I could now get up Castle Hill without pausing for breath. The sight of the old castle walls, golden in the sunlight, never failed to lift my heart, and the view over the town and the river from the top of the hill was lovely too. On other days, I took the footpath out to Windy Tor and gazed out across the vast expanse of Dartmoor, now ablaze in every direction with the bright yellow flowers of gorse.

But one place I stayed away from that week – and for the next week or so to come – was Moor View Lane. Whenever I thought back to the illusion I'd had of seeing Shane in the window of Bilberry Cottage, I felt sick with dread. I knew it must have been my fever, giving me hallucinations, but I just couldn't chase the fear away yet. And the other place I didn't visit was the office of the *Crickleford Chronicle*. I'd seen Matt, briefly, in the newsagent's on Town Square a couple of days after I came out of hospital, and thanked him effusively for his help in taking me there after I'd collapsed.

'You're welcome,' he'd responded. 'I hope you're feeling better now.'

But the exchange was stilted, as if we were two strangers who weren't very sure if we even liked each other. *Perhaps that's what we are, after all*, I thought to myself sadly. He'd helped me when I was in need, as I suppose anyone would have done, but it seemed clear that he didn't want any more to do with me since I'd refused to let him write about me in the paper. Well, in that case the feeling was mutual, I decided crossly. But feeling cross about it just made me sad. I hated to admit it, but I missed him. I should know better by now. I should have learned that I couldn't trust anyone, especially a man, particularly a good-looking one with a smile that made my heart miss a beat. No good could ever have come from it.

But apart from this, I was mostly happy. Lauren, Jon and little Holly treated me like one of the family. I loved living with them in Primrose Cottage, playing with Holly, chatting to Lauren while I helped her cook dinner, sleeping in my little blue bedroom. It could never be *quite* the dream home I imagined when I looked at Bilberry Cottage, though. For a start, the view from the windows was of the other, less appealing and more modern houses in Primrose Gardens rather than the expanse of Dartmoor I imagined seeing from the windows of Bilberry Cottage. And while it was an attractive and homely little house, it didn't have that 'chocolate box' prettiness of my favourite cottage.

My work helped to keep me happy too. It was so much fun looking after people's pets that it didn't even feel like work, but I'd taken Mary's advice seriously, and looked after the administrative side of my little business carefully. She'd already called on me a couple of times, while bringing Lauren more books to read, to check that I was managing, and offer me more help if I needed it. I'd given her Shakespeare books back, trying to talk about the weather, the state of the footpaths, Scrap the dog – anything other than the plots of the damned plays – and to my relief she didn't ask anything more at all about whether I'd enjoyed them, or offer to lend me more. Perhaps she'd forget, now, that we ever discussed literature. I hoped so.

And then, one day in May when I came home from looking after a rather spoilt Pekingese, Lauren greeted me at the front door with a worried look on her face.

'There's somebody here to see you, Emma,' she said. 'I tried to call you, but I presume you had no signal.'

My heart started to race.

'Who is it?' I whispered, the vision of Shane in the window of Bilberry Cottage coming back to me so forcefully that I had to hold onto the doorframe to stop myself from bolting back down the road.

But before Lauren could even answer me, the door to the lounge opened and there – her arms open ready to hug me, tears glistening in her eyes – was my twin sister.

'Kate!' I gasped.

And before either of us could manage another word, I was in her arms, both of us blubbing out loud from the emotional impact of being together again. My sister! It was so lovely to see her – even though I had no idea what she was doing here.

CHAPTER EIGHTEEN

Lauren took Holly into the kitchen, closing the door to give us some privacy, and Kate and I sat down on the sofa together, our arms still entwined.

'It's so good to see you,' I said. 'I'm sorry – it was such a surprise – I didn't even notice your car outside.'

'Lovely to see you too.' She smiled, but there was something in her eyes I couldn't quite work out. 'Are you OK, Emma?'

'I'm fine.' I frowned. She surely couldn't have heard about the cat bite incident? 'I mean, I miss you, obviously – and Mum and Dad – but it's all worked out so well down here. Lauren and Jon are so kind—'

'Yes, they do seem nice. I must say I'm relieved that you're staying with a decent family.'

'There's a huge *"but"* at the end of that,' I said. 'What is it, Kate? You obviously haven't just turned up here out of the blue like this because you fancied a break from the kids. Is everything all right at home? Are Mum and Dad OK?'

'Yes. Everything's fine at home. It's *you* we're all worried about,' she said, her voice now rising indignantly. 'You never phone, you don't respond to texts or emails—'

'But I explained, didn't I! I don't get a phone signal here, and the internet service is terrible. Anyway, I *have* texted you, loads of times. I have to walk into town to get a signal—'

'You haven't, Emma. You did at first, but you haven't lately. Not since you told us about this new business of yours.'

'Well, that's exactly why!' I was on the defensive now, but I felt guilty too, because she was right, of course. I probably hadn't been in touch with the family for weeks. 'I've been so busy! I've had one job after another, and some of them have been difficult. I didn't want to worry you, but I ended up in hospital because of one job. And it's not exactly nine to five—'

'You've been in hospital?' she shot back, her eyes wide. 'What happened? Why didn't you let us know?'

'Because it was only overnight. It was nothing, I'm fine. Like I said, there was no need to worry you. A cat bit me and it got infected, that's all.'

She stared back at me. 'A cat bit you? I thought you said it happened while you were working.'

'Yes, Kate. I was looking after a cat who's not allowed out, but the client's husband didn't close the door of her run, and—'

'You're looking after your clients' *cats*?'

'Yes!' I frowned. 'And dogs. And rabbits, and all sorts. I've got a budgie next week, and a hamster soon … what else did you think I'd be doing, as a pet sitter? What's funny, now?'

'You're a pet sitter?' she said, laughing. 'Oh my God, Emma, why didn't you say?'

'I did! You asked me what sort of business it was, and I know I took a while to get back to you, but I did eventually.' I hesitated. 'Didn't I?'

She bent down and picked up her bag, pulling out a sheet of paper.

'Listen to this,' she said. 'It's a printout of your email:

Dear sis. Don't worry, my business is going well. Some clients are lovely, just want me to cuddle them and stroke them. Others are a bit demanding and have a lot of excess energy if you know what I mean, ha ha, I go home exhausted. Got soaking wet with one of them out on Dartmoor, he's a big beast but very clever and does what I tell him. My favourite is Sugar, she's playful and really special. I think she loves me too.'

Kate stopped reading, folded the paper, and then looked up at me and we both burst out laughing. Within seconds we were falling into each other's arms again, on the verge of hysteria.

'*Got very wet with a big beast on Dartmoor*!' I repeated when I managed to catch my breath.

'Some just wanted you to *stroke* them!' She wiped tears of laughter from her eyes. 'You think *Sugar* is special and she loves you too!'

'Well, she is, and she does,' I conceded. 'But she's a Burmese cat.'

We both giggled again, and then I noticed the lounge door opening just a fraction, and Holly peeping round it at us, her eyes like saucers.

'Why are you screaming?' she said, promptly setting us off again.

I held out my arms to her and she came running to jump on my lap.

'Sorry, darling. I'm just telling Kate about some of the funny things the animals do, that's all. We're happy because we haven't seen each other for ages.'

'Mummy said Kate's your twin sister. I wish *I* had a twin sister. I keep asking Mummy to get me one but she says she can't. It's not fair.'

'Oh dear,' I said, glancing at Kate and trying to keep a straight face. 'Well, at least you've got me as a ... kind of ... *big* sister, haven't you.'

'Sorry, is she being a nuisance?' Lauren said from the doorway. 'I've been trying to keep her out of your way.'

'No, she's fine,' I said, laughing. '*We're* fine, now – I hope – aren't we, Kate?'

'Yes, despite you worrying the life out of everyone at home, you idiot,' she said, giving me a playful punch. 'We've all sent countless messages and emails asking you to please explain, but you're *useless* at keeping in touch.'

'Our phone signal and internet here are terrible,' Lauren started to excuse me, but I shook my head.

'No, it's my fault. I'm not very good at writing emails, and I ... um ... managed to give my family the wrong impression about my work. They didn't think it was animals I was involved with.' I glanced sideways at Holly and added quietly, 'I'll tell you later.'

'I'd better call our parents and give them some reassurance,' Kate said, getting to her feet and trying again to smother a laugh. She got her phone out of her bag and shook her head, remembering about the signal.

'Use our landline,' Lauren said at once. 'And, Kate, I've got dinner cooking, and I hope you're going to stay the night? Or as long as you want, of course. We haven't got loads of space, but the sofa's comfy, or there's a spare mattress you can put next to Emma's bed?'

'That's really kind of you. I'd love to stay tonight, thank you, but I'll need to get back to the kids tomorrow. My other half's taken time off to look after them.'

I hugged her again. 'I'm *so* sorry I've caused so much trouble,' I said, more soberly now. 'As usual,' I added with a sigh.

'Don't be daft. I haven't had such a good laugh in ages!'

Later that night, we moved the spare mattress that was kept under Holly's bed, into my little room and made up a bed for Kate next to mine. It only just fitted. I had to climb over her when I realised I hadn't cleaned my teeth, making us both start giggling again.

'I do miss you,' I said when I climbed back into bed. 'I miss sharing a room with you like this, like we did when we were kids.'

'I miss you too, you fool. I missed you terribly when you were in America, but when you came home it was just so awful for you, and so unfair that you had to go away again.'

I sighed. 'Well, it was my own fault, as always. If I hadn't reacted quite the way I did—'

'Emma, from what you told me, anybody would have reacted the same way. Shane deserved every bit of it. He was lucky you didn't kill him.'

'No, *I* was lucky I didn't kill him,' I said seriously. 'Or I'd have the police after me instead of just the paparazzi.'

'Well, I'm sorry, anyway, that it ended the way it did,' Kate said, giving my hand a little squeeze. 'I know how much you loved him.'

I sighed. 'Yes, I did, at first. But it was a stupid kind of love, wasn't it – one that made me leave my family and cut myself off from all my friends. And once he got *really* famous, in the States, I'm not sure I really loved him any more. I loved the lifestyle, if I'm honest. But now, living the way I am here in Crickleford, I look back and think what an idiot I was. It wasn't real, Kate. It was all just … like I was a character in a film. A rubbish film,' I added crossly.

'But you're happy here, now? Really? You don't miss anything about your life in America?'

'I love it here, honestly.' I lay back and smiled to myself in the dark. 'It's the complete opposite of that life. I've got nothing of my own now, except my work, but it's made me realise what matters. There's only one thing I miss from New York: Albert.'

'Your cat? Oh, Emma, I forgot about Albert. What happened to him? Has Shane kept him?'

'No. It took me ages to get an answer from either him or Leo – his agent – and when Leo finally replied, it was to threaten me with legal action if I tried to contact Shane again. I only wanted to know about Albert! I don't care if I never speak to Shane again, obviously. And all he told me was that Albert's gone to *a good home*. It had better be.'

'Can't you find out who's got him?'

'I don't know how.' I fell silent for a moment, then added quietly, 'Well, there's only one way I can think of. One person I could ask. But I've resisted it, so far.'

'*Her*?'

'Yes. Her. The bitch.' I snorted.

'You don't think *she's* got Albert?'

'No. She doesn't like animals. Or children. Or anyone who isn't as blonde, beautiful and rich as her, of course. But she'll probably know where Albert's gone. If I can only bear to lower myself to contact her.'

'Poor you,' Kate sympathised, reaching out for my hand in the dark. 'What a mess.'

'Well, I should have been like you, shouldn't I. Worked harder, got my exams, gone to college …' I said. Suddenly I felt like crying. 'Made Mum and Dad proud,' I added in a little voice.

'Emma,' she said gently, 'you've got to stop blaming yourself for everything. Mum and Dad love you, you know. When you have kids yourself one day, you'll realise what I understand now: you never stop loving your children, no matter what. Even when they empty a cup of milk all over your new rug,' she added, 'like Rose did last week.'

I gave a short little laugh, but didn't say anything. She squeezed my hand again and we fell asleep like that, like we used to when we shared our bedroom at home, hands stretched out across the gap between our beds. We always needed at least our fingertips touching, or we couldn't sleep.

In the morning, I held onto her for so long on the doorstep that she eventually had to gently disengage herself from my arms.

'I've got to go, honey. The kids—'

'I know. I'm sorry. Sorry you had to come and check me out.' I smiled. 'But I'm glad too. Will you come again? Bring Tim and the children? I'd love to see them. There are B&Bs in the town – you could stay for a week. Please say you will?'

'I'd like that. I'll talk to Tim about it, I promise. Perhaps later on in the summer. Bye, sis. Be good. Behave yourself with Sugar!'

I was glad she'd made me laugh before she drove off, otherwise I'd almost certainly have been crying.

CHAPTER NINETEEN

May drifted into a showery June, and the animals I was being asked to look after suddenly became as varied as the weather.

'I've got some fish next week,' I told Lauren when I was checking my Primrose Pets appointments diary one evening.

'Fish?' Lauren queried. 'What, in a tank?'

'No, Koi carp, in a pond. And a hamster in a couple of weeks' time. And later on in the summer I've got a booking for a budgie.'

'Well, it makes a change from dogs and cats, doesn't it,' she said. 'I presume you know how to look after all these different creatures?'

'Most people give me loads of instructions. The fish can't really be difficult, though, can they? I mean, they just swim around in their pond. I just have to feed them. I can't exactly take them out for walks, or sit and cuddle them.'

'Not that you'd want to! Yuck! Slimy!'

She pulled a face, and we giggled together, and I felt a bit better for a few minutes. But it didn't last.

I'd felt lonely since Kate had left. I couldn't help it – despite Lauren and I having become quite close now, nobody was quite like my twin sister, and although we'd spent so long apart, now that we'd been together again briefly I found myself missing her more than ever. I was throwing myself into my work with the animals in an effort to stop myself from brooding.

I'd already met the Koi carp before I turned up on my first day. The man who owned them, Gary, seemed keen to show them off.

'They're worth a lot of money, you know,' he told me proudly as I watched them swimming around the big pond in his garden. I wasn't surprised to hear that. Some of them were as big as a small cat. They came up to the surface, mouths open greedily, as he threw in handfuls of food to show me how much to feed them.

'Is that all I have to do?' I asked. It seemed a bit like money for old rope, but I didn't like to say that, of course.

'Well, you can talk to them if you like,' he said – and when I looked up at him in surprise, he laughed and I realised he was joking. 'Just make sure the level of the water doesn't go down. It's not likely to, unless the weather turns hot,' he said, and showed me how to top it up if necessary. 'Otherwise, that's it. But if you have any worries, call my mate – he used to keep Koi himself and he's a bit of an expert, so he'll be able to sort you out. I'll leave his number next to the fish food.'

I did wonder why he didn't ask his mate to look after the fish while he was away – I'd have thought he was better qualified. But perhaps he was too busy. That first day, after I'd fed the fish and spent ten minutes watching them swimming up and down the pond, I wandered around the garden, went back to the pond and spent another ten minutes watching them, looked at my watch, and wondered if it was really fair to charge him for half an hour's work: there was no way I could stretch it to any longer. Too late, I realised I could easily have done this job at the same time as another one. I'd know for next time, but now I'd blocked the whole week out of my diary, just to stand here gazing at a dozen big fish in a pond.

The next day, to pass the time, I invented names for all the fish. The big spotty one was Dot, the smaller, spotty one, Freckles. The one with stripes on his back was Tiger. The fat orange one was Tango, and the littlest one, Tiddler. Then I started to get confused because there were several that looked the same. Which one did I just call Nemo? Which was Bubbles? And did it really matter, as they obviously didn't respond to their names? Anyway, Gary might have named them something else.

The third and fourth days passed much the same way. I was getting so bored of watching them swimming around, I did actually start talking to them. And then, on the fifth day, I looked into the pond – and Freckles was floating on the top, looking rather too much like the previous night's dinner.

'Oh my God,' I whispered. 'You're dead, aren't you? What did I do wrong? And what the hell do I do now?'

I raced back into the utility room where Gary kept the fish food, and found the piece of paper where he'd written his mate's phone number. It was the first time I'd looked at it. And as soon as I did, my heart sank. How likely was it that there might be another Rob Montgomery in Crickleford? Not very likely at all, it turned out, when I called the number next to his name.

'I'll come straight round,' he said when I explained about the dead fish. 'I didn't realise Gary had left *you* in charge,' he added as if it was the most ridiculous thing he'd ever heard of. Perhaps it was.

He was there within minutes, and marched straight out to the pond.

'Pass me that net,' he said, and promptly lifted out the poor dead fish and inspected it closely. 'Doesn't appear to be diseased,' he muttered. He looked back at the pond. 'There's another one there looking a bit unhappy – see?'

He was right. Nemo was tilting to one side a bit and lying on the surface with his mouth open.

'Has Gary changed the water recently?' he asked.

I spread my hands. 'How would I know? You're supposed to be the expert. I didn't know you had fish, by the way,' I added.

'I haven't, any more. One member of the family being obsessed with animals is more than enough,' he said sourly. 'Look, I think the problem is that this water needs changing.

I'll test it, but my betting is that it's become too acidic. If you leave it like this, they'll all die.'

'What!' I gasped. 'But they're really expensive, aren't they! And I have no idea how to change the water. Gary never said anything about that.'

'I'll show you,' he said with a sigh. 'We'll change twenty per cent of the water today, another twenty per cent tomorrow, and so on, OK?'

'And then Gary will be back,' I said, with heartfelt relief. 'I hope he doesn't think it was my fault Freckles died.'

'*Freckles*?' Rob said, with a lift of an eyebrow in my direction.

'Yes. Well, I was bored, so I gave them all names. The one who looks a bit poorly is Nemo. The other spotty one there is Dot, and—'

'I get the picture,' he said with a laugh. 'Well, let's give Freckles a decent burial, first, then I'll show you where the outlet pipe is. OK?'

It was strange: I'd been so wary of him whenever I was looking after Sugar for Vanya, but there in Gary's garden, he seemed different: he was helpful, nice, almost charming. It wasn't until we'd finished topping up the pond again and I was thanking him for his help, that he started to show his other side again.

'You're welcome, sweetheart,' he said, putting an arm around my shoulders. 'Anything for my wife's favourite little cat sitter, eh? And what do I get in return?' he added in a horrible smarmy tone.

'I've said thank you.' I tried to shake off his arm, but he pulled me closer.

'A little kiss, at least, surely? To show we're still friends? Even though you dropped me in it with Vanya about the bloody cat getting out of her run.'

'I didn't *drop you in it*,' I retorted. 'It was your fault! I wasn't going to take the blame.'

'Fair enough. And I won't let you take the blame for *Freckles* when Gary asks me what happened here today. Unless you're mean to me, of course.' His arm moved to encircle my waist and he gave me a little squeeze.

'Get off!' I pushed him away, angry now. 'Nothing about this has been my fault and you know that. You said Gary obviously forgot to change the water before he left.'

'So you don't even want a little kiss and cuddle? We could go indoors – there's nobody here to disturb us. We'll just have a little sit down on the sofa …'

'No! I don't want anything like that. I thought I'd made that quite clear,' I said, conscious of sounding like a maiden aunt at a vicarage tea party. The truth was, of course, that I was struggling to control my own conflicting feelings. It had been a long time since I'd felt a man's arms around me like this, his hot breath in my ear, the excitement of being *wanted* – and I was lonely. But there was no way I was giving in. 'Please leave me alone,' I added more firmly, 'or I'll have to tell Vanya.'

He laughed, but finally let go of me. 'She wouldn't be surprised.'

'Maybe not, but that's between you and her. She's my client, and you're her husband, so you must be mad to think I'd be tempted—'

'Oh, but you are, aren't you,' he replied softly, with a very gentle touch of his finger on my cheek, making me shiver. 'But if you'd rather play with pussy cats and puppies, that's your loss. Pity though,' he added with a last lingering look at me before he turned to go, waving goodbye without a backward glance.

I took a deep breath and turned back to the pond:

'Men!' I said. 'Is it any easier, being a fish?' I asked Tiger as he swam up to the surface to blow bubbles at me. 'Do you have problems with the ladies ... or the men ... which are you, anyway, I wonder?'

Then I burst out laughing at myself. Rob was right, I was better off playing with pets than getting involved with a man, especially a married one who was completely off limits. But talking to a fish about its love life? That was slightly weird, even for me.

I managed the water changing business myself, the next couple of days, which at least kept me busy, and when Gary came back he was full of apologies.

'I meant to do it before I went away. I just got so busy ... I'm so sorry to have given you a problem.'

'No worries. Rob was ... very helpful. But I'm really sorry Freckles – I mean, the spotty fish – didn't make it. At least Nemo – um, the other one – recovered.'

He smiled at me. 'You gave them names too? I called him Smartie – the one that died. And that one there is Stripey.'

'Tiger,' I said.

'Right. The big orange one? I call him Mango.'

'Oh. Close. Tango.'

He nodded, looking pleased. 'That little one – Titch.'

'Tiddler.'

We laughed together. Who'd have thought it? I'd ended up quite liking those fish. What next? Tortoises? Owls? Stick insects? Primrose Pets was nothing if not versatile, I told myself with satisfaction as I made my way home.

I decided, that Sunday, to have a walk past Bilberry Cottage again. I wanted to cheer myself up and try to recapture my delight about living in Crickleford. The uneasy feeling I'd had since I'd imagined seeing Shane there had dissipated over the weeks, and now I just felt silly for taking too much notice of the hallucination, or whatever it was, that I'd had while I'd been feeling ill.

It had been raining all morning, but as I turned into Moor View Lane, the rain suddenly stopped and there was a break in the clouds. In the sunshine, the wet leaves on the trees gleamed, the flowers in the front gardens and the daisies and buttercups on the grass verges sparkled, and everything looked pretty and bright again. Feeling more cheerful, I hurried round the bend and stopped outside my favourite cottage, sighing with contentment. I knew it was strange to feel such a connection to a house that didn't belong to me and never could. But it just looked, somehow, *perfect* – so exactly what I imagined for myself. Perhaps it was a good thing I was never likely to go inside, I thought, as I stared at the two apple trees in the overgrown little garden and the stepping-stone path up to the blue front door. It might be absolutely awful – gloomy and dark with horrible wallpaper hanging off damp walls, chipped and stained bathroom fittings and an ancient unworkable cooker in a kitchen where every surface was impregnated with decades of grease. But try as I might, I couldn't talk myself out of my infatuation with the place.

I was so engrossed in my daydream that when I first glimpsed a movement in the front room of the cottage, I didn't even react. I blinked, and the movement became a figure, and for a minute I thought I was experiencing *déjà vu*.

'*Shane?*' I muttered to myself, my heart beginning to race. But this time, I wasn't ill, I wasn't feeling dizzy and feverish and about to faint in a heap. I looked a little closer, watched as the figure turned around, and gasped with surprise. It wasn't Shane, of course. It was Matt.

I didn't wait to see whether he'd noticed me loitering outside. I bolted back down the lane, annoyed with myself for doing so, but at the same time annoyed with Matt – for being there, in *my* special cottage, for looking so similar to Shane from behind that he'd scared the life out of me and … and this was the thing, I realised as I slowed down and walked back over Crickle Bridge with my heart rate finally calming down: I was annoyed with him for not telling me the truth! When I'd asked him why I'd seen him outside Bilberry Cottage before, taking photos, he'd seemed really awkward, and just made some excuse about liking the place. And hadn't I thought it was a coincidence that he just happened to turn up there on the day I collapsed outside? What *was* he doing there? Was he moonlighting as a decorator, or maybe doing the place up for a friend? Why lie about it?

I slowed down a little more. It had started to rain again, and I put up the hood of my jacket and sighed to myself. Who was I to talk about telling lies? I'd told everybody here, including Matt, a pack of them. I'd refused to tell him anything true about myself whatsoever – no wonder he'd given up on me. Perhaps it *wasn't* just because I wouldn't let him write a story about me that he seemed to have gone off me. Maybe he'd have liked to know more about me, and I didn't suppose it was much fun hanging around with someone who was a complete mystery, never revealing anything more personal about themselves than the colour of their hair. Come to think of it, even that was fake!

If only there was somebody I felt close enough to, some-body I could trust implicitly here in Crickleford, so that I could be myself with them and tell them the truth. But even Lauren would probably throw me out if she found out. Why would she want someone with such a whiff of scandal about them lodging in her lovely home, with her lovely family? Why would anyone! It was no good. I'd just have to carry on keeping my secrets to myself.

CHAPTER TWENTY

I tossed and turned in bed that night, wondering about Matt. Did he see me lurking outside Bilberry Cottage? Why hadn't he told me he was doing DIY or whatever, there? And more to the point, was I ever going to swallow my pride and ask him if we could be friends again? I missed his easy company, missed his smile and his teasing and his hugs. All through the following week, I looked out for him around the town, stared fixedly at the *Chronicle* offices when I walked past them, willing him to suddenly appear out of the door and give me a wave. If I saw him now, I decided, I'd try to make up with him. I'd apologise for refusing him his story, try to give him some sort of explanation – or should I say excuse – even if it meant telling more lies. But it was as if he'd gone into hiding.

I looked after Pongo the Alsatian again, enjoying my healthy walks in the sunshine and did a good job, this time, of chasing the vicious tabby cat out of the garden before it frightened Pongo. I chatted to the people I'd got to know in the town, enjoying their country accents and their shameless gossip. But every day when I went home, despite Lauren's cheerful chatter as we enjoyed the warmth of Primrose Cottage's little walled garden while Holly played outside, I felt an inexplicable emptiness. It felt like there was something missing, and it wasn't just about the ache for Kate and my family.

One evening I gave in and composed the email I'd been thinking of writing for weeks. It would have to sit in my 'drafts' folder, anyway, until there was a spark of internet here to send it on its way. It had been hard to know what to say, but I kept it short, if not sweet. This was the gist of it, very roughly:

Ezmerelda: (who the hell calls themselves Esmerelda, let alone with a 'z'?)

I'm only asking you because I'm sure you will know the answer and nobody else will. Where is my cat? Who has him, and are they looking after him? In spite of everything, I think you should let me know. I think you owe me that, at the very least. Please reply, and I will never bother you again.

Candice

OK, I suppose you could ask who the hell calls themselves Candice, too. Well, I did. For the entire time I lived in New York, I was Candice Nightingale. Living with someone as famous as Shane Blue, and hanging around with people like bloody Ezmerelda Jewell, who of course was a top model and didn't she just know it, I couldn't very well be plain boring Emma, could I. That was my mistake, or just one of them. I turned myself into somebody else, somebody whose image I could never live up to. If I'd stayed myself, silly little Emma from Loughton, perhaps I'd have survived the whole celebrity experience and come out of it unscathed. But no, I couldn't hack it, and in the end I turned back into Emma from Loughton anyway, in the most spectacular and ugly way possible.

I felt a bit better after writing the email. At least I'd finally done it, and perhaps at last I'd get some answers about my lovely Albert. If I knew he really was in a good home I could stop worrying about him, however much I'd always miss him.

My next assignment was a hamster called JoJo. I was looking forward to it. I liked hamsters – Kate and I had had one, briefly, when we were children, although I couldn't remember what had happened to it. As with the Koi carp, I didn't imagine there would be much work to do for JoJo. For a start, I knew he'd be asleep during the day, so once again I should have booked in another pet to look after during the same week. I only needed to care for JoJo in the evenings when he woke up. At least I'd be able to play with him then, after I'd fed him and cleaned out his cage. That was more than I could say for the fish!

I'd already met JoJo and his owner, Billie, who lived with her family in Castle Hill House, a very old detached house just below Castle Hill. Billie had shown me how to put my hand into JoJo's cage and let him come to me.

'He's very tame,' she said. 'My kids were desperate for a pet, and I'm not very keen on cats or dogs. I'm nervous of them, to be honest.' She shuddered. She struck me as a very nervy person in general. 'But JoJo's OK, he's cute,' she went on. 'Just remember when you clean his cage out, to put him in his hamster ball. He can roll around the room in it, without the risk of getting lost.'

The children, two boys aged about six and eight, were very noisy and boisterous, so I reckoned if JoJo was happy to be handled by them, he wouldn't mind me too much. He was a nice, cheeky little thing with bright eyes, a twitchy pink nose and an inquisitive expression on his face. I went to the house every evening after I'd had dinner, by which time he was awake and, after the first couple of days, he seemed to be getting used to me. He'd be sitting up, looking at me expectantly, hoping for a treat – a piece of apple or some sliced carrot or banana – as well as his normal pellets of hamster food.

When I cleaned his cage, JoJo seemed perfectly happy to be lifted out and popped into his hamster ball. In fact he seemed eager to get started on his exercise – as if the lounge was a great wide world that was his to explore, rather than the same old room he rolled around every time he was out of his cage. In fact he rolled the ball around so wildly, his little beady eyes bright with excitement, that it was great fun to watch him, and for a while I almost forgot the whole idea was for me to clean out his cage while he was out of it. He'd roll himself under the table, then out again and zip across the carpet at an astonishing speed, as if his life depended on clocking up as many hamster-miles per hour as possible. One evening while I was cleaning the cage, he actually bashed his ball really hard into the skirting board, giving a little squeak of surprise.

'Are you all right in there, JoJo?' I asked anxiously – but he must have been, because he was already rolling off in another direction, the glint of determination back in his eyes.

I brought in a bowl of soapy water, turned on the TV and started watching the first part of a new thriller series while I worked. Once everything had been washed and left to dry, I turned my attention back to JoJo. The ball had stopped rolling around after a while and I'd guessed he must have finally tired himself out.

'Come on, lazybones!' I sang out cheerfully as I bent down to talk to him. 'Let's get this ball moving again, shall we?'

I gave it a little push, and immediately realised something was wrong. The ball moved too easily. It felt too light. I knelt down and picked it up, and to my horror it was empty. No hamster. Ridiculously, I turned it upside down and shook it, as if I expected JoJo to come tumbling out. Then I saw it: a long crack in one side of the ball. It looked only just about wide enough for a spider to crawl through – but JoJo must have somehow squeezed himself out of it.

'Oh my God,' I muttered. 'Where is he?' I scanned the room, panic mounting. 'JoJo!' I yelled. 'Where are you? Come on, boy – look, I've got you a lovely slice of cucumber!'

I found myself whistling, like I would for a dog. But how else were you supposed to get a hamster to come back to you? And how far could they travel in ... perhaps half an hour? I stopped shouting and whistling, and listened carefully. I could hear a little squeak coming from the corner of the room, behind the television. I tiptoed over there, pulled a few things out of the way and knelt down carefully. *Squeak, squeak* – there it was again, but I couldn't see JoJo anywhere. I pulled back the corner of the rug, lifted the edge of the curtains and listened again. *Squeak, squeak.*

'JoJo!' I whispered. 'Where the hell *are* you?'

I stared around the room, trying to work out where a little hamster could be hiding. The room had a wooden floor – they looked like they might be the original floorboards, polished up and made to look attractive in a *vintage chic* kind of way – and expensive-looking rugs scattered around. But looking more closely, I could see there were gaps between some of the floorboards, and particularly between the floor and the skirting boards. Gaps that were, in places, wider than the crack in the broken hamster ball. I might not have been Einstein, but it didn't take much to work out from this that there was one obvious answer as to where my little fugitive might have gone.

'JoJo!' I called again, bending close to the widest gap – and the answering squeak told me I'd guessed correctly.

Relief washed over me. At least he was alive. At least I knew where he was. Phew!

But that was only half the problem. Now I had to coax him out. My adventurous little friend was probably quite happy exploring a whole new world under the floorboards, free from the constraints of his ball, and would doubtless be in no hurry whatsoever to be captured. I went to get the piece of cucumber I'd been trying to tempt him with, but then thought better of it.

'Don't be ridiculous, Emma,' I told myself. 'Cucumber hasn't got much of a smell to it.'

But how much of a sense of smell did hamsters have, anyway? I had no idea. I discarded the cucumber in favour of a slice of apple, wafting it across the gap in the floor, crooning about *a lovely juicy apple for JoJo* and getting nothing, not even a squeak now, in response.

Over the next couple of hours I tried everything Billie had left in the kitchen and in the fridge for the hamster's treats: banana, celery, red pepper, lettuce, even his special hamster choc drops. I knelt on the floor, begging and pleading with him until my knees hurt and my voice was getting hoarse. He gave the occasional little squeak in response, but even when I lay down and put my eye directly to the hole, I didn't get a single glimpse of him.

By half past nine that evening I was close to tears. I couldn't go home and leave him underground. I'd have to try to take up a floorboard – but how? I wasn't exactly experienced in DIY, never mind *un*-DIY-ing. What tools would I need? Would Billie or her husband have them here? Where could I get them from? I didn't want to call Lauren and Jon – it wasn't fair to disturb their evening, and little Holly would be asleep in bed. For some odd reason, there was just one person I wanted to call for help. In the end, I gave in and called him.

'Matt, I'm really sorry. I've got a problem.' I was trying not to sound too pathetic, but I was tired and worried, and felt guilty for letting my little charge get lost. 'I need to take up a floorboard or two, and I don't know how to do it. Have you got any tools?'

'Where are you?' he asked, sounding a bit bemused. 'What's happened?'

I wondered if he'd refuse to help me. After all, it wasn't as if we'd been on the best of terms recently. But as soon as I explained the situation, he said he'd come straight round.

'Don't worry, it won't be difficult. Keep talking to the little chap, won't you, to make sure he doesn't run off under there, to a different part of the house.'

This didn't exactly make me feel any better. While JoJo stayed reasonably close to the spot where he'd slipped down under the floorboards, I felt fairly sure we could get him out – somehow – but if he started exploring further afield, what was I supposed to do? Take up more and more boards until I found him? I lay down on the floor and started talking to him again.

'Listen, JoJo. You trust me, don't you? I thought we were friends. Please don't start running around down there. Stay right where you are, OK, or I'll be in dead trouble.'

He squeaked a couple of times. Maybe he understood Human. I needed to keep talking to him but what on earth could you talk to a hamster about for any length of time? Well, I supposed I could just tell him a story …

When the doorbell rang I nearly jumped out of my skin. I'd made myself comfortable on the floor with a couple of cushions and had been on the point of dozing off, mid-story.

'Thanks for coming,' I said when I opened the door to Matt. He was carrying a bag with some tools in it.

'Well, we can't have a poor little hamster stuck under the floorboards forever, can we,' he said lightly. 'Where did he get down? Show me the gap.'

We went back into the lounge, as I explained what had happened in more detail.

'I didn't notice his hamster ball was cracked,' I said. 'He did bash it into the skirting board but I can't believe it was enough to break the thing right open.' I pointed to where I'd left my cushions on the floor. 'See that little gap? It's hard to imagine it's big enough for him to have squeezed through, isn't it?'

'I see you've been sitting comfortably down there,' he said, with a flicker of a smile.

'I've been telling him a story.'

'Right.' The smile widened a little. 'A fairy story?'

'No. My life story, if you must know.' Too late, I realised this was a bit of a sore point. He was probably thinking I'd told the hamster a lot more than I'd been prepared to tell *him*! 'Not that it was very interesting,' I added quickly. 'It even made *me* fall asleep.'

He laughed out loud now and, although I was relieved, because it seemed he wasn't too cross with me any more, I put my finger to my lips and warned him: 'Ssh! We need to listen for JoJo squeaking.'

He nodded solemnly, and we both sat down on the cushions and waited in silence for a moment.

'JoJo!' I called gently. 'Are you still there?'

Squeak, squeak.

'Phew,' I whispered to Matt. 'Sounds like he's still in the same spot.'

'Must be waiting for you to carry on with your story,' he whispered back, a note of sarcasm in his voice. 'Right. Look, the floorboards are just nailed down. It won't be a problem at all to get this one up. But can you just keep talking to him while I do it? I'll be as quiet as I can, but I don't want him to get spooked and run away.' He hesitated, giving me a little grin now. 'If you want to carry on with your life story, I promise not to listen.'

'Nah. It's a bit boring,' I said. 'I'll sing to him instead.'

'Got any earplugs?' he joked as I started singing. I nudged him and pulled a face and he chuckled as he got up to get his tools. It felt good to be back on friendly terms with him. It was a few minutes, though, before I realised what I was singing. It was Shane's first number one hit – needless to say, my favourite song once upon a time. Not any more. I stopped singing abruptly and started reciting the Lord's Prayer instead.

'Are you religious?' he asked me, as he carefully pulled out the first nail.

'Not really. But I went to a Church of England school so we learnt it off by heart. *Thy kingdom come, Thy will be done ...*' I continued.

'Let's hope we don't need prayers to help us find this hamster,' Matt muttered as the next nail came out of the floorboard.

I frowned to myself. I couldn't believe I'd inadvertently started talking about my childhood. Before I knew it, I'd be telling him all about what happened at school and how I went off the rails as a teenager. Then it'd only be a short step to telling him about meeting Shane, and then it'd be too late. I really, really needed to guard my tongue around him if we were going to be friends again.

'There,' he was saying. 'That's the nails out. Now, I'll just prise the board out from this end. Can you still hear him squeaking? Hold his cage ready to catch him when I lift him out.'

But it seemed the prayers might be needed after all. As the floorboard was lifted clear, we just caught sight of JoJo's tail as he scampered off.

'He's only just out of reach,' Matt said, lying flat on the floor next to me and shining his phone's torch under the floorboards. 'He's sitting there staring at me. Can you sweet-talk him back again? Try a different prayer. Perhaps he's a Methodist.'

By midnight, we'd had to take up five floorboards. There was a gaping hole in the floor of my clients' home. But JoJo the runaway hamster, after leading us a merry dance moving from one spot to another as each board came up, was finally back safely in his cage, rewarded with all the bits of fruit and vegetables I'd been trying to tempt him with earlier. I put the broken hamster ball in the kitchen out of the way.

'Phew.' Matt wiped his brow. 'Put the kettle on, can you, Emma? I think we both deserve a cuppa before I nail these boards back down. I'll just retrieve my phone. I left it down there …'

He lay down and reached under the floorboards again. I went out to the kitchen to make the tea, and a couple of minutes later he joined me, carrying what looked like an old tin box.

'What have you got there?' I asked.

He set it down on the kitchen worktop. 'I've no idea. I was reaching for my phone when I found it. It looks pretty old and rusty, though.' He raised his eyebrows at me. 'Shall we have a look inside?'

He opened the lid. And we both gasped with surprise.

CHAPTER TWENTY-ONE

'So what *was* in the box?'

Lauren was sitting opposite me at the kitchen table the next morning. It was a Saturday and I'd slept in late, after being at JoJo the hamster's house with Matt until the early hours. I didn't even stir until Holly peered round my bedroom door and called me a lazybones, and I was still in my pyjamas now at ten o'clock, eating a late breakfast while I told Lauren all about the drama.

'About five hundred quid in cash, for a start!' I said. 'And some jewellery ...'

Lauren's eyes widened. 'What kind of jewellery? Gold, silver, pearls?'

'Yes. All of those. Plus what looked like a diamond and sapphire ring, and a brooch with ... I think ... rubies set in it.' I shrugged. 'I've no idea how valuable any of it is, of course.'

'But do you think the couple who live there know about this – that they hid it there?'

'Well, it's a good hiding place, but it seems very strange, particularly as they've got a safe! It's hidden behind a picture on the wall. Billie showed it to me, in case I needed the keys to the garage for any reason. She apparently always puts them – and absolutely everything else – in there when they go away. She seems a really nervy, jumpy kind of person. I can't imagine she'd feel happy living with a cache like that under the floorboards when she even locks her car keys and chequebook away in a safe.'

'No. How very odd.'

'It is. But that's not all …' I wiped the toast crumbs from my mouth and leant closer to her across the table. 'There were some old papers in the box too.'

We'd nearly missed the papers. They were folded at the bottom, under all the cash and jewellery. Matt pulled them out, handling them carefully – they were dog-eared and yellow with age. But when he opened them up and spread them out to read, he fell silent.

'What is it?' I asked him.

'Listen.' He lifted a sheet of faded and smudged close-typed text, which looked like a carbon copy of something written on an old-fashioned typewriter, and started to read it to me.

'*The haunting of Castle Hill House is well documented. Many cases have been reported over the years, including sightings of specific supernatural embodiments. I have documentation refer-ring to several appearances of a young man dressed in military uniform, with one arm severed at the elbow and a gaping hole in his neck, who floats across what would once have been the back parlour, moaning and calling out for "Florrie". Another apparition mentioned more than once is that of a small blonde child wearing a Victorian style of nightgown and carrying a candle, crying "Mama! Mama!" in a distressingly pitiful voice. Other reports detail voices calling through keyholes and a blood-curdling scream coming from the fireplace in the sitting room.'*

I glanced towards the fireplace. It was very late at this point – nearly two o'clock in the morning – and it was all too easy to imagine something spooky and unpleasant going on within these old walls, however much modernisation had taken place over the years.

'And it carries on in much the same vein,' Matt said, turning the page over and glancing at the next one. 'Ghosties, ghoulies, and things that go bump in the night ... sorry, are you getting spooked?'

'No,' I laughed, 'although the lady who lives here probably would be,' I added thoughtfully. But Matt was thumbing through the rest of the papers, his eyebrows raised in surprise.

'Look at this!' He held up the large and very tattered page he'd just unfolded. It was the front page of a newspaper. 'It's the *Crickleford and District Gazette* – I've heard it was the *Chronicle*'s rival paper here, till it went out of business, oh, probably about thirty years ago. And look at this headline: CRICKLEFORD HOUSE VISITED BY GHOSTS.'

'Oh my God. It was actually in the paper?' I exclaimed.

'Yes.' He was scanning the article quickly, beginning to look excited. *'Evidence has recently been uncovered of frequent* supernatural activities taking place in Castle Hill House, Castle Hill Road, Crickleford. Residents both past and present have reported seeing ...' He looked up at me, his eyes shining now. 'It repeats much of what's in the typewritten document. Emma, whoever wrote this, had it published on the front page of the local newspaper!'

'Yes.' I nodded. 'Well, it was quite a story, I suppose.'

'And it still is,' he said, very pointedly. He got to his feet, carefully folding the papers again. 'And enough years have passed since this was published – it's time for it to be resurrected!'

I dropped the gold bracelet I'd been holding, back into the box, and shook my head at him.

'No, Matt. You can't.'

'Why the hell not?' he shot back. 'Who's it going to hurt?'

'The people who live here, for a start! If they don't know about it, and I can't believe they do, it's going to be enough of a shock, when they come back from their holiday, to find out that all this has been under their floorboards, without it being shouted at them from the *Chronicle*! It's their home, Matt. You can't write about it without their permission.'

'I'll *get* their permission, then. I'll wait till they come home, fair enough, but they're bound to find it really interesting themselves—'

'I'm not so sure about that. Billie seems like a very anxious person. She'll probably be terrified. And they have two young children. It might upset her so much, she'll want to move house. I'm not convinced we should even be telling her about it.'

'What? We have to tell them, Emma! It was under their floorboards!'

'Obviously they need to know about the money, and the jewellery, but all this *ghost* stuff . . .' I hesitated. 'I don't know.'

'Tell the husband, then, if you think the wife's going to have an attack of the vapours!' He sighed and shook his head. 'I'm really starting to think you don't want me to write *any* big story for the paper, even if it's not about you.'

I didn't respond. We agreed to put the papers back in the box, and put the box in the safe until the family came home, and he went back to nailing down the floorboards.

'I'm going to talk to Billie's husband about it first,' I told Lauren now, as I spread marmalade on my second slice of toast. 'In case he wants to keep it quiet from Billie and the kids.'

'Yes, that makes sense. Wow. The things that go on in your own town that you don't even know about!' she said. 'But I'm glad the little hamster was safe.'

'Yes. That's the main thing.' I smiled. 'He's had the adventure of his life, though. I don't think it bothered him one bit – he really didn't want to be recaptured, the cheeky little thing. Matt thought it was hilarious.'

I hadn't thanked Matt properly, yet, for helping me. I felt bad now about the way we'd parted. We'd both been tired, and I knew he was cross and frustrated with me, but

I wasn't in the mood for an argument. Now, though, I was worried that we'd go back to not speaking to each other again, such a short time after things had felt better between us. I decided I'd get dressed and walk into town in the hope that Matt might be working a Saturday again in the *Chronicle* office.

The weather had suddenly turned hot: high summer had come to Crickleford and, with it, an influx of tourists. *Grockles*, I found myself thinking, and chuckled. I was starting to think like a local! There were groups of hikers in the Town Square, rucksacks on their backs, hiking poles clutched in their hands, taking a rest on the benches in the Square while they spread out maps and consulted their handheld sat-navs. Family groups strolled in the sunshine, looking in the shop windows and exclaiming about cute artefacts that were, of course, displayed precisely to attract them, their prices bearing no relation to their value except to the holidaymaker who wanted a 'Souvenir of Dartmoor'. I headed straight for the *Chronicle* office, but to my disappointment the only person there, a girl who was slumped at her desk, reading a magazine and looking bored, told me Matt wasn't working that day, and that they only had a *skeleton staff* on Saturdays. I presumed she was the skeleton that week.

At a loss now, I decided to go home and see if Lauren and Jon would like me to take Holly to the park. I had nothing to do until it was time for JoJo the hamster to wake up that evening. But as I passed Ye Olde Tea Shoppe, I glanced in the window and Annie waved to me, so I popped in for a quick coffee and chat. The place was heaving, and Annie was red in the face, yelling out orders to a guy who was helping her, and then yelling across the room to customers when their drinks and snacks were ready.

'You're a bit busy today!' I shouted above the noise of the customers' chatter, and she gave me a look, her eyebrows raised.

'Fair makes the gravy run,' she muttered.

'Sweat,' her helper translated for me with a grin. 'She's working up a sweat is what she means. My mum can't speak normal English, can you, Ma?'

Annie laughed and pretended to swipe him across the top of the head with a menu, telling him not to be so *forthy*.

'Cheeky,' he explained with another grin at me. 'I'm Kieran, by the way. I don't think we've met before. I help Mum out, in here, when I'm home from uni.'

'Emma,' I said. 'Pleased to meet you.' I smiled at him, thinking he looked a bit older than the usual university student – but already he was explaining:

'I worked in here full time when I first left school. Then I realised my mistake.'

'How do you mean?' I asked.

He shrugged. 'Well, I had no ambition back then. You know how it is, when you're a teenager, you just want to leave school, and earn enough money to have some fun with your mates. Then you grow up! I realised if I was going to make anything of my life I needed to go to uni and get a degree. I'm a mature student at Bristol. Doing engineering.'

I nodded, suddenly feeling sad. If I hadn't gone to the States with Shane, would I eventually have 'grown up' and done something sensible about my own life? Got an education, instead of being a dimwit? Probably not. I didn't have it in me.

'You OK?' Kieran asked, looking concerned.

'Yes, I'm fine,' I said. 'Sorry, I'm holding up the queue. I'd like a cappuccino, please.'

'She'll die of thirst afore you serve her, poor maid,' Annie scolded him. 'Chattering on about your bliddy university. She don't want to hear 'bout all that.'

'No, it's fine. I was interested, actually,' I said. 'But I'll let you get on.'

I smiled my thanks at Kieran and took a seat in the corner, where I listened with half an ear to the gossip going on at the next table about somebody's husband and someone else's girlfriend. After I'd drained the last of my cappuccino, I was getting up to leave when Kieran suddenly appeared at my table, ostensibly collecting crockery.

'Did you mean that, about being interested, or were you just being polite?' he asked quietly. 'Only you looked kind of upset when I started talking about being a student.'

'Oh, no, it's nothing,' I said, feeling embarrassed. Then I sighed. 'I suppose I just envy people like you – people who are clever enough to go to college and get a proper career.'

'Well, it's never too late, Emma,' he said, looking at me with his head on one side. 'I know it's a big financial burden these days, but it pays off in the end. There's nothing stopping you doing what I've done—'

'Yes, there is,' I said, shaking my head and taking a step towards the door. I looked back at him. He seemed a nice guy, with an open, friendly face and intelligent grey-green eyes. A mature student, studying for an engineering degree. How could he possibly understand how it felt to be too stupid to pass even the most basic school exams? 'I didn't even get any GCSEs,' I said quickly, looking away from him.

As I pushed my way through the group of hikers now coming down ye olde steppes into the teashop, I thought I heard him call after me '*Nor did I!*' – but, of course, I must have heard wrong. Or he was trying to be kind. Or patronising me. Whatever, I decided to steer clear of him in future. I didn't need comparisons with university students to make me feel any worse about myself. I had my own business now, I reminded myself crossly as I walked home, the sun burning the back of my neck. I was an *entrepreneur*. But still, the meeting had unsettled me and I wanted to just get home and take little Holly to the park. Spending time with her always cheered me up.

'Oh, Emma, I'm glad you're home,' Lauren said as soon as I'd closed the front door behind me. 'A letter came for you in the post this morning.'

'For me?' It was no wonder she'd sounded surprised. I'd had no mail whatsoever since I'd come to live here, and hadn't expected to get any.

'Yes, here it is!' She came out of the kitchen, waving an envelope at me.

I took it from her cautiously, as if I expected it to blow up in my face. I recognised the writing on the envelope straight away – it was Mum's. But when I ripped it open, there just was another envelope inside, a blue airmail one this time with an American stamp on it. My heart began to race and my hands were clammy as I carried it upstairs to open in private. There was one sheet of flimsy paper inside this second envelope, covered with the same hand-writing that was on the airmail envelope. It was written in a tiny, cramped script that blurred before my eyes. The signature was completely illegible. But I wasn't even going to try to read it yet – because the important thing was what had been folded inside the letter. It was a photo of Albert, my beloved cat.

CHAPTER TWENTY-TWO

I stared at the photo of Albert until my eyes ached. I even stroked him, in the picture, ran my finger over his head and his back, and tickled his tummy just the way he used to like. It was a while before I could think about anything other than the fact that he was here in front of me, even if only in a photograph. But eventually I started to look at the picture more closely. I didn't recognise anything about the background. He was lying on a stripy cushion, on a gold-coloured sofa. Behind him was a wall decorated in a floral wallpaper, with a glimpse of some stripy curtains that clashed with the wallpaper, and in front of the sofa was a brown floral carpet. There weren't any other clues as to whose home it was, but in a funny way I was pleased about the cushion. Whoever was looking after him was obviously happy for him to make himself comfy on the sofa. From what I could see, he seemed to be in good health. He looked calm and contented. That was very reassuring. I picked up the letter again and stared at the spidery writing, screwing up my eyes, trying to make out what it said.

'*Dear Candice*', it began. I grabbed the envelope again. Sure enough, it was addressed to '*Ms C. Nightingale*', and I could just about make out enough of the address to know it had been sent care of my parents' home. Of course, everyone I'd ever met in the States knew me as Candice. But . . . who'd sent it? Who, apart from Shane, could possibly have known my parents' address? And *he* certainly wouldn't be writing to me! Come to that, who among my acquaintances in America, would send a photo like this, in an airmail envelope, folded inside a sheet of tiny, cramped writing? Everyone I knew shared photos on WhatsApp or Instagram. Why wouldn't this person at least have sent it attached to an email? It made no sense to go to all the trouble of an airmail letter, not these days. It was . . . surely . . . something only an elderly person would do.

I sat up straight, staring at the illegible script again. That was it, of course – the old-fashioned decor in the room, the carpet, the curtains – Albert had been adopted by someone of my grandparents' age! I sighed. Despite the frustration of not knowing who they were, and the worry of how they'd got hold of my Loughton address, I did feel better now I'd figured this out. Surely a nice old lady or gentleman would love my darling Albert and look after him well, give him lots of affection and cuddles. An indoor cat was the perfect pet for someone of that age, after all. I tucked the photo under my pillow, together with the letter, and went back downstairs.

'Everything OK?' Lauren asked me. 'You looked a bit upset about that letter.'

'Oh, no, it's fine – it was just from my mum,' I said. 'She's sent me some photos of my sister's kids,' I lied smoothly. It was worrying how easily the lies came to me these days.

Lauren smiled. 'That's nice.' She looked at me slightly questioningly and I realised the normal thing would have been to show off the photos.

'Oh, they're not very good photos,' I said, turning away. 'In fact the children both had the chickenpox, that's why she was showing them to me. They wouldn't want the pictures shown around. Embarrassing for them – great big spots on their faces. Poor things, eh!'

Lauren nodded. She looked a bit surprised but seemed to accept it. 'Well, it's nice that your mum's been in touch, anyway,' she said.

I realised she probably thought it was odd that Mum and Dad never called, never wrote to me, hardly ever emailed or texted me. She couldn't be expected to know what it was like to be the black sheep of the family, the daughter everybody was glad to see the back of, the one who'd never given them anything but grief.

'Yes,' I said, trying to ignore the sudden pain in my heart, the certainty that most mothers, in forwarding on an airmail letter to their daughter, would surely at least put in a covering note. Just a few lines. Even just *'From Mum'* and a couple of kisses would have been nice. 'She said how much she misses me, of course,' I lied. 'I know she's desperate to have me back. But I prefer it here. I love being here in Crickleford.'

Lauren smiled again and said she was glad I was happy there. But I noticed her giving me the occasional odd little glance over the course of the next couple of days, and once again I wished I could confide in her.

My second week of looking after JoJo the cheeky hamster passed uneventfully, except that I'd had to find an old box to put him in while I cleaned out his cage now that the ball was no good – which rather cut down on his adventures and resulted in lots of squeaks of frustration. He seemed happy to see me every evening, though, but because it was always dark by the time I left, I'd often find myself thinking about the story of hauntings in the house. I wasn't sure what to make of it, and the place didn't really spook me, as long as I had the lights on. But I did wonder what Billie's husband would say when I told him, and whether he'd want to risk telling his wife.

Because I had the daytimes free, I spent quite a bit of time helping Lauren in the garden, ignoring her protests that she didn't expect me to be doing it. I'd never done any gardening before in my life, but the weather was fine and warm and I was enjoying being outside in the fresh air. Under her direction I was learning which were weeds and which were plants, and how to cut the lawn and sprinkle the cuttings on the flowerbeds to mulch down into the soil. One afternoon when I was trimming the edges of the lawn at the same time as trying to maintain a game of hide and seek with Holly, Lauren came out saying I had a visitor. It was Matt, and Holly seemed almost as excited to see him as I was.

'I'm sorry I sulked, about that ghost story,' he said, as soon as Holly had finally run off to play on her swing at the end of the garden. 'I haven't had a chance to talk to you about it since that night – I've been really busy covering loads of summer events. But you were absolutely right, of course. If you think the story might upset the lady, you must talk to the husband first and see what he says. When do they come home?'

'Saturday evening. I'd better go round and see them straight away, otherwise they'll beat me to it – open the safe and find all the stuff there.'

'Maybe you should bring it all home with you on Friday night, just in case.'

'Yes, that makes sense.' I smiled at him. 'I'm glad you're not annoyed with me. I do want you to have your chance of a story, but—'

'I wasn't annoyed with you,' he interrupted me. 'Just frustrated, I suppose. I shouldn't have taken it out on you. How about I make it up to you by taking you for that drive across Dartmoor I promised you? I'm free on Saturday.'

'Yes, I'd love that,' I agreed. I was so pleased about being friends with him again that I managed to block out my worries about him being a journalist – for now, anyway. 'Thanks, I'll look forward to it.'

Lauren brought out a tray of cold drinks, and the three of us chatted for a while about the possibility of ghostly goings-on at Castle Hill House.

'Perhaps someone died a violent death there, years ago,' Lauren suggested.

'Yes. It's usually a murder, isn't it?' I said, with a little shiver.

'But the document we found mentioned someone dressed in military uniform, with horrible wounds. Could they have been war injuries?' Matt suggested.

'Mm. And they mentioned a little child, crying for her mother.' I glanced from Lauren to Holly, who'd now gone back to whizzing back and forth on her swing and was singing a happy little song to herself. 'That was a bit upsetting,' I added quietly.

'Yes. But it's strange that the family who live there now haven't experienced anything, isn't it?' Lauren said thoughtfully.

'Perhaps they're not attuned to the supernatural,' Matt said, with a grin.

'You don't believe in it, do you,' I said. 'You think it's all – what? – someone's overactive imagination?'

'I wouldn't say I don't believe. But I'm sceptical, sure.' Just then, Romeo came rushing out of one of the bushes, haring across the lawn and jumping up in the air after a butterfly, making us all laugh. 'Cats are supposed to be linked to the paranormal world, aren't they?' he went on, as Romeo, looking offended with us for laughing at him, sat down with his back to us and started washing himself.

'So they say,' Lauren said, laughing again. 'They do have nine lives, after all!'

But I stayed silent, thinking about Albert. It hurt me to admit to myself that I was never going to see him again. I wondered if he even remembered me now. I supposed I should be grateful that I could at least enjoy the company of Lauren's two cats, to say nothing of all the pets I was looking after now.

For the next couple of days I looked forward eagerly to my day out with Matt, but, sadly, on the Saturday morning I pulled back my bedroom curtains to see the rain pouring down outside.

'The weather's broken,' Lauren said with a sigh. 'It was too much to hope for sunshine for longer than a week, here in Devon. We've had more than enough rain this year already, though.' She shook her head. 'Everyone's getting nervous about the river again.'

'What do you mean?' I asked, staring at the heavy clouds and rain-spattered windows.

'The part of town nearest the river gets flooded sometimes when there's been a lot of rain all year. Fortunately it's been a few years since it's happened, but it's always on our minds.'

'Oh yes – I noticed sandbags by the houses along the riverbank, back in the winter. It must be really worrying for those people.'

'It is. Fortunately we're OK over this side, we're on higher ground of course. But let's hope it doesn't happen.

Will you still go out for the day, though, Emma? The moor can be really dreary in the rain.'

'I don't know. Matt's supposed to be picking me up in an hour, but perhaps he'll want to reschedule.'

However, Matt was shocked at the very idea of cancelling the trip.

'We could wait all year for a dry day,' he pointed out. 'Come on, a bit of rain won't hurt us. It might even clear up by lunchtime.'

'Ever the optimist,' I muttered, as I ran from the front door to the car with my anorak over my head.

We headed slowly out of town, the rain beating down on the car windows, the wipers going double time. I wondered how I was going to see any of the scenery through the downpour. But before we'd gone very far the rain eased off a little. We turned onto a tiny narrow lane, passed across a little bridge over a stream and started to climb slowly uphill, and by now there was only a fine drizzle in the air. At first there had been a few cottages beside the lane, but now there was nothing – no habitation whatsoever, just miles and miles of moorland in every direction. Suddenly, there was a shaft of sunlight through the clouds, illuminating everything, making all the greens, yellows and purples of the landscape brighten up as if someone had switched on the lights.

'Oh, wow,' I said quietly. 'It's so beautiful.'

There was a little lay-by just ahead where someone, perhaps a group of walkers, had parked an old Land Rover, and Matt pulled in behind it. We got out of the car, ignoring the persistent light drizzle, and crossed the road to stare around us.

'That's Grey Tor,' he said, pointing towards a rock formation looming above the moor. 'And that stack of rocks in the distance – see? – is called the Giant's Nose. You'll see why when we get there. It's right on the top of Tinker Ridge. From there we can head down to Widecombe-in-the-Moor.'

'Oh, that's one place I have heard of,' I said.

'Yes. It's a bit touristy, unfortunately, but there's a nice pub there for lunch.'

'Sounds good!' I smiled at him as we shook the rain off our jackets and got back in the car. 'I knew Dartmoor was beautiful, but seeing it in this light, with the sun shining through the rain – it's just ...'

'Stunning,' he finished for me. He put the key in the ignition but didn't start the engine. He was watching me as I did up my seatbelt. 'I'm glad you like it,' he said softly.

'The colours are amazing. The yellow – is that gorse?'

'Yes. Though we usually call it furze.' He laughed and added, 'And I've heard some of the old folk call it Dartmoor Custard.'

'It's not edible though, is it?'

'Only for cattle. And the Dartmoor ponies, of course.' He was still smiling at me. 'We're bound to see some of them today.'

'Oh, yes. I've heard a lot about the ponies, I'd love to see them!'

He turned away to start the car, but still didn't move off.

'There's an old saying about gorse,' he said without looking back at me. *'When gorse is out of bloom, kissing's out of fashion.'*

'What's that supposed to mean?' I asked.

He looked back at me, smiling again. 'There's always gorse in bloom,' he said. 'Kissing's never out of fashion.'

I felt a sudden shock of excitement as he leant towards me. The kiss took me completely by surprise – I hadn't even realised I'd wanted it until it was happening, but within seconds I was melting into it. His hand caressed the back of my neck, his lips were soft and warm, the kiss gentle and lingering. I didn't want it to stop, but eventually I couldn't bear the discomfort any more.

'Ouch!' I said, pulling away slightly. 'The handbrake ...'

'Oh, sorry.' He took a deep breath and looked at me anxiously. 'I hope I'm not presuming too much?'

'No, you're not.' I leant over and gave him a quick peck on the cheek. 'Of course you're not. I enjoy being with you, Matt. I want us to be ...' I hesitated, and I saw his eyes cloud over.

'Friends?' he said. 'Just friends?'

I sighed. I wanted more than that, of course I did. But how could I begin a relationship with somebody when even our friendship was built on my secrets and lies? I'd never be able to tell him who I really was, not when his livelihood and his future depended on making everything he found out public knowledge. It wasn't fair.

'Can we take it slowly?' I suggested gently. 'It's not that I'm not interested. But . . .' I searched for a reason that would buy me some time to think how to manage this, without hurting his feelings. 'I'm getting over a break-up.'

'Oh yes – you did say you'd come here to get away from an abusive relationship. I should have been more considerate. Sorry.'

'It's fine.' I smiled. 'I just don't want to rush things.'

'I understand.' He reached for my hand and squeezed it. 'No wonder you don't want to talk about the past.' He looked into my eyes now, and to be honest all I really wanted to do was collapse into his arms and kiss him again. 'I'd like us to be more than friends, obviously. But I promise I won't put any pressure on you, until you're ready.'

We were both quiet as Matt drove us on across the moor. All I could think about was that kiss, how much I wanted to do it again, how much I wanted to be with him. And how unbelievable it was that I'd let myself have feelings like this for a *journalist*. I'd never liked them, never trusted them! Back in New York it had felt like Shane and I had had to run in and out of our apartment block, and dive in and out of our cars, almost every day to avoid the paparazzi constantly hanging around, with their cameras and their microphones, trying to get the latest picture, the latest snippet of gossip to twist a supposed story out of – the more salacious, ridiculous and frankly untrue the better. In the end I was scared even to open my mouth when any of them were around, I was so afraid of saying something stupid. I knew these people. They'd do *anything* for a story. How could I expect Matt to be any different?

We were climbing higher now, and gradually the clouds were lifting, the drizzle giving way to sunshine.

'You were right about it clearing up by lunchtime,' I said, to lighten the mood.

'Ah, well, I'm used to the Devon weather.'

'Have you always lived in Devon? You haven't got much of an accent. And – well, your surname's not very English.'

He was silent for a moment. I glanced at him, wondering if I'd offended him. I was just about to apologise when he responded:

'I'm half Italian. But I've never even been there. And anyway, we don't all speak like Annie, around here.'

I laughed, but I couldn't help thinking he'd seemed reluctant to answer. The thought came to me suddenly that perhaps *he* was hiding something about his past too. In a funny way, that would actually make me feel better.

When we came across a group of Dartmoor ponies grazing in the shelter of one of the huge rocky outcrops, Matt pulled over again and wound down the window.

'They're so cute,' I said. I was surprised by the variety of their colours – brown, cream, black, white. 'Although to be honest I was expecting to see more of them today.'

'They're an endangered species now – only a small number left, compared with the past. They've been on Dartmoor for thousands of years. I'm surprised you haven't seen any before. We do occasionally get one or two wandering into town. They shouldn't be handled or approached, though.'

'No, of course not – they're wild animals after all,' I said.

'Not really. They're not *tame*, of course, but they all belong to various people who live on the moor. Every year in September or October they hold *drifts*, where the ponies are rounded up and herded into fields where they get sorted out by their owners' brands, and given health checks, to make sure they're OK for the winter. It's tough out here on the moor but they're hardy little ponies.'

'That's really interesting.' I smiled at him. 'You're an excellent Dartmoor guide, you know.'

'At your service, madam,' he said, starting the car again. 'Are you ready for lunch now?'

The rest of the day passed quickly. It was easy to feel relaxed and happy in Matt's company, as long as I stopped thinking about his profession. I'd have liked the day to go on forever. As we finally headed back to Crickleford, I sat in silence, still enchanted by the scenery but struggling with my feelings. Would I ever be able to have a relationship again? Would I always think about Shane and my life in the States, whenever I got close to being with another man? How could I ever leave my past behind?

CHAPTER TWENTY-THREE

That evening, I went back to Castle Hill House, taking the tin box with me in a carrier bag. Billie was just trying to chase the two boys upstairs to bed.

'They're tired and overexcited after the flight and everything,' she said apologetically. 'I really need to get them settled. Can I leave Carl to sort out your payment? Was everything OK with JoJo?' she added with her usual anxious look.

'Yes, absolutely fine,' I said. 'Thank you.'

I was glad I had an excuse to talk to Carl, her husband, on my own.

'I didn't actually just come for my money,' I explained quietly as soon as Billie and the children were out of earshot. 'I needed to show you something.' I pulled the tin box out of the bag and placed it on the table in front of him. From the way he stared at it and looked back up at me blankly, it was obvious he'd never seen it before. 'I'm afraid everything wasn't *completely* fine while you were away. I had to ask a friend of mine to come and help me take up a couple of floorboards,' I said. 'JoJo's hamster ball is broken. I've left it in the kitchen for you to see. It doesn't look like much of a hole, but he managed to get out of it and went down that gap—' I pointed to the place, and Carl gasped.

'Oh my God, I'm so sorry you had all that to cope with,' he said.

'Not at all. I was just relieved we got JoJo back safely. But the thing is, my friend found this under the floorboards.' I passed him the tin box and he turned it around in his hands, frowning. 'I'm sorry, but we were kind of nosy,' I went on. 'We looked inside. And ... well, you'd better take a look for yourself.'

I sat back in my chair while he took the lid off the box. I was trying to listen out for the sound of Billie coming back downstairs. I could hear her talking in a constant, quiet voice and presumed she was reading the boys their bedtime stories. Downstairs, we were silent apart from the occasional whirr of JoJo's wheel as he took his usual lively evening exercise. Carl's eyes were almost popping out of his head as he thumbed through the pile of banknotes and let the gold necklaces and bangles drape across his fingers.

'I presume you didn't know about any of this,' I said.

'No. But the house has been in the family for more than a century. We inherited it from my Great Aunt Maud. She was a wealthy lady, the only daughter of a rich family who used to own a lot of land around here, so I can only presume this little lot belonged to her. Oh! Yes, I recognise this ring.' He suddenly looked quite overcome. 'She used to wear it all the time. I always liked looking at it when I was a little boy, the stones fascinated me. I think she stopped wearing it because it became too loose for her as she aged. How nice to see it again. And I remember this brooch too. What a find! Fancy it lying under our floorboards all this time.' He shook his head, and then picked up the papers that Matt and I had folded carefully underneath everything. 'What are these?'

'Well ...' I lowered my voice. 'I didn't want to talk to Billie about them without showing you first, because they're a bit, um, disconcerting.'

'Really?' He spread the typewritten sheet out in front of him and started reading. To my surprise, he began to smile, and by the time he'd read to the end of the page he was chuckling to himself.

'You don't think there's any truth in it?' I said. 'But look: it was even written up in the local paper.'

He opened out the old newspaper and immediately started chuckling again. I was completely perplexed.

'What – do you think it's just somebody's overactive imagination, then?' I said.

'Yes.' He grinned up at me. 'That's exactly what it is. Good old Great Aunt Maud! I'd heard this story when I was a boy, but I'd forgotten all about it. Apparently she was always up to mischief, had a wicked sense of humour and loved to play tricks on people. As a teenager it seems she drove her parents mad, always hanging around with the local kids and frightening visitors up at the castle by dressing up in sheets and making ghostly wailing noises.'

I smiled to myself, thinking of this high-spirited young girl whose wealthy parents would probably, in those days, have preferred her to sit quietly at home doing her sewing.

'And *this*,' Carl went on, jabbing a finger at the newspaper page, 'is what she did when she was a lot older and living here on her own. She never lost her sense of fun, you see. She just loved to tease people. Apparently this story really got everyone around here excited and worked up – but when people started wanting to come to the house to hunt for ghosts, she eventually had to admit it was all a practical joke.'

'Oh, I see.' I was relieved, really, for Billie's sake, but I couldn't help feeling just a tiny bit deflated too – just as, I supposed, the townspeople had felt back then. 'Were people annoyed with her?'

'She said most of them took it in good part. But some people actually took a lot of convincing that it wasn't real. That's why some of these ghostly stories still come up today, from time to time.'

'Do they? I haven't heard anything.'

'Oh yes. The story about the man in military uniform is repeated by tour guides up at the castle!'

Just then Billie appeared back in the lounge.

'I hope you're not scaring Emma with all these old ghost stories,' she said, smiling at me. 'They're all made up, Emma – mostly by Carl's great aunt!'

'So I hear.'

Carl pushed the tin box across the table towards his wife. 'Look what Emma found.'

By the time we'd filled Billie in about JoJo and the floorboards, it was nearly nine o'clock, but they wouldn't hear of me leaving until I'd had a drink with them to celebrate finding the jewellery, which I completely understood was more important to Carl, for sentimental reasons, than the money and the papers. They opened a bottle of wine and we shared it, toasting each other and also raising a glass to Great Aunt Maud and her quirky sense of humour. Billie seemed completely relaxed, and I came to the conclusion that it was mostly with her children that she became anxious. Perhaps she was just an overprotective mum. I was glad I hadn't had to scare the life out of her or make her want to move away from their lovely old family home.

'There's one more thing,' I said after we'd all drained our wine glasses. 'My friend – the one who actually found the box – works for the *Chronicle*.'

Carl immediately started laughing again. 'I bet he liked this story, didn't he?'

'Yes, he did! He wondered whether you'd mind if he wrote it up for the paper. I could get him to come round and talk to you, and you can decide how much you're happy for him to include.'

Billie glanced at Carl and they both smiled. 'Wouldn't your great aunt have loved that?' she said. 'The irony of it – after all this time, getting her story in the paper again!'

'She would have,' he agreed. 'Of course you can bring your friend round, Emma. And he can write whatever he likes. A lot of people around here still remember Great Aunt Maud and they'll think it's wonderful to see her story in print again. I'll hang onto these original papers, though. I'd quite like to have them framed! Your friend can make copies.'

'I'll make sure he explains that it's all made up, this time,' I said.

When I finally said goodbye to them, and to JoJo, Carl gave me some extra money on top of my payment.

'A little of my great aunt's money,' he said. 'Share it with your friend. It's made my day, coming home from holiday to find all these things. We'd never have known they were down there if it hadn't been for you.'

'Thank you, that's really kind. But really it was all because of JoJo,' I reminded him. 'Perhaps it's hamsters that have supernatural powers, not cats!'

I'd arranged to meet up with Matt the next day in The Star to tell him what had happened, and his eyebrows shot up with surprise when I gave him his half of the money Carl had shared with us. I expected him to be a bit disappointed that there weren't any genuine reports of hauntings to write about, but he was quite philosophical about it.

'It'll make quite a nice little story,' he said with a shrug. 'Not my big break, exactly, but I can describe how we found the box, and write about the great aunt's mischievous childhood, as well as telling the story of her ghost hoax.'

He picked up his drink and took a big gulp, suddenly looking away and sighing.

'What's wrong?' I said. 'It sounds really good. I bet the *Chronicle* readers will lap it up.'

'It's not that.' He was silent for a moment, looking down at his beer glass, then he suddenly looked up at me again and said, quietly: 'It's just – what you told me about that family, how they inherited the house from his great aunt – it's a little bit like my own situation.'

I waited, silently. He'd told me so little about himself so far, and I sensed that I was finally going to get a revelation about his family or his childhood. Something that seemed to be upsetting him to think about.

'I've recently inherited a house, too,' he said. 'From my grandparents. Well, it's a cottage, actually.' He looked up at me then, and immediately I saw it in his eyes. I understood.

'Bilberry Cottage,' I said, letting out a long breath. 'That's why I've seen you there. It's yours, isn't it? Why didn't you say?'

'I'm sorry. I … just didn't want to talk about it.'

'Oh.' I swallowed back my irritation. So he'd lied to me – said he just 'liked' the cottage. The day he took me to hospital, he'd told me that he'd just happened to be passing – when in fact he *had* been inside the cottage at the time. It hadn't been Shane, obviously, but it wasn't a hallucination either – it was Matt. He wasn't only doing the place up, he actually owned it! 'Don't you like it, then?' I couldn't help saying. 'Are you doing it up to put it on the market? I saw you taking photos of it.'

'No, I'm not selling it. Well, probably not. The photos were just for me to keep, for sentimental reasons. So that I can remember how it looked when my grandparents lived there. I suppose I'll move in, one day. When I've finished renovating it.'

'You don't sound too excited about it. If *I* owned a beautiful cottage like that, I'd be—'

'Well, there you go,' he cut me short. 'I'm sorry for not telling you. I just … don't like talking about it.'

I looked back at him in surprise. His face was a picture of misery.

'Did your grandparents only pass away recently?' I asked gently, reaching out to take his hand across the table.

He nodded, swallowing hard. 'I'm sorry, Emma. Can we change the subject?'

'Of course.'

We sat like that, in silence, holding hands across the table, for a few minutes while he managed to compose himself. I was sorry for his loss, of course, although I hoped he wasn't making things worse by bottling up his feelings. And how could I blame him now, for not telling me the truth about Bilberry Cottage? Not only was he battling his own grief, but I still hadn't told him a single iota of truth about myself or my background. I liked Matt, a lot. I'd been harbouring the hope it might develop into something between us, somehow, eventually, if I could ever get over my fear of journalists. But if he couldn't confide in me about things like this, and I couldn't confide in him about, well, *anything*, really – what was the point of it? There was no future for us.

And no future for me in my dream cottage, either, I thought to myself sadly as I eventually walked back home.

There was no point taking any more little strolls down Moor View Lane to stand outside and daydream about living there. It belonged to Matt, even though he didn't seem very happy about it. Was he really going to move in there? It would make sense; at present he rented a one-bedroom flat above the Chinese takeaway. Or would he finish smartening the cottage up and then put it on the market at a massively inflated price, as people tended to do after a renovation? Either way, it was going to be out of my reach. Not that it was ever seriously *within* my reach, but at least I'd been able to dream, and now I couldn't.

I sighed. The day had somehow gone sour. Matt had remained quiet and sad for the rest of our time together, and although he'd given me a little kiss on the cheek when we parted, I'd been aching to be kissed again the way we'd kissed in the car the previous day. It was my own fault, of course – it had been me who'd asked to take things slowly. Now I was wondering if it would be better if we didn't even see each other. I might be making a success of my little business here in Crickleford, but if I thought I could make a success of a relationship with a man, I was surely fooling myself.

CHAPTER TWENTY-FOUR

For the rest of the day I moped around at home, unable even to make the effort to play with Holly. I listened to Lauren and Jon chatting and laughing together in the kitchen as they cooked the Sunday dinner, and instead of the happy glow it normally gave me to be included in this nice little family, it gave me an ache in my chest that made me want to cry.

More than ever, I missed *my* family. I missed my sister. I wanted my mum to hug me, my dad to call me his little girl. Why had I messed everything up so badly with them? Would I ever be able to go back home? Would I ever have a life like Kate's, a proper life with my own home, a man who loved me, children of my own? It seemed so unlikely, all I could envisage was a future of being an eternal lodger in other people's homes, growing old and embittered, talking to myself and resenting other people's happiness.

'Snap out of it, girl!' I told myself crossly. Skulking around feeling sorry for myself wasn't going to change anything. Juliet was dozing on my bed, and I lifted her onto my lap, immediately feeling the wave of contentment that comes with stroking a cat, the warmth of her fur and the pleasant rumble of her purrs. 'You still love me, don't you, sweetie?' I said. 'You might not, though, if you really knew me.'

I pulled the photo of Albert out from under my pillow. It was already creased and dog-eared from so much handling.

'This is my own cat, my lovely Albert,' I told Juliet, showing her the picture. She looked at it and meowed, finally making me smile. 'You might have been friends if you could have met him,' I said. 'But I ... I left him behind. I did something very bad, you see, and I had to run away and leave poor Albert behind.' I wiped a tear from my eye. No use crying about it. It was all my own fault, after all.

Juliet yawned and stretched, and jumped down from my lap.

'You're right,' I said. 'It's time to stop moping around and go and help with the dinner.'

She meowed again and padded to the bedroom door, looking back at me to make sure I was following. I'd always known cats were empathetic. It felt like she'd understood every word I was saying.

The next day I felt better. I was back with Sugar, the pedigree Burmese, for the week. She'd become one of my favourites and I always enjoyed spending time with her. It was a pity I couldn't say the same for her owner's husband. Because of the occasion when he'd let Sugar outside and failed to close the door of her run, Vanya now wanted me to stay overnight at the house. And the fact that I was sleeping just along the hall from him seemed to give Rob the incentive to pester me as much as he possibly could.

'Please don't do that, Rob,' I told him as he leant over my shoulder, breathing into my ear, while I was dishing up Sugar's food.

'I thought you liked it,' he responded huskily. 'How about this?' He snaked his arm around my waist, giving me a squeeze that made me jump and spill some of Sugar's food on the worktop.

'I've already told you: if you keep on like this, I'll have to tell Vanya,' I said. I was trying to sound firm, but his nearness, his physicality and overt masculinity, were getting to me. Of course, it wasn't him I wanted, but the memory of being in the car with Matt, his fingers caressing my neck, his lips against mine, was still achingly fresh in my mind and my loneliness were intensifying the need for someone to hold me.

'Just leave me alone,' I said shakily, pushing him away.

But of course, he wasn't going to take no for an answer. For the next couple of days it was exhausting trying to keep him at arm's length. I found myself thinking that if he hadn't been married, and more particularly if his wife hadn't been my best client, I'd probably have given in by now, just to stave off the feeling of emptiness. But as it was, he was strictly off limits.

I tried to make sure I never ventured outside of my guest bedroom in my pyjamas, and took to carrying Sugar around the house as much as possible. He couldn't do much to me while she was in my arms – if he'd made me drop her, Vanya would never have forgiven him. It would probably have been a far worse crime, in her eyes, than any kind of infidelity on his part.

'Us girls have to stick together, don't we,' I whispered to Sugar, and she looked back at me adoringly, answering me with her usual loud meow, as if to say that being carried around like royalty was exactly what she deserved and expected.

But on the last day, needing the bathroom early in the morning, I opened my bedroom door to find Rob waiting outside, blocking my way.

'Excuse me,' I said quietly, but instead of moving away, he jostled me back into the room.

'Come on, Emma,' he said, holding my arm firmly as I stumbled backwards towards the bed. 'You know you want to.'

The reality of the situation banished any lingering temptation from my mind.

'No!' I said firmly. 'I *don't*. Let me go!'

He was holding me close to him, fumbling with my dressing gown, one hand already finding its way inside. I was losing my balance and knew that if I fell onto the bed he'd take it as a green light.

'Stop it!' I shouted, stumbling and putting my hand behind me on the bed to stop myself from falling – but he completely ignored me, using his strength to force me backwards while his hands continued to pull my dressing gown open. As I struggled frantically, turning my face away from him to stop him slobbering over me, I caught sight of the jewellery box on the dressing table. Instantly the memories came rushing back to me. That last day in New York. The shock, the violence, the angry words. The heavy jewellery box being thrown across the bedroom, finding its target. The screams. The hasty escape from the apartment, never to return.

'Get *off* me!' I gasped now. Rob was laughing at me, and he seemed to have no intention of stopping. I pushed at his chest as hard as I could. How could this be happening? How was it that men seemed to end up treating me so badly? He was on the point of overpowering me now and I knew there was no alternative. I slapped him, hard, across the face. He reared back in surprise, and I pushed him again, making him lose his balance just enough for me to get away from him. I grabbed my clothes off the end of the bed and ran to the bathroom to get dressed, locking myself in.

When I came back out, the house was in silence. His car was gone. I breathed a sigh of relief, but I was still badly shaken. Despite my anger with Rob, I found myself wondering if it was partly my own fault, whether it had been too obvious to him, despite how often I'd rejected his advances, that I was longing to be held and loved by someone. I would never actually have given in to his demands – he was a nasty piece of work, after all, who treated his beautiful wife with complete indifference. And not only was he married, he was about twice my age. He wasn't even my type, I thought miserably, thinking of Shane's dark good looks, Matt's beautiful brown eyes.

I fed Sugar, and was just about to take her out for some exercise in her outdoor run when I heard Rob's car pull up on the drive again. I stood in the kitchen, holding Sugar, breathing hard. I couldn't face him – I'd slapped him, hard, and although he'd deserved it, I had no idea how to handle being in his company for the rest of the day. Putting Sugar back down, I closed the door to keep her safely in the kitchen, and made a bolt for it out of the back door, and round the side of the house, where I waited until I heard Rob closing the front door behind him, and then ran down the road as fast as I could. Vanya would be home later, and I'd have to tell her what had happened. She might not want to use me any more – after all, she couldn't trust her husband around me now. I'd lose my best client, thanks to him, but it was only right that she knew what he'd done.

I wandered the streets for a while, trying to calm down, and this time when I found myself facing Moor View Lane, I decided that, despite what I now knew about Bilberry Cottage belonging to Matt, I still really wanted to have another look at it. I stood, sighing, outside the gate as usual, staring up at the windows. I half expected to see Matt inside; if I had done, I would have waved to him this time and perhaps he'd even have invited me in to look around. *Don't be stupid, Emma*, I chastised myself for the thought. How would that help? It was bad enough being in love with the outside of the cottage and knowing it was already taken, without giving myself the chance to fall in love with the interior too! Wasn't it enough of a problem, knowing that I'd fallen in love with Matt himself?

The thought made me actually gasp out loud. Love? Was that what it was? Had it really taken a run-in with that dirty old man Rob to make me realise how strongly I felt about Matt? I was pretty sure he wouldn't be feeling the same way about me. Wanting a relationship with someone was one thing, but *love* ... I doubted that had been part of the plan! I leant against the cottage gate and took a few deep breaths. I'd promised myself never to get into this kind of situation again, after Shane, and yet here I was only – what, six, seven months later? – in danger of making the same mistake all over again.

But it's not the same, is it, a little voice inside me pointed out. *He's nothing like Shane.*

'Maybe not, but he's a bloody journalist!' I said out loud to the little voice, and a lady walking past on the other side of the road turned to stare at me. Talking to myself now – they'd be locking me up soon if I wasn't careful.

Anyway there was no sign of Matt in the cottage, which perhaps was just as well. So I strolled back into town, feeling no better, worrying about Matt and whether we could ever be more than friends, worrying about Rob and whether Sugar was safe with him for another hour or so until Vanya arrived home. Should I go back and check, or would he just pounce on me again, despite the slapped face? I didn't feel like going home, so I just kept walking, back and forth beside the river, up and down Fore Street, up Castle Hill and back down again, trying desperately, but failing, to take my mind off the situation. As I approached The Star pub, I realised it was past their opening time, and I suddenly decided that what I really needed was a drink. I pushed open the door and stumbled, blinking, in the darkness of the bar after the bright sunshine outside. There were one or two men sitting on stools at the bar, but I ignored them as I ordered my glass of white wine and waited while the barmaid poured it.

'Well, if it isn't our little pet sitter,' a voice close to me said. A very unpleasant voice. I turned to look at him. He was a huge man, too large to sit properly on a bar stool so he was just balancing his backside against it and leaning on the bar. He looked around the bar, appearing to make sure he had everyone's attention, before giving a nasty, sneering laugh and going on, loudly: 'What have you been up to, then, as if I didn't know?'

'What do you mean?' I stammered, thinking immediately of Rob. Not that it was *me* who should feel any shame about the incident, I reminded myself. Even if I *had* given him the wrong impression, even if he had mistakenly thought I'd be up for more than a cup of tea with him – it was still his fault, his shameful behaviour. 'I've been looking after pets, of course,' I said more firmly, 'like I've always done.'

'That's not what I've heard,' he said. 'Not what I've heard from my friend and fellow parish councillor, Mr Rob

Montgomery.' He leered at me again. 'He tells a *very* interesting story about you, miss. Very interesting indeed!'

'Well, he's lying,' I said, my face on fire. 'So please stop embarrassing me in front of all these people—' I turned to see how many people were, in fact, sitting at tables in the darkness of the shabby little bar, their ears no doubt flapping in anticipation of some tasty morsels of gossip. I knew only too well that just the fact that I'd stayed at the house overnight while Vanya was away, would be enough to make people's tongues wag around here. Fortunately the pub wasn't too busy, but ...

And then I saw him. Already getting to his feet, coming over to join me, a puzzled expression on his face. I knew he'd been listening. I guessed he was going to tell this rude, fat man to leave me alone – but he'd be too late. The councillor's mouth was already open, he'd already begun to speak and I knew what he was going to say.

'I can assure you Rob Montgomery doesn't lie,' the fat man said calmly. 'And what he tells me—'

Matt was beside me now, giving me an uncertain half-smile, his eyebrows raised as if to ask if everything was OK. I looked at that dear, kind face, those beautiful brown eyes, those soft lips that had kissed mine so tenderly, and I wanted to cry. Without even realising it, I'd fallen in love with this man, and now I was going to lose him. Because what this horrible parish councillor was about to repeat was bound to be a completely untrue version of this morning's events that he'd heard from Rob. To save his own pride, to put me down and feel that he'd got even with me, Rob would probably have made it sound like I was a tease, who'd lured him, encouraged him, right up till the last minute and then rejected him – or worse, he might have pretended I'd been happy to participate. How would Matt feel if he heard stuff like that about me? And would he believe me if I tried to tell him what really happened? Or would it just turn him off me for good – just as I'd started to realise how very much he mattered to me?

CHAPTER TWENTY-FIVE

I closed my eyes, wishing the sticky, dirty floor of the pub would open up and swallow me before Matt had to hear what was coming. The fat parish councillor had been leering at me, his face red and sweaty with obvious delight about what he was about to divulge.

'What my good friend Rob Montgomery tells me,' he repeated slowly, for the benefit of everyone listening, 'is that you haven't always looked after people's pets at all. You had a very different past altogether, and it only took a little bit of research on the internet for Rob to find out about it. It seems you're hiding yourself away here to avoid some kind of scandal. It wouldn't surprise me if you're actually wanted by the police.' He nodded triumphantly, looking around the bar at the other customers, who had all fallen silent. 'You've told everyone here in Crickleford a pack of lies, haven't you, young lady? You're not who you say you are at all.'

I opened my eyes again in shock as soon as I realised what he was saying – that he wasn't, after all, telling some fictitious version of the incident earlier that day when Rob had assaulted me. This was even worse. If Rob had somehow managed to find out my real identity, there'd be hordes of journalists here anytime now. And Matt, of course, would know why I'd never come clean to him about my background. Was I going to lose him now, just as surely as if he'd heard some invented licentious story about me and Rob Montgomery? I looked around the bar, my heart thumping. Everyone was staring at me. Some people were frozen in surprise, drinks halfway to their mouths, others were whispering together behind their hands. And Matt … I hardly dared to look at Matt. When I finally did, it was the hurt in his eyes that made me catch my breath.

'It's not true,' I said, in a trembling voice. 'Rob's just spreading lies about me.'

The fat councillor snorted and shook his head dismissively. It was enough to make me angry, and the anger took over from my shock and fear, and gave my voice the strength it needed.

'And I can tell you exactly *why* he's doing it,' I said much more loudly, turning to look around the pub, 'if anybody's interested.' People looked down into their drinks now, embarrassed, but quite obviously still agog. 'He didn't get what he wanted from me, OK? Yes, that's right. I refused to sleep with him, and when he tried a bit too hard to … well, let's say to *persuade* me … I slapped him round the face.' There was a gasp from a woman at a nearby table. 'He knows I'm going to tell his wife,' I went on, 'and he's obviously feeling humiliated. That's why he's trying to make *me* look bad.' I paused for breath, glanced at Matt again, and went on quietly, 'I'm not wanted by the police. If anyone thinks I've run away from something, they're right – I ran away from a disastrous relationship. That's not so very unusual, is it?'

'Well, that's not what Rob told me,' the councillor spluttered. 'He says you've come here from New York, and you used to be—'

'Shut up,' Matt said quietly, stepping in front of me to square up to him. 'You heard her. Your so-called friend assaulted her. Nobody gives a monkey's what he said – or what you say, either. So why don't you just get your fat arse off that stool and bugger off before I land one right on your fat nose?'

There was a background of sniggering in the bar now.

'You threatening me, boy?' the big man said, lumbering to his feet and glaring at Matt, their noses almost touching.

'That's enough!' the barmaid shouted. 'Get out – go on, both of you. And you,' she added to me. 'I don't want any trouble in here. I'll call Fred down from upstairs if you're not gone in ten seconds.'

We didn't stay to find out who Fred was. Matt grabbed my hand and tugged me towards the pub door. The parish councillor came swaggering after us, still snorting with disgust, but he'd obviously had too many drinks to manage his huge girth unaided, and he was falling over chairs and tables and swearing loudly as the door closed behind us.

We walked in silence, Matt still holding my hand, striding quickly away from the town centre.

'Where are we going?' I gasped. 'Slow down!'

'Sorry.' He stopped, turning to look down at me, his face still etched with pain. 'I'm just ...' He ran a hand across his face. 'I just don't know what I'm supposed to make of all that, Emma. Is it true? Have you been lying to me? Lying to everyone?'

I felt a shiver run through me. Of course I'd been lying. I'd done nothing *but* lie, hadn't I, since I'd been living here. From the lies I told Lauren about having a job in a care home and having been ill, to the stories I made up for Mary about my house burning down and my parents looking after refugees – to say nothing of pretending I liked reading Shakespeare. Even stupid wannabe burglar Josh had been on the receiving end of my lies – that I'd got a daughter called Holly, that I had a boyfriend back in Loughton – and no doubt he'd spread them around the town. For God's sake, Matt himself seemed to have been taken in by my lie about being a fitness fanatic.

'I just kind-of bent the truth a little bit occasionally,' I admitted miserably, 'because I didn't want people probing into my background.'

'I know *that*,' he said dismissively, making me stare up at him in surprise. 'I've heard the rumours about the house fire and everything, Emma,' he went on, with a shrug. 'Probably everyone has. I just ignored them. I realised you were just trying to keep people from being too inquisitive. But *that* – what that guy in the pub was saying—'

'I told you,' I said quickly. 'Rob Montgomery's saying things to spite me. He tried it on with me, and—'

'I don't doubt that. Everyone around here knows what he's like. I'm sure his wife knows too. But all that stuff about finding out about you on the internet ... about you coming here from New York ... you looked scared out of your life when that great slob of a man came out with all that stuff about you.' He started to walk again, more slowly this time. We turned down beside the River Crickle and he led me to a bench on the riverbank, where we both sat staring at the water. 'What did he mean?' he asked quietly without looking at me. 'You said you'd escaped from a bad relationship, but what else haven't you told me? You're not *really* in trouble with the police, are you? Tell me, Emma! I ... look, I care about you, you must realise that. I really do. I just want to help you.'

I sighed and leant my head against his shoulder. It was so tempting to finally unburden myself to someone, but whether I followed my heart or my head, I couldn't help thinking it would end up badly. Matt was a journalist, and how many times had he already told me he was looking for a big story? If I told him, and he acted on it, I'd definitely have to leave Crickleford and start all over again. Even if he didn't, it seemed likely I'd have to do that, now the story seemed to be leaking out. But when I looked up at Matt, and saw the tender expression on his face, the concern in his eyes, I convinced myself he wouldn't betray me; he wouldn't let me down. I'd found someone who really cared about me – hadn't I?

'No, I'm not in trouble with the police,' I told him, letting out a long breath. 'But there is stuff I haven't told you. I'm sorry. It's actually the press that are after me.'

He frowned. 'The press? Who? Why?'

'Because ... look, I haven't exactly been lying about who I am. My name *is* Emma Nightingale, but that's not what I've been called for the past five or six years.'

'I don't understand. Were you married? Did you take his name – this guy you've run away from?'

'No. While I was living in New York I changed my name to Candice.'

'Candice? That's ...' He stopped, his eyes suddenly widening. 'Candice *Nightingale*?' He actually jumped to his feet, as if I'd touched him with something hot. 'You're telling me you're ... oh my God, I can see it now! How did I not see it? I knew your hair colour wasn't natural, obviously, but ... oh my *God*! You're a redhead. You're Candice bloody Nightingale.' He stood there, shaking his head at me, while I waited, holding my breath. In the scheme of things, it was so unimportant it was ridiculous, but I wanted to ask him how he knew my Cheeky Chestnut hair colour wasn't natural.

'Sit down,' I pleaded eventually. 'You're making me feel nervous.'

He sat, a few inches away from me this time, as if my notoriety might be something infectious that he wanted to avoid catching.

'So you're a famous American "It" girl, and somehow you've wound up here in a little Devon town, hiding from the paparazzi,' he said. He shook his head again. 'Why do I suddenly feel like someone's going to jump out at me from a hedge in a minute and say *April Fool*?'

'It's *August*,' I reminded him in a little voice. 'And I'm not American – I don't have an accent, do I? I'm English. I only went to America because of Shane.'

'Shane Blue.' He nodded slowly. 'Of course. I read somewhere that you'd broken up with him. Then you ... just seemed to disappear from the scene.' He looked at me sideways. 'So the paparazzi were looking for you, and you were here, all the time, masquerading as an innocent little pet sitter.'

'I didn't want anyone to know. I had to leave my parents' home because they were invaded by reporters. I just wanted to lie low until all the fuss died down.'

'So you did lie to me.' He gave me a hurt look. 'Even though you must have known how I felt about you. Couldn't you have trusted me, Emma?' He gave a little snort of a laugh. 'Is that even what I should be calling you? Would you prefer me to say *Candice*?'

'No! I don't want to be known as that stupid, vacuous person I used to be, ever again! You said I was famous, but what for? Just for being Shane Blue's girlfriend, that's all! Just for wearing the latest fashions and hanging out with the rich, stupid, New York set who followed Shane around. I was a nobody, Matt! I hate that person I was. I ...' I grimaced and shrugged. 'I hate the life I had back then,' I went on. 'I've never been as happy as I've been since I came here to Crickleford and started working as a pet sitter. Since I met you,' I added very quietly.

'Even though I'm a journalist?' he retorted, and I shrunk away from his harsh tone. 'I suppose you realise I could blow your cover now? I could write a story about who you really are, and where you're hiding out. It'd be a game-changer for my career.'

'You wouldn't do that,' I whispered.

'Why shouldn't I? I could have understood, Emma, if you'd confided in me from the start. But I can't believe the way you've kept up this act with all the people in the town who've tried to be kind to you, to accept you and pay you for looking after their pets. How could you do that? I suppose even Lauren and Jon don't know who you really are?'

I shook my head miserably, and he exhaled in disgust and got up, brushing down his jeans as if he'd been contaminated.

'I haven't been hurting anyone,' I said, in a pathetic little voice. 'I just didn't want to be found. If you knew how it all ended, in New York ...'

'But I don't, do I, because you didn't want to tell me. Oh, keep your silly little secrets to yourself, *Candice* – frankly I'm not interested. I'm not the kind of journalist who goes running after two-penny celebrities. I'd rather write about something more worthwhile. To be honest, the story I wrote about the pretend haunting of Castle Hill

House was a far better one than anything I could write about you.'

'Don't go, Matt,' I said as he turned away. 'Please, let me explain. I'm sorry if I've hurt you. I never meant to. I love you!'

For a moment he hesitated, looking up at the sky. I counted the seconds. If I got to ten and he was still standing there with his back to me, I'd go after him. I'd put my arms round him and he'd say he forgave me. I'd tell him everything. He'd understand. But I only got to seven before he walked away.

PART 4

NO PLACE LIKE HOME

CHAPTER TWENTY-SIX

August turned into September. The little town seemed to be steaming with heat under a thundery sky. Even when it rained, which it was now doing with increasing regularity, it was still hot. When the sun did occasionally deign to come out, everywhere was bathed in a strange orange light, the wet streets shining, the trees dripping and glistening as they began to shed their leaves. Then the dark clouds rolled in again, blotting out the sun's light but not its intensity.

The weather matched my mood. I found it difficult to sleep, and just as difficult to drag myself out of bed in the mornings. I walked my doggie charges, petted my cats and stroked an assortment of rabbits and guinea pigs while I waited, expecting at any moment for everything to blow up in my face – for the story of whatever Rob had found out to spread through the town and for the paparazzi to descend on Crickleford in their hordes. For the people to turn against me, for Lauren to throw me out. For Matt to turn up ... or not. For a headline in the local paper to scream out his betrayal of me ... or not. The uncertainty, the fear, the heartbreak about Matt was as unsettling and exhausting as the stormy atmosphere. Over and over again I replayed the scene where Matt walked away from me. Should I have run after him? Would it have made any difference? Was he really angry enough with me now to ruin my new life by exposing me?

I was terrified of meeting up with Rob again; but I knew I had to talk to Vanya, in case Rob had told her an untrue version of what had happened on my last day at the house. Finally, a week later, I walked down that long driveway twice, my legs shaking, turning back and hiding behind a tree, before finally plucking up the courage to go on, all the way to the house, where I breathed a sigh of relief when I saw the garage door up, Rob's car missing. As far as I knew, he never walked anywhere other than on the treadmill at his gym.

'Don't worry, Emma,' Vanya said, as soon as she opened the front door to me. 'He's not here. And he won't be back.'

I blinked at her. How did she know I was frightened of confronting her husband? And why would he not be back?

She smiled at me as she put the kettle on, but the smile didn't quite reach her eyes. I wasn't sure what was coming. Sugar was winding herself around my legs and I bent to stroke her absent-mindedly, my thoughts chasing each other through my exhausted brain.

'Last week, on your last day here,' she began, still in the same calm but strange tone, 'you left in a hurry, didn't you?'

'I know. I'm sorry. I know I left Sugar on her own with Rob but it was only for a little while, and ...' I dipped my head and took a deep breath. 'I need to tell you—'

'It's OK.' Her voice, suddenly, was surprisingly, unusually, gentle. I looked up. She put a hand on my arm. 'I know what happened. Well, I guessed. You left your pyjamas and dressing gown in the bathroom, the bed unmade. That's not like you. And I knew you wouldn't just walk out and leave Sugar.'

'He came into my room!' I blurted out. 'He ... he tried ...' I gulped and stood up straighter. It wasn't my fault, I reminded myself. 'He assaulted me, Vanya. I promise I didn't do anything to encourage him.'

'Of course you didn't, love. And believe me, you're not telling me anything I don't already know. I just wish you'd confided in me sooner so that I could have stopped him straight away.' She smiled thinly as I stared at her in surprise. 'Oh, he's always stopped his little dalliances as soon as I've found out. He pretends he stays with me because he doesn't want the scandal of a divorce, but really it's because I keep him in the style to which he's become accustomed. He acts as if he's something rather special, doesn't he, but in fact he's been out of work for the past two years. He pretends to be working from home, while he's ... well, occasionally doing stuff for the parish council, but mostly just messing about playing computer games or looking at porn sites.'

I gasped, and she gave a little dry laugh. 'I suppose you thought he was working. Well, he was a sociology teacher at the local high school, until he got dismissed for texting dirty messages to a couple of the sixth form girls. All this posturing and posing about being on the parish council is just him trying to make himself feel better, because in fact he's just a rather pathetic piece of nothing.'

I flinched, but she carried on, apparently completely self-possessed.

'I asked him why you'd gone, of course – and he stuttered and stammered and tried to hide his red cheek—'

'Oh yes!' I put my hand to my mouth. 'I'm sorry. I slapped him ... it was the only way I could stop him.'

'Don't apologise.' She gave me that strange smile again. 'He had it coming to him. Anyway,' she added, picking up her beloved cat and laying her face against the fur of Sugar's little head, 'he'll have to get over his fear of the *scandal*, because I'd already decided to divorce him, after the last girl he tried to seduce. It's quite funny really,' she said. 'He hardly ever succeeds, you know. He's not unattractive, so I suppose it's his obnoxious personality that puts the girls off.'

'I ... I don't know what to say,' I stuttered. I took a deep breath. 'So has he actually moved out now, then?'

'I've thrown him out, yes. Not before time. So you won't let this put you off from coming back to look after Sugar again, will you? I couldn't bear to lose you now. Sugar adores you.'

It was bizarre. She seemed completely unmoved about splitting from her husband – her only concern was making sure her cat would be looked after while she was jetting off to her high-powered business meetings.

I finished my coffee, trying to calm my fears. It didn't appear that Rob had told Vanya anything else about me. So presumably, apart from his horrible friend in the pub (who I hoped had been too drunk to remember any of it), he hadn't told anyone else either. He may not have had time, of course. Apparently she'd given him his marching orders as soon as she guessed what had happened, and was so keen to see the back of him that she was paying the rent on a room in a B&B down in Paignton for him, as a temporary measure. He'd left the parish council and left Crickleford, hopefully for good. He surely had enough to worry about now – with a bit of luck, he wouldn't have time to think any more about me and my background.

'I'll have to sell the house, of course,' Vanya was saying sadly, looking around her. 'He'll expect a share. But I've got my furry baby. That's all that matters.'

I admired Vanya, even if I was still a little nervous of her. She was a strong, dignified woman, as well as being beautiful and successful. I understood why she wouldn't want someone like Rob in her life any more. But although I loved Sugar too, I knew her obsession with the little cat was somewhat over the top. Which had come first – his despicable behaviour, or her rejection of him in favour of her furry baby? Well, as long as I never had to see Rob again, I didn't care.

I'd still heard nothing from Matt, though. For the first few days, I'd gone to hang around outside the *Chronicle* office – despite my fear that even as I stood there, he'd be inside, writing his killer story about my past. I'd even stood for ages outside the door to his flat a couple of times, too scared to ring the bell. And of course I'd walked up and down Moor View Lane staring at Bilberry Cottage. I was desperate to see him, to try to sort things out between us, but conversely worried about what I might find out. If he *was* writing that story, it would mean the end – for me in Crickleford, and for us. That's if we hadn't already reached the end.

Eventually I stopped looking for him. Apart from the fact that I was beginning to feel like a stalker, as time passed and I looked fearfully every week at the *Crickleford Chronicle* as well as all the national papers, half expecting to see my name splashed across the front page, I began to believe that he wasn't going to betray me. Perhaps he might still care about me – might have forgiven me? But even if he had, I sensed that I'd have to give him time. How much time did he need? I spent hours lying on my bed, staring at my photo of Albert, cuddling Romeo and Juliet, or listening to little Holly's chatter and keeping her amused by playing games with her, waiting for my fear of exposure to recede, waiting for my heart to mend.

There'd been a big change at Primrose Cottage. Holly had started school at the beginning of the month, and told anyone who'd listened that she was now a *big, grown-up girl*. It seemed that there was now nothing her parents or I could tell her that she didn't already know.

'Mrs Jones told us that,' she'd say about anything we discussed. 'Mrs Jones knows about everything.' Her heroine was apparently pretty, funny and kind, as well as being the fount of all knowledge.

'I feel a bit redundant,' Lauren admitted one day while we were preparing dinner together. 'I'm going to talk to the school about working afternoons as well as mornings. I'm lucky that my job fits in perfectly with Holly, of course, and I do love being a teaching assistant. But sometimes I wish I could do something more challenging – perhaps train to be a proper teacher. But ...' She shrugged apologetically. 'I only got a handful of GCSEs at school.'

'You're cleverer than me, then,' I said. 'I didn't get any.'

'But you *are* clever, Emma,' she said, looking at me in surprise. 'Look at you! Running your own business! Being clever isn't *just* about passing exams.'

Wasn't it? I pondered this when I lay in my bed that night, as usual trying to get to sleep while the rain pattered against the window of my little room. Nobody had ever said that to me before. I'd always just assumed I was as stupid as the other children at my school said I was.

One afternoon, I was upstairs in my bedroom, with some time to spare before going back to the house where I'd been looking after a rather annoying budgie, when I heard Mary arriving with her latest supply of books for Lauren. Lauren put on the kettle and started to chat with Mary about her career ambitions. I lay back against my pillow and, as I often did, pulled out the photo of Albert and the letter that had come with it. It was now completely creased up, and I smoothed it out and stared at it, making my eyes go funny in the vain hope that some of the words might suddenly jump out at me, when there was a tap at the door and Mary appeared in the doorway.

'Sorry to disturb you, Emma,' she said. 'Lauren's just making a cup of tea and she asked me to call you to see if you wanted one. You obviously didn't hear me.'

'I must have been miles away,' I said, quickly folding up the letter.

'You looked as if you were struggling with that,' she commented, giving me a smile.

'Oh ... um, yes. I don't seem to be able to see properly.' I screwed up my eyes and blinked a couple of times. 'It must be my eyesight.'

'Do you need glasses?' She looked concerned now. 'You should make an appointment at the optician's.'

'Oh, no, it's just ... I've probably got something in one of my eyes.' I rubbed them and started to put the letter back under my pillow. 'I'll come downstairs – thanks, Mary. I'd love a cuppa before I go out.'

She stayed in the doorway, looking at me with her head on one side.

'Would you like me to read that for you?' she asked, slightly cautiously. 'I mean – not if it's anything personal, obviously. Or would you prefer to wait until your eye's better?'

I hesitated for a minute. 'It's very spidery writing,' I said. 'Really small and cramped. I doubt you'd be able to—'

'Well, I'll have a try, anyway, if you'd like me to. I've got very strong glasses,' she said with another smile.

Again I hesitated, the letter in my hand. It would be so good to know who had Albert, wouldn't it? But what else might be in the letter? What might it give away to Mary about me, about my identity and my past life?

As if she could read my mind, she said quietly as she sat down next to me on the bed: 'Whatever is in the letter, Emma, it'll be between you and me, I promise. And anyway, I'll forget it as soon as I've read it. My memory is shocking.'

I laughed. 'I'm sure it's not. But – well, OK, then. It might take a while for my eye to get better, I suppose. Thank you.'

As soon as she started to read, I realised how odd it must seem.

'The letter's dated July,' she said, looking up at me. Of course it was. I'd had it under my pillow for two months, unable to read a word of it.

'Oh, it must have got lost in the post,' I said.

'Mm, must have done. Well, anyway – it starts: *Dear Candice ...*' She paused, glancing at me again.

I shrugged awkwardly. 'It's a nickname some of my old friends used to call me.'

'Oh, I see. So: *Dear Candice, You don't know me, but ...*'

By now my face was burning. I felt like grabbing the letter back from Mary but she was already ploughing on, peering through her glasses at the scratchy writing.

' ... my name is Dorothy Mason and I'm Shane's grand-mother. Not that I'm proud of that fact, and by now I'm sure you'll agree with me. He's never behaved like a grandson to me, nor has he ever been a good son to his mom or his dad. In fact the rest of the world may worship him but as far as I'm concerned, he's a disgrace to the family. I'm only sorry you ever got involved with him, dear, as I'm sure you're a good girl at heart and it's just dreadful the way he treated you. I'm not so old that I don't see what's going on in the papers, all the scandals with the other women. It made my blood boil, I just wanted to disown him. I'm glad you've finally got away.

'Anyway I wanted to let you know I have your dear Albert here with me in my little home. I'd like to say my grandson redeemed himself a little by giving him to me, but in fact it wasn't him. It was a girl with a strange name – Emerald or Esme? – you'll have to forgive my bad memory. Very thin. Very shrill voice. Too much make-up. She brought Albert round to me in a basket and said you'd run off and left him, and although she didn't want him herself, she didn't want to leave him with Shane. It seemed that despite appearances, she had a little bit of common sense at least, because she realised Shane would have neglected him. She told me he'd said he didn't care what happened to Albert, but had suggested giving him to me because "old women always like cats". Wouldn't have hurt him to bring the cat to me himself, would it, as I haven't seen him for years on end, but that's Shane for you.

'I told the thin girl to get me your address so that I could let you know Albert is being well looked after. He's a beautiful cat. I realise you couldn't have taken him with you, and I'm *sure you're missing him. Please be assured I will love him on your behalf. I enclose a picture, so that you can see for yourself how well he has settled down with me. I hope your life will be happier from now on. With kindest regards, Dorothy.'*

Mary stopped reading but continued to look down at the letter for a moment. She folded it, handed it back to me and finally looked up at me. The tears were trickling down my face – I'd stopped caring about what on earth Mary might be thinking about the contents of the letter, almost as soon as she'd started reading, I was so overcome with emotion by everything Shane's grandmother had said.

'She seems like a lovely lady,' Mary said quietly. She didn't ask any questions. She just looked into my eyes and went on, very gently, 'I'm sure she'll love your cat and care for him.'

'Are you two coming down for your tea?' Lauren yelled up the stairs. 'It's getting cold!'

'Coming!' Mary called back.

'Thank you, Mary,' I said, wiping the tears from my eyes.

'You're welcome,' she said – and she folded me into her arms and gave me a hug. I wondered how much she'd understood, how much she might have guessed. But before I could say any more, she added softly: 'And of course, I've already forgotten every word.'

CHAPTER TWENTY-SEVEN

Of course, I knew exactly who the 'thin girl' in Dorothy's letter was. Ezmerelda. Otherwise known as The Bitch. Ezmerelda Jewell, my so-called best friend in New York, the top model who took me under her wing when we first arrived *on the scene*, fresh from our simpler, happier life at the beach house in California. Ezmerelda, who shaped me, dressed me, taught me which hairdresser, which manicurist, which orthodontist to see, which personal trainer and which stylist to use, to change myself into a celebrity, to make myself more like her. To be slim, beautiful and perfect, like all the other girls we hung around with. To lose my real identity and become a clone of herself, so that she'd hang on my arm and tell people we were *sisters*.

Why had I gone along with it? What the hell had I seen in her? Perhaps it was the *sister* thing. I missed Kate, of course, and maybe I was trying to replace her, but if so, it was an insult to my sweet, clever, funny twin. Ezmerelda was nothing special: she was just as stupid as me, but she had poise and self-belief; she believed in her own hype. She thought she was wonderful, and sadly all the hangers-on, the press, the fashionistas and society photographers, as well as the world's impressionable teenagers and even plenty of older women who should have known better, agreed with her. She pretended to love me like a sister, but it was all fake. Under the veneer, behind all the make-up and the plastic surgery and the fixed smile, she was just a nasty, deceitful, cheating bitch.

So what was I to make of Dorothy's letter? It seemed Ezmerelda had, at least, had the decency to take Albert to someone who'd care for him – which was more than I could imagine Shane doing. And although she hadn't responded to my email asking if she knew where Albert was, at least she'd given Dorothy my home address. Was she actually too *ashamed* to contact me herself? Was it even possible for her to feel shame? Or was it just that she couldn't be bothered?

I shook my head, trying to stop the flow of my thoughts. I didn't want to be grateful to Ezmerelda, even though I was relieved beyond measure to know that Albert was being loved and looked after by someone who seemed so kind. I didn't want to feel indebted to that bitch or even waste another minute of my life thinking about her. She was in my past, like Shane, and that was bad enough. I wasn't that person any more.

I tried instead to work hard at my new little business. I reasoned that if I filled my diary with as much pet sitting as possible and concentrated on building my reputation, I could block out the ache in my heart that overcame me whenever I thought about Matt. With each passing day, I felt more confident that there hadn't been any spread of gossip about me from Rob Montgomery or his horrible friend, and even if my personal life had taken a nose dive, Primrose Pets was going from strength to strength. The budgie I'd been looking after had been a first for me, and as with the Koi carp, I'd made the mistake of thinking it was going to be a doddle.

'I only have to top up his seed and water containers every day, clean his cage and let him have a fly around in the room,' I'd told Lauren on my first morning. 'It's a bit like looking after the hamster – I really can't see what I'm being paid for.'

'Well, you will have to be careful while he's flying free,' she'd warned me, 'or it *will* be like the hamster, you'll lose him somewhere! For goodness' sake close all the doors and windows! They haven't got an open fireplace, I hope?'

'They said not. But surely he'll just flutter onto a window-sill or something and sit there till I've cleaned the cage?'

'I don't know. I've never had a budgie.'

'Can we have one, Mummy? Please?' Holly squealed immediately, and Lauren and I both laughed when she added, with a puzzled expression, 'What *is* a budgie?'

'It's a bird, Holly,' I explained. 'And you really can't have pet birds when you've got cats.'

'Oh.' She sighed. 'Poor Romeo and Juliet. They might have liked it.'

'I'm sure they would,' Lauren whispered to me, and we'd exchanged a grin. I then had to pacify Holly by drawing a picture of a budgie for her to colour in.

But like the fish and the hamster, in fact, looking after the budgie had turned out to be different from what I'd expected. On my first day with him, I let myself into the house (a neat, modern little town house in a cul de sac just off Fore Street) and immediately stood stock still with fright in the hallway. Somebody was shouting at me from the living room.

'Close the bloody door, can't you? Born in a field, were you? Close the bloody door!'

As I'd already closed the front door, I walked cautiously towards the door of the room where the voice was coming from, wondering if a cantankerous aged relative had been left at home without anyone telling me. But there was nobody in there, apart from the green and yellow budgie sitting in his cage staring back at me.

'Close the bloody door,' he squawked again. *'Are you stupid or something?'*

'No, I'm not!' I retorted, which was a bit ridiculous really. 'I've closed the bloody door, thank you. And I presume you're Sid.'

I hadn't met my new charge before starting the job, and although I obviously knew budgies talked, I was completely unprepared for this little chap's vocabulary. From shouting about the bloody door, he went on, during the course of the two weeks I was in charge of him, to insult me at every turn, with the worst language possible. I was shocked to realise that his very pleasant, very pregnant owner, or perhaps her partner, must talk like this at home on a regular basis for the bird to have learnt these words. They must have a very volatile relationship!

Not only that, but far from sitting quietly on a windowsill while I cleaned his cage, Sid used his free-flying exercise breaks to dive-bomb me from the top of the curtain rail, before finally settling on my head and castigating me for being lazy, stupid and *up to no good again*. When, on the last day, he accused me of wanting *sex, sex, sex, morning, noon and night*, I didn't know how on earth I was going to face his owners on their return. But to my amazement he fell silent as soon as they arrived back at the house, turning his face to the wall as if he were sulking.

'How has he been?' the young woman, Karen, asked eagerly, peering into his cage.

'Er ... Fine. But very vocal,' I said, without looking at her.

'Oh my God.' She covered her face with her hands. 'Has he been swearing?'

'Yes. Quite a lot, actually.'

'I should have warned you. I'm so sorry. I thought we'd cured him of it. We just tell him to be quiet and behave himself, and he stops.'

'That's right,' her partner Mike agreed, looking chastened. 'In fact he hasn't done it now for ages, so it didn't occur to us that he'd start again, with someone new looking after him. I'm so sorry. How embarrassing.' He glanced at me and added: 'Oh! I hope you didn't think he learnt all that stuff from *us*!'

'Well ...'

'Oh, no!' Karen exclaimed. 'We got him from our next-door neighbour. The old boy had passed away and his wife didn't want to keep Sid, so we said we'd have him. He's only been with us for a few months. We've been trying to teach him nicer things. Nursery rhymes,' she added with a happy smile, 'so he can recite them when our baby's born.'

'*Jack and Jill went up the hill*,' Sid squawked obediently. '*Polly put the kettle on.*'

'Good boy, Sid!' Karen crooned.

I stared at him in disbelief. The little bugger! He was masquerading as a *good boy* to keep on their right side, but I knew better! I could only hope he'd behave once they had a baby in the house.

Of course, I'd been keeping Lauren amused during the whole of this time, recounting the things that naughty bird had been saying to me, and she fell about laughing when I told her the outcome of Karen and Mike's return.

'You could write a book about all your experiences with these pets, Emma,' she said.

I knew it was just something people said. But I shook my head, giving a sour little laugh.

'Me? No. I can barely even write an email.'

But laughing about Sid the budgie had cheered me up a bit. And I'd been looking after a cat during the same two weeks – I'd finally got myself better organised, always booking in another pet, if I could, during any period when I had a fish, bird, hamster, tortoise or anything similar that wasn't a full-time job. As a result, Primrose Pets was becoming a more efficient business as well as a more lucrative one. Of course, there were odd days when I had no work, but this happened less and less often as I was getting repeat bookings and recommendations from satisfied customers.

During the daytime, I'd been correct in hoping that being busy would help, a little, to keep my mind off Matt. Unfortunately, though, during the long nights when I tossed and turned and struggled to sleep, that didn't help at all.

The rain had continued for most of September, but finally at least the stifling and sticky heat gave way to cooler, showery weather, and most people – those who weren't nursing a broken heart anyway – were apparently sleeping better. Feeling aimless and lost on an odd Saturday afternoon between pet bookings, I ducked into Ye Olde Tea Shoppe to escape yet another heavy downpour and was surprised to see Annie's son Kieran helping out again.

'I thought you'd have gone back to uni by now,' I said as he served me my tea and a scone.

'I go back on Monday,' he said and, registering my surprise, explained that university terms were shorter than school ones.

'Oh, I see.' I nodded. 'I didn't realise. My sister's the only person I know who's been to university, and I was ... well, I'd left home before she started. So I don't know the first thing about it.'

He looked at me in silence for a moment, and just as I was about to go and find a table, he said quietly: 'I hope I'm not interfering, Emma. But when we talked about it before, you said something about not going to university because you didn't pass any exams at school.'

I shrugged. I remembered the conversation, and how it had made me feel. It had taken me right back to my days as a school failure and dropout, and afterwards I wished I'd never talked to him about it.

'That wasn't why I didn't go,' I said, turning away. 'I didn't go because I had other things to do.'

Other things? I thought to myself. *What, like moving into Shane's bedsit?* I didn't want to talk to Kieran about university and exams now, any more than I had a few weeks before. I started to walk away, but tripped on the uneven olde floore. My tea sploshed over the rim of the cup onto the nearest table, and my scone flipped off its plate to land, buttered side down, in the middle of a lady's salad. She jumped, knocking over her glass of white wine, and her friend squealed as the cold wine splashed down her blouse, getting up so quickly she knocked over her chair.

'Oh, no – I'm so sorry!' I said, and grabbed a handful of serviettes off the counter, which I threw at the lady who was flapping at her wet blouse with her hands. 'Sorry!' I said again to the other lady, trying at the same time to mop up the spilt tea and grab my scone off her lettuce. It dripped mayonnaise across the table and I felt my face burning up with embarrassment. By now everyone was staring at me and all I really wanted to do was run back up ye olde steppes and never come back.

'Now then, my lovelies, don't 'ee be gettin' all in a miff about that,' Annie boomed, appearing from behind me with a damp cloth, with which she quickly wiped down the table, and a small towel, which she offered to the lady with the damp blouse. 'All sorted now? Sit theself down, maid,' she added to me with a wink, 'and we'll make 'ee another leak of tea direckly.'

Kieran was already holding out a plate to me, with a fresh scone on it, but he caught my eye and began to smile.

'Maybe I should bring it over to your table for you,' he said, and suddenly we both started to laugh.

'I'm not usually so clumsy,' I said, apologising again to the two ladies whose lunch I'd bombarded. Well, I might as well stay after all, now everyone seemed to have got over the disturbance and gone back to their conversations. I found myself another table and allowed Kieran to bring my replacement scone and tea to me.

'I hope I didn't upset you,' he said quietly, 'talking about university just then. I only mentioned it because, well, I tried to tell you before, and I don't think you heard me: *I* didn't pass my exams at school, either. But of course,' he added quickly, 'I'm not saying university is for everyone, and if you didn't go because you had more important things to do with your life, then good for you.'

I blinked at him. More important? I'd done *nothing* with my life, had I? Not until I came here to Crickleford and started my pet-sitting business, anyway.

'How come you went to uni without passing any exams?' I asked him.

'I took an access course for mature students.' He shrugged. 'I messed around at school, completely flunked year ten, and left without any qualifications. When I started to regret it, I went to Mrs Field for help. Do you know her?'

I shook my head.

'She was the special educational needs co-ordinator at our school. Now she's a private tutor.'

'Oh. Do you mean Mary?' I guessed suddenly.

'Mary Field, that's right. She tutored me for a couple of years and got me through the access course. She's a brilliant teacher.' He grinned. 'Still, you don't need to know about that. You've done really well from what I hear, running a successful business on your own. Why would you need anyone's help? I just wanted to explain.'

I smiled. 'Thanks, Kieran. And I'm sorry if I was a bit abrupt. I suppose I do get defensive when people talk about exams and careers and stuff. I never wanted to go to uni. But I do feel pretty stupid sometimes.'

'I'm sure you're not. Exams aren't everything, Emma, but if you ever felt like you needed to, well, catch up on anything you didn't learn at school, I'd definitely recommend Mrs Field.'

'I do know Mary as it happens. She helped me set up my business. And she gave me some Shakespeare plays to read, but ...' I hesitated, 'I couldn't quite get into them.'

'No, I can imagine!' He laughed. 'I couldn't, either.' He started to walk back to the counter, then paused and turned back to me again, adding quietly, 'Look, if you'd like to chat some more when I'm back from Bristol again at Christmas, maybe we could get together for a drink or something?'

'Oh! Thank you, but I'm ... um ... well, I might be already seeing someone ... I don't know ... It's kind of complicated ...' I stammered, blushing.

But he was laughing and shaking his head.

'Sorry, Emma, I didn't mean that the way it came out. I wasn't asking you out. I've got a partner, actually, but he's quite OK about me having girl friends!'

'Oh!' I said again, and started laughing too. 'My fault for misunderstanding. Of course I'd like us to be friends, Kieran, and yes, it'd be nice to chat some more to you when you're home again.'

'Great. And don't be offended. I'm sure I'd fancy you if I wasn't gay,' he joked.

His mum was by now bellowing across the room for him to stop gossiping and give her a hand, and I watched him go with a smile. It was true that I could do with a friend. Lauren was lovely, but I missed having someone of my own age to chat to. Most of all I missed Matt, of course, but I had no idea what was going to happen to our relationship even if we did get back together. Perhaps a gay best friend was the nicest thing I could hope for with a man, after all.

CHAPTER TWENTY-EIGHT

But as it happened, I finally bumped into Matt the very next day. I was outside the newsagent's, trying to persuade the little dog I was now looking after to wait nicely, with her lead fastened to the metal ring that was fitted to the shop wall for that purpose, while I popped in to buy a paper. Trixie, who was a sweet but quite feisty spaniel-terrier cross, wasn't having any of it. As soon as she realised she'd been tethered, she let rip with a barrage of complaint, first barking and then beginning to howl.

'That's what happens when you try to leave someone,' said a familiar voice behind me. I stopped, my hand still on Trixie's head where I'd been stroking her to try to calm her down. 'They howl with misery.'

For a moment I fought with the urge to turn round and throw myself into his arms. I'd missed him so much, longed so much for the moment when I might see him again. But then my disappointment with him came to the forefront. Why hadn't he sought me out to try to patch things up?

Did he really not want anything to do with me, now he knew who I was? Was that how little he cared about me?

'*I* didn't leave *you*,' I pointed out without turning round.

'Well, theoretically you did,' he said mildly. There was a smile in his voice. 'Emma left me, and out of the blue this bird called Candice turned up.'

At this, I swung round to face him.

'I'm *not* Candice!' I said, so vehemently that Trixie started whimpering and a lady passing by in the street nearly jumped out of her skin. 'I'm not that person any more. I never was, not really – I was just putting on an act.'

'OK.' He was speaking gently now, the smile gone, his eyes looking into mine. 'I believe you. I do. And like I said in my messages, I'm sorry for walking away, but it was a shock—'

'Messages? What messages?'

He frowned. 'You didn't get them? Are you serious? But I texted you the very next morning. And then I emailed you – twice. When you didn't reply, I assumed you didn't want any more to do with me, after the way I reacted. I thought about coming to see you, but I didn't want to behave like a stalker—'

'Nor did I! I wanted to come and see *you*, but I thought, if you needed more time to come to terms with it, or if you ... just didn't want to see me any more ...' I tailed off and looked down at the ground, 'I'd just have to accept it.'

'But, Emma, I wouldn't do that. It was a shock, yes, finding out you weren't who you said you were. I was hurt

309

that you didn't confide in me, and I suppose I felt a bit stupid. But it didn't change how I feel about you. I said all that in my messages.'

'Oh.' So he *did* still want to see me? 'I'm sorry. I thought I'd told you, we've got a lousy phone signal at home. And the internet's just as bad.'

'Don't you check your phone for messages when you're in town? Or use your mobile data to pick up your emails? What sort of a celebrity are you?'

I glanced at him, cross again for a minute, but I could see he was teasing. And I deserved it.

'No, I don't check very often. It suits me to be out of touch, these days, and anyway I don't usually get many messages.' I sighed. 'I've just got out of the habit of looking at my phone much at all, since I left the States. I didn't have any friends over there. Not real friends. My sister texts me occasionally, but ... well, she's busy. And my family are better off without me in their lives.'

'I'm sure that's not true,' he said, looking at me sadly.

'Oh, it is. I upset everyone by coming home, and ruined their Christmas. The paparazzi swarmed round the house. We had to sit with the curtains drawn and all the phones off, whispering to each other. And now someone's found out about me, I suppose it's going to happen all over again here in Crickleford.'

'It might happen eventually, I suppose. But not yet.'

I frowned. 'I don't know how you can say that. I mean ... well, maybe *you* won't say anything, or write anything, but—'

310

'Of course I won't!' He paused and glanced at Trixie, who was now whining constantly to be set free. 'Look, were you on an urgent mission to get something from the shop here? I could hold onto the dog while you go in. Otherwise, shall we take a stroll, and I'll explain what I've been doing while I was waiting for you to reply to my messages.'

I laughed. 'OK. Trixie would appreciate the walk, I'm sure. I was only going to get a paper. To check ... you know. I've been checking every day whether they've found out where I am.'

'Well, so have I,' he said matter-of-factly as I unclipped Trixie from the shop wall and we started to walk down towards the river. 'And not just in the papers.' He smiled at my puzzled expression. 'I'll explain as we walk.'

It was raining again, but we both had cagoules on and, to be honest, I was so happy to be in Matt's company once more, I don't think I'd have cared if it had hailed, snowed and thundered. Trixie calmed down as soon as I gave her the length of her lead and she could run ahead of us, sniffing into the undergrowth and stopping from time to time to investigate other dogs' scents. Then she trotted on again, her tail wagging with excitement, her ears erect with the expectation of more interesting smells around the next corner.

'She's cute,' Matt said. 'Whose is she?'

'Oh, she belongs to Mr and Mrs Barton. They live in that house there, right on the riverbank,' I said, pointing it out to him.

We both stood in silence, staring at it. Although I'd been there in the house, looking after Trixie, for a few days now, I hadn't actually seen it from this angle before. The river was literally lapping at its walls.

'I didn't realise it was quite *so* close to the river,' I said.

'It isn't, normally. The river's as high as I've ever seen it. Haven't you heard everyone around here talking about it? They're really worried it's going to burst its banks. It's only happened once before, in my lifetime anyway, but we've had so much rain ...' He shook his head. 'If it carries on like this, I think we could be in trouble, seriously.'

Sure enough, looking at the riverbank now I could see what he meant. Some of the houses nearest the river already had sandbags against their doors. I gazed around me, trying to gauge the impact it could have if the river did overflow its banks.

'This part of town could all end up underwater, couldn't it?' I said quietly. 'Should I have sandbags for the Bartons' house?'

'Yes, probably, as a precaution anyway.' He glanced again at the house, and added, 'I'll get you some for their doors. And if it becomes necessary, I'll help you move as much as possible upstairs.'

I wondered if I ought to try to contact them. They were a nice couple who'd just retired, and were celebrating with a holiday of a lifetime in Australia. It would be awful to worry them if it turned out to be unnecessary, and even worse if they felt they ought to interrupt their holiday and come rushing home.

'Wait for a few more days,' Matt advised, 'and we'll see whether the water level's starting to go down.'

I loved that he was saying *we*. I instinctively tucked my free arm through his, and then started to worry that I might be assuming too much. But he smiled down at me and squeezed my hand.

'So: tell me,' I said as we strolled on through the rain. 'What did you mean when you said you were checking, but not just in the papers?'

'Well, there's no point being a journalist if you can't use your contacts, is there?' He laughed. 'It was easy enough to pretend I was interested in doing a follow-up on the whole "Where is Candice Nightingale?" story. You know, poor little unknown hack working on a minor Devon newspaper that nobody in Fleet Street's even heard of – desperate to get his big break by fair means or foul—' He stopped, giving me an ironic little smile. 'Which isn't far from the truth. Apart from the fact that, in this instance, I can't actually do it.' He squeezed my hand again. 'My heart won't let me.'

My own heart skipped a beat at this. Did he mean it? Could I believe him? I so much wanted to trust him. Surely, if he'd been going to betray me, he'd have done it by now?

'So who are they? Your contacts?' I said.

'People working on the nationals. Some I knew from uni. Others I met down here in the West Country, and they've since moved on to better jobs in London. Some I've spoken to in the past about various stories. I managed to find at least one contact on every national paper.' He turned to smile at me now. There was rain dripping off his hood onto his nose, running down his cheeks so that he looked as if he was crying. 'Nobody has heard anything yet, Emma. If anyone here in Crickleford had spread the word, these guys would have got hold of it by now, trust me. Your trail's gone cold.'

When I hugged him, the rain from his cagoule dripped over my face.

'Thank you,' I breathed. 'I can't tell you what a relief that is.'

'You're welcome. And the more time that goes by without anyone from the national press finding you, the less interesting the story will be. You'll be old news.'

There was a time, back when Shane made his first recording, when I actually longed for fame. Now, being *old news* sounded like a dream come true. What Matt had done for me – not only passing up the chance of writing my story himself, but also talking to all those contacts on the national press, without revealing what he knew – was surely proof in itself that I could trust him. He didn't hate me for my deception. We were still friends. But was that all we were? It had been me, after all, who'd said I didn't want a relationship with him. And even though, in a fit of panic when I thought I was losing him after the revelation about my identity, I'd told him I loved him, how could I blame him if he didn't want to risk being hurt or rejected by me again? Rather than risk embarrassing or upsetting either of us, I decided to make do, for now, with just enjoying his company again.

We walked on, neither of us speaking for a while. And just as I was thinking that poor Trixie looked more like a drowned rat than a dog, and I ought to be getting her home, Matt suddenly turned to me and said:

'I should be apologising to you, anyway. I haven't been fair to you. I got all upset and self-righteous about you not telling me the truth about yourself, but I've kept stuff about myself back from you as well. The thing is, it's hard for me to talk about it. Most people around here don't realise who *I* am, either.'

I looked at him through the pouring rain. His lovely, warm brown eyes were gazing into mine, his expression apologetic.

'What?' I said. 'Are you really someone famous as well? Who? Don't tell me – Ed Sheeran? Have you coloured your hair too?'

He laughed softly. 'No. Nothing like that. I can't compete with your fame. I was born Matteo Sorrentino.'

'Oh. I presumed Matt was short for Matthew. But, of course, your father's Italian.'

'Yes. And the thing is, most people around here don't realise I'm the little boy who used to come and stay here in Crickleford every summer with my grandparents. They were my mum's parents, so their surname was different, you see. My parents split up when I was just a baby, and my dad went back to Italy.' He sighed. 'I never knew him, and I don't want to. He walked out on my mum and that was enough for me. Mum and I lived in Plymouth, and life was hard for her. She had to work long hours to support us both. But eventually she got together with this new guy, Jim, who frankly wasn't interested in me. So I used to spend every school holiday here in Crickleford with Nan and Grandad, as well as a lot of weekends. I was ...'

He swallowed a couple of times and shook his head, struggling to go on. I took hold of his hand and held it tight. 'I was very close to them,' he said eventually. 'I just lived for my holidays here in Crickleford. But as I grew up, I came less often. My own life took over – you know how it is with teenagers. I went away to uni, then I got my first job, on the *Plymouth Daily News*. Mum was too involved with Jim – he's my stepdad now – to occupy herself very much with her own parents. And I ... I just didn't keep in touch with them as much as I should've done. I kept thinking there'd be plenty of time left. And as it turned out there wasn't.'

'They both died around the same time?' I said.

He nodded. 'Together, actually. In an accident. They were crossing Fore Street from behind a parked van. Hand in hand. A motorbike came out of nowhere and knocked them both flying. They died at the scene, still holding hands.' He took a deep breath and swallowed again before going on. 'That should have been a comfort to me, the fact that they died together, but I just couldn't ... still can't, really ... get the horror of it out of my mind. The pictures in my head, of the accident, never leave me, whether I'm awake or asleep. It haunts me. They left Bilberry Cottage to me because they knew I loved it so much. It used to be home to me, my second home. But after they'd gone, I couldn't bear going inside.'

'I'm not surprised. That's awful, Matt – so sad. But they wanted you to have it!'

'I know. And I owed it to them to come back here to Crickleford and sort it out. It was a bit neglected – they were old, after all. They'd let it get shabby and run down. The kitchen and bathroom were old-fashioned, the décor was dated, the paintwork was peeling and the carpets, well, they were thin and worn. I got the job on the *Chronicle* here, but I knew the only way I could cope with living in the cottage would be to change it completely first. I had everything ripped out and started making it into a completely different place.'

'Yes. I can understand that.' I nodded. 'And I'm sure your grandparents would have understood, too.' I wanted to add that I'd love to see what he'd done to the interior of the cottage, to tell him how much I'd always loved it from the outside, but obviously it wasn't the right time. 'I'm glad you've been able to tell me about them, Matt.'

'I think it's helped me, talking to you about it,' he replied quietly. 'It ... just feels right, somehow, sharing it with you.'

I thought about hugging him again, but we were both dripping rain from everywhere by now and the result would probably have drowned us both. Trixie broke the moment by shaking herself thoroughly, rainwater flying in every direction.

'I'd better take her home,' I said, and we turned to walk back into town. I gave Matt a quick kiss on the cheek as we said goodbye.

'I'll bring some sandbags round to the Bartons' house later, OK?' he said. 'And I'll be watching the river level, but call me if you need my help. You know, using that strange new-fangled invention you seem to have forgotten about – the mobile phone?'

I was laughing to myself as I took Trixie back indoors. It was good to feel happy again, good to know that there didn't seem to be any paparazzi after me yet, and even better to know Matt was OK with me. But when I looked out of the back window of the Bartons' house, I felt a tremor of anxiety, watching the river water lapping at the banks. I gave Trixie a warm bath and rubbed her dry, enjoying the little dog's woofs of pleasure, then dried myself off and changed into the spare clothes I now always brought with me on any canine assignments. A couple of hours later Matt returned with the sandbags, as promised, and showed me how to pile them against the Bartons' doors when I left for the evening.

'It's stopped raining for now,' he said, looking up at the sky. 'So keep your fingers crossed.'

As I walked home in the twilight, I noticed that several other people living in this area near the river were also outside putting sandbags up against their houses. They nodded to me as I walked past.

'Can't be too careful,' one man said, shaking his head ruefully. 'It's fifty-fifty now whether she goes back down or whether she bursts her banks. A couple of dry days and we'll be all right. Otherwise, we're stuffed.'

I shivered with anxiety again. I really didn't like the idea of trying to move all the Bartons' furniture upstairs, even with Matt's help. But if that was what was needed, that was what we'd have to do.

CHAPTER TWENTY-NINE

The next day I left early in the morning to return to the Bartons' house. I was greeted by a shocking sight: the water had risen higher still, and people in the neighbouring house called out to me that they were moving furniture and belongings upstairs.

'It b'aint gonna leave off,' the woman said, nodding at the black clouds above us. 'She be 'bout to burst 'er banks,' she added, nodding at the river now. And I had to agree, it did now look pretty certain.

I went indoors to pacify little Trixie, who was up at the window barking her head off in alarm at the goings-on next door, and after feeding her I called Matt. He'd apparently just woken up, but agreed straight away to come over and help me move things upstairs.

'We'll do the best we can,' he said when he arrived, glancing anxiously at the level of the river. 'But we won't be able to get everything up the stairs. Have you called the Bartons?'

'I've sent them a text message. I didn't really want to alarm them into cutting their holiday short. I said not to worry, that I'd got help and we'd deal with it.'

'I imagine they'll come home anyway.' He sighed. 'OK. Let's get this rug rolled up and carried upstairs first, and take the chairs. The sofa might have to stay here and take its chances.'

By mid-morning, with the help of the neighbours, we'd moved everything we possibly could. The rain had eased off and we all felt as if we were balancing on a knife edge – almost as if one single raindrop more, at any moment, and half the town would be flooded. I took little Trixie for a long walk to calm her down – she was so upset by all the disruption to her home – then after I'd eaten my lunch I went home to Primrose Cottage briefly to collect some night things and to tell Lauren that I was going to stay overnight at the Bartons' house.

'Are you sure?' she said. 'You could always bring the poor little dog here for the night, instead.'

'Thanks, but no, I think I ought to be there. Trixie and I will stay upstairs, obviously, and at least I can raise the alarm if the worst does happen.'

I spent most of the afternoon and evening pacing the top floor of the house uneasily, watching the sky for signs of further rain, and even when I did eventually go to bed in the spare room, with Trixie snuggled down in her bed next to me, it was a long time before I managed to get to sleep. Sometime after midnight, I was woken up by the sound of torrential rain against the window. I jumped out of bed and rushed to look out. To my horror, instead of the familiar landscape of the riverside path, the road, the fields and moorland beyond the river, all I could see was water: it was as if a huge lake was slowly swallowing everything up.

'Come on, Trixie,' I said, lifting the little dog into my arms. 'Time to leave.'

I'd taken the precaution of bringing welly boots back with me, as well as warm clothes and my raincoat, and I put these on quickly over my pyjamas. I knew the flood-water could rise fast and there was no time to lose. Carrying Trixie, with my bag on my back, I carefully opened the back door of the house, which was further from the river, so the water in the back garden wasn't too deep yet. I closed the door after me and piled sandbags against it, before banging on the neighbours' windows to wake them up and get them moving.

I switched on the torch on my phone. Water swirled all around me, but a little distance down the lane, the humped shape of Crickle Bridge rose out of the flood, and I headed for this now, ringing Matt's number as I went. I could see something moving on top of the bridge, but it wasn't until I was closer that I realised what it was.

'Matt, there's a *pony* stranded on the bridge,' I said. 'The road up to the bridge is flooded. He can't get away.'

Within a few minutes Matt had joined me and we waded through the flood to the top of the bridge.

'It's one of the little Dartmoor ponies,' he said quietly. 'He'd be able to swim, for sure, but he's probably too frightened. Have you called the police?'

'Yes.' I peered through the darkness at the pony. He was keeping his distance from us, shying away, whinnying and pawing at the air in distress. 'They said they're coming, and they've called a vet.'

'Well done. I'm glad you got out in time,' he said. We were both keeping our voices very low so as not to panic the pony any further. 'The flood's rising fast. Are the neighbours out of their house?'

'Yes. They were following me. They're going to her sister's place up the hill, on the other side of town.'

We waited in silence, beginning to shiver in the cold. Trixie wriggled in my arms, giving little whimpers.

'I should have taken her back to Primrose Cottage,' I said, stroking her gently to try to calm her.

'I don't think you're going to be able to, now,' Matt said. 'The water's getting deeper by the minute.'

I looked around me with the help of my torch. He was right. We'd soon be completely stranded. But a few minutes later we heard the sound of a motor dinghy approaching. Flashlights cut a huge arc of light across the water and someone called through a loudhailer to let us know they were the police. I wanted to shout at them to shut up, but it was too late. The pony, now completely terrified, tossed his head, galloped from one side of the bridge to the other and before either of us could reach him to try to stop him, jumped into the water and began to thrash around wildly.

'He's out of his depth!' I cried.

'He can swim,' Matt reminded me. 'But he's frightened and panicking.'

A burly man in a yellow waterproof was already jumping out of the boat.

'I'm the RSPCA vet,' he said. 'Somebody called about a pony in distress. I take it that's him in the water there?'

'Yes, he's terrified,' I said. 'But he's only just gone into the water.'

'OK.' A police officer had now jumped out of the boat to join us. 'We've got ropes. We'll take it from here, love.'

'Be careful,' I begged him as he and his colleague began to uncoil their rope in the dinghy. 'He's so frightened.'

'There's a lifebelt here!' Matt shouted, grabbing it from the parapet of the bridge.

'Thanks, that could help,' the vet said. 'Don't worry. Once we've got him secured, we'll tow him to dry ground. We'll take it very slowly so he can swim behind us. I'll sedate him and check him over once we've got him to safety, before we release him back onto the moor.'

'Can we help at all?' Matt asked.

'Thanks, but you've been a great help already just by calling us so quickly,' one of the police officers said. 'You need to get yourselves to safety now. We've got room for one of you on board – we'll take you to the rescue centre that's been set up at the church hall. There'll be another police dinghy here soon and we'll keep doing return trips till you're all safe.'

'All?' I said, looking around me. To my surprise, a small group of people had joined us on the brow of the bridge while the drama with the pony was going on. They were all cold and shivering from wading through the floodwater to get to this high point. As the pony's rescue operation was swiftly completed, the small crowd clapped and cheered. The pony had tired himself out and once the lifebelt was around him, with the rope attached to the boat, he seemed to have given in and was lying in the water looking nervous but subdued.

'Get yerself and the little dog into the boat, my lovely,' said a voice from the darkness. It was Pete, the Bartons' next-door neighbour. 'You've done a good job there, getting help for the pony. You're shivering cold.'

'No, I'm fine,' I said. 'I can wait for the next boat.' I nodded at Pete's wife, Shirley, who was looking pale and shocked in the light of my torch. 'You go with them, Shirley. But if you could take Trixie with you, that would be good.'

She climbed into the boat, and I passed her the trembling little dog. We watched the dinghy head back out across the water, the pony floating behind.

'He'll have a story to tell the other ponies when he gets back up on the moor,' Matt joked, putting an arm around me to try to warm me up.

'I'm glad he's going to be OK.'

'Me too. And hopefully everyone on this side of town has got out of their houses.' Matt was shining his flashlight across the water. 'What about that house on the other bank?' he said suddenly.

'An elderly couple live there,' Pete said. 'Stan and Madge. I tried calling them earlier but didn't get a reply. They must've got out. It's all in darkness.'

'Everywhere's in darkness now,' Matt pointed out. 'The power's gone.'

I stared at the little house on what was normally the other bank of the river. The floodwater was halfway up the door now and lapping at the downstairs windowsills.

'Matt, what if they *haven't* got out?' I said. 'If they're elderly, they might not have been able to. And they probably couldn't get to their phone, if it's downstairs.'

'We'll have to tell the police when they get here in the next boat.'

'We don't know how long they'll be!' I protested.

'The Bartons have got a rowing boat,' Pete suddenly remembered. 'It's tied up, under a tarpaulin, round the other side of their house – if it hasn't been washed away by the force of the water. You probably didn't notice it.'

Without thinking, I started to run across the bridge back towards the Bartons' house, but before I'd even got back down to street level again, the water was pouring in over the tops of my wellies, making it almost impossible to wade. I stopped and pulled them off, abandoning them in the water, and ploughed on in my socks.

'Emma! Wait!' Matt was shouting after me. 'For God's sake, wait for the police. Pete's calling them again now, he'll get them to prioritise that house.'

'No, we have to try, at least,' I called back. 'The police will have so many calls to emergencies tonight, it could be ages before they get there.'

I waded on through the water, waving my torchlight in front of me. I was soon aware of Matt following behind me. As he caught up with me he grabbed my arm and helped to pull me along. The water was well past my knees now, and I was getting tired, but when we reached the Bartons' house I was relieved to find the boat, exactly where Pete had described, around the side of the house. The tarpaulin had gone but because the boat was now afloat on the floodwater, it was gradually working loose from its lashings, making it easier to set free. The oars were clipped to hooks high up on the wall, and within moments Matt had lifted them down. With a grunt of exertion he helped me into the boat before heaving himself over the side.

'Can you row?' I asked, a little late in the day.

'I sure can. Grandad taught me, here on the river, when I was a little lad.' He started to pull out into the flood. 'The water's still rising,' he added anxiously. 'You should have stayed on the bridge.'

'Maybe I should have, but it's too late now. Besides, if those people are still in their house and need our help, it'll be easier with two of us.'

As we approached the other side of the river, I could see a flickering light in an upstairs window.

'A candle,' I said. 'They *are* still there, Matt. Thank God, at least it looks like they're upstairs.'

'Yes. Look at the water level here,' Matt said worriedly. 'It's up past their downstairs windows now.'

We were having to raise our voices to hear each other above the swish of the oars and the rushing of the river, and just then the window where we'd seen the candlelight was flung open and a man's face appeared at the window.

'Is someone out there?' he called in a frail voice. 'Help us, please, we can't get out.'

'OK, sir,' Matt yelled. 'We're just outside your house now. Is there anyone downstairs?'

'No,' the man shouted back. 'It's just me and my wife, but she's got bad arthritis. The water's halfway up the stairs and rising.'

'We can't get them out, on our own, if the wife's disabled,' I said to Matt.

'I know.' He nodded grimly. He was having enough trouble keeping the boat steady. 'Don't worry, sir,' he yelled back regardless. 'We're here to help.'

I'd been constantly trying to call 999, and finally I got through. 'The police rescue boat is on its way now,' I said to Matt as I ended the call. 'They're saying we shouldn't be out here.'

'OK, but at least we can try to reassure these poor people,' he retorted. 'Look how fast the water's rising.'

It was only a few more minutes before a voice came through a loudhailer out of the darkness:

'Stay out of the water. Get to high ground. The floodwater is rising. I repeat, get out of the water and up to high ground.'

Our little boat began to rock dangerously in the swell of the approaching rescue craft.

'The police are here now, sir,' Matt called up to the man at the window.

'Thank God!' he said, his voice shaking. 'The water's reached upstairs now.'

'We ought to row back to shore,' I said. 'They're telling everyone to get to high ground. Look at the bridge!'

The small group of spectators on the Crickle Bridge – what we could still see of it, above the water – were now clinging to the parapet and looked like they wouldn't be able to wait much longer for their own rescue. Sure enough, the voice came through the loudspeaker again now:

'People on the bridge – stay where you are, another boat is on its way. I repeat, stay where you are. Do not attempt to swim through the floodwater!'

'Can we help?' Matt shouted as the police dinghy drew level with us.

'You shouldn't even be out here,' the nearest officer replied, shining his flashlight at us. 'But as you are, then yes, put these on—', he threw two lifejackets across to us, 'and perhaps you could help to keep our boat steady against the house with yours, while we bring these people out. We'd have been better off with the helicopter, but Crickleford's not the only town flooded tonight and we're all struggling with the number of callouts. Do you know how many people are in the building? Is everyone accounted for?'

The next twenty minutes or so were a precarious night-mare. We were cold and wet, but hung on for grim life to keep both boats steady while the trained officers, with their rescue equipment, helped the elderly couple into lifejackets and out through their window into the boat. Thankfully by now a further boat had turned up to rescue the people on the bridge.

'Thanks for your help,' one of the officers shouted to us finally when Stan and Madge were seated in the rocking rescue launch. 'We'll take this lady and gentleman to safety. I advise you to follow us. A rescue centre's been set up, higher up in town.'

We slowly followed them out of the flood, relief beginning to wash over me along with a sudden weariness. Suddenly I heard a familiar sound.

'There's a cat out here somewhere,' I said, staring around me with the flashlight. 'Oh, look, Matt, there it is, in that tree. The poor thing's terrified!'

Only the top of the tree was showing above the water, and the cat was clinging to one of the highest branches, yowling its head off. Matt pulled alongside the tree, grabbing hold of a branch to steady the boat while I stood up, wobbling dangerously, and reached up for the cat.

'It's that big tabby from the farm!' I exclaimed as I grabbed hold of him and fell back into the boat, his weight nearly overbalancing me.

'The one who bit you? Be careful, then, for God's sake.'

'But look at him, Matt!' I laughed. 'He's quaking with fear! What a difference from the spiteful thing he normally is!' I settled the frightened cat on my lap and to my amazement he nuzzled up against me, mewing gently now and hiding his head under my arm. 'All right, you daft thing,' I said, suddenly feeling sorry for him despite everything. Perhaps he just wasn't a very happy boy. Maybe the farmer was too busy to show him any affection and just left him to his own devices, roaming around town upsetting people.

Even in the animal world, I guessed bullies were mostly just unloved and unhappy.

By now we'd arrived at the part of the river where the banks were steeper, and the ground higher above sea level, as the town spread upwards towards the peak of Castle Hill. Matt must have been exhausted from rowing, but he moored carefully at the side of a footpath leading to the church hall, where the rescue centre had been set up, before helping me out with the tabby cat. There was a round of cheers as we stumbled, dripping water, into the warmth of the hall. It was the early hours of the morning, but it seemed nobody was asleep. Townspeople were out in their dozens to help those of us who'd been caught in the floods. Apparently the story of Matt and I borrowing the little rowing boat to try to help the elderly couple was already spreading like wildfire, to say nothing of the story of the pony's rescue – and when they saw the big cat I was carrying, who was now clinging to me like we were best friends, they all fell about laughing.

'That'll learn 'im,' Pete said. 'Mebbe he won't wander so far from home in future!'

People were handing out blankets and mugs of hot soup, and at the same time patting us on the back and calling us heroes.

'You've got your big story now,' I said to Matt after I'd handed my bedraggled and shivering feline charge over to someone with a towel and a carrying basket, who promised to call his owner. We sat huddled together under a blanket, sipping our soup.

'Yes, I suppose I have. But I've realised that's nowhere near as important to me as you are,' he said softly, leaning in to plant a gentle kiss on my forehead. 'When I saw you wade off that bridge into the floodwater back there, I nearly had a heart attack. You didn't stop to think about how dangerous it was.'

'No. Sorry, I suppose it was a bit stupid. But I'm still glad we were able to help.'

Remembering how the cold water had rushed around my legs, I started to shiver again despite the blanket and the soup.

'Come on,' he said, taking the mug out of my hands. 'You're chilled right through, aren't you? You're coming home with me. You can call Lauren on the way, to let her know you're safe. Will she pick the little dog up?'

Trixie, who'd been brought to the rescue centre on the first boat, had rushed over to bark around my legs as soon as she'd spotted us and was now desperately trying to climb onto my lap.

'Yes, I'm sure she will. Holly will love that!' I smiled. 'But why are we going back to your flat?'

'We're not. We're going to Bilberry Cottage,' he said quietly. 'It's about time I started calling it home.'

CHAPTER THIRTY

Before I could even register my surprise, Matt had grasped my hand and we stumbled, both of us half asleep on our feet, the short distance to Moor View Lane.

'Is the cottage finished now, then?' I asked as we turned into the lane. 'Ready for you to move in?'

'It's been pretty much finished for a while,' he admitted. 'I've just been ... too pathetic to move in, on my own. With the memories. My grandparents ...'

'I understand,' I murmured. 'If you'd rather I didn't—'

'No. You're coming in,' he said firmly. He opened the cottage gate, and then the front door for me. 'Come on, princess. Your palace awaits.'

I giggled. I was almost too exhausted to put one foot in front of the other, but nothing could diminish the thrill I was feeling at finally walking into my dream cottage. Matt switched on the light, and the cosy welcome of the cream-painted little hallway made me feel like I was settling into a soft, snugly duvet.

'Get out of those wet clothes,' he told me, steering me towards a pretty blue and white bathroom. 'Have a hot shower. There are towels in the airing cupboard, and I'll get you one of my jumpers and a pair of joggers. They might be a bit big,' he added with a grin. 'You see – I've even brought half my stuff round here, but I've still been putting off moving in.'

I didn't need telling twice about the shower. The relief of finally feeling warm and dry was wonderful. When I came out of the bathroom, Matt had changed into dry clothes too and was in the lounge, where a wood fire was now burning and candles had been arranged on every available surface.

'Sit here,' he said, urging me towards a cream sofa piled with cushions in different shades of red. I leant back and drank in my surroundings – the timbered ceiling, the oak-clad fireplace, the wooden floor scattered with thick cream and red rugs, the heavy deep crimson curtains – while Matt went around the room lighting all the candles.

'It's beautiful,' I breathed. 'It's just ... just exactly how I'd have done it myself.' How I'd imagined it, in fact, when I'd dreamed so often of living here. But I couldn't tell him that. It had been Matt's beloved grandparents' cottage, and although I knew he must still be finding it hard to live with the memories, it was his home, not mine.

He was in the corner of the room now, taking a bottle out of a cabinet and pouring two large glasses of red wine. He pressed a button on a controller and the room filled with soft music.

'This'll help you sleep,' he said, handing me one of the glasses of wine and sitting down next to me with the other. 'I've put a hot-water bottle in the bed for you. The main bedroom's at the back of the house – it overlooks the moor.' Just as I'd imagined, I thought with a sigh. 'It's as quiet as the grave here, so you won't be disturbed. You can sleep till midday if you like.'

'But what about you?' I blinked at him, wondering sleepily if he intended to share the bed with me, and if he did, whether I'd be able to stay awake long enough to enjoy it! 'You must be tired too,' I added.

'I'll sleep down here on the sofa. I haven't furnished the spare bedroom yet.'

'OK.' I couldn't rouse myself to argue – the warm bed in the quiet bedroom sounded so good.

He moved closer to me and put his arm around me. He was warm now, and smelt of coconut shampoo. I took a gulp of my wine and rested my head on his shoulder.

'You've been so good to me,' I muttered. 'If you knew what a terrible person I really was, you wouldn't—'

'Shush,' he said, laughing. 'You're not terrible, are you, just for living the high life in New York? Besides, you've said yourself, you're not that person any more.'

'But ...' I sighed, 'if you knew what I did, when I walked out on Shane ... why the press were so desperate to hear my side of the story ...'

He looked sadly into my eyes. 'Did he treat you badly?'

I nodded. I didn't really want to go into it, but surely I owed him more of an explanation.

'Not at first, of course. But when he got really famous, and all the girls were after him, the most beautiful girls on the New York scene throwing themselves at him – well, you can imagine. He lost interest in me. I knew there were other girls, of course—'

'He must have been mad,' Matt whispered.

I shrugged. 'But finding out he was sleeping with my so-called best friend was really hurtful. I should've left him then, but, stupidly, I still hung on.'

'Not much of a best friend, was she.'

'No. Ezmerelda bloody Jewell – everyone's favourite top model. I just call her The Bitch now, though,' I added with a faint grin.

Matt whistled. 'No wonder the paparazzi want your story.'

'Yes. Well, like I say, I didn't behave very well when I finally ended it.' I took a deep breath. Could I do this now? Finally share it with someone – the dreadful memory, to say nothing of the shame? 'I found him in bed – in *our* bed – and not just with Ezmerelda, despite him having promised me he wasn't seeing her any more. No, they were having a threesome. The other woman was another model. A really stuck-up cow who'd never liked me.'

I swallowed, closed my eyes, trying to dispel the image, burnt onto my brain, of the three of them, naked, looking back at me as I stood in our bedroom doorway staring at them in horror. Shane and Ezmerelda at least managed to look slightly guilty. Ezmerelda grabbed a sheet and tried to cover herself. Shane started babbling about it not being what it looked like – as if it could have been anything else. But *she*, the other one, Gloriana Glee, who'd always looked down her nose at me and was now getting slowly off the bed, flaunting her perfect naked body as she shimmied brazenly across the room to me – she was *sneering* at me.

'*Thanks for the loan of your boyfriend, darling,*' she drooled, trailing a finger down my arm. '*Hope we haven't messed up the bed too much. Oh – and you might need to buy some more of this.*' She picked up an empty bottle from my dressing table and showed it to me. '*I seem to have used it all up.*'

It was my favourite perfume. *Candice.* Yes, I'd even named myself after it. The stupid thing was that it wasn't a particularly expensive brand, but a small bottle of that perfume was the first birthday present Shane had ever bought me, back in the days when we had nothing. Nothing but each other. The fact that this … this nasty, gloating, *tramp* … had used it, *stolen* it, from my dressing table before getting into *my* bed with *my* Shane, suddenly galvanised me into the kind of action I'd never before imagined myself capable of. Screaming abuse at her, I grabbed her by the hair and, taking her by surprise, threw her back onto the bed. She'd existed on a diet of rocket and quinoa for so long, she couldn't have weighed more than seven stone.

'*Oh! You bitch!*' she bleated at me.

Bitch? *Me*? Huh! If they wanted to see bitchiness, I'd show them some. For a start, she was mistaken if she thought I'd only have one bottle of my favourite perfume. I opened my dressing table drawer, took out two more and marched over to the bed.

'*Like my perfume, do you?*' I yelled at her. '*Well, have some more!*'

I whipped the top off the first bottle and tipped it straight over her stupid head. As she screamed and covered her eyes, I repeated the process with Ezmerelda. Ignoring their yells, pleas and frantic scrambling to get off the bed and away from me, I turned back to the dressing table and grabbed everything I could get my hands on.

'*Need to brush your hair after all that activity? Here you go!*' – I threw a heavy onyx hairbrush at Gloriana's head. She ducked, it hit Ezmerelda square on the shoulder and I screeched with maniacal laughter.

'*Want to borrow some of my make-up, as well as my boyfriend – my* fiancé *actually, not that you'd care? OK, be my guest.*' The heavy cosmetics bag caught Gloriana full in the face and she gasped and fell back on the bed again.

'*Oh dear, do you want some nice cold cream to stop that stinging?*' – the jar hit its target with a satisfying thud and, as the girls both squealed and ducked and tripped over their knickers in their haste to dress, I squirted them with moisturiser, emptied hand cream over them, threw bottles of hair spray and tubes of gel, combs, nail varnish and even a mirror, until there was almost nothing left on the dressing table.

'*And as for you,*' I said, finally turning to Shane and laughing at the stunned expression on his face, '*you might as well have* this *back.*' I pulled off my engagement ring and chucked it at him. '*And why not have* all *this jewellery you've lavished on me over the years. Give it to your trollops! I don't want it! I don't want YOU!*' It took both hands for me to heave my jewellery box across the room. Shane didn't move fast enough, and although he shielded his face, he yelled in pain as the box made contact with his hands. It'd probably be a while before he picked up a guitar again. Shame.

'So now you know why the press *really* want to talk to me,' I finished. Matt had turned away. I wondered if he was disgusted with me. 'I suppose I'm lucky it's just the press, and not the police.'

To my surprise, there was a muffled snort in response. He looked back at me, his hand over his mouth.

'I'm sorry. I've been trying not to laugh. Come on, Emma, it's hilarious! So you might have hurt them a bit. Well, good for you! They deserved it! You came out of that situation with your head held high—'

'Actually, no. I crept out with my tail between my legs and nothing but my Hermès handbag, my passport and phone, and the clothes I stood up in,' I corrected him. 'And I sold the bag when I realised I had nothing else.' But I was smiling now. 'Aren't you shocked?'

'Not in the slightest. If anything, I love you even more for it.'

Love? He'd said the 'L' word, and didn't even flinch and correct himself? Still, I supposed people loved their friends, didn't they. There were different types of love. People loved their dogs, too, and their cats and rabbits, and hamsters and—

'I said I love you, Emma Nightingale,' he repeated, softly, stopping me in my litany of animals and making me stare at him in amazement. 'In case you didn't hear me.' And, drawing my face towards his, he began to kiss me. I felt myself melting. The warmth of the fire, the flickering light of the candles and the glow inside me from the red wine, were combining with the rush of pleasure from feeling Matt's arms around me and his lips on mine. Aerosmith were singing their hearts out in the background about not wanting to fall asleep, not wanting to miss a thing and suddenly the words seemed to be my words, saying exactly how I felt about this moment.

'Don't sleep down here on the sofa, Matt,' I whispered when we paused for breath. 'I don't want to be on my own upstairs. I'm not even tired any more.'

'Are you sure?' he whispered back.

In response I got to my feet, pulling him up with me. He enveloped me in his arms again and this time when our lips met it was with a fiercer passion. He began to lead me out of the lounge, still kissing me, both of us stumbling in our haste, and our unwillingness to let go of each other.

Suddenly, there was a burst of light from outside. We broke off, staring at each other, confused.

'Lightning?' I guessed.

Another flash. We were both standing stock still in the lounge doorway. Frowning, Matt walked back towards the window, where he'd only drawn the curtains halfway across. He pulled them completely open – and immediately we were both almost blinded by a volley of flashing lights. At the same time, someone began hammering on the front door.

'Candice!' a voice shouted.

And another: 'Come on, Candice, come out and talk to us!'

'Who's your new man, Candice? Does Shane know about it?' yelled another.

'Oh my *God*,' I gasped. 'Matt, they've found me.'

Shaking, I fell back onto the sofa, covering my face with the cushions ridiculously – as if cushions could hide me from the vultures outside.

'Quick, out the back way,' Matt said, pulling me to my feet again. 'Come *on*, Emma – *quickly*! Before they find their way round the side.'

He got hold of my raincoat and pushed my arms into it, grabbed a black woolly hat from a peg in the hall and pushed it firmly onto my head, and manhandled me through the kitchen to the back door. I was almost too shocked to move of my own accord, but when I heard a shout from outside – '*They're going out the back!*' – I jumped into action. Together we dived out of the cottage's back door into the darkness of the garden. I could still hear shouts of '*Candice! Come out and talk to us, Candice!*' from the street, and then I became aware of more voices joining in the hullabaloo.

'What's going on? What's all this noise about?'

'What's happened? Who's Candice?'

'Candice *Nightingale*!' I heard one of the reporters yell back. 'She's in there! She's been hiding out here in Crickleford, posing as a pet sitter, apparently.'

'Do you know her?' one of the other journalists was asking people now. Is she one of your neighbours? Emma Nightingale, she's been calling herself. Has she looked after your pets?'

'All the neighbours are out there,' Matt whispered to me as he locked the back door after us. 'They've all been woken up.'

'Yes.' I felt another wave of panic. I'd already recognised the voices of Pat, Pongo the Alsatian's owner, and her next-door neighbour Hattie. Now they knew about my deception, it would be all over town in no time.

'Come on! This way,' Matt urged me, grasping my hand and tugging me after him. He obviously didn't want to use a torch, and the path through the little plot was rough and overgrown. When we reached a gate in the far corner of the garden, he opened it and ushered me through. 'It's a footpath – it leads back to the top end of Fore Street but it's quite overgrown all the way,' he said quietly. 'It doesn't get used much. Hang onto my hand.'

It was a relief to be away from the shouts of the reporters and the flashes of their cameras, but my heart was still going nineteen to the dozen as we made our way through the brambles and mud of the little narrow path in the darkness. When I finally glimpsed the shapes of the Fore

Street shops in the distance I slowed down, breathing heavily. There was nobody following us. It was quiet and dead in the town, the streetlights still out.

'You won't go back to the cottage yet, will you?' I asked Matt. My heart was aching with regret and guilt. I'd ruined everything for him. His beautiful home would probably feel violated now.

'No. I'll give them a while to give up and go away. But you'd better not go back to Primrose Gardens yet, either. One of the neighbours in Moor View Lane might just blab about where you live. Come back to my flat.'

'No.' I stopped, put my hands up to his shoulders and rested my head against his chest just for a moment. 'No, Matt. I can't. It won't work. They won't give up, not now they know I'm in Crickleford. They'll find me; you know that as well as I do.'

'So, what are you going to do?'

'Move on.' I swallowed back my tears. I had no choice. It wouldn't be fair to Matt or to anyone else here, for me to stay now. 'I'll go home quickly now and pack – hopefully I can be in and out before they get there. I'll leave a note for Lauren. I'm so sorry, Matt. It's all my own fault.'

'No, it isn't!' He grasped my hands, trying to hold onto me. 'Stay, Emma! Brazen it out. Once they've got their story, they'll go.'

'But that's not the point.' I reached up, traced the line of his lips tenderly with my fingers and kissed him lightly before turning away. 'I've lied to everyone here in Crickleford. When it all comes out, they'll hate me. How can they trust their pets with me – someone who throws hairbrushes and mirrors at people's heads? I can't stay, Matt.'

'Don't do this!' he said hoarsely. 'I love you!'

'I love you too,' I said. 'That's why I'm not staying to ruin your life.'

CHAPTER THIRTY-ONE

I ran across the road, pounded down the deserted streets, tears pouring down my cheeks, not even stopping for breath until I reached Primrose Cottage, where I quietly let myself in and tiptoed upstairs to my little blue bedroom and packed my things. Within ten minutes I was in the kitchen, writing a note to Lauren and Jon saying I was sorry – sorry for lying, sorry for running away – but that I didn't want to bring them trouble. They'd soon find out why. I asked them to please cancel the pet-sitting bookings that were in my diary, to apologise on my behalf and, of course, to take little Trixie back to the Bartons when they returned from Australia. I guessed they were already on their way back. I added some special kisses for Holly – my tears splashed onto the paper as I wrote them. Then I folded some money into the note, to make up for letting them down with the next month's rent and thank them for sorting out my business. I propped it against the kettle with my door keys, gave Trixie a little goodbye stroke, and went cautiously back out into the street.

It had stopped raining, and the sky was beginning to lighten. Ironically, it looked like it was going to be a nice day. All the way into town, I kept glancing behind me and into the shadows, but the only people around were heading to the church hall with more items of comfort for those who'd been forced out of their homes. From the Town Square, I phoned for a taxi. The driver had to take a circuitous route out of town to avoid the floods, and the journey seemed to take forever, and cost a fortune. But by the time dawn had broken I was at Newton Abbot Station, boarding the first train of the morning heading back to London.

As the train sped across the country, I called my sister and asked her if I could sleep on her sofa that night. I couldn't risk going to my parents – the paparazzi would be turning up there as soon as they realised I'd left Crickleford. And it wouldn't take long for them to track down my sister's address, either, so I'd have to move to a B&B the following day.

'What on earth's happened, Emma?' she asked. 'What's gone wrong? I thought you were so happy there.'

'I was.' I sniffed and swallowed back tears. The young couple sitting opposite me were staring at me. If I mentioned the press pursuing me, their eyes would be out on stalks and before I knew it they'd have worked out who I was. 'I'll tell you when I get there,' I said miserably. 'I'm sorry, Kate. Sorry for worrying you, all over again.'

After I'd hung up, I closed my eyes and tried to sleep, but despite having been awake for most of the night, my brain refused to wind down. I couldn't stop turning everything over and over in my mind. The floods, the pony, the elderly couple we helped to rescue, the people cheering. Matt looking after me, taking me back to Bilberry Cottage, listening to my terrible story but still telling me he loved me. Oh my God – he said he *loved* me! I wiped my eyes again. I was running away from the man I loved! But how could I have stayed, now that the press had found me? They'd hound everyone in town for what they knew about me.

Then I sat up with a start, my eyes wide open again. How *had* those reporters found out about me? Matt had told me he'd spoken to someone on every national paper, and that none of them knew where I was. '*Your trail's gone cold*,' he'd said. Was he *lying* to me? Oh, please God, don't let him have been lying! Surely, if he'd been going to betray me, he'd have written the story himself, wouldn't he, not passed it on to the national press? I could only think that one of the journalists he'd spoken to had been suspicious about his phone call, and finally guessed that I was in Crickleford with him. And anyone in town could have quite innocently told someone who was looking for Matt, about Bilberry Cottage.

By the end of the journey I was worn out from thinking about it all. Getting the Tube from Paddington to Liverpool Street and then changing lines for Loughton was the last straw. As usual there were no empty seats and I was so exhausted I'd have sat on the floor if there'd been room between people's feet. When my sister opened her front door and took me straight into her arms, I instantly burst into tears.

'I'm so sorry,' I sobbed. 'I've messed up again. Yet again. I've let everyone down.'

'Of course you haven't,' she soothed me, taking me through to the lounge and sitting me down. 'Whatever's happened, you could never let us down. We're your family and we love you.'

This, of course, just made me cry all the more. By the time I'd managed to give Kate the full story, I felt completely drained. She heated me up some soup and sat by me, insisting I finished it all, and although I hadn't wanted it I realised it had been a long time since I'd eaten.

'Now I'm going to pick the children up,' she said when I'd finished, explaining that she'd left them with a friend so that she could give me her full attention. 'And when I come back, I expect you to be asleep. On my bed. I insist. I'll close the curtains, shut the door and keep the kids away from you until you wake up.'

Too tired to argue, I allowed myself to be tucked up for a nap as if I were a child myself. I fell asleep almost instantly, but woke up to loud 'shushes' from Kate, when she came back with the children. I lay there for another hour, pretending to be asleep, and when I eventually went downstairs it was to hugs and kisses from little Jeremy and one-year-old Rose, who had apparently been told by their mum that Auntie Emma was feeling very poorly and needed lots of rest.

I stayed for two days, by which time I felt stronger but no less depressed. I'd spoken to my parents, and the disappointment in their voices had made me feel sick. Apparently the house had been surrounded by reporters again, as I'd guessed, but they were drifting away now that they seemed to have accepted that they'd lost me.

'Stay with us,' Kate and Tim had both urged me. 'They don't know where we live.'

'They'll find out. It's lovely of you to offer, but it's not fair,' I said, and I found myself a room in a nearby B&B, which would do until I could get a job and rent a flat.

Would I have to move away again, somewhere new? The thought made me feel sad and weary. Would it ever end, this running away? Had I been naive to think I could settle in Crickleford and make a new life for myself without ever being found out? That I could run my own business and be a success – me, stupid little Emma who failed all her exams, who failed her family, failed in her relationship, even failed to be a success as a celebrity! Let's face it, I thought, I was just a failure all round, always had been. I'd have given anything to stay in Crickleford, to stay with Matt. My heart ached with a real, physical pain every time I thought about him. But I'd even failed *him*, in the end. Running away seemed to be the only thing I was good at.

October seemed a long month. The weather was getting colder, my heart getting heavier and heavier. I had countless texts and emails and missed calls from Matt, saying it was safe for me to go back, that the paparazzi had left Crickleford. But how could I go back, after all the stupid lies I'd told, how I'd deceived everyone and let them down. It would be ridiculous, anyway, to believe Matt would still want me back after I'd run away from him like that. Eventually the calls and messages stopped, so I guessed I'd been right. He'd given up on me, and I couldn't blame him.

I couldn't seem to stop feeling sorry for myself. I let the Cheeky Chestnut hair colour grow out, and surprise, surprise, nobody in Loughton recognised the drab, pathetic woman with unkempt red hair as the glamorous Candice Nightingale. I hardly even recognised myself. I got a job in a supermarket, stacking shelves. It was boring, punishing work but I felt like it served me right. I couldn't be both-ered to look for a flat, staying instead in the same tatty B&B, where my dingy room overlooked a council car park. At night I dreamed of my little blue room in Primrose Cottage, of Castle Hill and Fore Street, the views across Dartmoor, the river, the Town Square and, of course, Bilberry Cottage. I'd wake up with nothing to look forward to, no hope in my heart. My life was going nowhere. Of course, I was suffering from depression, but I didn't recog-nise it as such. I just felt worthless.

It was a little dog who finally started to bring me out of the darkness. Kate called me one day during the last week of October to tell me she and Tim had got a rescue puppy, and of course I couldn't resist going round to see him. Jeremy was rolling around on the floor with this little scrap of fur, not much bigger than my hand, and it was impossible to look at him without smiling.

'We've called him Casper. We don't know what breed he is – well, what *mixture* of breeds,' Kate said, laughing, as she lifted the puppy up and passed him to me for a cuddle. But over the next couple of weeks, as he grew bigger, I saw a definite resemblance to both Trixie, and Scrap, Mary's Cairn terrier.

'He could be a cross between a spaniel and a terrier,' I said as I went with Kate and the children to take Casper for his first outdoor walk after he'd had his second vaccination. I'd been spending more time with them since they'd got the puppy. 'He reminds me of a couple of the dogs I used to look after.'

'You must miss them,' she said, glancing at me sympathetically, and I just nodded. The less I had to talk about my time in Crickleford, the easier I found it. But Kate didn't give up that easily. 'You must miss *him*, too,' she added.

I'd told her about Matt, that first day I arrived back in Loughton, and she knew I'd ignored all his emails and phone calls.

'It's too late,' I said, shrugging as if it didn't matter. 'He's probably found someone else by now.' I knew she could see through my flippancy. My heart still felt as if it was going to break, just talking about him.

But with the puppy to take for walks, which meant I saw more of the children too, my mood gradually lightened a little. November passed, with its damp, dreary chill in the air, and its sudden onslaught of far-too-early Christmas tat in all the shops, and to my surprise my mum told me how much they were looking forward to me spending Christmas with them.

'It won't be like last year,' she said. 'That's all behind us now.'

I wasn't sure whether that made me feel better or worse.

And then, one Saturday morning during the first week of December, she called me to say something had arrived at their house for me in the post.

'But nobody knows your address,' I said, puzzled. 'It's not from America, is it?' I added, going suddenly hot with panic.

'No, love. It's a brown envelope, quite a big one. Come and pick it up – it might be important.'

As soon as I saw the envelope I knew it was from Matt. My heart began to pound. What was this? It was too big to be a letter, and even if he'd wanted to write to me, how had he got my parents' address? I thanked Mum but declined her offer to sit down in her lounge and open it while she made me a cup of tea. I needed privacy for this.

I went back to my grotty room at the B&B, sat on the bed and stared at the envelope, turning it over and over in my hands for several minutes before finally ripping it open. And then I froze. Inside there was just a single item: a folded page of a newspaper. I recognised it straight away as the *Crickleford Chronicle*. I made out only the headline before screwing up the page and throwing it across the room:

CRICKLEFORD'S SECRET CELEBRITY

He'd betrayed me. After all his promises, making me believe he'd never write about me, never tell anyone who I was! How could he do this to me? He'd obviously known that now it was out there, in the press, all the national papers would pick it up and the whole thing would blow up again. He'd done *this* to me – after telling me he loved me! Was this some kind of *revenge* on his part? Was he so angry and upset with me for not taking his calls or responding to his messages that he'd wanted to spite me?

If Kate hadn't called me at that precise moment, suggesting we take Casper for a walk, the newspaper page would probably have stayed screwed up in the corner until I put it in the bin. But when I answered the call she obviously picked up on the tone of my voice.

'What is it?' she asked sharply. And when I couldn't answer, she said: 'I'm coming straight round.'

I let her in, and just pointed at the ball of paper.

'He's written about me. And sent it to me, just to make it even worse.'

She picked up the paper and straightened it out, glancing at the headline.

'Have you read it?' she asked. We exchanged a look. 'Sorry,' she said quietly, and started to read it to herself.

'Oh, don't bother,' I told her impatiently. 'It's obvious what it'll be about. How I lied to everyone in Crickleford,

and the awful thing I did when I found Shane in bed
with—'

She held up her hand to silence me, still reading.

'You've got it all wrong,' she told me, looking up at me
with wide eyes. 'Listen.'

And she proceeded to read it to me. And it changed
everything.

CHAPTER THIRTY-TWO

'*CRICKLEFORD'S SECRET CELEBRITY*,' Kate read out. '*Crickleford residents have been surprised to learn that a popular resident of the town, who has endeared herself to the community since arriving here almost a year ago, turns out to have a very distinguished background.*'

'See?' I groaned immediately. 'He's *betrayed* me, Kate!'

She just shushed me and carried on:

'*Emma Nightingale, who has been running a successful pet-sitting business in Crickleford, was hoping to hide the fact that in her previous life she was none other than well-known celebrity Candice Nightingale, ex-partner of the rock singer Shane Blue. Emma had hoped, following the breakdown of her unhappy relationship with the star, to start afresh in Crickleford, living a simple and anonymous life, but unfortunately it seemed there were people within our community who had other ideas. Having managed to uncover her identity, someone appears to have found an inexplicable satisfaction from contacting the national press to let them know she was here among us.*

'This spiteful action resulted in Emma being pursued by the very publicity she'd tried so hard to avoid. Sadly, it happened just as she'd made herself a heroine to the people of this town by her courageous and selfless actions during the floods earlier this year, when she helped with the rescue of two elderly people, a stranded Dartmoor pony and a frightened cat. Sadly, once she'd been found by reporters from national newspapers, she felt she had no option but to leave Crickleford, and has yet to return.

'"I can't understand why anyone would want to spoil things for Emma like that," said her landlady and good friend Lauren Atkinson, 35, of Primrose Gardens. "She's a lovely girl and nobody here cares what she used to be. My little girl has been inconsolable since she left."'

I gasped. Holly was inconsolable? And it was my fault! But there was more. Kate glanced at me and read on:

'"Emma's not just a wonderful pet sitter, she's a very special person," said Pat Wilkins, 56, of Moor View Lane. "She took no credit at all for her bravery when she caught an intruder in my house while she was looking after my dog Pongo. I hope she'll come back. Pongo loved her and so did I."

'"I'm very fond of Emma,' Mary Field, 68, of Church Hill, told me sadly. "She worked hard at setting up her business and her new life here in Crickleford. If only she'd realised how popular she is with everyone here, she wouldn't have left. People won't turn against her just because she hid her identity. She had good reason, and we don't blame her at all."

'I spoke to many other residents and pet-sitting clients of Emma's, who all described her in similarly glowing terms. But *perhaps the last word should go to Vanya Montgomery, 45, of Woodside House, who told me with tears in her eyes: "I really don't know what I'll do without Emma. I can't trust anyone else with my beautiful pedigree cat. Emma was absolutely wonderful with her. I have a very good idea who was behind this leak to the national press and I'm furious that poor Emma felt she had to run away from all the people in Crickleford who love her, just because of some idiot's venomous actions. Please tell Emma through your paper that we don't care about her past. We want her back. We need her back!"*

'*It seems the people of Crickleford are unanimous in their outpouring of love and regret. Candice Nightingale may have been a celebrity, a so-called "It" Girl with a past that's irresistible to some journalists on the national papers. But to us, she's not Candice, she's just Emma Nightingale of Primrose Pets. This newspaper has only one thing to say to those journalists: Print what you like. Nobody cares. It's yesterday's news. And if you come to Crickleford again looking for Emma when she comes back, as we hope she soon will, you are sure to be chased out of town. This newspaper says: "Come back Emma! Crickleford needs you."*'

Kate stopped reading, but I sat staring at her with my mouth open. I couldn't speak. I'd never felt so completely gobsmacked in my life.

'Wait!' she said suddenly, although I hadn't moved an inch. 'There's something scrawled on the bottom of the page.' She screwed up her eyes, and then looked up at me, smiling. 'It says: "But I need you the most. I love you. Matt."'

'Oh,' was all I could say. I opened and shut my mouth and said it again: 'Oh!'

'You'd better get back there, girl,' Kate said briskly, but I heard the wobble in her voice. 'There's a whole town that needs you, by the sound of it. And a man who needs you even more.'

'Do you think he really means it?' I asked in a little voice.

'Of course he does!' She sounded exasperated now. 'What planet are you on? Look what this guy's done for you! He must have known you could have hated him for *outing* you like this, but he's taken the risk – God knows he tried hard enough to persuade you first, with all those calls and messages.' She sighed and put her arm round me. 'Look, he obviously knows the story's going to come out sooner or later, Emma,' she went on more quietly. 'So he's decided to get in first. To pre-empt it all. To show you in a positive light, so that nobody's in any doubt how popular you are in Crickleford, no matter what.'

'Not to everybody. Not to the bastard who told the tabloids about me,' I said. Vanya's comments in the article had, of course, left me in no doubt about who'd given me away to the press. So much for me supposing Rob had other things on his mind and wouldn't think any more about my background. Perhaps he blamed me for his divorce and had decided to get even.

'But who cares what he thinks?' Kate said. 'If it's that pervy guy you told me about. You said he'd left town now that his wife's divorcing him, and hopefully it'll be for good. Look, there might be some salacious gossip for a while, especially if the details of your break-up do come out.' She grinned, and to my own surprise I smiled back. Was it really so important any more if that story broke? If people gossiped? As Matt had said, who could blame me, really, for what I did to Shane and those two bitches? 'But from what Matt's written here,' Kate went on, tapping the paper with her finger, 'I think most of the community will defend you to their last breaths.'

'I do miss Crickleford,' I admitted now, my heart beginning to race. 'And, well, of course I'd be lying if I said I didn't miss Matt the most.'

'So call him, Emma, for God's sake. Get yourself back there. Why waste any more time?' She hugged me and spoke close to my ear now, the way we whispered in the night when we were teenagers sharing our room. 'I'll miss you, you know that. I always miss you. But I've never seen you as miserable as you've been since you ran away from Crickleford. It's where you belong, where you've made your life now. Matt's been patient, but he might not wait forever.'

That was enough to get me motivated. I called him as soon as Kate had left. It had been so long since I'd spoken to him, I felt like a teenager making a call to her first crush.

'Emma. At last!' he said softly, as soon as he answered the phone.

'Matt, I got the paper, your story, I don't know what to say,' I gabbled, too flustered to make much sense.

'Are you annoyed? Was it a mistake?' he asked anxiously.

'Of course not. It's brilliant. Thank you so much. But you've made me sound like a ... well, like somebody much nicer than I really am.'

'Somebody nicer than Candice, perhaps. But exactly like Emma Nightingale.' He paused for a moment and then went on, 'I didn't know what else to do. You wouldn't answer my calls or anything – have you got that bloody phone turned off? You can't tell me you haven't got any signal up there in the Big Smoke.'

'Loughton isn't quite the Big Smoke,' I said, laughing. 'Speaking of which, how did you find out my address?'

'I'm a journalist, Emma,' he said. 'I can find out anything. Actually I'd made up my mind to come to Loughton myself if you still didn't get in touch. And I might just do that anyway,' he added. 'I can't wait any longer. There's an Emma-shaped hole in my life here.'

'You've no need,' I said, smiling from ear to ear now. 'I'm coming back to Crickleford. There's ... a Matt-shaped hole in *my* life, too.'

'You're coming back? Yay!' He sounded just like a little kid getting excited about Christmas. 'I can't wait! When?'

'As soon as I've said goodbye to my dead-end job. I suppose they'll expect a week's notice.'

'Another whole week?' He groaned. 'It'll be the longest week ever. I've been so grumpy with everyone here since you left. I've had two official warnings from my editor about not smiling enough.'

I laughed. 'No you haven't!'

'No. Actually, I've got a promotion. Senior reporter. And I've been offered the chance to write a column for the paper every week too – something I've always wanted, Emma. It's going to be called "Crickleford People". The boss liked the way I wrote about you, so every week I'm going to interview someone different; a Crickleford character, you know? I'm starting with Annie from Ye Olde Tea Shoppe.'

'A good place to start!' I agreed. 'Oh, Matt, I'm so pleased for you. Congratulations.'

'Never mind about all that. Just hurry up back!'

I didn't need telling twice. I was working at the supermarket that evening, so I handed in my notice straight away. I'd already told my landlady I was going. And when I called on my parents the next day to tell *them*, their response was both instant and surprising.

'Next Sunday? OK, we'll drive you,' my mum said.

'Yes,' Dad agreed. 'It's about time we came down to see Crickleford again.'

'And it's about time we met this young man you've been moping over,' Mum added with a wink at me. 'Don't worry, we'll only stop for a cup of tea or whatever and come straight back. We won't want to get in the way of your reunion.'

'Oh!' I said, somewhat bewildered by the whole idea of them coming down with me. 'Well, thank you, but of course you should stay over. I'll ask Matt to book you into one of the pubs, though. Lauren hasn't got another spare room.'

I'd been half expecting Lauren, when I'd called her with the news that I was going back, to say that she'd let my room out to someone else. '*Oh, no, I haven't bothered,*' she'd said, which seemed a bit odd. '*I'm really glad you're coming back. The room's still waiting for you.*'

And so it was agreed. The following weekend I hugged my sister and her family goodbye yet again, with a promise to see them at Christmas. And with my bag stowed in Dad's car, we were on our way. Every nerve in my body was quivering with excitement. I could hardly believe I was leaving home again ... leaving home, and yet *going* home. Home to Crickleford.

CHAPTER THIRTY-THREE

As we drove over the Crickle Bridge and into the town, the memories of the night of the flood came rushing back to me. There must have been a massive clean-up operation here after the floodwater retreated. I felt bad now for deserting all my friends in Crickleford, when I could have been helping them. The Bartons' house looked as spick and span as before, the river flowing past at its normal level and their rowing boat presumably back under the tarpaulin I could see at the side of the house. I wondered who'd taken it back, but that must have been the least of their worries when the poor things got back from their Australian holiday to find their house half submerged.

It was the middle of the afternoon, and already people had their Christmas lights twinkling in their windows. There was a big Christmas tree in the Town Square, an illuminated Father Christmas, and a *HAPPY CHRISTMAS!* banner strung between the lampposts.

'Nothing seems to have changed here since we came for holidays when you were children,' Dad commented as he drove slowly down Fore Street.

'No,' I said, smiling happily out of the car window at the familiar shops, now all decorated up for Christmas shopping; the church, where another Christmas tree had been erected with some lights that seemed to be flashing rather erratically; the pub, with festive lights around its windows; the castle up on the hill, floodlit in the December gloom; the Olde Tea Shoppe ...

'Oh!' I said suddenly. 'What's going on here? Stop a minute, Dad – is that some kind of *protest* meeting or what?'

There was a large crowd of people outside the tea shop, spilling into the road, some of them holding flags or banners of some sort. As we pulled over, my phone rang.

'How far away are you now?' Matt asked. It was the tenth time he'd called to ask the same question.

'We're here!' I squawked. 'I mean, we're in Fore Street – but what's happening at Annie's?'

'Oh! You got here sooner than we thought. Stay where you are!' he yelled in response, and hung up. I stared at my phone, puzzled.

'Is everything OK?' Mum asked.

And then the door was flung open and Matt was there, beside me, trying to pull me out of the car before I'd even managed to undo my seatbelt. And while I was still being clasped in his arms, there was a huge shout from the crowd milling around outside the tea shop.

'She's here! Welcome back, Emma! Welcome back to Crickleford!'

WELCOME BACK EMMA! screamed the banner being waved by Annie and her son Kieran, who was presumably home from Bristol Uni for Christmas. I stared around the crowd. There was Lauren, with Jon and Holly, waving flags. Mary was standing next to them, with little Scrap on his lead barking with excitement, and with her, her neighbour Jackie, owner of the two rabbits. There was Pat Wilkins talking to her neighbour Hattie, as Pongo the Alsatian strained at his lead between them. There was Vanya Montgomery, on her own and looking pleased about it. There were Gary of the Koi carp pond, and Billie and Carl whose hamster I nearly lost. Behind them were Karen and Mike whose budgie had insulted me – Karen holding their new-born baby. And on the edge of the crowd were Mr and Mrs Barton, accompanied by Stan and Madge, who we'd helped to rescue in the flood. I even spotted young Josh, my erstwhile burglar, in the crowd, looking surprisingly smart and grown-up. I hoped his brush with the law – and with Pongo – had given him enough of a fright to teach him a lesson. Perhaps he'd be let off with a fine and community service and he'd stay out of trouble in future. As the crowd called my name and cheered and clapped, and someone started to sing a rather out-of-tune rendition of 'For She's a Jolly Good Fellow', Annie approached me and gave me a hug.

'You're right welcome back in Crickleford, maid. We've opened up special for you on a Sunday. This your folk?'

She pointed at my parents, who were still sitting in the car, looking stunned. 'Bring them inside the warm for a leak of tea and some stay-stomach. I've got some scones fresh out the oven.'

'Mind ye olde steppe, Mum,' I warned her as I took her arm and led her to the door. I caught Annie's eye and we both laughed.

'This is all ... just absolutely amazing,' I told Matt when we were settled at a table with everyone chattering around us. 'Did *you* organise it?'

'With the help of a few of your friends here,' he said, smiling down at me. 'Now do you see how much people here care about you?'

'But I lied to them,' I said, hanging my head. 'I made up so many stupid stories.'

'Do you think we didn't realise that?' It was Lauren, sitting down opposite me now and smiling broadly. She looked different, somehow, in a way I couldn't quite put my finger on. 'We knew you didn't want to tell us every-thing about yourself, Emma. We respected that.'

'I kind of guessed that your house hadn't burned down,' Mary said quietly from behind me. She lowered her voice a little more. 'And your parents seem very nice. But I had a feeling they probably hadn't taken in refugees.'

'We all knew you had a secret,' Vanya joined in, touching my hand gently. 'But we didn't care what it was. We guessed you'd tell us when you were ready. We just loved the way you looked after our animals.'

'And how you stopped the burglaries,' said Pat. 'With Pongo's help, of course,' and young Josh blushed and looked at the floor.

'And the way you got stuck in and helped when the floods came,' called out Pete, the Bartons' neighbour, and there was another round of applause.

'And the way you ... just fitted in here,' someone else said. 'Almost as if you were Devon born and bred.'

'Not like you were one of those stuck-up folk from up-country,' another voice chimed in, 'begging your pardon, Emma's mum and dad.'

Everyone laughed, including Mum and Dad, and they all toasted me with their cups of tea.

'I really thought you'd have let my room to someone else by now,' I said to Lauren when it had quietened down again. 'I'm glad you didn't, obviously, but—'

'I was going to explain that later, Emma,' she said. 'We're thrilled to have you back – that goes without saying. But we weren't bothering to let the room out to anyone else. I'm sorry to say my dad finally passed away back in October—'

'Oh, Lauren, I'm so sorry,' I said, but she shook her head quickly.

'It wasn't a tragedy. He'd been very poorly for a long time, and he'd had enough. I lost him years ago, really. But – this sounds bad, but you see, it meant we weren't so desperate for money, to pay the care home fees.'

'I can understand that,' I said.

'And anyway, eventually we're going to need that little bedroom. Not for a while, don't worry. And the baby can sleep in our room to start with, obviously.'

'The *baby*?' I yelled, and I jumped up to hug her, nearly sending everyone's tea and scones flying. 'Oh, that's wonderful – congratulations!'

'I'm going to be a Big Sister,' Holly announced solemnly. 'The baby will need me, because it won't know *anything*.'

'Of course,' I said, giving her a hug too. 'Big Sisters are very important people.'

'The baby has to wait inside Mummy's tummy until the summer,' she explained. 'I'll be *five* by then. Babies take ages to get borned. I know that because Mrs Jones told us.'

I'd forgotten about the all-knowing Mrs Jones. I grinned at Lauren. 'You'll have to tell me when you need me to vacate the room, then.'

Matt nudged me. 'I'm rather hoping you might decide that for yourself. Before too long. There's a space in the wardrobe in Bilberry Cottage just waiting for you to fill it up with your things. And a space in the bed next to me,' he added in a whisper, making me shiver, 'just waiting for you to warm it up.'

'Are you serious?' I whispered back. 'We're not even … officially … *together* yet.'

'Soon fix that.' He grinned.

'What be you two lovebirds whispering about now?' Annie interrupted us, to a chorus of wolf-whistles, making me redden and giggle.

And so the welcome-home tea party went on, for another hour or so, with more people coming in, and everyone chatting and laughing and hugging me. It was overwhelming. It was wonderful.

'I've been meaning to ask you,' Matt said suddenly, as Kieran was doing the rounds with slices of Annie's special chocolate cake. 'You did *read* my story in the *Chronicle*, did you?'

'Of course I did!' I said without thinking.

'How?'

He looked at me, waiting, smiling gently, and the café seemed to fall silent around us. My parents were watching me. Kieran paused at our table and touched my arm, giving me a little nod of understanding.

I looked down at the table. There had been enough lies. It was time to be honest now – about everything. 'Well, OK, my sister read it to me,' I said, my voice shaking a little. 'How did you know—?'

'Haven't you told him, Emma?' Mum said. 'There's no shame in it.'

'Oh, isn't there, Mum?' I shot back at her. 'No *shame*, in not being able to read properly, at my age? How do you think that makes me feel? Knowing what a failure I am, how disappointed you and Dad have always been in me?'

'We're not ... we never were ...' she protested.

'Really? You never compared me to Kate? Never thought how strange it was to have one clever, perfect twin and one complete and utter waste of space like me? Sorry, I don't believe you.'

There were some mutterings around us. People suddenly pretended to find their scones and cakes very interesting. The afternoon felt spoilt, and as usual it was my fault.

'Sorry,' I began, looking down at the table again. But my dad had already begun to speak, loudly and clearly enough for everyone to hear.

'We were *never* disappointed in you, Emma. We were *worried*. Everything was so hard for you, having dyslexia. All the extra lessons, all the private tuition and extra homework – it helped, but only a bit. You were fed up with it all by the time you left primary school, and who could blame you? You just gave up, refused to try any more. *You* compared yourself to Kate, *we* didn't. We just worried about you. It was your behaviour we were disappointed in, not you. Never you. We loved you. We'll always love you, no matter what you do – but right now, we're *proud* of you.'

'Proud?' I whispered, tears starting in my eyes.

'Couldn't be prouder,' Mum said, grabbing both of my hands in hers. 'Starting again, the way you have here, after that idiot rock singer caused you so much heartache? Making your own way, making yourself so popular here in Crickleford? Of *course* we're proud of you.'

'Running her own business has been an amazing achievement for someone who can't read or write very well,' Mary said, smiling at me.

'*You* knew?' I said. I looked around the room. Everyone was nodding. 'You *all* knew?'

'Sorry, Emma, but it was fairly obvious,' Lauren said apologetically. 'You always made excuses not to read to

Holly. And, well, I saw your spelling. When you wrote on the shopping list.'

'You borrowed children's books out of the library,' young Josh said, his cheeks red with the embarrassment of talking to me.

'And the way you got so defensive about university,' Kieran joined in, 'it just didn't ring true that you were too stupid, or too busy doing something more important than getting a good education. I guessed it must be something like this.'

'I thought at first you were just not very good at spelling,' Mary said gently. 'I noticed the notes you were making when you came to me for help setting up the business. But then ... all those fibs about reading Shakespeare ...'

I blushed, mortified, but she was laughing. 'It was naughty of me to try to catch you out, but I just wanted you to admit to me that you couldn't read properly, so that I could try to help you. And then, when it was so obvious you couldn't read that letter about your cat, when it was plainly so important to you – I felt so sorry for you.' She paused. Everyone had fallen silent. 'But from that letter,' she went on, 'I realised you'd lived in America and wanted to keep it a secret for some reason, and I respected that. I knew I had to back off.' She laughed. 'Anyway, I'd never even heard of Candice Nightingale! I'm not interested in celebrities.'

'Nor am I any more!' I said with feeling. There was a ripple of laughter in the café. I clasped Mary's hands, trying to find words to thank her for everything. For caring. For understanding. For keeping my secret. But Matt was already speaking.

'I guessed about your reading and writing when I saw you staring at messages on your phone as if they were written in Greek,' he said. 'And the ones you sent me, not that there were enough of them—' He winked at me. 'Well, I could understand them but sometimes they were ... kind of weird. In a nice way!' he added quickly, squeezing my hand.

'But I use predictive text!' I said crossly. 'Surely spelling doesn't matter so much, with spell checks on computers and so on.'

'That's true to some degree,' Mary said. 'But sometimes they do get it wrong. The wrong *too* or the wrong *their* and so on.' I looked at her blankly and she gave me another smile. 'I could try and help you, Emma. I've had lots of experience with adult literacy. I promise I won't expect you to read Shakespeare.'

I smiled, and she grinned back. 'Sorry,' she said, laughing.

'It was my fault for not being honest,' I said. 'And thanks for offering to help me, but ... I think I'm happy just being a pet sitter.' I looked around the room. 'If you all still want me?'

'Of course we do!' everyone chorused.

I took a deep breath and turned back to my parents. I owed them an apology. They'd suffered enough because of me.

'I'm sorry for snapping at you. And for being such a worry to you when I was growing up,' I said. 'And since.

With Shane, and the paparazzi and the publicity and everything.' They were both shaking their heads as if none of it
mattered. 'And thank you,' I went on, 'for everything. You've
got no idea how much it means to me, you being here,
meeting all my friends.' I spread my hands, looking around
the café, and then back at them both. My voice cracked
as I finished: 'Hearing you say you were proud of me. It's
what I've dreamed of, my whole life.'

We hugged goodbye soon after that. Matt had booked
them a room at The Riverboat Inn, and they were going
to leave early the next morning for the long drive back.

'You'll still come home to us for Christmas, won't you?'
Mum asked me anxiously.

'Of course I will.' I glanced at Matt and added: 'Will it
be OK to bring a Plus One?'

'Absolutely!' she said calmly, with a smile at him. 'Our
family has needed a friendly journalist on our side for a
long time.'

'Actually, it might be Plus Two,' he said, suddenly giving
me an anxious look. 'There's something I haven't had a
chance to tell you yet.'

'Oh God – what?' I stared at him. Was this where he'd
tell me he had a wife or child hidden away somewhere?

'I've bought a dog,' he said. 'I needed someone to keep
me company after you left. She's really sweet and friendly,
but the thing is, it seems our local pet sitter won't be here
over Christmas, so I'd have to bring her with us. Would
you mind?' he asked Mum.

'Of course not. We all love dogs.'

'I wonder if Kate will bring their puppy round. They could be company for each other,' I said. 'How exciting that you've got a dog!' I added, putting my arms round Matt. 'What's her name?'

'Candice,' he said – and then laughed at the look on my face and said. 'Just joking. It's Lola. It just seemed to suit her somehow.'

'Lola. Aw, I can't wait to meet her!'

After my parents had gone, I looked around the room again at all these people, these Crickleford people who had taken me into their hearts, and trusted Primrose Pets with their animals, despite the stories and lies I'd told them, not even caring who I really was, or whether I could read or write properly. Maybe in time I *would* accept Mary's offer. Maybe one day, with her help, I'd be able to read Matt's new column in the *Chronicle* for myself. I could even write a letter back to Shane's grandmother in the States to ask how Albert was. I smiled as I thought how much she might appreciate that.

But, for now, all I wanted to do was to curl up with Matt on the sofa in Bilberry Cottage and dream of a future together – a future I'd never have dared to dream of when I walked away from my life as Shane Blue's vacuous little arm-candy.

'It was stupid of me, thinking I could hide the dyslexia,' I said, but he put his fingers on my lips and shook his head.

'Why does it even matter?' he said. 'I love you just as you are, Emma Nightingale.'

And as he finally kissed me, another cheer went up around us from everyone in Ye Olde Tea Shoppe.

'Gawd love us!' Annie exclaimed.

'Get yerselves a room!' someone joked.

'Good idea,' Matt said, looking into my eyes. 'How about coming back to Bilberry Cottage with me now? Lola's waiting to meet you.'

I linked arms with him and, waving goodbye to all our friends, we headed out into the dark of the winter afternoon. I had a definite feeling I was going to be spending a lot of time at Bilberry Cottage from now on. And, much as I loved animals, it wasn't all going to be about meeting Lola!

If you enjoyed

THE PETS AT PRIMROSE COTTAGE

Leave a review online

Follow Sheila on Facebook/SheilaNortonAuthor and on
Twitter @NortonSheilaann

Keep up to date with Sheila's latest news on her website
www.sheilanorton.com

Make sure you've read Sheila's other novels . . .

Sheila Norton

The Vets at Hope Green

A new start
A fresh chance at love...

Sam has always dreamed of working with animals . . .

But her receptionist job in a London vets is not hitting the spot. Unsure whether a busy city life is for her, she flees to her Nana Peggy's idyllic country village.

But despite the rolling hills and its charming feel, life in Hope Green is far from peaceful. On first meeting Joe, the abrupt and bad-tempered local vet, Sam knows she must get him on side, but that is easier said than done . . .

With her dream close enough to touch, will she get there or will events conspire against her?